Microalloying in Steels

PROFESSOR J.J URCOLA

(1946-1996)

Proceedings of the International
Conference on

Microalloying in Steels

7th-9th September 1998
Donostia-San Sebastián,
Basque Country, Spain

Editors:

J.M. Rodriguez-Ibabe
I. Gutiérrez
B. López

Centro de Estudios e Investigaciones Técnicas de Gipuzkoa (CEIT)
and
Escuela Superior de Ingenieros, Universidad de Navarra

ttp TRANS TECH PUBLICATIONS LTD
Switzerland • Germany • UK • USA

Volumes 284-286 of
Materials Science Forum
ISSN 0255-5476

Distributed *in the Americas by*

Trans Tech Publications Inc
PO Box 699, May Street
Enfield, New Hampshire 03748
USA

Phone: (603) 632-7377
Fax: (603) 632-5611
e-mail: ttp@ttp.net
Web: http://www.ttp.net

and worldwide by

Trans Tech Publications Ltd
Brandrain 6
CH-8707 Uetikon-Zuerich
Switzerland

Fax: +41 (1) 922 10 33
e-mail: ttp@ttp.ch
Web: http://www.ttp.ch

Printed in the United Kingdom
by Hobbs the Printers Ltd,
Totton, Hampshire SO40 3WX

EDITORS' FOREWORD

Since the Middle Ages steel takes an important place in the industrial development of the Basque Country, which has long been reputed for shipbuilding and heavy manufacturing. Actually, a broad set spectrum of factories from small family industries to large steel producers is concentrated here and with an annual production of 6 million tons manifests one of the highest steel production densities in Europe.

The chosen subject: "microalloying in steels" is of great importance to local steel industries with which Prof. J.J. Urcola Galarza felt engaged to collaborate until his untimely death. The team of people working in close collaboration with him, on steel research, participating in the same inquietude, decided to organise an international symposium on microalloying, following his initial idea. It has been the aim of the organising committee to dedicate this Symposium and the actual proceedings to his memory.

The technology of microalloying is now, at the end of the 20th century, widely accepted due to the attractive balance of properties achievable with the correct control of the interaction between steel chemistry and processing. Microalloying is applied to a broad spectrum of products and has two main goals: improve the mechanical properties and save-costs. To cover new development and new trends, invited speakers are presenting papers on thermomechanical treatments, phase transformation, modelling, mechanical properties and alloy design.

The editors acknowledge gratefully the assistance and support provided by the organisations and governmental agencies listed as sponsors. Particular thanks are due to the advice and dedication of the members of the International Committee and very specially to Prof. C.M. Sellars for his enthusiastic encouragement when the event was no more than a simple project. Sincere thanks are also due to the invited speakers and authors of contributed papers.

The Editors

International Scientific Committee

E. Anelli	CSM (Italy)
H.K.D.H. Bhadeshia	Cambridge Univ.(U.K.)
A.J. DeArdo	Pittsburgh Univ. (USA)
B.J. Duggan	Hong Kong Univ. (H K)
M. Fuentes	CEIT (Spain)
J. Gil Sevillano	Navarra Univ.(Spain)
P.D. Hodgson	Deakin Univ. (Australia)
B. Hutchinson	SIMR (Sweden)
J.J. Jonas	McGill Univ. (Canada)
P. Karjalainen	Oulu Univ. (Finland)
M. Korchynsky	STRATCOR (USA)
A.Köthe	IFW (Germany)
G. Krauss	Colorado S. Mines (USA)
V. Leroy	CRM (Belgium)
B. Mintz	City Univ. (U.K.)
D.J. Naylor	British Steel (UK)
C. Ouchi	NKK Corp.(Japan)
R. Priestner	UMIST (U.K.)
J.M. Rodriguez-Ibabe	CEIT (Spain)
C.M. Sellars	Sheffield Univ. (UK)
S. van der Zwaag	TUDelft (Netherland)

Organising Committee

Dr. J.M. Rodriguez-Ibabe
Dr. I. Gutierrez
Dr. B. López

CEIT and Universidad de Navarra

Sponsoring Organisations

- Diputación Foral de Gipuzkoa
- Dpto de Industria, Agricultura y Pesca, Gobierno Vasco
- Dpto de Educación, Universidades e Investigación, Gobierno Vasco
- Comisión Interministerial de Ciencia y Tecnología (CICYT), Madrid
- Directorate General XII of the European Commission

FOREWORD

The Professor J. Javier Urcola Symposium

The "Microalloying in Steels: New Trends for the 21st Century" International Symposium was conceived primarily as a vehicle to mark the career of Professor J. Javier Urcola, who died in a tragic accident in the summer of 1996. Tributes have already been made for Javier Urcola´s distinguished service to the University of Navarra and to education - particularly for his contribution to the teaching of Material Science and engineering in Basque - and for his earnest dedication to the advancement of scientific and technical *euskera,* the ancient language of his home Country; to that end he published two excellent text books and numerous enlightening and educational articles on physical metallurgy and related topics.

Javier Urcola had a deep understanding of metallurgy, which he was able to impart to those who had the privilege of being his students. He provided them with wise guidance and unreserved friendship, and here lies perhaps the key to his success in setting up what is known as his "Virtual School of Metallurgy", so active in his home Country and overseas, particularly in Chile, Colombia and Argentina where so many of his former Ph.D. students are now successful practitioners in the metallurgical trade.

He was endowed with a rich personality. He possessed an uncommon intuition and a special ability to grasp the essence of an engineering problem, two features that were matched with an unusual sagacity to deliver practical solutions. He had a unique ability of observation for natural phenomena, supported by a rigorous academic training. Also, he was a passionate researcher. With all these ingredients, the outcome of his work could be nothing short of the very highest standard of research, carried out in collaboration with an enthusiastic team consisting of past and present Ph.D. students: Isabel Gutiérrez, Jose Maria Rodríguez Ibabe and Beatriz López are among the former. Working with Javier Urcola was great fun. He was frank, warm and extremely easy to get along with. He had a contagious vitality which extended to those that worked with him, and beyond. It is, perhaps, a tribute in itself, that his younger colleagues were able to take up the mantle and maintain a Research Group, of which they, themselves, and CEIT can be justifiably proud.

Javier Urcola pursued research in a wide range of topics. He started off his research, at the Faculty of Engineering of the University of Navarra on its campus at Donostia/San Sebastian (where later on, in 1991, was appointed Full Professor of Metallurgy), working on diffraction of induced substructures in F.C.C. single crystals, followed by a doctoral degree in engineering (Dr. Ing.) He took up thereafter investigations into the solid phase transformations of Fe-based alloys. Later, he moved on to Sheffield where, under the guidance of Mike Sellars, he carried out a very fine piece of work on strain rate transients during high temperature deformation of ferritic stainless steels, Aluminium and Al/Mg alloys and obtained a Ph.D. in metallurgy.

On returning to CEIT, he simultaneously launched two research lines - one dealing with powder metallurgy and the other with the thermomechanical processing of steels. Throughout this period, up until his tragic death, he enjoyed the confidence of industrialists and fellow researchers in European Universities and R&D institutes, which allowed him to win the financial support of companies - to carry out contract research - and the sponsorship of the European Union Brite, ECSC and CRAFT programs. The encouragement he gave to this approach - a core concept in a contract research organisation such as CEIT - and the success that followed contributed to his promotion to the position, in 1993, of Head of its Materials Department, a position thoroughly deserved and one he pursued with his usual vigour and intellect. Colleagues and friends will miss the characteristic two-handed manipulations with the rolling up of his tie -in the rare occasions he felt it was justified to put one on- as he listened to a speculative rational of some experimental results or a tale of great mountaineering feats.

However, it is out of respect for Javier Urcola´s reputation as an eminent engineer that his many friends and colleagues are gathered here today, in recognition of his contributions to the advancement of steel processing and, more specifically, to the area of microalloying and its associated thermomechanical processes. As convenors, we are gratified by the delegates response: by so warmly entering into the spirit of the occasion. We take this opportunity to thank both speakers and delegates for their whole-hearted support for the meeting and its prime purpose of honouring the outstanding stature of Javier Urcola both as a human being and as a gifted metallurgist.

No meeting of this nature would be possible without considerable financial sponsorship or without the patronage of significant bodies. Their names are listed overleaf but we would like here to express special thanks to Mike Sellars -champion and mentor of Javier Urcola- for his most helpful assistance in working out the concept of this symposium.

Prof. Manuel Fuentes
General Director of CEIT

Donostia/San Sebastián, March 1998

Table of Contents

1. Invited Papers

2. Hot Working

3. Phase Transformation

4. Mechanical Properties

5. Modelling

6. Medium/High Carbon Steels And Forging

7. Thin Slab And Hot Ductility

8. Cold Deformed And Annealed Steels

INVITED PAPERS

Materials Science Forum Vols. 284-286 (1998) pp. 3-14

Effect of Interpass Time on the Hot Rolling Behaviour of Microalloyed Steels

J.J. Jonas

Department of Metallurgical Engineering, McGill University,
3610 University St., Montreal, H3A 2B2, Canada

Keywords: Rod Rolling, Strip Rolling, Mill Logs, Mean Flow Stress, Interpass Time, Microalloyed Steels, Dynamic Recrystallization, Metadynamic Recrystallization, Static Recrystallization, Strain-Induced Precipitation, Torsion Testing

Abstract

The three ranges of industrial interpass time (i.e. short, intermediate and long) are described and compared. These correspond approximately to the finishing stands of rod mills, strip mills, and plate or reversing mills. The concept of the T_{nr} or "no-recrystallization" temperature is reviewed. The possible types of interaction between strain-induced precipitation and softening by means of static, dynamic and metadynamic recrystallization are then characterized. With the aid of mill logs obtained from a number of rolling mills as well as of laboratory torsion tests, it is shown that when the interpass time is short, as in rod rolling, neither strain-induced precipitation nor static recrystallization is possible. Under intermediate interpass time conditions, as in strip rolling, two types of behaviour can be seen. One corresponds to the long interpass time type of response associated with plate rolling (i.e. pancaking or CCR). The other follows the rod mill pattern of strain accumulation, leading to the initiation of dynamic and then metadynamic recrystallization. The first requires the presence and the second the absence of strain-induced precipitation.

Dedication

Before embarking on the main topic of my presentation, I would like to make a few commemorative remarks about J.J. Urcola and his character and work. On each of my visits to San Sebastian, the most recent of which took place just a few weeks before his untimely death, Javier received me with great warmth and affection. On one of these occasions, he presented me with a red Basque beret, which I very much treasure, and on the most recent, with a bottle of Pacharan, the well-known Basque digestif. These simple examples of his legendary generosity are typical and his wide circle of friends and associates can provide further instances, I am sure. He will be sorely missed.

Born in Tolosa, here in the heart of the Basque country and near San Sebastian, he was inordinately proud of his native region and never tired of introducing its scenic glories to his visitors. He was a gifted story-teller, and very knowledgable about the history of Guizpocoa and of Euskal Herria. This pride extended to his technical and academic interests, and he worked tirelessly to build up the Materials Department of the CEIT, eventually becoming its Director. Others will attest to his contributions to the development of metallurgical expertise in the local industries. In my own case, our shared interests were focused on Thermomechanical Processing and on the arcane mysteries of the Torsion Test. These very topics will be discussed in some detail during the various sessions of this conference. In this way, those of us gathered here will pay homage to his memory.

1. Introduction

Although thermomechanical processing is most widely applied to plate products, there is increasing interest in extending the principles developed for these materials to other types of applications, such as long products and strip. All three categories of steels can be based, not only on the use of Nb additions, but on those of Mo, Ti and V as well. In order to do so, however, it is important to keep in mind some of the basic differences that apply to the three types of process; these arise in turn from the characteristics of the equipment involved. Although there are other parameters of interest, the most important one from the present point of view, and the one that is addressed in this paper, is the interpass time.

The interpass time is particularly important because of two reasons: i) it determines the extent to which dynamic recrystallization (in addition to the more conventional static recrystallization) plays a role; and ii) it also determines whether the precipitation of microalloy carbonitrides is likely to occur. Such precipitation, if it takes place, can then interact with the two recrystallization mechanisms. Alternatively, in its absence, recrystallization will proceed without impediment. Through the interaction between recrystallization and precipitation, the interpass time is therefore able to affect both the rolling loads and microstructures that are developed during processing.

In order to examine the possible interactions in more detail, it is useful to begin with a brief review of the possible ranges of interpass time and then to follow this examination with the effect of the interval itself on the types of interaction.

2. Interpass Times and Types of Recrystallization/Precipitation Interaction

2.1 Ranges of Interpass Times There are three distinct ranges of interpass time, which are illustrated in Table 1. These arise directly from the geometries of the processes involved. Reversing mills, it can readily be seen, have times of around 10 seconds. Such intervals are long enough to permit recrystallization at elevated temperatures; they also allow for the occurrence of carbonitride precipitation when the temperature is well below (e.g. 100 °C below) the solubility limit of the particle of interest. Above or near the solubility limit, it is clear that no interaction is possible between precipitation and recrystallization; under these conditions, recrystallization can proceed without hindrance.

Table 1. Comparison of hot rolling mills in terms of their ranges

of strain rate and interpass time

Mill Type	Strain Rate Range (s^{-1})	Interpass Time (s)
Reversing Mills (plate, roughing, etc.)	1-30	8-20
Hot Strip Mills	10-200	0.4-4
Wire Rod Mills	10-1000	0.010-1

The other extreme in industrial interpass times is exemplified by the finishing stands of rod mills. Here, because of the short distances between stands and the high rates of rotation of the rolls, the times can be as short as 10 to 50 milliseconds [1,2]. Under these conditions, it is clear that only limited precipitation can take place, or perhaps even no precipitation at all. Thus, there is little possibility for precipitation and recrystallization to interact. As there is not much time for static recrystallization either, there is considerable scope for the retention of strain, i.e. for work hardening to accumulate. This can lead in turn to the initiation of dynamic recrystallization, followed in general by metadynamic or post-dynamic recrystallization. The interactions between strain accumulation and the three types of softening mechanism mentioned above are of considerable interest in the present context and will be considered in more detail below.

We turn, finally, to the intermediate case of moderate interpass times. These fall somewhat above or somewhat below one second. Although this range is involved in the production of considerable quantities of steel, the behaviour under these conditions has received much less attention than in the case of plate rolling (long interpass times). As can be expected, the interplay of strain accumulation, precipitation, and the various types of recrystallization is considerably more complex than in the two simpler examples described above. Because of the often delicate balance between the different mechanisms, the extent to which one or another dominates is also much more sensitive to small changes in chemistry and in the processing conditions. The clarification of these phenomena is evidently a topic that has much potential for future research. Progress in understanding the mechanism interactions will then lead to improved rolling mill models as well as to the more accurate prediction and control of properties.

2.2 The No-Recrystallization Temperature or T_{nr} The full understanding of the interactions introduced above involves a clear definition of the T_{nr} or "no-recrystallization" temperature, the basic elements of which are illustrated in Fig. 1. In this diagram are presented the results obtained from a laboratory simulation of plate rolling carried out on a hot torsion machine [3]. After reheating this particular Nb steel (0.45 %Mn, 0.3 %Mo, 0.04 %Nb, 0.08 %Ti, 0.06 %V), the sample was cooled at about 1 °C/s, so that, given the 30 s. interpass times employed, the deformation temperature decreased by about 30 °C from "pass" to "pass". It can be seen that the rolling load increased gradually at first, due entirely to the continuous decrease in temperature. After about nine passes, the rate of load change increased perceptibly. This is because strain accumulation, i.e. the retention of work hardening, took place at temperatures below those corresponding to pass no. nine. Mechanistically, it is evident that precipitation had begun to take place in this particular interpass interval and that it proceeded to a degree that was sufficient for it to impede further recrystallization. This type of test can thus be used to determine this "slope change" temperature, which therefore represents the temperature below which recrystallization is retarded to a significant degree by the occurrence of "strain-induced" precipitation.

The trends suggested in Fig. 1 can be discerned more clearly in Fig. 2, which was constructed from a set of flow curves similar to those of Fig. 1. Here the mean or average flow stress developed in each pass is plotted against the inverse absolute temperature. This type of plot can be used, not only to determine the T_{nr}, but also to define the A_{r3} and A_{rl} temperatures that apply to a particular steel when it is being deformed under conditions that are representative of the rolling mill and schedule of interest. In this way, such torsion tests can also be used to define the temperature limits for intercritical or indeed for warm rolling [4].

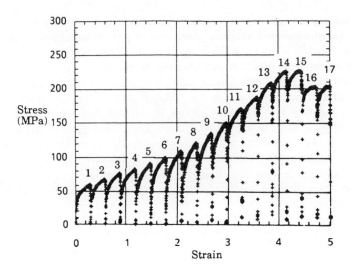

Fig. 1. Stress-strain curves obtained by torsion simulation of the hot rolling of a 0.45%Mn-0.3%Mo-0.04%Nb-0.08%Ti-0.06%V steel. The initial and final passes represent roughing and finishing, respectively, with an exit temperature of 745 °C [3].

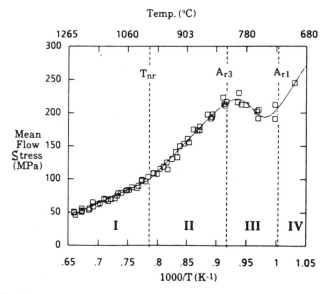

Fig. 2. Mean flow stress as a function of 1000/T, where T is the absolute pass temperature (combined data of four tests [3,4].

2.3 The Interaction Between Recrystallization and Precipitation Recrystallization/precipitation interactions of the type introduced above are of interest in many branches of ferrous metallurgy. In the case of classical controlled rolling which, in the present context, is the reversing mill or long-interpass-time case, it has long been known that the occurrence of NbCN precipitation during finish rolling prevents recrystallization and in this way leads to "pancaking" of the austenite grains [5].

The grain flattening attributable to pancaking, i.e. to rolling below the T_{nr}, is in turn primarily responsible for the fine ferrite grain sizes, and therefore the high fracture toughnesses, that are produced by this type of rolling. In this category of process, precipitation takes place *before* recrystallization and the latter mechanism is therefore prevented from operating. "Pancaking" thus involves two conditions: i) that rolling be carried out below the T_{nr}; and ii) that precipitation occur before recrystallization.

Other categories of interaction are also possible, as summarized in Table 2. For example, the opposite relationship, i.e. *recrystallization before precipitation*, is a common occurrence. It applies to the long interpass time processes as long as rolling takes place above the T_{nr} and there is no strain-induced precipitation. Here the T_{nr} is again significant: it defines the temperature above which there is no strain-induced precipitation. In the case of plate (i.e. long interpass time) rolling, recrystallization before precipitation is the basic paradigm for RCR processing, also known as recrystallization controlled rolling. However, these conditions can also apply to the short and medium interpass time processes, although in these cases there is, of course, less time for both strain-induced precipitation and static recrystallization. The absence of precipitation makes it harder to arrest recrystallization; conversely, the short interpass times make recrystallization *less* likely. The net effect can only be determined experimentally at the moment; it is primarily dependent on the extent of solute retardation, which replaces the function of precipitate pinning under these conditions.

Table 2. Relations between recrystallization and precipitation for the three ranges of interpass time

Time Interval	T range with respect to T_{nr}	Role of strain-induced precipitation	Relation between recrystallization and precipitation
Long	below	presence required	precipitation before static or dynamic recrystallization
Long	above	absence required	static recrystallization before precipitation
Intermediate	above or below	presence or absence both possible	several combinations possible
Short	below	absence required	no static recrystallization; dynamic recrystallization before precipitation

We turn now to the case where the interpass time is too short to allow either static recrystallization or strain-induced precipitation. (The first condition corresponds to rolling *below* the conventional T_{nr}, which no longer applies in the present case; the second corresponds to rolling *above* the conventional T_{nr}!) This leads to strain accumulation, at least at first. The point of interest here is that the occurrence of strain-induced precipitation retards, not only static recrystallization, but dynamic recrystallization as well. When it is *absent*, the stage is set for the initiation of dynamic recrystallization. Once dynamic recrystallization has been initiated, it is generally followed by post-dynamic (metadynamic) recrystallization during the succeeding pass interval.

Most rolling mill models make allowance for static recrystallization and include relations describing the softening kinetics. Such expressions generally take into account the current temperature, the prior or accumulated strain and strain rate, as well as basic differences in steel chemistry. However, when dynamic and post-dynamic recrystallization take place, they affect both the rolling load as well as the microstructure. Thus a complete model must allow for the effects of these mechanisms as well. This is an area of much current research [6-10], and it appears reasonable to expect that appreciable progress will be made on this front.

Now that the three basic types of interaction between recrystallization and precipitation have been introduced and defined, it will be useful to consider some examples of these interactions for the short and intermediate interpass time processes. The case of long interpass times (classical controlled, i.e. plate, rolling or CCR) will be omitted here as it has frequently been described.

3. Examples of Rolling Processes in which Dynamic and Metadynamic Recrystallization Play Important Roles

3.1 Short Interpass Times:- Rod Rolling Rod mills typically have four sections, referred to as the i) roughing; ii) intermediate; iii) prefinishing; and iv) finishing stands. The interpass times are generally well above one second in the roughing stands (and the temperatures are relatively high), so that the behaviour in this section of the mill is typical of that of all roughing mills in that static recrystallization takes place. This corresponds to rolling above the T_{nr} in the absence of strain-induced precipitation, i.e. to RCR rolling. In the intermediate and prefinishing stands, the interpass times are somewhat longer and somewhat shorter than one second. Thus the behaviour here is expected to be fairly complex, resembling that observed in strip mills, see Table 1. This case will be considered in more detail below.

The number of passes employed in the finishing stands depends on the dimensions of the final product. In the case of 5.5 mm rod, for example, there may be ten sequential reductions [11]. Under these conditions, the following set of interpass times has been cited as typical.

Pass No.	1	2	3	4	5	6	7	8	9	10
Time to Next Pass (ms)	79	61	47	36	28	23	18	14	11	817

The above times fall below 50 ms and decrease to as low as 11 ms in seven out of nine cases (the final number refers to the elapsed time while travelling to the "laying head"). Under these conditions, it is can be expected that little static recrystallization will occur in plain C or Nb steels and that, for similar reasons, there will be little CN precipitation in microalloyed steels.

This can be seen to better effect in Fig. 3, taken from a paper by Hodgson and co-workers [12]. The flow curves were determined at 900 °C in a torsion simulation of the rod rolling of a 0.04 wt%C-0.3 wt%Mn plain C steel. The test was performed at a constant strain rate of 3 s^{-1} and the following interpass times were used: i) 2s after a single simulated roughing pass; ii) 0.33s between the first and second prefinishing passes (no "intermediate" passes were applied); iii) 1.9s after prefinishing; and iv) 30 ms for all the finishing passes. It is evident that there is nearly complete softening (i.e. nearly complete recrystallization) after holding intervals of 1.9 and 2s,

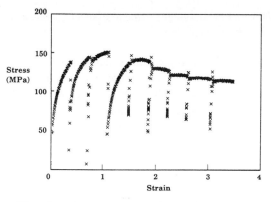

Fig. 3. Torsion simulation flow curves determined at 900 °C for a 0.04%C-0.3%Mn steel. The last six passes, which collectively exhibit dynamic recrystallization, make up the finishing schedule [12].

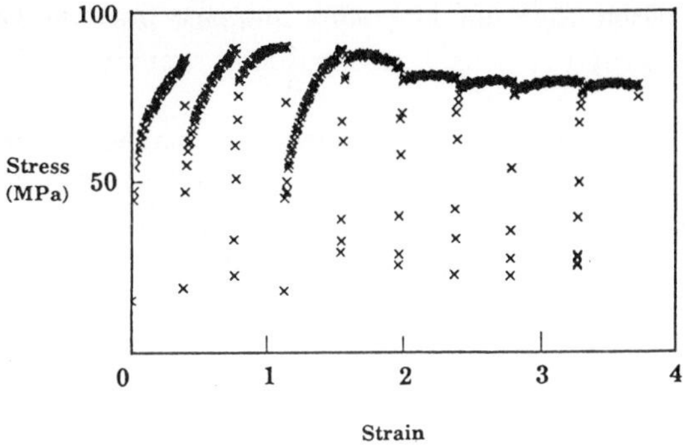

Fig. 4. Torsion simulation flow curves for a 0.03%Nb microalloyed steel tested at 1100 °C. It can be seen that the behaviour is similar to that of the C-Mn steel tested at 900 °C [12].

while there is little or no softening (recrystallization) when the interpass times are well below 1s. Of particular interest is the overall shape or "envelope" of the finishing flow curves. This is characterized by the typical single peak appearance of the "grain refinement" type of dynamic recrystallization flow curve [13].

Somewhat similar results were obtained from a simulation carried out on a 0.03 wt%Nb microalloyed steel, see Fig. 4. In this case, the test was carried out at 1100 °C, i.e. *a full 200 °C above* the temperature applicable to the test of Fig. 3. This temperature is well above that at which precipitation can be expected in this particular steel. Thus the Nb would normally be in solution and some solute drag is expected to be playing a role. Once again, despite the high temperature, little softening is observed when the interpass intervals are well below 1s.

3.2 Correcting for the Differences Between Laboratory and Industrial Strain Rates The results illustrated in these two figures are not fully representative of rod rolling because of the differences in strain rate between the laboratory tests (3 s^{-1} in the above two examples) and the mill. Those applicable to the ten-pass schedule quoted above are listed for reference in the table that follows [11].

Pass Number	1	2	3	4	5	6	7	8	9	10
Strain Rate (s^{-1})	84	71	121	151	314	221	498	421	868	782

As a consequence of the higher mill strain rates, there is more work hardening, so that softening is more rapid under industrial as opposed to laboratory conditions. More interpass softening therefore takes place industrially for a given interpass time. However, allowances can be made for this effect by using the following "law of similarity" for metadynamic recrystallization [11]:

$$t_{lab} = t_{mill} \left(\frac{\dot{\varepsilon}_{mill}}{\dot{\varepsilon}_{lab}} \right)^n$$

Here t_{lab} is the laboratory interpass time equivalent to a mill interval of t_{mill}, where $\dot{\varepsilon}_{lab}$ and $\dot{\varepsilon}_{mill}$ represent the laboratory and mill strain rates, and n is an exponent in the range *0.5 < n < 1.*

Two further problem arise. The first is that the flow stress is also strain rate sensitive, so that simulated mean flow stresses must be corrected by a factor of $(\dot{\varepsilon}_{mill}/\dot{\varepsilon}_{lab})^m$ to produce an estimate of the separation force [6]. The second is that the peak of the stress/strain curve is attained at a strain that is itself sensitive to strain rate [13]. As dynamic recrystallization is initiated in the vicinity of the peak, the strains applied in a given simulated "pass" must be adjusted so that whether strain accumulation takes place on the one hand or dynamic and post-dynamic recrystallization on the other is truly representative of the industrial schedule.

Such simulations have been carried out for a 0.07 wt%C-0.3 wt%Mn plain C steel, leading to the flow curves illustrated in Fig. 5. Here it can be seen that the retention of work hardening is essentially complete during the first and second interpass intervals. Dynamic recrystallization is then initiated in pass 3, followed at first by some softening and then by almost complete softening after pass 4. Strain accumulation also takes place in passes 5 to 7, followed by another cycle of softening and strain accumulation. This sequence of events is summarized in Fig. 6, where the mean flow stresses have now been corrected for the continuous increase in strain rate. The two cycles of softening are readily seen.

The accuracy of the model has also been verified by quenching samples and measuring the austenite grain size after 19 passes of rolling [14]. Such comparisons indicate that the grain sizes obtained from a dynamic/metadynamic model are in close correspondence with observations; by contrast, those based on a model that invokes static recrystallization alone do not lead to realistic predictions.

Although not discussed here in any detail, tests such as those illustrated above involve drawing a distinction between the occurrence and effects of static as opposed to metadynamic recrystallization. This arises because the kinetics of static recrystallization are particularly sensitive to the accumulated strain, while those of metadynamic recrystallization are not. Conversely, those

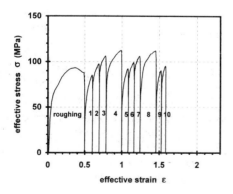

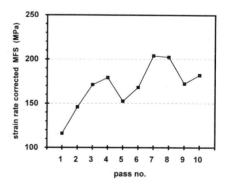

Fig. 5.   Stress vs. strain for the torsion simulation of the rod mill finishing schedule carried out using the strain rate corrections described in the text and in Ref. 6.

Fig. 6.   Strain rate corrected mean flow stress vs. pass number for the rod mill simulation of Fig. 5 [6].

of the metadynamic variety depend critically on the prior strain rate, while the static ones do not. The former are also considerably less sensitive to temperature than the latter [12]. Because of these distinctions, modelling of the microstructural and softening mechanisms acting during rod rolling requires careful determination of the appropriate kinetics [7]. It is also why analysis of the metallurgical events taking place during rod rolling has not proceeded as rapidly as in the case of plate rolling.

Despite these experimental difficulties, it is clear from tests such as those outlined here that there is strain accumulation over several consecutive passes. Then, because of the absence of strain-induced precipitation, dynamic recrystallization is initiated. In the general case, conventional *static* recrystallization does not take place, but there are exceptions. When the interpass time is long enough for appreciable post-dynamic recrystallization, there can be significant softening, leading to behaviours similar to that illustrated in Fig. 6.

3.3 Intermediate Interpass Times:- Strip Rolling The types of interaction outlined above also apply to strip mills. Here again there is strain accumulation, followed as the case may be by the initiation of dynamic and metadynamic recrystallization. The principle difference concerns the time available for strain-induced precipitation. As the total elapsed time from the first to the last (e.g. fifth, sixth or seventh) pass is about ten seconds, and therefore long enough to permit precipitation, the potential is there for the latter mechanism to interfere with both the dynamic and post-dynamic softening processes.

As discussed above, there are many experimental difficulties involved in the laboratory simulation of high strain rate (i.e. short interpass time) processes and strip rolling is no exception to this rule. However, unlike the case of rod rolling, where no separation force data are available by the nature of the process, there are copious amounts of mill log data which can be analyzed to provide some insight into the relative importance of the various metallurgical phenomena. Some examples of this type of approach to the analysis of strip rolling will now be reviewed.

The approach described here based on the analysis of mill logs was developed by Siciliano and Minami [8,9]. They considered data from six industrial mills and investigated the behaviours of both plain C as well as Nb microalloyed steels. Only the latter materials will be discussed here as the presence of Nb in solution in this case leads to the significant solute retardation of recrystallization and therefore to considerable potential for strain accumulation. Two typical examples of "pancaking", i.e. strain accumulation, in the strip mill are illustrated in Fig. 7 [9]. The two steels selected here contained 0.03 and 0.08 wt%Nb in combination with 0.06 wt%C. The other alloying elements of importance were 0.65 wt%Mn and 0.115 wt%Si in the first instance and 1.25 wt%Mn and 0.325 wt%Si in the second.

It is evident from the two diagrams that there is good agreement between the Williams model predictions (here the Hodgson model predictions should be disregarded as inapplicable) and the mill log data. The important point to be noted is that there is strain accumulation over all seven passes and that neither static recrystallization, on the one hand, nor dynamic/metadynamic, on the other, takes place. The *absence of any kind of recrystallization* has been attributed by the above authors to the occurrence of strain-induced precipitation, which is accelerated in turn by the relatively high levels of silicon present in the two steels under consideration. Thus, strip rolling in this case follows the well-known paradigm associated with plate rolling, i.e. with CCR.

In other steels, the above authors reported behaviours that stood in sharp contrast to that displayed in Fig. 7 above. Some of their observations are reproduced here in Fig. 8. The first

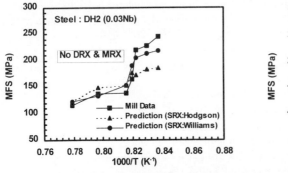
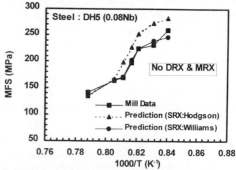

Fig. 7. Comparison of measured MFS values with those predicted by
two different mill models (only the Williams model is relevant here).
Although these models allow for static recrystallization, none takes
place here, leading to strain accumulation instead [9].
a) 0.06%C-0.03%Nb-0.65%Mn-0.115%Si steel
b) 0.06%C- 0.08%Nb-1.25%Mn-0.325%Si steel.

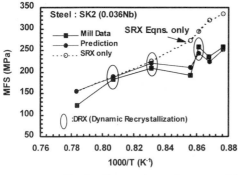
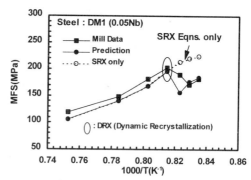

Fig. 8. Comparison of measured MFS values with those predicted by a
mill model that allows for dynamic and metadynamic recrystallization
[9]. The latter two softening processes are responsible for the drops in
MFS observed.
a) 0.09%C-0.036%Nb-1.33%Mn-0.06%Si steel
b) 0.06%C-0.050%Nb-0.70%Mn-0.07%Si steel.

material contained 0.09 wt%C, 0.036 wt%Nb, 1.33 wt%Mn, and 0.06 wt%Si; the second had the
following composition: 0.06 wt%C, 0.05 wt%Nb, 0.70 wt%Mn, and 0.07 wt%Si. It can be seen
that the conventional static recrystallization (SRX) model overpredicts the rolling loads; this is
because it calls for strain accumulation to take place, as in the previous example. (Although the
static recrystallization spreadsheet allowed for softening by static recrystallization, the model did
not detect the occurrence of any such recrystallization under the rolling conditions employed.)
However, when allowance is made for dynamic and metadynamic recrystallization, good agreement
is observed. The difference in behaviour was ascribed to the absence of strain-induced precipitation
in this case, which was attributed in turn to the much lower Si levels that were present in the second
category of steels.

    Thus the strip mill behaviour seemed to depend, in the cases analyzed, on the Si level, and to
a lesser extent on the Mn level as well. Moderate and high Si concentrations appeared to favour the

occurrence of NbCN precipitation, while low levels were consistent with the absence of precipitation. This was attributed to the effect of Si on the C activity, enabling precipitation to take place under the rolling conditions of Fig. 7, but not those of Fig. 8. When precipitation occurs, the strip mill displays "plate mill behaviour"; conversely, in the absence of precipitation, the strip mill displays "rod mill behaviour". Thus, "intermediate" interpass time processes seem to follow the "long" interpass time model in some instances and the "short" interpass time model in others. These phenomena clearly call for further investigation, particularly with regard to the effects of steel composition on the interactions described above.

A further example of "rod mill behaviour" in strip mills will be given here. This is taken from a recent paper by Kirihata and co-workers [10]. It is of interest because their steels contained little or no Nb, but substantial concentrations of Cr (around 1 wt%) and Mo (up to 0.97 wt%), as well as an addition of 0.46 wt%Ni in one case. Thus these materials displayed the type of behaviour expected when recrystallization is controlled by solute drag as opposed to precipitate pinning. Two examples of the results obtained are presented in Fig. 9, again derived from an analysis of the mill logs. The steel in question here contained 0.47 %C, 0.66% Mn, 0.98 %Cr, 0.97 %Mo, 0.46 %Ni, 0.12 %V, and 0.016 %Nb, all in wt%. The point of interest is that the behaviour does not follow the "retention of work hardening" (i.e. plate mill or strain-induced precipitation) model, while significant softening events take place, as called for under the rod mill conditions of Figs. 5 and 6, as well as the strip mill conditions of Fig. 8. As before, these are attributed to the occurrence of dynamic recrystallization followed by metadynamic recrystallization.

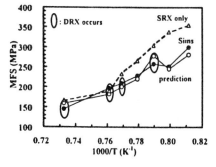 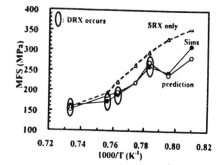

Fig. 9. Comparison of measured MFS values (determined from mill logs) with those predicted by the above dynamic/metadynamic mill model [10]. As in Figs. 6 and 8, in the absence of static recrystallization, the departures from strain accumulation can be attributed to the occurrence of these two softening mechanisms.

4. Conclusions

1. When short interpass times are involved, as in rod rolling, there is insufficient time for the occurrence of static recrystallization in the interstand interval. Thus strain accumulation (the retention of work hardening) takes place. In a similar manner, the times are too short for strain-induced precipitation. Under these conditions, the absence of precipitation makes possible the initiation of dynamic recrystallization, followed by greater or lesser amounts of metadynamic (post-dynamic) recrystallization. Thus, both the rolling loads as well as the microstructures are controlled by the latter two softening processes.

2. When intermediate interpass times are involved, two kinds of behaviour are possible. As long as strain-induced precipitation does not intervene, the rolling load follows the rod mill pattern. This is the case in the presence of appreciable concentrations of alloying elements in solution (e.g. Cr, Mo and/or Ni) or when NbCN is prevented from precipitating by the use of relatively high levels of Mn in association with very low ($<$ 0.1 %) Si contents.

3. Conversely, when strain-induced precipitation takes place within the mill stands, the rolling load follows the "pancaking" pattern that is typical of plate mills, i.e. of classical controlled rolling or CCR. This type of behaviour has been observed in the presence of Si levels above 0.1%.

Acknowledgements   The author is grateful to the Alexander von Humboldt Foundation for a Research Award and to Professor Guenter Gottstein, Director of the Institut fuer Metallkunde und Metallphysik of the RWTH-Aachen for the provision of laboratory facilities. He is also indebted to the co-workers cited above for many stimulating discussions. Finally, he would like to thank the Canadian Steel Industry Research Association and the Natural Sciences and Engineering Research Council of Canada for supporting the investigations that made this presentation possible.

References

1.   J.J. Jonas and C.M. Sellars: Thermomechanical Processing: Sir Robert Honeycombe Commemorative Symposium; eds. J.A. Charles, G.W. Greenwood and G.C. Smith, The Institute of Materials and the Royal Society, London, 1992, pp. 147-177.

2. J.J. Jonas: "Materials into the 21st Century", University of Cape Town Special Issue, Mater. Sci. and Eng., A184 (1994), pp. 155-165.

3.   F. Boratto, S. Yue, J.J. Jonas and T.H. Lawrence: Proc. Int. Conf. on Phys. Metall. of Thermomechanical Processing of Steels and Other Metals (Thermec-88), Tokyo, Japan, June, 1988, pp. 519-526.

4. F. Boratto, R. Barbosa, S. Yue and J.J. Jonas: ibid., pp. 383-390.

5. T. Tanaka, N. Tabata, T. Hatomura and C. Shiga: in "Microalloying '75", Proc. Int. Conf. on HSLA Steels, Union Carbide Corp., New York, 1977, p. 107.

6. T.M. Maccagno and J.J. Jonas: ISIJ Int., 34 (1994), pp. 607-614.

7. T.M. Maccagno, J.J. Jonas and P.D. Hodgson: ISIJ Int., 36 (1996), pp. 720-728.

8. F. Siciliano Jr., K. Minami, T.M. Maccagno and J.J. Jonas: ISIJ Int., 36 (1996), pp. 1500-1506.

9. K. Minami, F. Siciliano Jr., T.M. Maccagno and J.J. Jonas: ISIJ Int., 36 (1996), pp. 1507-1515.

10. A. Kirihata, F. Siciliano Jr., T.M. Maccagno and J.J. Jonas: ISIJ Int., 38 (1998), pp.187-195.

11. P.R. Cetlin, S. Yue, J.J. Jonas and T.M. Maccagno: Metall. Trans. A24 (1993), pp. 1543-1553.

12. P.D. Hodgson, J.J. Jonas and S. Yue: in Proc. Int. Conf. on the Processing, Microstructure and Properties of Microalloyed and other Modern HSLA Steels, ed. A.J. De Ardo, Iron and Steel Society of the Metallurgical Society of AIME, Warrendale, PA, 1992, p. 41.

13. T. Sakai and J.J. Jonas: Acta Metall., 32 (1984), pp. 184-209.

14. J.J. Jonas: in Proc. of Int. Conf. on Thermomechanical Processing in Theory, Modelling and Practice; Swedish Society for Materials Technology, Stockholm, 1997, pp. 24-34.

E-mail address: JohnJ@Minmet.Lan.McGill.CA. Fax number: 1-(514)-398-4492.

Materials Science Forum Vols. 284-286 (1998) pp. 15-26

Microalloyed Strip Steels for the 21st Century

A.J. DeArdo

Basic Metals Processing Research Institute, Department of Materials Science and Engineering, University of Pittsburgh, Pittsburgh, PA 15261, USA

Keywords: Microalloyed Steels, Microstructure, Precipitation, Recovery, Recrystallization, Texture, Thermomechanical Processing, Ultra-Low Carbon Steels

Abstract Over the last quarter-century, microalloyed steels (MA) have attained a premier position in the spectrum of high-quality steel products. Recent history has shown that microalloyed steels are used when high performance and low cost are essential. In the early days of MA steels, around 1970, they were little more than modifications of existing carbon steels. The most popular examples of the early MA steels were products intended for linepipe, strip and forging use. As a better understanding of the behavior of microalloying elements evolved, especially the synergy with hot deformation, the later MA steels started to take on characteristics quite different from both carbon and even earlier MA steels. The evolutionary improvement in MA steels will undoubtedly continue well into the 21st century. This evolution will be aided by understanding how MA elements (MAE) act in contemporary steels. The purpose of this paper is to discuss some recent advances in our understanding of modern MA strip and sheet steels. It is hoped that a better understanding of the various metallurgical phenomena will not only lead to better steels in the future but also aid in the more realistic modeling of these materials.

1. Background

It is well-known that microalloying elements are added to structural steels for three principal reasons: (i) to refine the austenite grain size during rolling and, therefore, to aid in refining the ferrite grain size after transformation; (ii) to lower the transformation temperature, thereby also refining the ferrite grain size and increasing its dislocation density; and (iii) to possibly impart precipitation hardening or solute hardening [1-15]. Since numerous, detailed review papers have been written covering the alloy design and thermomechanical processing of these steels [18,31-33,36-38], only certain aspects of these topics will be discussed here. What follows is focused on the strip mill rolling of two grades: (i) high strength MA strip ($350 \leq YS \leq 420$ MPA), intended for use either as-hot rolled or after cold rolling and annealing and (ii) Ultra-Low Carbon (ULC) strip intended for the cold mill.

Hot strip mill rolling can be summarized as slab reheating near 1250 °C, followed by rough rolling from 250 to 25 mm, followed by finish rolling from 25 to 2 mm, followed by water cooling to a designated temperature, followed by coiling to RT. The roughing passes are usually separated by fairly long interpass times while the finishing passes by very short times. This means that the γ will undergo repeated recrystallization during rough rolling very probably leading to a refined grain size as the transfer bar enters the finishing stands. Although there can be extensive precipitation in austenite depending upon the processing conditions, the combination of very short interpass times and the presence of at least a portion of the MAE in solution means that the last few passes in the finishing train occur below the T_5, the recrystallization stop temperature (Figure 1). Hence, the γ will be "pancaked" as it begins to transform during cooling. The crystalline defect structure of this metallurgical condition, characterized as high Sv [33], a parameter described below, will have numerous sites for ferrite nucleation, and will be at least partially responsible for grain refinement of the final ferrite [31-33,36-38]. As will be discussed below, this ferrite grain refinement is important since it is responsible for about 80-90% of the final yield strength in a 350 MPa grade hot rolled strip.

The addition of elements such as niobium, titanium and vanadium to the early pearlite-reduced steels was used to facilitate this austenite conditioning process [29-32] as well as to lower the transformation temperature [36-38] and to possibly provide precipitation hardening of the ferrite [24]. Exhaustive research has shown that the effectiveness of austenite conditioning can be quantified through the use of the parameter S_V, the total interfacial area of near planar crystalline defects such as grain boundaries, deformation bands and incoherent twin boundaries. Since it is these defects that act as nucleation sites for ferrite during transformation, larger values of S_V were found to be associated with finer ferrite grains, Figure 2. It is now well understood that fine ferrite grains are a result of high S_V and low Ar_3 temperatures [36-40].

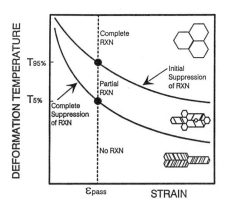

Figure 1 - Schematic illustration of austenite microstructures resulting from various deformation conditions [41]. T_{95} and T_5 are the temperature for 95% and 5% recrystallization, respectively.

Table 1 Typical Characteristics of Hot Rolling Processes

Mill	Plate	Finishing Stands of Strip Mill	Bar
Pass Reduction, %	15	50	25
Strain Rate, Sec^{-1}	20	100	100
Interpass Time, Sec.	20	1	1
Finish Temp, °C °F	750-950 1382-1742	900 1652	1000 1832
No. Passes	15	6	8

The characteristics of various types of hot rolling are quite specific to the actual practice, and perhaps also to the precise mill of interest. A summary of typical, average rolling parameters used for simulation purposes is shown in Table 1 for various types of hot rolling.

When γ is isothermally hot deformed and held to simulate passes in an industrial practice such as one of those described above, and then quenched to RT to observe the resulting microstructure, the pattern of behavior which is found is shown schematically in Figure 1.

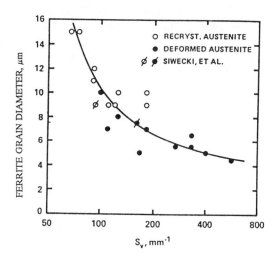

Figure 2 - Ferrite grain sizes produced from recrystallized and unrecrystallized austenite at various S_v values [31].

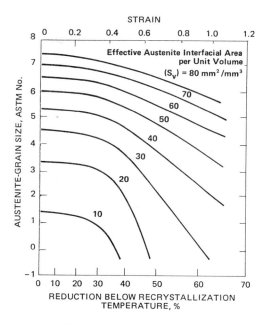

Figure 3 - Variation of effective interfacial area (S_V) with rolling below the recrystallization temperature for the 0.03% Nb steel. [33]

When the deformation occurs above T_{95}, a recrystallized γ grain structure results. However, when the pass strain occurs below T_5, a pancaked microstructure is observed. Passes which take place between these two temperatures lead to detrimental mixed grain structures.

It is apparent from the phenomenology shown in Figure 3 [33] and from the quantification of such data

$$Sv = 2/D_\gamma, \qquad\qquad \text{for } T > T_{RXN} \qquad (1)$$

and

$$Sv = (1 + R + 1/R)/D_\gamma + 0.63(\%R - 30) \qquad \text{for } T < T_{RXN} \qquad (2)$$

where R is the average aspect ratio of deformed austenite grains [31,42], that large values of S_V can be achieved through: (i) refining the starting γ grain size, (ii) increasing the grain boundary area per unit volume by changing the grain shape, e.g., pancaking, and (iii) increasing the density of intragranular crystalline defects such as deformation bands and/or incoherent twin boundaries. Hence, large values of Sv can be achieved through γ grain refinement, γ grain pancaking, or a combination of the two. The first contribution can occur only through control of the reheating practice and by applying most of the passes above the temperature T_{95} described in Figure 1. The second contribution can occur only through applying most of the passes below T_5, described in Figure 1. In this second case, the more total deformation applied below T_5, the higher would be the Sv caused by this approach, i.e., pancaking.

Two comments can be made regarding the general behavior of γ shown in Figure 1. The first is that the diagram will vary with the type of process being considered, even for the same steel and reheating conditions. This is a result of the difference in mill characteristics shown in Table 1 and in the response of the material to these differences. The second is that the overall behavior shown in Figure 1 represents the result of the competition between the driving forces trying to eliminate the excess crystalline defects induced during the processing and the retarding forces that slow or retard this change. Obviously, when the driving force exceeds the pinning force and sufficient diffusivity of pertinent atomic species exists, the movement of defects leading to their reduction or elimination can proceed. The migration of dislocations and low and high angle boundaries resulting in lower numerical and energetic densities are, in fact, what is meant by recovery, recrystallization and grain coarsening. However, when the retarding force exceeds the driving force, or when insufficient atomic mobility exists, then the structural restoration processes will not occur in the time available. For static restoration, which is the predominant mode possible during industrial processing, the time available is the interpass time shown in Table 1.

Thermomechanical processing is normally considered to start with the charging of billets, blooms or slabs into a reheat furnace. In this case, the major events of interest are the grain coarsening that occurs during reheating and the degree of recrystallization or pancaking that takes place during rolling. However, there is another aspect of TMP that is often overlooked, i.e., the high temperature straightening of freshly solidified steel that has been processed through a curved mold caster. The very large austenite grains that exist in this temperature range often lack the hot ductility required by the straightening process, especially in microalloyed steel. The combination of large austenite grains and the presence of fine microalloy precipitates result in a lack of hot ductility and the occurrence of surface cracking. These effects are shown in Figure 4 for steels microalloyed with Nb and/or Ti , and similar results are found for steels containing V [17]. Hence, the application of hot charging, hot direct rolling and continuous strip production (CSP) for improved productivity are processes that are all susceptible to the phenomena shown in Figure 4 and must be considered in the application of these technologies. Fortunately, these hot cracking problems can be eliminated through the combination of a hypostochiometric addition of titanium to suppress high temperature precipitation of carbonitrides and the control of the cooling practices on the caster.

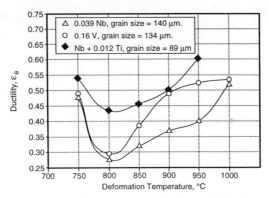

Figure 4 - Hot ductility curves for V, Nb, and Nb-Ti steels. [17]

2. Optimization of Composition for Strip Rolling

Rolling on a hot strip mill is quite different than on a plate mill. The widths are narrower, the rolling speeds are higher, the reductions are greater and the total rolling times are much shorter. In a typical strip mill the time from slab drop out to exiting the last finishing pass is only about one to three minutes. Comparable times are five to ten minutes for plate or structural rolling. Since nearly all of this time is interpass time, static hardening and softening processes are expected to be much more important and variable in plate and structural rolling than in strip rolling.

High Strength Strip

Since high strength strip produced as hot rolled band is used as a structural material, the properties of interest are strength, ductility and toughness. Therefore, rolling to achieve a large Sv is a primary goal of the alloy design and rolling procedure. During strip rolling, the lack of adequate time between passes means that most of the MAE stays in solid solution during rolling. This has been confirmed in two recent studies of strip mill simulations conducted on HSLA steels containing Nb in the range 0.02-0.03 wt% [43,44]. In these studies, a portion of the total Nb was found as strain induced precipitation in specimens quenched immediately after the last finishing pass. The analysis of the strength observed in the as-coiled condition revealed that about 80-90% of the observed strength of 350 MPa can be attributed to the combination of grain size and dislocation strengthening, depending upon the Mn level and coiling temperature. HSLA steels with higher levels of MAE, e.g., ≥ 0.05 Nb with or without 0.04 Ti, have shown evidence of very extensive precipitation hardening, Figure 5 [45]. A comparison of the Nb and the Nb + Ti steels reveals a much higher strength in the latter. The higher strength found in the Nb + Ti steels indicates that the lower strength found in the Nb steels might have been caused by some strain induced precipitation of NbNC in austenite during high temperature rolling. The suppression of precipitation NbNC in the Nb + Ti steels was achieved by the precipitation of TiN. Hence, nearly all of the dissolved Nb would have been retained in solution in austenite during rolling and available for precipitation hardening of the ferrite during cooling in the Nb + Ti steels. This most probably was responsible for the higher yield strengths observed in these steels [45].

Although only a small amount of strain-induced precipitation was found in the low Nb steels mentioned above [43,44], the as-rolled austenite was nevertheless pancaked. This indicated that the retardation of recrystallization was caused by both precipitation and Nb in solid solution. Pancaked prior austenite in hot strip is not rare, and MAEs in solution are known to retard static softening, as shown in Table 2 [46] and Figure 6 [47].

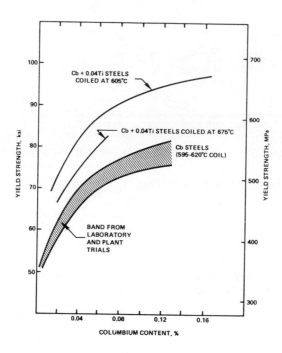

Figure 5 - Summary of mill-trial results for hot-rolled Nb-Ti sheet steels. [45]

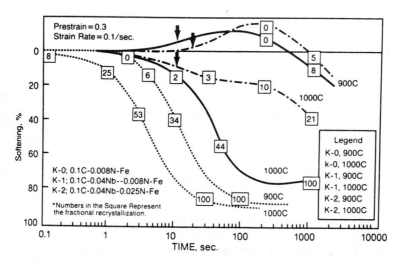

Figure 6 - Static restoration behavior of three steels at 900 and 1000°C. Arrows indicate the start of strain induced precipitations. [47]

Table 2 Solute Retardation Parameters for Static Recrystallization [46]

Element (0.1%)	SRP (0.1at%)	SRP (0.1wt%)
V	12	13
Mo	33	20
Ti	70	83
Nb	325	222

The influence of MAE on recrystallization stop temperature, T_5, shown in Figure 7, is applicable to plate rolling, with rather long interpass times. The effects observed are probably caused mainly by particle pinning with some minor contribution by solute drag. There is no similar plot illustrating the influence of solute MAE on T_5. Since the pinning effect by solute drag is substantially smaller than that of particles, the slope of a curve of the types shown in Figure 7 might be expected to be somewhat smaller. Nevertheless, a substantial amount of Nb in solution could be expected to increase the T_5, especially since this Nb would almost certainly be located at the austenite grain boundaries [48]. The presence of 0.04 at % Nb was found to increase T_5 by about 250°C for the plate simulation shown in Figure 7. How much this Nb would raise the T_5 for a strip simulation where a portion of the Nb remains in solution is not known, but it can be probably safely assumed to be substantial.

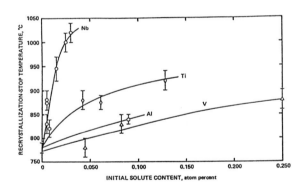

Figure 7. The increase in recrystallization temperature with increase in the level of microalloy additions in a 0.07C, 1.40Mn, 0.25Si steel [49].

Ultra-Low Carbon Steels

Ultra-low carbon steels (ULC) or interstitial-free (IF) steels are used in applications where excellent formability is paramount. Since formability appears to be improved with reduced solute carbon contents in ferrite, these steels are first vacuum degassed to carbon and nitrogen levels near 20-30 ppm, and are then microalloyed with Ti and/or Nb to further reduce these solute levels. These extremely low interstitial levels also meant that these steels were ideally suited for processing on continuous annealing lines, where the rapid post-annealing cooling precludes the precipitation of cementite that normally can be found in batch annealed AKDQ steel.

The first generation of these steels was fully stabilized with yield strengths near 150 MPa. While these steels have shown excellent formability, i.e. deep drawability, their low yield strengths have proven to be detrimental in both handling and weight reduction. To overcome these difficulties, higher strength or second generation versions of these early steels have been developed. These newer steels have yield strengths in the range 200-300 MPa, again with good formability. The higher strengths exhibited by these newer steels have been achieved by either phosphorus additions or by understabilizing followed by bake hardening.

Stabilization What follows is a discussion of the alloy design and TMP of first generation ULC steels stabilized with either Ti or Ti + Nb. The role of TMP in these steels is altogether different from that for high strength strip. The optimum combination of composition and TMP for ULC steels should provide the proper levels of: (i) stabilization of C and N, (ii) solute Nb, and (iii) hot band texture.

Recent work on stabilization in ULC steels has shown that there can be six versions of these steels, depending upon the composition, Figure 8 [50]. Each version is defined by the precipitation reaction expected during rolling or the precipitates expected in the hot band.

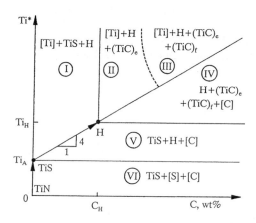

Figure 8 - Simplified stabilization map for Ti-containing ULC steels. Subscripts e and f indicate carbides that form as epitaxial and as free standing particles, respectively. [50]

Early work in this field assumed that the following reactions would occur with falling temperature [51]:

$$Ti + N = TiN, \tag{3}$$
$$2Ti + S + C = \tfrac{1}{2}\,Ti_4C_2S_2, \tag{4}$$

and

$$M + C = MC \ (M{=}Ti,\,Nb). \tag{5}$$

Recent work has offered a different view of these reactions [50,55]

$$TiS + Ti + C = \tfrac{1}{2}\,Ti_4C_2S_2 \tag{6}$$

A new view of carbon stabilization has emerged that is based upon the in situ transformation of TiS to $Ti_4C_2S_2$ by an intercalation process whereby Ti and C are consumed by the reaction, equation 6. This view is based on the role of sulfur in the formation of $Ti_4C_2S_2$ or H-phase [50], and is shown in Figure 8, which is plotted for a given S level. Figure 8 is based upon two points, the first describing TiS and the second describing H in Ti-C space. The line connecting these two points is equation 6. All compositions lying above the line are fully stabilized, assuming equilibrium, while those below the line are not. Clearly, all stabilized compositions are not equivalent in terms of the precipitation reactions and precipitates formed. Steels designed to fall in Region I will have all of its carbon stabilized by H. Steels containing higher S can stabilize more carbon using this reaction. As the carbon content exceeds the $C_{H,}$ this carbon is removed from solution as MC, first by epitaxial MC at low excess C and then by free standing MC at high excess C. The selection of composition can be expected to have a dramatic effect on mill loads during rolling and recrystallization temperatures during annealing following cold rolling.

The influence of stabilization routes on rolling behavior is shown in Figure 9 which shows the alpha and gamma fibers for as-coiled hot band [53,54]. The MC stabilized hot band, where H has been explicitly suppressed, shows a broad maximum extending from near (223)[1$\bar{1}$0] to (001)[110], whereas the two H stabilized steels show peaks near (111)[1$\bar{1}$0]. This difference can have important implications during cold mill processing and in the final product [53,54].

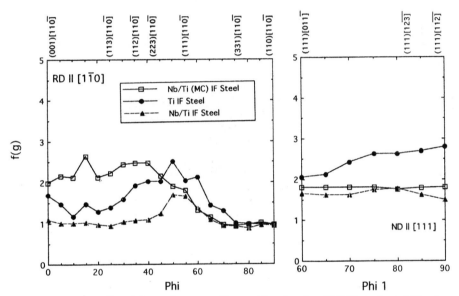

Figure 9 - The alpha and gamma fibers of hot bands of ULC steels stabilized by either $Ti_4C_2S_2$ or MC. [52]

One reason for this difference may lie in the variation in the carbon content during hot rolling of austenite that results from the two stabilization mechanisms. This difference is shown schematically in Figure 10. Another reason, recently observed, may lie with the effect caused by differences in as-reheated grain size.[18,54]

When Nb is added to a steel whose composition lies above the stabilization line, especially in Regions I and II, it remains largely in solution in the final ferrite. However, when it is added to a steel falling below the line, the Nb acts as a stabilizer by forming NbC precipitates. In the former case, detailed studies have indicated that about 60% of the total Nb remains in solution on the ferrite grain boundaries. A typical example, taken from atom probe field-ion microscopy studies of an as-coiled ULC steel lying in Region II, is shown in Figure 11 [50].

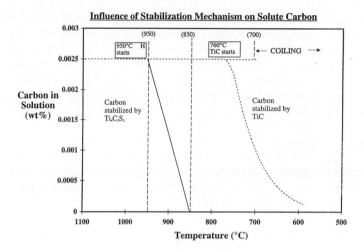

Figure 10 - Illustration of solute [C] as a function of temperature. [18]

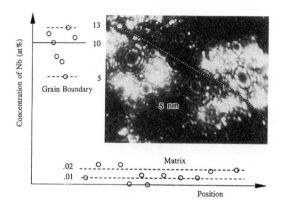

Figure 11 - The concentration of Nb on grain boundaries, subgrain boundaries and within the ferrite matrix, measured by APFIM, from hot band and coiled Ti+Nb ULC steel [49].

The Nb in solution on the ferrite grain boundaries appears to have interesting beneficial effects in terms of improvements in texture development, formability, powdering resistance, and cold work embrittlement without increasing the recrystallization temperature during annealing [54].

3. Summary and Conclusions

Microalloyed steels are complex, sophisticated engineering materials whose ultimate performance depends upon the synergistics of alloy design, deformation processing and controlled cooling through to be done in both optimizing known grades and in developing grades of the future. This information and knowledge has been shown to be and will continue to be vital to the modeling efforts which can be so beneficial to the steel industry.

References

1. Proc. Microalloying 75 (Washington, DC), Union Carbide Corp., New York, 1977.
2. Proc. Hot Deformation of Austenite (Cincinnati), TMS-AIME, New York, 1976.
3. Proc. HSLA Steels: Technology and Applications (Philadelphia), ASM, Metals Park, OH, 1984.
4. Niobium (San Francisco), TMS-AIME, Warrendale, PA, 1984.
5. Proc. Thermomechanical Processing of Microalloyed Austenite (Pittsburgh), TMS-AIME, Warrendale, PA, 1984.
6. Proc. High-Strength Low Alloy Steels (Wollongong, NSW), South Coast Printers, Port Kembla, NSW, 1986.
7. Proc. Accelerated Cooling of Steel (Pittsburgh), TMS-AIME, Warrendale, PA, 1986.
8. Proc. HSLA Steels (Beijing), TMS-AIME, Warrendale, PA, 1986.
9. Proc. Microalloyed Forging Steels (Golden), TMS-AIME, Warrendale, PA, 1987.
10. Proc. Accelerated of Rolled Steel (Winnipeg), Pergamon Press, New York, 1988.
11. Proc. Processing, Microstructure and Properties of HSLA Steels (Pittsburgh), TMS-AIME, Warrendale, PA, 1988.
12. Proc. Microalloyed HSLA Steels (Chicago), ASM, Metals Park, OH, 1988.
13. Proc. HSLA Steels (Beijing), TMS-AIME, Warrendale, PA, 1992.
14. Proc. Processing, Microstructure and Properties of Microalloyed and Other Modern HSLA Steels (Pittsburgh), ISS-AIME, Warrendale, PA, 1992.
15. Proc. Low Carbon Steels for the Nineties (Pittsburgh), TMS-AIME, Warrendale, PA, 1993.
16. D. N. Crowther and B. Mintz, "Influence of Grain Size on Hot Ductility of Plain Carbon Steels," Mater. Sci. and Eng., 2, (1986), 951-955.
17. J. Y. Fu et al., "On the Hot Ductility of Continuously Cast Microalloyed Steels," Proc. 8th Processing Technology Conference: The Effect of Microalloys on the Hot Behavior of Ferrous Alloys (Dearborn), (ISS-AIME, Warrendale, PA, 1989), 43-50.
18. A. J. DeArdo "The Physical Metallurgy of Thermomechanical Processing of Microalloyed Steels," Thermec '97, (Wollongong, 1997), in press.
19. C. I. Garcia and A. J. DeArdo, "Structure and Properties of ULCB Steels for Heavy Section Applications," Ref. 12, 291-300.
20. C. I. Garcia et al., " The Physical Metallurgy of Ultra-Low Carbon Bainitic Steels for Plate Applications," Trans. Iron Steel Soc. AIME, (1991), 97.
21. E. O. Hall, Proc. Physical Soc., Series B, 64B, (1951), 747.
22. N. J. Petch, "The Ductile-Cleavage Transition in Alpha -Iron," Proc. Fracture (Swampscott), (Wiley, New York, 1959), 54-67.
23. W. B. Morrison and J. A. Chapman, Phil. Trans. Roy. Soc., 282A, (1976), 289.
24. D. Webster and J. A. Woodhead, "Effect of 0.03%Nb on the Ferrite Grain Size of Mild Steels," J. Iron Steel Inst., 202, (1964), 987.
25. F. B. Pickering, Ref. 1, 9.
26. N. S. Pottore et al., "Interrupted and Isothermal Solidification Studies on Low-Alloy Steels," Metall. Trans. A, 22A, (1991), 1871.
27. C. I. Garcia and A. J. DeArdo, "Formation of Austenite in 1.5% Mn Steels," Metall. Trans., 12A, (1981), 521.
28. C. I. Garcia and A. J. DeArdo, "Formation of Austenite in Low Alloy Steels," Proc. Solid-Solid PhaseTransformations, (TMS-AIME, Warrendale, PA, 1982), 855.

29. M. E. Fine, <u>Phase Transformations in Condensed Systems</u>, McMillan, New York, 1964.
30. D. J. Swinden and J. H. Woodhead, J. Iron Steel Inst., 209, (1971), 883.
31. G. R. Speich et al., "Formation of Ferrite From Controlled Rolled Austenite," <u>Proc. Phase Transformations in Ferrous Alloys (Philadelphia)</u>, (TMS-AIME, Warrendale, PA, 1984), 341-390.
32. A. J. DeArdo, "Ferrite Formation From Thermomechanically Processed Austenite in HSLA Steels," Ref. 6, 20.
33. I. Kozasu et al., "Hot Rolling as a High-Temperature Thermo-Mechanical Process," Ref. 1, 120-135.
34. J. D. Grozier, "Production of Microalloyed Strip and Plate by Controlled Cooling," Ref. 1, 241-250.
35. K. Amano et al., "Moderation of Controlled-Rolling by Accelerated Cooling," Ref. 7, 349-365.
36. A. J. DeArdo, "Influence of Thermomechanical Processing and Accelerated Cooling on Ferrite Grain Refinement in Microalloyed Steels," Ref. 7, 97.
37. T. Tanaka, "Overview of Accelerated Cooled Steel Plate," Ref. 10, 187-208.
38. A. J. DeArdo, " Accelerated Cooling: A Physical Metallurgy Perspective," Can. Met Q., 27, (1988), 141.
39. E. E. Underwood, <u>Quantitative Metallography</u>, (McGraw-Hill, New York, 1968), 77.
40. L. J. Cuddy, "Grain Refinement of Nb Steels by Control of Recrystallization During Hot Rolling," Metall. Trans. A, 15A, (1984), 87-98.
41. E. J. Palmiere, University of Pittsburgh, Unpublished Research, 1990.
42. C. Ouchi, T. Sampei, and I. Kozasu, Trans Iron and Steel Institute of Japan, V. 22, (1982), 214.
43. M. Hua, Unpublished Research, University of Pittsburgh, 1996
44. V. Thillou, M. Hua, C. I. Garcia, C. Perdrix, and A. J. DeArdo, In this Conference, 1998.
45. P. E. Repas, "Control of Strength and Toughness in Hot-Rolled Low- Carbon Manganese-Columbium-Vanadium Steels," Ref. 1, 387-396.
46. M. G. Akben et al., Acta Metall., 29, 1981, p. 111.
47. O. Kwon and A. J. DeArdo, "Interactions Between Recrystallization and Precipitation in Hot-Deformed Microalloyed Steels," Acta Met., 39, (1991), 529.
48. E. J. Palmiere, C. I. Garcia, and A. J. DeArdo, "The Influence of Niobium Supersaturation in Austenite on the Static Recrystallization Behavior of Low Carbon Microalloyed Steels," Metall. Mater. Trans. A, 27A, (1996), 951-960.
49. L. J. Cuddy, "The Effect of Microalloying Concentration on the Recrystallization of Austenite During Hot Deformation," Plastic Deformation of Metals, (Academic Press, New York, 1975), p.129.
50. M. Hua, C. I. Garcia, and A. J. DeArdo, Metall. Mater. Trans. A, 28A, (1997), 1769.
51. D. O. Wilshynski-Dresler, D. K. Matlock, and G. Krauss, <u>Int. Forum for Physical Metallurgy of IF Steels</u>, (ISIJ and Nissho Iwai Corporation, Tokyo, 1994), 13.
52. R. M. Fix et al., "Mechanical Properties of Ti-V Microalloyed Steels Subjected to Plate Rolling Simulations Using Recrystallization Controlled Rolling," Ref. 8, 219-228.
53. L. J. Ruiz-Aparicio, M. Hua, C. I. Garcia, and A. J. DeArdo, "Thermomechanical Processing of Interstitial-Free Steels," <u>Int. Conf. on Thermomechanical Processing</u>, (Stockholm, Swedish Society for Materials Technology, 1997), 407-414.
54. L. J. Ruiz-Aparicio, C,I. Garcia, A. J. DeArdo, "Effect of Initial Grain Size on Transformation Textures of Ultra-Low Carbon Steelsl", <u>Proceedings of 39th Mechanical Working and Steel Processing Conference, International Symposium on Flat Rolled Products</u>, (Iron and Steel Society of AIME, Warrendale, PA., 1998) in press.
55. G. Tither, C. I. Garcia, M. Hua, and A. J. DeArdo, "Precipitation Behavior and Solute Effects in Interstitial-Free Steel", <u>International Forum for Physical Metallurgy of IF Steels</u>, (ISIJ and Nissho Iwai Corporation, Tokyo, 1994), 293-322.

Correspondence: A. J. DeArdo's e-mail address is deardo@engrng.pitt.edu

Materials Science Forum Vols. 284-286 (1998) pp. 27-38
© *1998 Trans Tech Publications, Switzerland*

Modelling the Austenite Decomposition in Steel on a Physical Basis

S. van der Zwaag

Laboratory of Materials Science, Delft University of Technology and Netherlands Institute for
Metals Research, Rotterdamseweg 137, NL-2628 AL Delft, The Netherlands
E-Mail: S.vanderZwaag@stm.tudelft.nl

Keywords: Phase Transformation Kinetics, Modelling, Grain Size Distribution, Interface Mobility, Surface Tension, Early Growth, JMA Kinetics

Abstract.

This paper presents some recent models [1-3] for modelling the $\gamma \rightarrow \alpha$ transformation kinetics of Fe-C alloys on a physical basis. The models illustrate the effect of the austenite grain size distribution and the ferrite nucleation density, the effect of surface tension during early growth and the effect of a finite interface mobility during later stages of growth. Although these models are not yet capable of modelling the transformation behaviour of lean steels let alone that of HSLA steels, the concepts presented here may be useful in the development of more advanced models applicable to real steel grades.

1. Introduction.

Notwithstanding its long history and its empirical background steel technology has developed into a high tech industry, with many fundamentally new process routes being developed even now. Inherent to high tech industries is the need for detailed models describing all relevant processes taking place during production. These models can either be semi-empirical and closely linked to the production facilities or more physico-chemical in nature and linked to fundamental parameters. While the first approach is far more suitable for product and process optimisation, the second approach, in combination with creative technological thinking, has advantages when evaluating not-yet existing processes and material grades. Furthermore, fundamental models can indicate which parameters should be incorporated in, or be deleted from, the more empirical models. Of course, if the product complexity increases, as it does in modern steel technology and steel grades, it becomes increasingly difficult to build and validate such models.

Modelling the transformation behaviour of complex microalloyed steels, the subject of this conference, just from physical principles is too complex at present. However, in recent years several models for the transformation behaviour of simpler steels have been developed which may form the basis of such future models. These models concentrate on three important issues in physical modelling the transformation kinetics : the proper description of the starting structure, the nucleation and early growth phenomena and the growth kinetics. In none of these models the effect of precipitates is taken into account explicitly, but the models are such that their effect can be incorporated at a later stage.

2. The geometry of the austenite grain and the nucleation site distribution.

The transformation kinetics are determined not only by thermodynamics but geometrical features, the distribution of nucleation sites over the austenite grain and the grain geometry itself, play an important role too. A highly suitable approach to modelling the austenite microstructure consists of the construction of randomly chosen Voronoi cells (also known as Wigner-Seitz cells). The very construction of a Voronoi tessellation of space shows a strong resemblance with the evolution of a real microstructure, during which more or less randomly distributed nuclei grow to form grains. The tessellation of such grains is space filling and possesses a natural capriciousness. The realistic presence of grain edges and corners is of importance, since it has been shown experimentally that these specific sites in the structure are governing the nucleation behaviour [4,5]. Nucleation mainly takes place heterogeneously, *i.e.* at specific sites in the structures, at which a lower activation energy occurs. It is a statistical process; all sites in the structure have a certain probability to induce nucleation. Observations by both Huang [4] and Zurek [5] on partially transformed steel samples have shown that nucleation preferentially takes place, in order of increasing potency, at grain boundaries, edges and corners. For an adequate modelling of the transformation process, it is therefore of importance that a model structure contains a realistic density of grain boundaries, edges and corners.

In model calculations, however, due to the random shapes of the Voronoi cells, a large number of cells should be considered, which in practice is too computer-time consuming. In order to overcome this practical problem without discarding the connection to the Voronoi tessellation, a single geometrical body, the tetrakaidecahedron, has been proposed as a regular, but typical, representation of the Voronoi cells [1,6].

It can be shown that the geometrical characteristics of a lognormal volume distribution of tetrakaidecahedra and those for a Voronoi-cell tessellation are highly comparable [1,6], hence the tetrakaidecahedron is a suitable shape for testing the effect of geometrical assumptions on the ensuing overall transformation kinetics. In order not to obscure the geometrical effects we performed calculations assuming isothermal massive transformations, i.e. the growth rate is constant in time and unlike the usual diffusional transformation behaviour independent of the degree of transformation. This approach enables the normalisation of the transformation times as a reduced time t^* can be defined which equals the diameter of the tetrakaidecahedron divided by the growth rate. Now the effect of the number of active nuclei per tetrakaidecahedron, N_n, invariably located at the grain corners, on the transformation kinetics can be modelled. The influence of the nucleation site density on the transformation curves is shown in figure 1, which also contains the transformation curve calculated for a spherical austenite grain with instantaneous nucleation all along the surface.

Clearly, the effect of the geometry is significant. In contrast to the spherical model, the calculations on the tetrakaidecahedron result in sigmoidal transformation curves, which is consistent with the well-known Johnson-Mehl-Avrami (JMA) kinetics. In particular, the transformation rate at the start of the transformation is very sensitive to the model geometry. This is due the large γ/α-interface surface area at the start of the transformation described by the spherical model, compared to the (more realistic) small area of the ferrite nuclei in the tetrakaidecahedron. As the ferrite-volume change rate is the product of the interface surface area and the interface velocity, the spherical model predicts a much higher transformation rate at the start of the transformation than the tetrakaidecahedron. At somewhat higher ferrite fractions, the ferrite grains in the tetrakaidecahedron impinge and continue to grow in what is more similar to a shell-like manner. Therefore, for relatively large numbers of nuclei N_n the shape of the tetrakaidecahedron curve approaches the shape of the spherical-model curve for larger ferrite fractions.

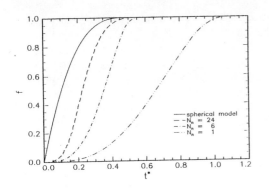

figure 1: the isothermal transformation kintics for various nucleus densities per tetrakaidecahedron. The solid line applies to the spherical model.

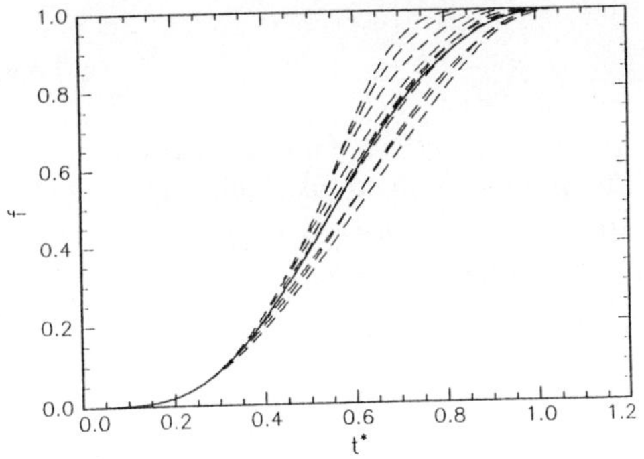

figure 2: the isothermal transformation kinetics for relative nucleaus positions (N_n=2)

The parameters that influence the exact shape of these curves are not only the nucleus density, but also the position of the nuclei in the tetrakaidecahedron, as becomes clear from figure 2. Figure 2 shows, for $N_n = 2$, the dependence of the transformation curves on the relative positions of the nuclei. In general, the further apart the nuclei are, the faster the transformation proceeds, since overlap effects occur only at the last stage of the transformation. When confronting these curves with experimental transformation curves, the solid line in figure 2, giving the average transformation curve, is of importance.

The calculations presented so far have all been performed on a single grain. However, also the austenite grain-size distribution is an important parameter for the transformation kinetics. We will now present simulations in which three kinds of distributions have been taken into account: the mono-disperse distribution (a single grain size), the normal distribution and the lognormal distribution. These distributions apply to the size parameter a, representing the dimensions of the austenite grains. In all calculations the number of nuclei per austenite grain is $N_n = 1$. The size of the austenite grains strongly influences the transformation curves. Consequently, the introduction of a grain-size distribution results in a broadening of the transformation curves. This phenomenon is illustrated in figure 3, in which the transformation curves are presented for the three distributions under consideration. The normal and lognormal distributions have an average diameter that is equal to the mono-disperse grain size and a standard deviation of 0.19 (normalised to the average grain size). Figure 3 shows that, although the average grain size is the same in all three cases, the transformation curves for the normal and lognormal distributions are not only broader than for the mono-disperse distribution, but also shifted to longer times. This shift can be ascribed to the presence of larger austenite grains in the distributions, especially for the lognormal distribution, since it is the largest grain that determines the total transformation time. As the number of austenite grains per volume element scales with one over the average of a^3, this value is different for the three distributions. The lognormal distribution has the smallest number of austenite grains per volume element. As in the simulations the number of nuclei per austenite grain was kept constant. This corresponds to a lower nuclei density and thus to a slower transformation.

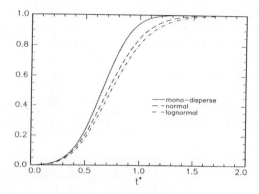

figure 3 : the effect of the grain size distribution on the transformation kinetics for the case of 1 nucleus per tetrakaidecahedron.

As is manifest from figure 1 the transformation kinetics for the tetrakaidecahedron model resemble those of the well known Johnson-Mehl-Avrami equation [7,8]

$$f(t) = 1 - exp\left(-(kt)^n\right)$$

(1)

in which k is a rate factor that is related to the growth rate of the new phase and to the density of active nuclei, and in general will be temperature dependent, and n is the so-called Avrami-exponent. In order to investigate the correspondence of the tetrakaidecahedron geometry with a prevailing transformation model, the transformation curves calculated for the tetrakaidecahedron geometry are fitted using the JMA-equation (12). The rate factor k and the Avrami-exponent n are used as fitting parameters. Figure 4 shows the results for the transformation curves using $N_n = 1$, $N_n = 6$ and $N_n = 24$. It can be seen that the JMA-equation and the simulations show high similarities in their graphical results. It was found that the fitting parameters depend rather sensitively on the number of nuclei per tetrakaidecahedron and their relative position (figure 5).

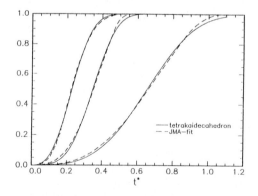

figure 4 : the effect of the nucleus density on the transformation kinetics and the JMA fits.

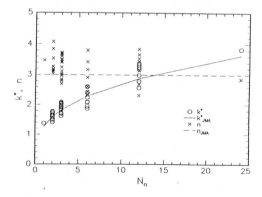

figure 5: the Avrami coefficient as a function of the nucleus density and relative position.

These simulations indicate that the conclusion of a change in the physical transformation mechanism may not be drawn purely on the basis of a change in the parameter values for JMA fits to experimental data. Furthermore, the simulations indicate that great attention must be devoted to the original austenite grain size distribution and the number of nucleation sites per austenite grain when analysing experimental transformation kinetics data.

3. Nucleation and early growth phenomena.

The process of diffusion controlled phase transformations can be divided into three steps: 1. The formation of a nucleus with a size equal to the critical size, i.e. the nucleus is sufficiently large to decrease the energy of the alloy by growth. 2. Early growth during which the surface tension effects are significant. 3. Further growth of the phase when surface tension effects are no longer significant. Few experimental data on the nucleation process are available because of the small scale (~10 nm) of the stable nuclei and their transient nature. No experimental techniques are available for measuring the nucleation behaviour in situ, although the newest grades of TEM with point element analysis facilities may help in unravelling the physics and the chemistry of the process. It is this inaccessibility that has resulted in nucleation still being described by the classical nucleation theory [9,10], which although probably basically correct, contains parameters whose value can not be determined unambiguously or separately. The second step of the diffusion controlled phase transformation involves curvature dependent interfacial concentrations as surface tension effects contribute to the Gibbs free energy curve. During the third step the interfacial concentration is commonly assumed to be constant and equal to that of both phases in thermodynamic equilibrium. This section focuses on the second step.

For very small particles the Gibbs free energy of the constituent phase is increased due to the small radius of curvature of the particle. Due to the change of the Gibbs free energy, the solid solubilities at the interface are no longer constant and change continuously during the transformation. Hence the standard diffusional growth models do not apply to the early stage of the ferrite growth. However, a model for the early growth kinetics can be derived as follows.

The critical nucleus size, R^*, depends on the surface energy, γ, the molar volume of the second phase β, V_{mol}^β, and the driving force for the nucleation of the new phase β, $\Delta G_{\alpha\beta}$ by the following relationship [9]:

$$R^* = -\frac{2\gamma\, V_{mol}^{\ \beta}}{\Delta G_{\alpha/\beta}} \qquad (2)$$

To determine the parameter $\Delta G_{\alpha\beta}$ the parallel tangent construction is used. The energy for the formation of a nucleus of size R is [12]:

$$G^\beta_{\ nuc} = \int_0^{n^*_\bullet} G^\beta_{\ m}(p^\beta)\,dn = \int_0^{n^*_\bullet} (G^\beta_{\ m}(p^o) + \frac{2\gamma}{R}V_m^{\ \beta})\,dn \qquad (3)$$

where n is the number of moles of the β phase per unit volume. The pressure p^0 refers to the actual pressure in the alloy at the nucleation site if no nucleus were present. As the increase of the number of moles of the second phase per unit volume is given by $dn=4\pi R^2 dR/V_m^\beta$, substitution of this into Eq. (3) and subsequently carrying out the integration, yields:

$$G^{\beta}{}_{nuc} = n^{*}G^{\beta}{}_{m}(p^{0}) + 4\pi\gamma\,(R^{*})^{2} \tag{4}$$

For the critical radius it follows from Eq. (2): $\gamma = \dfrac{p^{*}R^{*}}{2}$. Substitution into Eq. (4) yields:

$$G^{\beta}{}_{nuc} = n^{*}G^{\beta}{}_{m}(p^{0}) + 2\pi p^{*}(R^{*})^{3}{}^{*})3 \tag{5}$$

For the critical volume of the nucleus, $V^{*}=4\pi(R^{*})^{3}/3$, $V^{*}=n^{*}V^{\beta}{}_{m}$. Hence, the Gibbs free energy needed for the formation of a nucleus of the critical size is :

$$G^{\beta}{}_{nuc} = n^{*}G^{\beta}{}_{m}(p^{0}) + \frac{3}{2}\cdot p^{*}V^{\beta}{}_{m}n^{*} \tag{6}$$

The above expression represents the energy required for the formation of a nucleus with the critical size. As the Gibbs free energy depends on the particle radius, R, the common tangent construction of the Gibbs free energies of both phases yields particle size dependent solid solubilities in both phases, i.e. $c^{\alpha} = f(R)$, with c^{α}, R and $f(R)$ respectively denoting the solid solubility of the alloying element in the matrix, the particle size and a function of R. Assuming a spherical β particle to grow into a finite cell and applying the correct boundary conditions, the interfacial position and hence the growth rate can be calculated by numerical integration. The results of such calculations will now be demonstrated for the case of ferrite formation from supersaturated austenite under isothermal transformation conditions.

Firstly, the influence of the initial matrix concentration on the growth kinetics has been investigated starting with an initial particle radius of 1.001 times the appropriate critical radius. The common physical input parameters are listed in table 1. The critical radius, interfacial concentration and particle concentration at t = 0 are listed in table 2.

Table 1: The physical parameters used in the calculations corresponding to Fig. 3.
D is the diffusion coefficient of carbon, R_c the cell size, G_i^j the free energy of i atoms in j phase.

D	R_c	$^{0}G^{\alpha}{}_{A}$	$^{0}G^{\alpha}{}_{B}$	$^{0}G^{\beta}{}_{A}$	$^{0}G^{\beta}{}_{B}$	$R \cdot T$	$V^{\beta}{}_{m}$	γ
$m^{2}\,s^{-1}$	m	kJ mol^{-1}	kJ mol^{-1}	kJ mol^{-1}	kJ mol^{-1}	kJ mol^{-1}	m^{-3}	J m^{-2}
10^{-10}	$5 \cdot 10^{-4}$	2	20	20	1.5	15	10^{-4}	1

Table 2: The matrix concentration, critical radius, interfacial concentration and the particle concentration corresponding to the cases of non-zero surface tension in Fig. 3.

c^{m} (mass %)	R^{*} (m)	$c^{\alpha}(t = 0)$	$c^{\beta}(t = 0)$
0.30	$7.3145 \cdot 10^{-8}$	0.2999	0.8300
0.29	$8.5076 \cdot 10^{-8}$	0.2899	0.8231
0.28	$1.0218 \cdot 10^{-7}$	0.2800	0.8159
0.27	$1.2877 \cdot 10^{-7}$	0.2700	0.8082
0.26	$1.7578 \cdot 10^{-7}$	0.2600	0.8001
0.25	$2.8142 \cdot 10^{-7}$	0.2500	0.7916
0.24	$7.3800 \cdot 10^{-7}$	0.2400	0.7826

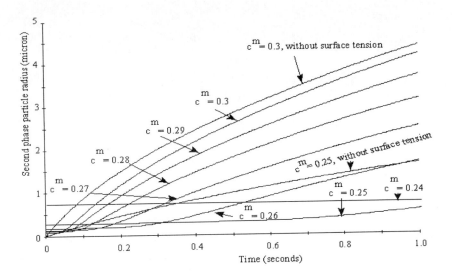

Fig. 3: The particle radius as a function of time for different initial matrix concentrations.

From Fig. 3 it can be seen that the growth rate increases and that the apparent incubation time decreases with increasing initial matrix concentration, i.e. with an increasing degree of supersaturation. For the lower initial matrix concentrations the driving force for nucleation is lower, causing a larger critical radius. To explain the difference in the behaviour of the curves in Fig. 3, two effects have to be considered:

1. From Eq.2, it is clear that the increase of the Gibbs free energy per unit increase of the particle radius is higher for small particles. This implies that the change of the interfacial concentration is large for a unit change of the particle radius.

2. A second consideration concerns the increase of the surface of a spherical particle during growth. The two correlated effects both give rise to a lower growth velocity for larger particles and thus yield a larger incubation time of growth for larger critical radii, which corresponds to a lower degree of supersaturation. This degree of supersaturation can also be considered as a measure for the undercooling. However, for all curves in Fig. 3 the same diffusion coefficient has been applied.

Analyses such as these might be used to understand the effect of intentional or accidental grain boundary segregation on the transformation kinetics.

Secondly, to elucidate the influence of the surface energy, γ, on the early growth kinetics, the particle radius has been plotted in Fig. 4 as a function of time for various surface tensions and hence for various critical particle radii.

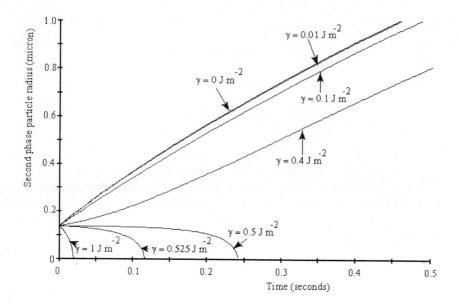

Fig.4: The particle radius as a function of time. The initial matrix concentration is 0.25 mass. % for all curves.

All curves correspond to an initial particle radius of 0.14 μm (which is the critical radius for a surface tension of 0.495 J/m^{-2}) and an initial carbon mass fraction of 0.25 % in the matrix. The further input data are as listed in table 1. It can be seen that the growth rate is considerably higher for low surface energies and finally approaches the case without any effects from surface tension. For the higher values of the surface tension, i.e. $\gamma = 0.50$ to 1 J m^{-2}, the initial particle size is smaller than the critical radius. For these cases the particle is unstable and therefore the particle shrinks.

4. The growth kinetics of ferrite.

Most of the growth models for allotriomorphic ferrite formation are based on the local equilibrium concept. Implicit in this model is that the austenite-ferrite interface moves at such a rate that the net interface velocity is determined by carbon diffusion only and the interface behaves as a volatile, carbon transparent film. However, there are various reasons to suspect that this assumption may not always be correct. Firstly, in-situ TEM observations of moving austenite-ferrite interfaces [12] in pure Fe-C alloys have shown that local atomic arrangements of as yet unspecified nature determine the local growth rate significantly. Secondly, in the case of IF steels with no interstitial carbon being available, the local equilibrium model suggests an infinitely high transformation rate while in practice the rate is high but clearly finite. Hence the interface itself must have a finite mobility. Thirdly, in the case of HLSA steels interface pinning by precipitates is an important rate reducing factor. Again the transformation rate is no longer determined by the carbon diffusion only, but by the (effective) interface mobility too. (a mixed mode).

The transformation rate in Fe-C alloys under conditions where both the carbon diffusion and the interface mobility play a role can be modelled as follows (assuming the α phase to be represented by and infinite x-z plane, moving in the x direction):

The carbon rejected by the ferrite and transferred into the austenite causes a carbon mass flux $J_{\alpha \to \gamma}$ across the interface, given by

$$J_{\alpha \to \gamma} = \left(c^\gamma (x_{int}) - c^\alpha (x_{int}) \right) \cdot v, \tag{7}$$

where $c^\gamma(x)$ and $c^\alpha(x)$ are the carbon concentrations of austenite and ferrite as a function of the position x, respectively. x_{int} represents the position of the interface. The interface velocity is given by v. Carbon diffusion in the ferrite is assumed to take place at such a rate that the carbon is homogeneously distributed. This assumption is justified by the diffusion coefficient of carbon in ferrite [13] being larger than that in austenite [14]. In the austenite a diffusive carbon flux J_D occurs, described by Fick's first law:

$$J_D = -D \cdot \left(\frac{dc^\gamma}{dx} \right), \tag{8}$$

where D is the diffusion coefficient of carbon in austenite. The net carbon flux ΔJ at the interface is given by the difference of the two fluxes:

$$\Delta J = \left(c^\gamma (x_{int}) - c^\alpha \right) \cdot v + D \cdot \left(\frac{dc^\gamma}{dx} \right)_{x = x_{int}}. \tag{9}$$

This equation reflects the interaction between carbon diffusion and interface mobility that is taken into account in the mixed-mode model of the ferrite growth. Note that the net carbon flux ΔJ is not necessarily zero at all stages of the transformation. A non-zero value for ΔJ results in a change in the carbon concentration at the interface. The development of the carbon concentration at the interface during the growth is thus determined by both the interface velocity and the diffusive flux. In turn, both these quantities depend on the carbon concentration at the interface. Equation (9) therefore gives the essentials of the present mixed-mode approach.

In the evaluation of the mixed-mode concept, information on the diffusivity is readily available [14], but the interface velocity v poses a potential problem. According to the theory of thermally activated growth [15], the interface velocity is determined by the free energy gain of the system. However, since the transfer of the interstitial carbon is expected to be much faster than the lattice change of the iron, we assume that the chemical potential difference between the Fe-lattice in the ferritic and in the austenitic state constitutes the free energy gain that determines the interface velocity. The interface velocity is given by [16]

$$v = M \, \Delta\mu_{Fe}, \tag{10}$$

where M is the interface mobility and $\Delta\mu_{Fe}$ the chemical potential difference of austenitic and ferritic iron, which is given by

$$\Delta\mu_{Fe} = \mu_{Fe}^\gamma (c^\gamma) - \mu_{Fe}^\alpha (c^\alpha). \tag{11}$$

The chemical potentials $\mu^{\gamma}{}_{,Fe}$ and $\mu^{\alpha}{}_{,Fe}$ depend on the local carbon concentrations in austenite and ferrite, respectively. Both c^{γ} and c^{α} are to be evaluated at the α/γ-interface. The simplicity of eq. (4) implies that a multitude of structural aspects that influence the mobility, like the degree of coherency of the interface, pinning effects, build-up of stresses, or solute drag, are united in a single mobility parameter M, which is therefore to be interpreted as an effective interface mobility. In our work, the chemical potentials are described using the regular solution sublattice model, originally due to Hillert and Staffansson [17], using the parameters as determined by Gustafson [18]. The mobility of a disordered α/γ-interface has recently been estimated from an analysis of the heat effects accompanying the $\gamma \rightarrow \alpha$ transformation in low-Mn Fe-Mn alloys [16]. For low-alloy compositions, it was found to be essentially independent of the alloy composition. The intrinsic interface mobility M is described by an Arrhenius-type temperature dependence:

$$M = M_0 \exp\left(-\frac{E}{RT}\right) \qquad (12)$$

From the work in Ref. [11], for the activation energy E the value 140 kJ/mole is used, and the pre-exponential factor M_0 is taken as 58 mm·mole/(J·s).

Using this approach in a suitable numerical routine in a transformation model regarding the austenite as a simple infinite slab of finite thickness, the carbon concentration profile ahead of the moving interface and the effective transformation velocity can be calculated.

The development of the carbon concentration profile during the transformation of a Fe-0.20 mass%C alloy at 1025 K is shown in figure 5. The carbon concentration profile builds up during a considerable period of time. The slowly increasing c^{γ} at the interface causes the interface velocity to decrease continuously during a substantial part of the transformation. Due to this build-up of the carbon concentration profile, the austenite grain size explicitly plays a role in the growth kinetics as modelled by the mixed-mode model. When the interface approaches the middle of the austenite grain, the increased carbon concentration due to the rejection of carbon by the growing ferrite stretches out over the entire remaining austenite. This soft impingement affects the growth kinetics. Upon further growth the carbon concentration approaches the equilibrium concentration and the velocity drops to zero.

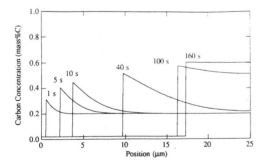

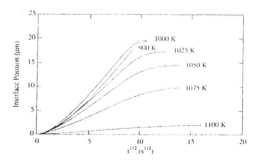

figure 5 : the development of the carbon concentration profile with progression of the transformation

figure 6 : time dependent position of the interface (see text)

The temperature dependence of the growth kinetics is rather complex. First, the driving force for interface migration, $\Delta\mu_{Fe}$, depends strongly on the temperature, as well as on the carbon concentration at the interface. In general, this aspect causes the growth rate to *increase* upon decreasing temperature, *i.e.* upon larger undercooling. Secondly, the interface mobility itself *decreases* with decreasing temperature according to eq.(12), and, thirdly, the carbon diffusivity in austenite also *decreases* upon decreasing temperature. An additional temperature-dependent factor is the temperature dependence of the equilibrium fractions of austenite and ferrite. Therefore, the temperature dependence of the growth kinetics cannot be summarised in a single relation. For the case of a Fe-0.20 mass%C system, the interface position at different temperatures is given in figure 6. For the austenite grain size $d = 50$ μm is used, and the final positions of the interface reflect the equilibrium state. In figure 6 the interface position as a function of $\sqrt{t}$ shows a distinct linear part, especially in the curves at low undercoolings. At large undercoolings this range of parabolic growth is less pronounced. This is due to the increasingly long initial phase of sub-parabolic growth, during which the carbon pile-up at the interface takes place. The deviation from parabolic growth in the final stage of the curves is due to soft impingement effects.

It is particularly interesting to compare the adequacy of the mixed mode model and the local equilibrium model with respect to the experimental growth data from Bradley *et al.* [19] given in figure 7 which shows both the thickening and the lengthening data for Fe-0.23 mass%C as a function of $\sqrt{t}$. It is clear that the local-equilibrium model strongly overestimates the growth rate.

figure 7 : lengthening and thickening kinetics for a Fe-0.23 mass % C alloy. Experimental data by Bradley et al [20]. Curves as predicted by the mixed mode transformation model. Straight lines apply to local equilibrium model.

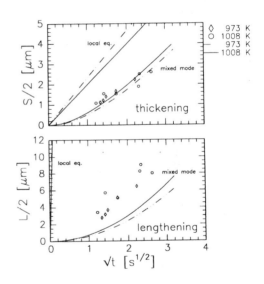

Whereas in the thickening data the difference is moderate, the lengthening data show a marked difference. In fact, the grain-boundary diffusion that plays a role in the lenghtening is so fast that a flat carbon concentration profile exists in the austenite throughout the transformation. This means that the growth kinetics for lengthening is essentially interface-controlled, and the composition at the interface is far from the local-equilibrium value. In the kinetics of the thickening, the difference is less pronounced, since the slower diffusion reduces the effect of the limited interface velocity. Figure 7 shows that the presently used mobility parameters, which were previously obtained for Fe-Mn alloys [16], give a reasonable account for the effect of the interface mobility for Fe-C alloys. More research on determining the effect of alloy composition on the interface mobility is currently under way.

5. Concluding remarks.

Physical modelling the transformation behaviour of HLSA steels would involve various new components in the transformation model : first of all, the precipitation kinetics of the Nb, V and T carbonitrides. Subsequently, the retarding effects of these precipitates on the recrystallisation and the recovery process. The combined effects of these processes would lead to an austenite which has more features than the austenite modelled in the present work. Using that refined austenite description the nucleation behaviour should be reconsidered, with precipitates and strain bands as additional nucleation sites. Furthermore, the driving force for the transformation is increased due to the energy stored in the dislocation networks. Finally, the growth behaviour will undoubtedly be affected with a complex balance between solute drag and grain boundary pinning. However, the current mixed mode growth model would enable in principle such an extension.

6. Acknowledgements.

The author gratefully acknowledges the contributions of his collaborators dr.ir. J. Sietsma, dr.ir. G. Krielaart, dr. M. Onink, dr.ir. F. Vermolen, ir. Y. van Leeuwen and drs T.A. Kop in developing the models presented here. This research program is supported by Koninklijke Hoogovens R&D, the Dutch Ministry of Economic Affairs through their IOP-Metalen program and the Netherlands Institute for Metal Research (NIMR).

References

[1]   Y. van Leeuwen, S.I. Vooys, J. Sietsma and S. van der Zwaag,. Submitted to *Acta Mat.* (1998)

[2]   G.P. Krielaart, J. Sietsma and S. van der Zwaag. *Mater. Sci. Eng* A237 (1997) 216-223

[3]   F. Vermolen. PhD thesis, Delft University of Technology (1998)

[4]   W. Huang and M. Hillert: *Metall. and Mater. Trans. A*, 1996, vol. 27A, pp. 480-483

[5]   Ch. Zurek, E. Sachova, H.P. Hougardy: Abschlussbericht Project COSMOS, Max-Planck-Institut für Eisenforschung GmbH, Düsseldorf, Germany, 1993.

[6]   T. Gladman. *The physical metallurgy of microalloyed steels*. The Institute of Materials (1997), London, UK

[7]   M. Avrami: *J. Chem. Phys.*, 1939, vol. 7, pp. 1103-1112.

[8]   M. Avrami: *J. Chem. Phys.*, 1940, vol. 8, pp. 212-224.

[9]   P.A. Porter and K.E. Easterling in: "Phase transformations in metals and alloys", Chapman&Hall, London, (1991)

[10]   J.S. Kirkaldy and D.J. Young, in: "Diffusion in the condensed state", The Institute of Metals, London, (1987)

[11]   Mats Hillert in: "Lectures on the theory of phase transformations", ed. H.I. Aaronson, AIME, New York, (1977)

[12]   M. Onink, C.M. Brakman, F.D.Tichelaar, E.J. Mittemeijer and S. van der Zwaag. J. Mater. Sci 30 (1994) 6223-6228

[13]   J. Ågren, Acta Met. 30 (1982) 841.

[14]   J. Ågren, Scripta Met. 20 (1986) 1507.

[15]   J.W. Christian, *The Theory of Transformations in Metals and Alloys*, 2nd ed. part 1, Pergamon Press, Oxford (1981), p. 476.

[16]   G.P. Krielaart and S. van der Zwaag, Mater. Sci Techn. (1998) in press

[17]   M. Hillert and L-I. Staffansson, Acta Chem. Scan. 24 (1970) 3618.

[18]   P. Gustafson, Scan. J. Met. 14 (1985) 259.

[19]   J.R. Bradley, J.M. Rigsbee and H.I. Aaronson, Met. Trans. 8A (1977) 323.

Materials Science Forum Vols. 284-286 (1998) pp. 39-50

Alternatives to the Ferrite-Pearlite Microstructures

H.K.D.H. Bhadeshia

University of Cambridge, Department of Materials Science and Metallurgy,
Pembroke Street, Cambridge CB2 3QZ, UK
E-Mail: hkdb@cus.cam.ac.uk www.msm.cam.ac.uk/phase-trans

Keywords: Structural Steel, Ferrite, Pearlite, Bainite, Microalloyed, Acicular Ferrite, Welding

Abstract

Structural steels based on a mixed microstructure of allotriomorphic ferrite and pearlite have a well–established history of cost–effectiveness and reliability. The purpose of this paper is to review the possibility of alternatives to this microstructure for large scale applications. Controlled–rolled bainitic steels, accelerated cooled steels, ultra–low carbon bainitic steels and inoculated steels are discussed in this context.

Introduction

I had the privilege to work with the late Professor Urkola Galarza for about two years; we even shared in the teaching of steels for three wonderful days set in his Basque Country. He had a passion for steels – with Rodriguez Ibabe he wrote what must be the only book on the metallurgy of steels in the Basque language [1].

Professor Urkola's dedication to steel is easy to understand; the material is remarkable in its complexity and use. There are three stable allotropes that occur in nature, with a body– centered cubic, hexagonal close–packed and cubic–close packed crystal structure. Two further allotropes (face–centered tetragonal and trigonal) can be created artificially, and there are at least six different magnetic transitions in solid iron. Just the addition of carbon then leads to a large variety of new possibilities including the classical ferrite, Widmanstätten ferrite, pearlite, bainite and martensite microstructures.

The most popular microstructure in the context of structural steels has undoubtedly been a mixture of ferrite and pearlite. Typical chemical compositions are given in Table 1 together with an indication of the usual mechanical properties. Fig. 1 shows that the main effect of microalloying (and the associated thermomechanical processing) is to refine the microstructure. The niobium carbonitrides that form also strengthen the ferrite by interphase precipitation or by strain–induced precipitation [2–5].

Pickering [4] has listed some of the major applications of these steels: oil and gas pipelines, drilling rigs, production platforms, ships, pressure vessels and tubing, earth–moving equipment, heavy goods vehicles and automobile components, high– rise buildings, bridges, transmission towers, fuel and other storage tanks and reinforcement bars for concrete. A few of the attributes that are consistent with these applications include the low cost, ability to produce a variety of forms, weldability, fabricability, reliability under extreme conditions (such as fire and earthquakes and other *force majeures*).

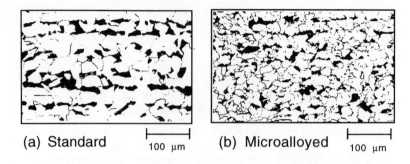

Fig. 1: Optical micrographs of banded ferrite–pearlite microstructures in (a) standard, (b) niobium microalloyed structural steel.

| Type | Composition / wt.% | | | | Stress / MPa | | | Elongation % | C_V (−20 °C) / J |
	C	Si	Mn	Nb	σ_y^l	σ_y^u	UTS		
Standard	0.11	0.21	1.24		285	320	480	36	55–90
Microalloyed	0.11	0.30	1.40	0.034	395	430	525	32	100–190

Type	Ferrite %	Pearlite %	Ferrite grain size	Pearlite interlamellar spacing
Standard	76	24	15 μm	0.22 μm
Microalloyed	78	22	9 μm	0.22 μm

Table 1: Typical ferrite–pearlite structural steels in both the standard and niobium microalloyed conditions.

It seems unlikely that the dominant position of the ferrite–pearlite steels will be challenged by alternative steel microstructures, let alone other materials! Nevertheless, we shall describe rivals based on bainite because such steels be produced using a similar production route and without any additional heat treatments.

Strength

The most elementary engineering specification is strength, which is easily obtained by transforming the austenite at ever decreasing temperatures (Fig. 2) and by increasing hardenability. And there are many other strengthening mechanisms available. Unfortunately, toughness does not necessarily increase with strength. The weldability may also deteriorate because the heat that flows into the steel causes the region adjacent to the fusion boundary to austenitise and more remote regions to be tempered. Two potential problems arise in this heat–affected zone (HAZ): (i) the excessive use of alloying elements can cause the formation of untempered hard phases in the fully austenitised region; (ii) there may be excessive softening in the tempered regions.

The tendency to soften depends on how far the microstructure is from equilibrium (Table 2). Large deviations naturally lead to greater rates of softening because the driving force for tempering is the energy stored within the material; the heat simply provides the thermal activation needed to bump the structure to lower energies. Higher strength steels are therefore more likely to be difficult to weld.

Steels that are stronger than 1000 MPa yield are important in certain applications, but the biggest markets are for lower strength varieties where the total alloy content rarely exceeds 2 wt.%. The

Phase Mixture in Fe–0.2C–1.5Mn wt.% at 300 K	Stored Energy / $\mathrm{J\,mol^{-1}}$
1. Ferrite, graphite & cementite	0
2. Ferrite & cementite	70
3. Paraequilibrium ferrite & paraequilibrium cementite	385
4. Bainite and paraequilibrium cementite	785
5. Martensite	1214
6. Mechanically alloyed ODS metal	55

Table 2: The stored energy as a function of microstructure, relative to the equilibrium state defined as a mixture of ferrite, cementite and graphite. The phases in cases 1 and 2 involve a partitioning of all elements so as to minimise free energy. In cases 3–5 the iron and substitutional solutes are configurationally frozen (for martensite even the interstitial elements are frozen). Case 6 refers to an iron–base mechanically alloyed oxide–dispersion strengthened sample which to my knowledge is the highest reported stored energy prior to recrystallisation [6].

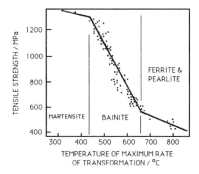

Fig. 2: Variation in the tensile strength of structural steels as a function of the temperature at which the rate of transformation is greatest during continuous cooling heat treatment [7].

alloy design then has to ensure sufficient hardenability to achieve the desired microstructure whilst at the same time avoiding the formation of hard martensite. Irvine and Pickering therefore developed a series of low–alloy, low–carbon steels, containing small amounts of boron and molybdenum to suppress allotriomorphic ferrite formation. Boron increases the bainitic hardenability and hence reduces the need for other solute elements which promote the formation of martensite. Steels like these (Alloy 1, Table 3) are found to transform into bainitic microstructures with very little martensite using simple, normalising heat treatments.

The other alloys listed in Table 3 are examples of some recent commercial bainitic steels. Although the steels listed appear to be similar, this observation is misleading because they have quite different mechanical properties because of the prominent role of the trace element concentrations, and because there are significant differences in the processing routes.

The steels developed by Irvine and Pickering exhibited reasonable combinations of toughness and

No.	C	Si	Mn	Ni	Mo	Cr	Nb	Ti	B	Al	N	Steel Type
1	0.100	0.25	0.50	–	0.55	–	–	–	0.0030	–	–	Early bainitic
2	0.039	0.20	1.55	0.20	–	–	0.042	0.015	0.0013	0.024	0.0030	Rapidly cooled bainitic
3	0.081	0.25	1.86	0.20	0.09	–	0.045	0.016	–	0.025	0.0028	Rapidly cooled bainitic
4	0.110	0.34	1.51	–	–	–	0.029	–	–	–	–	Rapidly cooled bainitic
5	0.020	0.20	2.00	0.30	0.30	–	0.050	0.020	0.0010	–	0.0025	ULCB
6	0.028	0.25	1.75	0.20	–	0.30	0.100	0.015	–	0.030	0.0035	ULCB (Cu 0.3, Ca 0.0004)
7	0.080	0.20	1.40	–	–	–	–	0.012	–	0.002	0.0020	acicular ferrite, TiO_x (Oxygen 0.0017)
8	0.080	0.20	1.40	–	–	–	–	0.008	0.0015	0.038	0.0028	acicular ferrite, TiB
9	0.080	0.20	1.40	–	–	–	–	0.019	–	0.018	0.0050	acicular ferrite, TiN

Table 3: Typical compositions (wt.%)of bainitic steels. The alloys appear similar in composition but the differences in carbon and trace element concentrations, and processing, are important.

strength, but in time proved to be unexciting when compared with the best of quenched and tempered martensitic steels. Nevertheless, the physical metallurgy principles established during their development are now being applied towards a new generation of bainitic steels, in which the emphasis is on further reductions in carbon and other solutes, together with advanced processing in order to refine the microstructure. The range of bainitic alloys which might reasonably be regarded as alternatives to the ferrite–pearlite microstructures are now discussed in greater depth.

Control–Rolled Bainitic Steels

The strengthening of iron by reducing its grain size is attractive because it gives a simultaneous improvement in the strength and toughness. This has led to the development of an impressive technology for reducing the austenite grain size and hence the scale of any transformation microstructure. In *controlled–rolling*, the ingots are reduced by hot–rolling at high temperatures making the austenite recrystallise many times before the finish–rolling temperature is reached, giving a fine austenite grain structure prior to transformation. Any grain growth during the hot–rolling process is hindered by the use of *microalloying* additions such as niobium, vanadium or titanium. These elements have a low solubility in austenite and are added in small concentrations ($\simeq$ 0.03-0.06 wt.%) to form stable carbides or carbonitrides which impede grain growth during hot deformation and subsequent cooling. It is also possible to continue deformation so that the austenite grains are flattened (pancaked) just before transformation because this leads to an even finer final microstructures.. A typical chemical composition of a steel suitable for control–rolling to a ferrite and pearlite microstructure is Fe–0.08C–0.3Si–1Mn–0.03Nb–0.004N wt.%.

The controlled–rolling process described above has now been adapted for bainitic alloys. There are two ways in which a bainitic microstructure can be obtained:

(i) The cooling rate can be increased to allow the austenite to supercool into the bainite transformation range.

(ii) Hardenability can be modified, thereby avoiding process changes.

This second option is discussed first; rapid cooling will be described in the next section. Alloying elements such as manganese are boosted in order to increase the possibility of forming bainite at the expense of allotriomorphic ferrite. Unlike conventional steels for control–rolling, TiN particles (of size $\simeq 0.02\mu$ m) are induced to precipitate during solidification and subsequent cooling to ambient temperature. The precipitation of fine TiN is stimulated by increasing the cooling rate of the molten steel so that the alloys are best produced by continuous casting. Slabs of the material are then reheated to a relatively low reaustenitisation temperature of 1150 °C, the particles inhibiting austenite grain growth. Reductions in grain size are obtained by repeated recrystallisation during control–rolling. The particles also help to produce a more uniform grain structure than is obtained in conventional control–rolling treatments. Finish rolling is carried out at a temperature where recrystallisation does not occur, resulting in deformed and pancaked austenite grains which then transform to a fine bainitic microstructure on further cooling.

The details of the steelmaking process are important in determining the final properties of control–rolled bainitic steels. The higher quality steels are dephosphorised, desulphurised and vacuum degassed prior to casting. Typical concentrations of phosphorous and sulphur after these treatments are 0.015 and 0.0015 wt.% respectively. In circumstances where formability and uniform ductility are important, the steel is usually treated with calcium which has the effect of fixing sulphur and of modifying the shape of the sulphide inclusions.

Rapidly Cooled Control–Rolled Bainitic Steels

Bainitic microstructures can be generated in control–rolled steel by changing the cooling rate during processing, rather than by increasing hardenability. This is desirable when weldability is important, in order to avoid the formation of martensite in the heat affected zone of the welded plate. The technology of rapid cooling during controlled–rolling is not as trivial as it sounds, given the speed of production, the kinetics of transformation, the need to avoid shape distortion and the need to achieve uniform cooling rates. Considerable progress has nevertheless been made and the rapidly cooled steels described here are commercially available. It is worth noting that rapidly cooled steels are often referred to as "accelerated cooled steels".

Pipeline and Plate Steels

There is a general demand for a reduction in the wall–thickness and an increase in the diameter of pipelines for gas transmission. Thinner walls permit faster and less troublesome girth welding operations, thereby reducing costs. The thickness of the steel used can be reduced by increasing the strength but without sacrificing toughness or weldability. When thickness considerations are not paramount, an increase in strength has the further advantage that the gas can be transmitted more efficiently under increased pressure ($\simeq 10$ MPa).

It is found that if, after controlled–rolling, the steel is cooled from the austenite phase field at a rate which prevents the formation of large quantities of allotriomorphic ferrite, but which is low enough to avoid martensitic transformation, then a fine–grained microstructure which is a mixture of allotriomorphic ferrite and bainite is formed. Such a microstructure has an increased yield strength and good toughness. The cooling rates involved are higher (10–40 °C s^{-1} over the temperature range 800–500 °C) than those appropriate for normal control-rolled steel processes (Fig. 3). The accelerated cooling is achieved in industrial practice by the use of water spray curtains directed on either side of the hot plate (which could be some 15 mm in thickness) in a manner designed to ensure uniform cooling and to minimise distortion. The rapid cooling of thick plate requires different technology with much more careful control of water pouring in order to cope with the large thickness and width of the plates and the relatively slow rate at which the steel moves through the mill.

The process leads to a more refined microstructure in which the bainite platelets contribute to strength and toughness via their fine grain size. For a given strength level, the properties can therefore be achieved using lower alloying concentrations, with the concomitant advantages of reduced cost and in some cases, increased weldability.

Although most investigations of accelerated cooled steels have relied on light microscopy which lacks

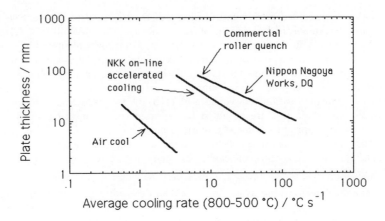

Fig. 3: An illustration of the average cooling rates associated with the manufacture
of steels for structural applications [8].

the necessary resolution, a detailed investigation [9] has demonstrated that the microstructure of these
rapidly cooled steels does indeed consist of a mixture of ferrite and bainite. The bainite consists of
sheaves of platelets of submicron thickness, as compared with the equiaxed grains of allotriomorphic
ferrite of size approximately 5 μm. It is also found to have a much higher dislocation density of
1.7×10^{14} m^{-2}, compared with 0.4×10^{14} m^{-2} of the allotriomorphic ferrite. In fact, the dislocation density of the allotriomorphic ferrite in rapidly cooled steels is known to be about four times
larger than in other steels containing ferrite, possibly because of plastic deformation by the bainitic
transformation which occurs after the allotriomorphic ferrite [10]. It is as a consequence of the lower
transformation temperatures involved, that the overall microstructure of these steels is found to be
more refined relative to the conventional control–rolled steels. The volume fraction of bainite can vary
from about $0.2 \rightarrow 1.0$ depending on the alloy chemistry and cooling conditions. Typical compositions
for accelerated–cooled alloys are given as Alloys 2-4 in Table 3. Of these, Alloy 2 is the leanest and
can be expected to contain the least quantity of bainite.

The production of the steels is not a continuous process of casting and controlled–rolling followed
directly by accelerated cooling. Instead, cast ingots are first allowed to cool to ambient temperature
and then reheated for the thermomechanical treatment. This ensures that the very coarse austenite
grain structure which evolves during ingot cooling is disrupted by transformation to ferrite. Hence,
the processing involves the reheating of thick ingots to 1150 °C, followed by rolling during cooling
of the ingot to 740 °C, with the total reduction in thickness being more than 600%, followed by
accelerated cooling at 20 °C s^{-1} to around 450 °C before allowing natural cooling. This treatment
changes the microstructure of the plate from the normal mixture of ferrite and pearlite to a ferrite–
bainite microstructure giving a better combination of mechanical properties. The tensile strength
achieved is typically 700 MPa which is about 50–70 MPa higher than that of conventional control–
rolled steels and the Charpy impact toughness can be an impressive 160–200 J at −20 °C. The extra
strength is attributed to the fine size of bainite plates, although Morikawa *et al.* [11] have demonstrated
that the strength of the allotriomorphic ferrite also increases with the accelerated cooling, probably
because of the dislocation density increase described above. The steels also do not exhibit sudden
yielding, although the relevance of this to pipeline applications is not clear [12].

Process Parameters

There are a number of controllable factors which can significantly influence the properties of the final
product [9, 12, 13]. For example, a high ingot reheating temperature allows more of the niobium
carbonitrides to dissolve in the austenite; the niobium carbonitrides may subsequently precipitate

during the $\gamma \rightarrow \alpha$ reaction to give fine dispersions within the ferrite ("interphase precipitation"), leading to an enhancement of its strength.

The temperature at which the rolling operation finishes is critical in the sense that it should leave the final austenite grains in an unrecrystallised, pancake shape. This ensures a further degree of refinement of the microstructure obtained after transformation and avoids the undesirable recrystallisation texture of austenite. If the finish rolling temperature is too low (below $Ar3$), then the allotriomorphic ferrite will deform, leading to an increase in strength but at the expense of toughness. For this reason, the deformation should be restricted to the austenite phase. The finish rolling temperature (T_R) also influences the variation in mechanical properties through the thickness of heavy gauge plates (Fig. 4). The surfaces, where the cooling rates are the largest, tend to be harder when compared with the central regions of the plates. The differences diminish as T_R is reduced because rolling deformation becomes focused at the plate surfaces, which consequently tend to transform more rapidly; this counteracts the effect of the higher surface cooling rates [14].

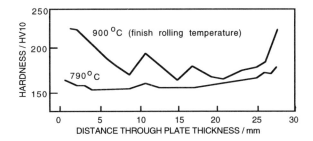

Fig. 4: The effect of the temperature at which rolling is completed, on the variation in hardness of a Fe–0.16C–0.63Mn wt% accelerated cooled steel containing bainite [15].

The steels used in the accelerated cooling operations have a hardenability which is so high as to prevent the completion of the bainite reaction during continuous cooling to ambient temperature. Large amounts of martensite are found in the microstructure under some circumstances, leading to a drop in toughness and a significant distortion of the plate product. As a consequence, the cooling has to be arrested by cutting off the water sprays at temperatures ranging between $600 \rightarrow 450\,^{\circ}\text{C}$, depending on the steel chemistry and the exact cooling conditions. The subsequent slower cooling rate allows the bainite transformation to proceed to a larger extent, leaving only very small amounts of residual austenite which may transform to martensite on further cooling. Another related problem has been found in alloys with a relatively high hardenability, typically those low–alloy steels with more than about 1.4 wt.% manganese. When produced using controlled–rolling and accelerated cooling, the yield strength decreases even though the tensile strength does not [Fig. 5a, 16]. This is because martensite replaces bainite as the dominant hard phase, and as in dual–phase steels, yielding becomes a gradual process as the stress is raised. Although this has clear advantages for applications involving forming operations, the lowering of yield strength is a disadvantage for pipeline and heavy plate fabrications where the design thickness is calculated using yield criteria.

A two–stage accelerated cooling process has therefore been developed to enhance the chances of forming bainite instead of martensite, while at the same time retaining the high cooling rate required to refine the allotriomorphic ferrite that forms first (Fig. 5b). After thermomechanical processing while the steel is in the austenitic condition, it is cooled rapidly ($25\,^{\circ}\text{C}\,\text{s}^{-1}$) through the ferrite temperature range in order to obtain the fine ferrite grain size, but the cooling rate is then reduced to about $3\,^{\circ}\text{C}\,\text{s}^{-1}$ over the temperature range where bainite forms, giving a greater opportunity for transformation to proceed before the martensite–start temperature is reached. The temperature (T_F) at which the forced cooling is stopped to allow the steel to air cool in the second stage of the process is also important. The

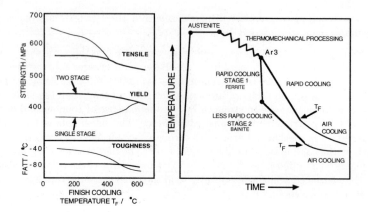

Fig. 5:   (a) The relationship between the tensile and yield strength, and toughness of accelerated cooled steels as a function of the temperature at which the forced cooled is stopped, for the single and double stage processes. (b) Schematic illustration of the thermomechanical cycles associated with the two–stage accelerated cooling process [17].

mechanical properties are found to be much less sensitive to T_F for the two–stage process presumably because much of the bainitic transformation is completed at a relatively high temperature during the second stage (Fig. 5a). The process is found to be successful in raising the yield stress of the steel when compared with the conventional accelerated cooling procedure.

Nishioka and Tamehiro [18] have emphasised that a general problem with the accelerated cooled steels is that the toughness and microstructure are not maintained in the heat– affected zones created after welding. On the other hand, the strength of these rapidly cooled steels is some 50 MPa greater than that of conventional control–rolled plates. For those applications where this excess strength is not needed, the carbon–equivalent of the steel could in principle be reduced further by the removal of alloying elements, thereby improving the weldability (Fig. 6). A further difficulty is that the process is not suitable for heavy gauge plates (20–30 mm thick) since it is then impossible to ensure uniform cooling throughout the depth of the samples [12]; it is often the case that the central regions of such samples transform instead to a ferrite and pearlite microstructure.

In control–rolled steels, heavy gauge plates which are cooled slowly after rolling, have a tendency to develop a coarse ferrite grain structure at the surfaces. This appears to be a consequence of the recrystallisation of ferrite grains deformed by rolling in the $(\alpha + \gamma)$ phase field, although the fact that recrystallisation only happens at the surface implies that the deformation must have been inhomogeneous. Accelerated cooling has the added advantage that it inhibits this recrystallisation at the surface [13].

Segregation

Control–rolled steels are made using the continuous casting process and therefore suffer from pronounced chemical segregation approximately along the mid–thickness of the plate. The resulting microstructure can be related directly to variations in the manganese concentration, which can reach twice the average value along the centre of the plate. Ferrite forms first in the manganese–depleted regions causing the carbon concentration of the manganese–enriched regions of austenite to increase. This is turn exaggerates the hardenability of the manganese–enriched regions, which then transform into bands of hard microstructure.

These hard bands are susceptible to hydrogen cracking. An advantage of the accelerated cooled

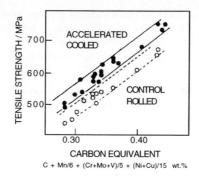

Fig. 6: Relationship between the carbon equivalent and tensile strength for conventionally produced control–rolled steels (CR) and accelerated cooled steels (AC); the latter have a mixed microstructure of ferrite and bainite [13].

steels is that the resulting microstructure of bainite and ferrite is experimentally found to be less sensitive to solute segregation, when compared with the gross banding effect observed with ferrite–pearlite microstructures [9,13]. The transformation in rapidly cooled steels is suppressed to lower temperatures, where nucleation is possible in all regions including the solute enriched zones, giving a more uniform distribution of carbon (Fig. 7). The resulting lower hardness in the segregated zone makes the steel less susceptible to hydrogen–induced cracking.

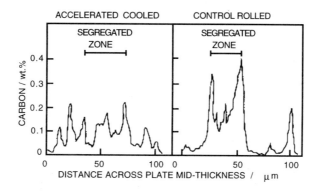

Fig. 7: Distribution of carbon concentration in the segregated zone for conventional control–rolled and rapidly cooled steel plates [19].

Ultra–Low–Carbon Bainitic Steels

Improvements in weldability brought about by reductions in carbon concentration can lead to major economies in fabrication, especially if the welding can be carried out without any preheat. Hence the development of the *ultra–low carbon bainitic* (ULCB) steels (Alloy 5 and 6, Table 3) with carbon concentrations in the range 0.01–0.03 wt.% [20, 21]. This level of carbon is sufficient to react with microalloying additions but not sufficient to allow the formation of martensite. The reduction in martensite at low carbon concentrations leads to an improvement in toughness without any undue loss of strength (Fig. 8). The carbon concentration should not fall below 0.01 wt.% since niobium carbide (and TiN) is necessary to ensure the development of a fine austenite grain structure during controlled–rolling. The niobium and titanium additions are also known to be effective in suppressing the precipitation of $Fe_{23}(CB)_6$ during control–rolling [22]. Only free boron is useful in enhancing hardenability. In other respects the ULCB steels are similar in concept to the Irvine and Pickering alloys; they contain boron and molybdenum or chromium to enhance bainitic hardenability, and titanium to getter nitrogen.

Boron can cause the hardness of the heat affected zone associated with welding to increase, but fortunately, its effect in ULCB steels is small because of very low carbon concentration, which also permits the greater use of substitutional solutes (Mn, Ni) compared with the Irvine and Pickering steels.

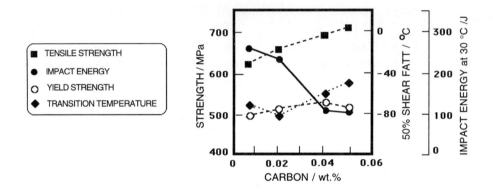

Fig. 8:   Variations in mechanical properties as a function of carbon concentration in thermomechanically processed ultra–low–carbon steels [21]. The fracture assessed impact transition temperature (FATT) begins to increase as the carbon concentration falls below about 0.02 wt.%.

The solubility of niobium in ordinary steels is limited by their relatively high carbon concentrations, most of the niobium being present as carbonitrides. By contrast, a larger fraction of the niobium remains dissolved in ULCB steels, which also contain Ti to combine with free nitrogen. Their response to niobium additions is therefore different from that of conventional microalloyed steel. The free niobium can significantly influence the transformation behaviour [23,24]. The finish rolling temperature for ULCB steels has a large influence on the soluble niobium in the final steel (Fig. 9a). Less niobium remains in solution as the finish rolling temperature is reduced because there is then a greater opportunity for strain–induced precipitation.

The variations in dissolved niobium affect the microstructure in air–cooled ($1\,^\circ C\,s^{-1}$) ULCB steels (Fig. 9b), perhaps because niobium strongly retards the formation of allotriomorphic ferrite, leading to an increase in the fraction of bainite. The effect is masked during accelerated cooling (15–$20\,^\circ C\,s^{-1}$) because the transformation to allotriomorphic ferrite is avoided (Fig. 9b).

ULCB steels are strong (500–620 MPa), tough and weldable. They have been designated for use in high strength line pipe in Arctic or submarine environments.

Inoculated Acicular Ferrite Steels

Acicular ferrite is intragranularly nucleated bainite. Whereas bainite contains packets of parallel ferrite plates, the acicular ferrite microstructure is much more chaotic with plates lying on many different planes. Such clusters of nonparallel platelets lead to an enhancement of toughness since, unlike bainite, any propagating crack frequently encounters plates in different crystallographic orientations.

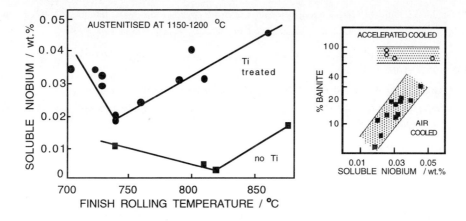

Fig. 9: Soluble niobium in ULCB steel [21]. (a) Variation in the soluble niobium concentration as a function of the finish rolling temperature; (b) variation in microstructure as a function of the finish rolling temperature.

The difference in microstructure arises because bainite nucleates from austenite grain *surfaces* thus allowing the development of packets, whereas acicular ferrite plates nucleate from *point* sites which are usually non–metallic inclusions.

Given the more desirable microstructure of acicular ferrite, it is now possible to inoculate steels with selected oxide particles in order to induce the formation of acicular ferrite for improved toughness [18,25,26]. Some examples of titanium oxides and nitride inoculated acicular ferrite steels are given in Table 3 (Alloys 7–9).

There appear to be considerable difficulties in the production of such steels so it seems unlikely that they will see many applications.

CONCLUSIONS

Ferrite/pearlite steels are outstanding on any performance criterion which includes cost and reliability. They will certainly not be threatened by the alternative microstructures discussed here. On the other hand, the novel microstructures do offer a better set of properties for high performance applications where their additional cost is not an issue.

Acknowledgments: My sincere thanks to the organising committee for this opportunity to pay a tribute to the late Professor Urkola. I would also like to thank Mike Lord for help with some thermodynamic calculations using the MTDATA program and SGTE database. I am very grateful to Shiv B. Singh for checking and commenting on the manuscript.

[1] J. M. Rodriguez Ibabe and J. J. Urkola Galarza: *Altzairuen Diseinurako Metalugia Fisikoa,* published by Elkar, (1993) 1–480.

[2] F. B. Pickering: *Physical Metallurgy and The Design of Steels,* Applied Science Publishers, London, (1978) 37- -100.

[3] R. W. K. Honeycombe and H. K. D. H. Bhadeshia: *Steels: Microstructure and Properties,* 2nd Edition, Edward Arnold (1995) Chapter 10.

[4] F. B. Pickering: *Constitution and Properties of Steels,* VCH publishers, eds R. W. Cahn, P. Haasen and E. J. Kramer, Germany, (1992) 339–399.

[5] T. Gladman: *Physical Metallurgy of Microalloyed Steels,* Inst. of Materials, London, (1997) 1– 360.

[6] H. K. D. H. Bhadeshia: *Materials Science and Engineering A,* A223 (1997) 64–77.

[7] K. J. Irvine, F. B. Pickering, W. C. Heselwood and M. Atkins: *Journal of the Iron and Steel Institute,* 195 (1957) 54–67.

[8] J. H. Gross, R. D. Stout and E. J. Czyryca: *Welding Journal,* 74 (1995) 53–62.

[9] M. K. Graf, H. G. Hillenbrand, P. A. Peters: *Accelerated Cooling of Steel,* P. D. Southwick, TMS–AIME, (1985) 349–366.

[10] A. J. DeArdo: *Accelerated Cooling of Rolled Steel,* eds G. E. Ruddle, A. F. Crawley, Pergamon Press, Oxford, U. K. (1988) 3–27.

[11] H. Morkawa and T. Hasegawa: *Accelerated Cooling of Steel,* P. D. Southwick, TMS–AIME, (1985) 89–96.

[12] L. E. Collins, R. F. Knight, G. E. Ruddle and J. D. Boyd: *Accelerated Cooling of Steel,* P. D. Southwick, TMS–AIME, (1985) 261–282.

[13] H. Tamehiro, R. Habu, N. Yamada, H. Matsuda and M. Nagumo: *Accelerated Cooling of Steel,* P. D. Southwick, TMS–AIME, (1985) 401–414.

[14] T. Tanaka: *Accelerated Cooling of Rolled Steel,* eds G. E. Ruddle, A. F. Crawley, Pergamon Press, Oxford, U. K. (1988) 187–208.

[15] S. Tamukai, Y. Onoe, H. Nakajima, M. Umeno, K. Iwanaga and S. Sasaji: *Testu–to–Hagane* 67 (1981) 1344.

[16] C. Shiga, K. Amano, T. Enami, M. Tanaka, R. Tarui and Y. Kushuhara: *Technoloyg and Applications of High Strength Low Alloy Steels,* ASM International, (1983) 643–654.

[17] K. Amano, T. Karomura, C. Shiga, T. Enami and T. Tanaka: *Accelerated Cooling of Rolled Steel,* eds G. E. Ruddle, A. F. Crawley, Pergamon Press, Oxford, U. K. (1988) 43–56.

[18] K. Nishioka and H. Tamehiro: *Microalloying '88,* TMS–AIME, (1988) 1–9.

[19] H. Tamehiro, T. Takeda, S. Matsuda, K. Yamamoto and N. Okumura: *Trans. Iron Steel Inst. Japan* 25 (1985) 982–988.

[20] H. Nakasugi, H. Matsuda and H. Tamehiro: *Steels for Line Pipe and Pipeline Fittings,* Metals Society, London, (1983) 90.

[21] K. Hulka, F. Heisterkamp and L. Nachtel: *Processing, Microstructure and Properties of HSLA steels,* ed. A. J. DeArdo, TMS–AIME, (1988) 153–167.

[22] H. Tamehiro, M. Murata, R. Habu and M. Nagumo: *Trans. Iron Steel Inst. Japan* 27 (1987) 130–138.

[23] G. I. Rees, J. Perdrix, T. Maurickx and H. K. D. H. Bhadeshia: *Materials Science and Engineering A* 194 (1995) 179–186.

[24] C. Fossaert, G. Rees, T. Maurickx and H. K. D. H. Bhadeshia: *Metallurgical and Materials Transactions A* 26A (1995) 21–30.

[25] M. A. Linaza, J. L. Romero, J. M. Rodriguez–Ibabe and J. J. Urcola: *Scripta Metall. and Materialia* 29 (1993) 1217–1222.

[26] M. A. Linaza, J. L. Romero, J. M. Rodriguez–Ibabe and J. J. Urcola: *Scripta Metall. and Materialia* 32 (1995) 395–400.

Materials Science Forum Vols. 284-286 (1998) pp. 51-62
© *1998 Trans Tech Publications, Switzerland*

The Role of Microstructure in Toughness Behaviour of Microalloyed Steels

J.M. Rodriguez-Ibabe

CEIT and ESII, P° Manuel de Lardizábal 15, E-20009 San Sebastián, Basque Country, Spain

Keywords: Toughness, Ductile-Brittle Transition, Cleavage, Facet Size, Effective Surface Energy, Non-Metallic Inclusions

ABSTRACT

The micromechanisms involved in cleavage propagation in the brittle and ductile-brittle behaviour of microalloyed steels are analysed. The cleavage process is divided into three different steps which must take place dynamically in order to succeed: nucleation of a microcrack in a microstructural feature, propagation of the microcrack across the particle-matrix boundary and finally, propagation across the matrix surmounting high angle boundaries. The microstructural parameters involved in the three steps are considered, together with their effective surface energy values. Depending on testing temperature, the microstructural feature controlling cleavage fracture changes. In relation to the ductile-brittle transition, the main parameter is the volume fraction of coarse microstructural units surrounded by high angle boundaries.

1. INTRODUCTION

In a significant number of steel applications toughness is of great engineering significance, together with a good strength level. Whilst it is possible to control the latter by different mechanisms and its prediction, by microstructural and process characterisation, is quite good, the different conditions to obtain good toughness values gives rise to complications. For example, vanadium microalloyed medium carbon steels with ferrite-pearlite microstructures have been produced with the equivalent as-forged or as-rolled strength as the quenched and tempered steels, but their toughness has not usually been as good. In a similar way, the strength of pearlitic steels has been improved by V addition, but toughness is not improved in the same way. In the case of as-forged and as-rolled microstructures, different approaches have been considered in order to improve their toughness, the most relevant being the refinement of the austenite grains (Ti microalloying) and the application of accelerated cooling [1, 2]. Nevertheless, their application to new products will depend on a better control of toughness quality.

In contrast to considerations of strength, toughness modelling can become very complex, particularly when the brittle-ductile regime is considered. In this situation, there is a competition between ductile and brittle mechanisms and the control of the micromechanisms intervening in the brittle process can contribute to a significant improvement of the toughness response of the material. This paper will focus on the micromechanisms controlling the cleavage initiation and propagation process, without taking into account the behaviour of the material in a completely ductile manner.

The process of brittle cleavage fracture has been attributed to the nucleation of a microcrack followed by its propagation into the surrounding matrix. This process can be divided into three different steps [3, 4]. In the first step (Fig. 1) a microcrack nucleates in an appropriate microstructural feature (a grain-boundary carbide, a non-metallic inclusion) which fractures easily in a brittle manner. The second step consists of the propagation of the microcrack to the surrounding matrix, given that the local stress exceeds a certain critical value. The third critical step, for crack progression through the matrix, arises at high angle boundaries which act as obstacles and force the microcrack to change the microscopic plane of propagation in order to accommodate the new local crystallography.

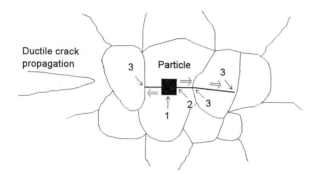

Fig. 1. Scheme of the different steps necessary for cleavage fracture (in ductile-brittle transition). A sharp microcrack nucleates in some microstructural feature (step 1) and propagates across particle-matrix (step 2) and matrix-matrix (step 3) boundaries under the action of a tensile stress.

The minimum stress (critical cleavage stress) necessary to surmount the obstacles is related to the effective surface energy and the size of the feature (particle or grain size) through the Griffith formula [3]. In the case of a penny-shape microcrack the equation takes the form:

$$\sigma_{im} = \left(\frac{\pi E \gamma_{im}}{\left(1 - \nu^2\right) d_i} \right)^{1/2} \tag{1}$$

where i can be either p or m, when referring to particle-matrix (p-m) and matrix-matrix (m-m) respectively, d_i corresponds to the particle size for d_p, and grain size for d_m and γ_{pm} and γ_{mm} are the corresponding effective surface energies. Once the crack has traversed several barriers, its length, together with the local stress state, is sufficient for catastrophic propagation to failure. The process must be dynamic to succeed [4, 5] and, in consequence, it is necessary that the three steps take place immediately one after the other.

The critical cleavage stress has been experimentally determined for a wide range of microstructures [6 - 12]. This value and the microstructural and energetic parameters are currently being used in the models developed to predict brittle fracture processes. Nevertheless, the problem arises in the measurement of γ_{pm} and γ_{mm} effective surface energy values and in the identification and quantification of the microstructural parameter controlling the process.

In microalloyed steels, both parameter types (microstructural and energetic) can be modified by adequate microalloying additions and thermomechanical processing, suppressing the cleavage process for a given temperature. Nevertheless, sometimes microalloying itself can make a contribution to inducing new microstructural features able to nucleate microcracks [13]. In this situation matrix microstructure becomes more relevant from the point of view of toughness improvement. In this paper an analysis of the role of the microstructure in the different steps of brittle fracture will be considered.

2. NUCLEATION OF MICROCRACKS

The first step of the cleavage process is the nucleation of a microcrack. McMahon and Cohen [3] first demonstrated that cleavage fracture in ferrite grains was associated with a broken carbide particle located somewhere in the grain or at the surrounding boundary. They defined that for the onset of ferrite cleavage, rupture of the carbide is essential, and for this, prior plastic deformation of the matrix is necessary. Later, in different low-carbon weld microstructures inclusion-initiated cleavage fractures were identified too (non-metallic inclusions rich in Ti and Si [14], complex inclusions with Mn, Ti, Si and S [15]). In electric melted low carbon Si-killed structural steels, complex Al, Si and Mn oxides (sometimes together with Ca oxides) have been identified as cleavage nucleation sites [16]. In the case of eutectoid steels, complex inclusions appeared as cleavage nucleants [9, 10]. The effect of TiN particles in promoting cleavage fracture has been observed too in Ti non-microalloyed conventionally cast steels (or blooms obtained by continuous casting) with Ti contents of ~20 ppm, present as a residual element originating from the scrap [18]. Sometimes the cleavage initiation particle is a consequence of a microalloying addition. This is the case of low and medium carbon Ti microalloyed steels [13, 17, 18], with coarse TiN particles (>1 μm). Ti is added to the steel to control the grain size through the pinning effect on grain boundaries by a fine distribution of TiN precipitates. However, in most cases, when Ti microalloying is used, it is difficult to avoid some degree of Ti precipitation from the liquid in the form of coarse TiN particles (> 1 μm). These particles are ineffective in pinning the grain boundaries and are detrimental from the point of view of toughness [19], acting as cleavage fracture nucleation sites. One of the common characteristics of all the inclusion types, reported as being active nucleation sites, is the very strong particle-matrix bonding that they exhibit. This bonding is a consequence of the mismatch between the non-metallic particle and the matrix. Taking into account the work of Brookbank and Andrews about the residual stresses associated with different thermal expansion coefficients [25], the most deleterious particles are calcium aluminates, alumina, TiN and some silicates.

In all the above mentioned cases, the nucleation of the cleavage is related to some microstructural feature that fractures under an acting (tensile) stress, promoting the formation of a microcrack. However, this cannot be considered as the unique mechanism responsible for the initiation of the brittle fracture. In bainitic V microalloyed steels [20] and pearlitic steels [9, 21, 22], for example, the cleavage fracture can propagate from a ductile defect (voids for example) nucleated under the action of a plastic deformation. In these circumstances, the cleavage fracture initiation would not be only a stress induced process, but also relevant to the applied strain level.

In relation to the geometry of the particles able to nucleate a brittle process (see Table 1), all types of shapes can be identified: globular (oxides in weldments, hard complex oxides), elongated (spessartite in Si killed steels) and cuboids (TiN inclusions). From the point of view of size, in agreement with equation (1), the coarsest particles would be the most deleterious. Nevertheless, as is discussed later in this work, the influence of the size of the particles can only be considered when taking into account the availability for the microcracks, thus produced, to propagate across the matrix. Finally, it is obvious that the volume fraction of these brittle particles will have repercussions on their effect on the toughness; nevertheless, this influence is difficult to quantify. In the case of normalised ferrite-pearlite microstructures, some relationships have been proposed between the grain boundary carbide thickness and the ITT transition temperature [26]. In a similar way, the carbide size distribution and their volume fraction have been taken into account for the toughness modelling in bainitic steels [27]. In the case of inclusions, the quantification of their effect on toughness becomes very complex and their influence on shifting the ITT temperature is, in general, unknown. Fig. 2 shows the influence of a very small volume fraction of Si-Al-Mn complex oxides on the ductile-brittle transition zone of the

Charpy curve for two V microalloyed structural steels with similar chemical composition, conventional mechanical properties and microstructural parameters (d_α and pearlite volume fraction).

Table 1. Second phase particles reported as being active on the nucleation of microcracks

Cracking of a microstructural feature		
FEATURE	STEEL	REFERENCE
carbides	ferritic steels, bainitic steels quenched and tempered steels	[3, 21]
complex inclusions	low-carbon welds, eutectoid steels,	[5, 8]
Al_2O_3-MnO-SiO_2, Al_2O_3-MnO-SiO_2-CaO	Si killed carbon and V microalloyed steels	[16]
TiN particles (>1µm)	Ti microalloyed steels	[13, 17, 18]
martensite-austenite films	HAZ, HSLA steels	[23]
Microcracks emanating from voids		
MnS elongated inclusions, ductile voids in matrix	pearlitic steels, bainitic steels, acicular ferrite steels	[9, 20, 22, 24]

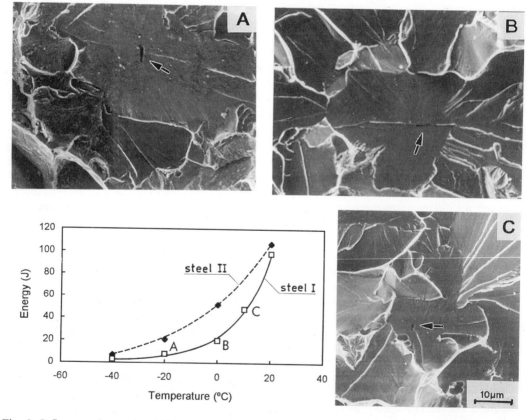

Fig. 2. Influence of complex Al-Mn-Si-Ca oxides in the shifting the ductile-brittle transition part of Charpy curve to higher temperatures in V microalloyed structural steels. Steel I has a higher volume fraction of these oxides than steel II, these particles being responsible for cleavage initiation (see fractographs at the origin of cleavage fracture) [16].

3. PROPAGATION OF MICROCRACKS INTO THE MATRIX

Once the microcrack (or the defect) has been nucleated, it is necessary to consider the second step: its successful brittle propagation into the matrix. According to equation (1), the minimum stress σ_{pm} required depends on the microcrack size and on the γ_{pm} effective surface energy. It is clearly evident that a high value of γ_{pm} or a small size of the nucleant microstructural feature requires a high stress value for the crack to progress. In consequence, at this step of cleavage propagation, both parameters must be considered.

One of the most common procedures for measuring the value of the γ_{pm} effective surface energy is by testing notched four point bending specimens at 77K. An extensive study has been carried out with different Ti, Ti-V, V and Ti-B microalloyed low and medium carbon steels with ferrite-pearlite, bainite, martensite and tempered microstructures [12, 28]. The cleavage origin (on the fracture surface of the specimens) was identified, the size of the cracked particle responsible for cleavage nucleation was measured and the local cleavage fracture stress (σ_F^*) determined. In Fig. 3, σ_F^* stress is plotted against $\phi/\sqrt{a_{min}}$, where ϕ is the ellipticity correction and a_{min} the minimum dimension of the broken particle. In agreement with equation (1), the straight line slopes correspond to the γ_{pm} effective surface energy, the majority of the experimental points lying between 7 and 20 J/m^2. It is worth emphasising that, due to the dynamic character of the microcrack nucleation and propagation, σ_F^* stress is always $\geq \sigma_{pm}$ and, as a result, the true value of γ_{pm} must be lower that the calculated data from the experimental measurements. Taking into account this characteristic, from the data of Fig. 3, an upper bound of $\gamma_{pm} = 7$ J/m^2 has been defined, this value being in a good agreement with the results obtained for a wide range of steels [9, 14]. This energy is usually considered to remain constant (or change only very slightly) with temperature.

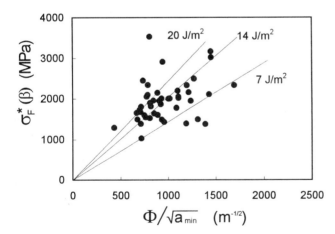

Fig. 3. Local cleavage fracture stress values plotted against the reciprocal square-root of the minimum particle size responsible for cleavage nucleation. Data correspond to Ti, Ti-V, V and Ti-B microalloyed low and medium C steels with different microstructures (ferrite-pearlite, bainite, martensite and tempered martensite) [12, 28].

Another important characteristic of step 2 is the size of the microcrack (broken particle size). Taking into account equation (1), a coarser particle would produce a larger defect and, in consequence, it would be easier for the microcrack to surmount the particle-matrix barrier. Nevertheless, in some microstructures the coarsest particles are not the most deleterious. This behaviour is related to a Weibull volume effect and is well documented for coarse particles (> 1 μm) and high strength matrix microstructures (martensite and tempered martensite) [12, 18]. As a consequence of a Weibull volume effect, large particles can break at lower acting stresses than small particles, and if this occurs at stress values lower than σ_{pm}, the crack will be arrested at the particle-matrix interface and will blunt, losing its capacity to promote cleavage fracture. As a consequence of this behaviour, for a given steel, the nature of the particles responsible for cleavage fracture can change as a function of the strength of the matrix which, in turn, is directly related to the different microstructures that can develop. For example, while coarse TiN particles promote cleavage fracture in a ferrite-pearlite microstructures, in the same steel with a martensitic matrix, smaller TiN particles and principally carbides are the microcrack nucleation sites [11].

In step 2 the misorientation angle, β, between the normal to the first formed cleavage facet and the direction of the applied tensile stress is another relevant parameter. For microcrack propagation from the particle to the matrix, it is required that at least one of the normals to the {100} crystallographic planes of the grains, surrounding the broken particle, should be close to the tensile stress component application direction. Taking into account that the relationship between the macroscopic and the local stress normal to the cleavage plane is $\cos^2\beta$, as β increases a higher fracture stress is required to propagate the microcrack to the matrix. In practice, β is found to be relatively low, as can be observed in Fig. 4, where experimental data obtained on 30 specimens of different microalloyed Ti and Ti-V steels with microcracks nucleated at TiN coarse particles are shown [12].

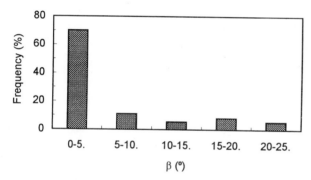

Fig. 4. Histogram of distribution of measured β angles between the normal to the first facet plane and the direction of the acting macroscopic tensile stress for different Ti and Ti-V microalloyed forging steels [12].

The crystallographic orientation relationship between the particle and the surrounding matrix can play a significant role in promoting or avoiding cleavage fracture. One of the ways to arrest the microcracks at step 2 is to develop a fine acicular ferrite structure. Taking into account the nucleation process of acicular ferrite [29, 30], the probability of having good parallelism between the cleavage plane of the particle and those of the matrix is very low and, in consequence, the microcracks stop at the interface and blunt, acting as nuclei for void formation [17] (Fig. 5 [31]).

Finally, step 2 of the cleavage process is considered to be the controlling step at very low temperatures (the zone corresponding to the completely brittle behaviour in the lower plateau of the Charpy curve) [11, 12]. In contrast, as will be shown later, in the ductile-brittle regime, matrix-matrix boundaries become, in the majority of situations, the most relevant step in the cleavage propagation process.

4. MATRIX-MATRIX BOUNDARIES

Once the first cleavage facet is formed, the successful propagation of the brittle microcrack depends on the trespassing of high angle matrix-matrix boundaries. In this situation, as happens with step 2, the two parameters that must be taken into account are the γ_{mm} effective surface energy and the size of the microcrack D (grain size).

In contrast with γ_{pm} energy, γ_{mm} experimental measured data are very scarce in the bibliography. Independently of the difficulty with its measurement, a lot of research work has been focused on the completely brittle region temperature, where step 2 becomes the controller, γ_{mm} energy being less relevant. Linaza et al. [11, 12] have proposed two approaches to estimate γ_{mm}. In the completely brittle behaviour temperature range, the dimensions of the first cleavage facet formed after the cracking of the brittle particle were measured on the fracture surfaces of four point bending tests performed at 77 K with Ti, Ti-V and Ti-B microalloyed steels. If this step is considered as the controller of the catastrophic failure, in a similar way as performed for step 2, the local cleavage fracture stress can be plotted against $\phi/\sqrt{D_{min}}$, where D_{min} is now the minimum size of the first cleavage facet. The data obtained are shown in Fig. 6 [12, 28] and the straight line slopes correspond to γ_{mm} effective surface energy. Taking into account that, in any case, at this low temperature a grain boundary has been observed preventing brittle propagation, that is, $\sigma_F{}^* > \sigma_{mm}$ (the process is controlled by step 2), from Fig. 6 an upper limit of $\gamma_{mm} = 50$ J/m^2 can be determined. In a similar way, San Martin [28] obtained an upper limit of 100 J/m^2 for a temperature of -100°C for a V microalloyed steel.

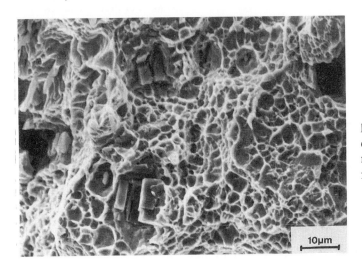

Fig. 5. Broken TiN particle in a ductile matrix. Ti-V microalloyed medium C steel with acicular ferrite microstructure [31].

In the ductile-brittle transition temperature range another approach has been considered [11, 12]. In different Ti, Ti-V and V microalloyed low and medium carbon steels, in the fracture surface of fatigue precracked specimens, brittle cleavage islands, totally surrounded by ductile fracture, were observed in the ductile crack propagation zone (see example in Fig. 7). In these islands the particle responsible for the cleavage initiation was identified (in the majority of the cases coarse TiN particles). It is clear that cleavage fracture was nucleated at a particle and propagated across the particle-matrix interface, but subsequently arrested at a matrix-matrix boundary. If the dimensions of the island are measured, and the stress value at which the cleavage process starts and stops is known, it would be possible to estimate the γ_{mm} energy (a lower boundary value).

The real stress value at which the islands are formed is unknown. Nevertheless, the following analysis shown schematically in Fig. 8 could be considered for estimating the stress:

- $\sigma_p \geq \sigma_{pm}$, that is, the particle breaks and cleavage develops.

- assuming $\gamma_{pm} = 7$ J/m^2, it is possible to determine the stress necessary to break the particle, as long as its size is known, with the help of equation (1). Considering then $\sigma_p = \sigma_{pm}$, the latter can be deduced.
- when $\sigma_{mm} > \sigma_p \geq \sigma_{pm}$ the cleavage facet is isolated, thus:

$$\sigma_{mm} = \left(\frac{\pi E \gamma_{mm}}{(1-\nu^2)D}\right)^{1/2} > \sigma_{pm} \qquad (2)$$

where D is the size of the cleavage island. Based on the assumption that σ_{pm} is uniformly acting all over the cleavage island, a lower bound limit for γ_{mm} can be estimated.

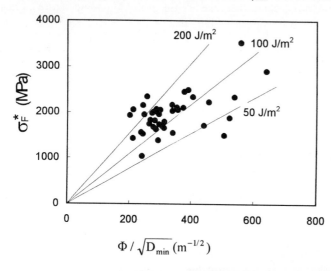

Fig. 6. Local cleavage fracture stress values plotted against the reciprocal square-root of the minimum first cleavage facet size. Data correspond to Ti, Ti-V, V and Ti-B microalloyed low and medium C steels with different microstructures (ferrite-pearlite, bainite, martensite and tempered martensite) [12, 28].

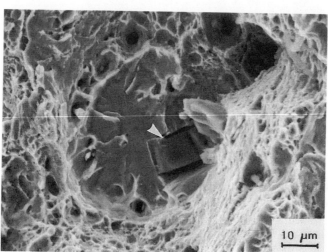

Fig. 7. Isolated cleavage island arrested by grain boundaries and completely surrounded by ductile features. The crack was initiated in a TiN particle and propagated across one cleavage facet before arresting (Ti-V steel with ferrite-pearlite microstructure).

Following this procedure, Linaza et al. [12] determined, from a total of 25 isolated islands, a value of 200 J/m² for room temperature tests, with Ti and Ti-V microalloyed medium C forging steels. More recently, for the case of a V microalloyed steel with a normalised microstructure, San Martin [28] determined, at -50°C, a lower bound limit of 500 J/m² (from 22 identified islands). For the temperature of -70°C, due to the small quantity of islands (the fracture was predominantly brittle), it was only possible to determine that the energy would be higher than 60 J/m². These values, together with those estimated as being upper limits at lower temperatures, are shown in Fig. 9.

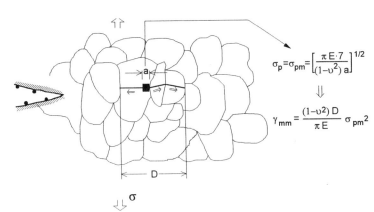

$$\sigma_p = \sigma_{pm} = \left[\frac{\pi E \cdot 7}{(1-\upsilon^2)a}\right]^{1/2}$$

$$\Downarrow$$

$$\gamma_{mm} = \frac{(1-\upsilon^2)D}{\pi E}\,\sigma_{pm}^2$$

Fig. 8. Schematic illustration of the estimation of a lower bound limit for γ_{mm} by analysing isolated cleavage islands in the ductile-brittle regime.

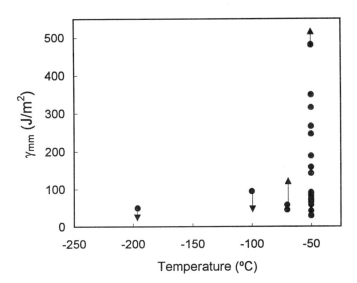

Fig. 9. Variation of γ_{mm} energy with temperature for a V microalloyed steel with a normalised microstructure. For temperatures lower than -100°C the data correspond to upper limits and for higher temperatures they are lower limits of the real γ_{mm} value [28].

Fig. 9 indicates that there is a significant increase of γ_{mm} with temperature. Taking into account equation (1), this means that, from a determined temperature value, matrix-matrix boundaries become more and more difficult to trespass and, in consequence, microcracks nucleated at brittle non-metallic inclusions can be arrested in step 3. It is worth emphasising that this change (related to the plastic contribution to γ_{mm}) is responsible for the relevant role of step 3 in the ductile-brittle transition, while for very low temperatures, into the brittle range, the cleavage process is controlled by step 2. In comparison with Fig. 9, the data of Linaza et al. show that there are large differences in γ_{mm} energy between different steels.

From the microstructural point of view, some parameters can be considered to define the m-m barriers, as listed in Table 2. In the case of low carbon steels with predominantly ferrite microstructures, the ferrite grain size corresponds to the cleavage facet size. In ferrite-pearlite medium C steels, the microstructural barriers that can stop brittle microcracks are "ferrite units" composed of ferrite plus pearlite with the same ferrite orientation [13, 19]. In a similar way, in pearlitic microstructures the unit with the same ferrite crystallographic orientation is the relevant parameter [32, 33]. In bainitic steels the packet size is the m-m barrier [34, 35] and in acicular ferrite steels the plate [36].

Table 2. Microstructural units involved in the 3rd cleavage propagation step.

Unit	Matrix	Reference
ferrite grain size	ferrite	[3]
ferrite unit (α + pearlite with the same α orientation)	ferrite-pearlite	[13, 19]
unit with same ferrite orientation	pearlite	[32,33]
packet size	bainite	[34, 35]
plate size	acicular ferrite	[36]

The size of these microstructural units will depend on the previous austenite grain size prior to transformation and on the number of nucleation sites for the transformation. Both parameters can be controlled, to a certain extent, by microalloy additions and thermomechanical treatments. Considering the temperature dependence of γ_{pm} and γ_{mm}, the critical "grain size" necessary to arrest microcracks at m-m boundaries will change with temperature, as shown in the scheme of Fig. 10 [12]. Since control of cleavage fracture may be by step 2 or step 3:

$$\text{step 2: } \sigma_{pm} > \sigma_{mm} \rightarrow \frac{\gamma_{pm}}{a} > \frac{\gamma_{mm}}{D} \rightarrow \frac{D}{a} > \frac{\gamma_{mm}}{\gamma_{pm}} \tag{3}$$

$$\text{step 3: } \sigma_{pm} < \sigma_{mm} \rightarrow \frac{\gamma_{pm}}{a} < \frac{\gamma_{mm}}{D} \rightarrow \frac{D}{a} < \frac{\gamma_{mm}}{\gamma_{pm}} \tag{4}$$

then, for a given T_1 temperature (Fig. 10) a critical value of γ_{mm}/γ_{pm} which turns out to be equal to $(D/a)_1$, can be defined. Consequently, for a given particle distribution, if the grain size is lower than D_1, cleavage fracture will be controlled by step 3 and the refinement of the microstructure will improve the toughness of the steel due to the arrest of the cracks at m-m boundaries. In contrast, if the grain size is coarser that D_1, step 2 will control cleavage fracture and m-m boundaries will not be able to arrest the brittle processes.

However, it has to be considered that, in opposition to what happens in the case of strength, brittle cleavage is a "weakest link" controlled process. In the ductile-brittle transition, the brittle fracture of a specimen (blunt-notch or fatigue precracked specimen) commences at the microstructurally weakest zone. It is worth emphasising that this zone will be composed of a microstructural feature able to nucleate a microcrack and by a coarse "grain" (unit). Usually the first cleavage facet is coarser than the mean "grain" size. As a result, the value of the mean "grain" size is not a sufficient parameter, from the point of view of toughness in ductile-brittle behaviour, it being essential to know the size distribution.

Independently of the control of the volume fraction and nature of particles and the features able to nucleate microcracks (steps 1 and 2 of brittle propagation), control of the m-m boundaries implies

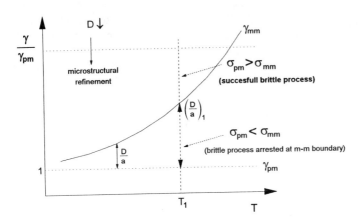

Fig. 10. Scheme of the relationship between γ_{pm} and γ_{mm} energetic barriers and the corresponding microstructural feature controlling cleavage propagation as a function of temperature [12].

the achievement of a fine and homogeneous microstructure in the whole volume of the material. This condition becomes the key factor in many industrial applications where some fraction of coarser grain sizes, not relevant from the point of view of strength, impair toughness values in the ductile-brittle regime. This can be the case with ferrite-pearlite and low carbon Ti microalloyed steels that suffer incomplete TiN grain growth control during hot forging or rolling [37], and in Nb microalloyed steels dependent upon the austenite grain size before the T_{nr} temperature and the degree of accumulated deformation after this temperature [38, 39]. In the same way, the effect of intragranular ferrite nucleation at MnS inclusions (high sulphur steels), used as a microstructure refinement procedure, can lead to uncertain results if the process has not taken place in the whole volume of the material.

5. CONCLUDING REMARKS

The brittle fracture of microalloyed steels proceeds through three differentiated steps:
1. Creation of a sharp microcrack in a critical microstructural feature, mainly at second phase particles.
2. Progression of the microcrack through the particle-matrix interface.
3. Progression of the microcrack through the matrix-matrix boundaries.

At very low temperatures, in the completely brittle regime, the cleavage process depends on step 2. In contrast, at higher temperatures in the ductile-brittle transition, step 3 controls cleavage propagation. This change is a consequence of the strong variation of γ_{mm} energy with temperature. Given that high angle boundaries can stop and blunt the microcracks nucleated at second phase particles, it is necessary to consider a microstructural unit surrounded by high angle boundaries (ferrite grain size, bainite packet, ferrite unit, ...) which becomes the main parameter at step 3. Thermomechanical treatments, controlling previous austenite grain size and transformation products, can refine the size of this microstructural unit. However, due to the weakest link character of the brittle process, the volume fraction of coarser units is more relevant than the overall mean unit size.

DEDICATION

This paper is dedicated to the memory of Prof. J.J. Urcola, our late head of the Materials Dept. of CEIT, a very good friend and responsible for our interest in steel research and development, and for beginning, pushing and developing many of the topics to which our research work is now devoted.

REFERENCES

1. D. Naylor, Ironmaking and Steelmaking 16 (1989), p. 246.
2. M. Korchynsky and J.R. Paules, SAE Technical Paper Series 890801 (1989).
3. C.J. McMahon and M. Cohen, Acta Metall. 13 (1965), p. 591.
4. J. F. Knott, Reliability and Structural Integrity of Advanced Materials, ECF9, S. Sedmak et al. eds, vol. 2, EMAS (1992), p. 1375.
5. J.H. Tweed and J.F. Knott, Acta Metall. 35 (1987), p. 1401.
6. P. Bowen, S.G. Druce and J.F. Knott, Acta Metall. 34 (1986), p. 1121.
7. P. Bowen, S.G. Druce and J.F. Knott, Acta Metall. 35 (1987), p. 1735.
8. D.J. Alexander and I.M. Bernstein, Met. Trans. 20A (1989), p. 2321.
9. J.J. Lewandowski and A.W. Thompson, Met Trans. 17A (1986), p. 1769.
10. J.J. Lewandowski and A.W. Thompson, Acta Metall., 35 (1987), p. 1453.
11. M.A. Linaza, J.L. Romero, J.M. Rodriguez-Ibabe, and J.J. Urcola, Scrip. Met. Mat. 32 (1995), p. 395.
12. M.A. Linaza, J.M. Rodriguez-Ibabe and J.J. Urcola, Fatigue Fract. Engng. Mater. Struct. 20 (1997), p. 619.
13. M.A. Linaza, J.L. Romero, J.M. Rodriguez-Ibabe and J.J. Urcola, Scrip. Met. Mat. 29 (1993), p. 451.
14. P. Bowen, M.B.D. Ellis, M. Strangwood and J.F. Knott, Fracture Control of Engineering Structures, ECF6, vol. 3, EMAS (1986), p. 1751.
15. D.E. McRobie and J.F. Knott, Mater Sci. Technol. 1 (1985), p. 357.
16. I. San Martin and J.M. Rodriguez-Ibabe, CEIT internal report (1998).
17. M.A. Linaza, J.L. Romero, J.M. Rodriguez-Ibabe and J.J. Urcola, Scrip. Met. Mat. 29 (1993), p. 1217.
18. M.A. Linaza, J.L. Romero, J.M. Rodriguez-Ibabe and J.J. Urcola, 36th Mechanical Working and Steel Processing Conference Proceedings (1995), p. 483.
19. M.A. Linaza, J.L. Romero, I. San Martin, J.M. Rodriguez-Ibabe and J.J. Urcola, Microalloyed Bar and Forging Steels, C.J. Van Tyne et al. eds. (1996), p. 311.
20. A. Echeverria, M.A. Linaza and J.M. Rodriguez-Ibabe, this volume.
21. A.R. Rosenfield and D.K. Shetty, ASTM STP 856 (1985), p. 196.
22. T. J. Baker, F.P.L. Kavishe and J. Wilson, Mater. Sci. and Technol. 2 (1986), p. 576.
23. B. Mintz, A. Nassar and K. Adejolu, Mater. Sci. Technol. 13 (1997), p. 313.
24. M.A. Linaza, Ph.D. Thesis, Univ. Navarra (1994).
25. D. Brooksbank and K. W. Andrews, JISI, 206 (1968), 595-599.
26. B. Mintz, G. Peterson and A. Nassar, Ironmaking and Steelmaking 21 (1994), p. 215.
27. A. Martin, I. Ocaña, J. Gil Sevillano, M. Fuentes, Acta Metall. Mater. 42 (1994), p. 2057.
28. I. San Martin, Ph.D. Thesis, Univ. Navarra (1998).
29. I. Madariaga and I. Gutierrez, Script. Mater. 37 (1997), p. 1185.
30. I. Madariaga and I. Gutierrez, Met. and Mat.Trans. 29A (1998)
31. M.A. Linaza, J.L. Romero, J.M. Rodriguez-Ibabe and J.J. Urcola, Thermomechanical Processing [TMP]^2, B. Hutchinson et al. eds (1997), p. 351.
32. J.M. Hyzak and I.M. Berstein, Metall. Trans. 7A (1976), p.1217.
33. D.J. Alexander and I.M. Bernstein, Metall. Trans. 13A (1982), p. 1865.
34. J.P. Naylor and P.R. Krahe, Metall. Trans. 5A (1974), p. 1699.
35. P. Brozzo, G. Buzzichelli, A. Mascanzoni and M. Mirabile, Met. Sci. (1977), p. 123.
36. M. Díaz, I. Madariaga, J.M. Rodriguez-Ibabe and I. Gutierrez, J. Construct. Steel Res. 46 (1998)
37. J.I. San Martin, M.A. Linaza, J.M. Rodriguez-Ibabe and M. Fuentes, Thermec'97, T. Chandra and T. Sakai eds. (1997), p. 265.
38. R. Bengochea, Ph.D. Thesis, Univ. Navarra.
39. Y. Kamada, T. Hashimoto and S. Watanabe, ISIJ Intern. 30 (1990), p. 241.

Materials Science Forum Vols. 284-286 (1998) pp. 63-72

The Production and Mechanical Properties of Ultrafine Ferrite

P.D. Hodgson[1], M.R. Hickson[2] and R.K. Gibbs[3]

[1] School of Engineering and Technology, Deakin University, Geelong, Australia

[2] BHP Research - Melbourne Laboratories, Mulgrave, Australia

[3] BHP Steel, Western Port, Hastings, Australia

Keywords: Thermomechanical Process, Steel Strip, Ultrafine Ferrite, Strain Induced Transformation, Mechanical Properties, Low C Steel, High C Steel, Microalloyed Steel

ABSTRACT

A new thermomechanical process has been developed to produce ultrafine (1μm) equiaxed ferrite grains in hot rolled steel strip. This process is remarkably simple and is applicable to a wide range of steel chemistries, including low and high carbon and microalloyed steels. Strips are reheated to produce a coarse austenite grain size, then rolled in a single pass at or just above the austenite to ferrite transformation temperature. It is suggested that the observed refinement is due to strain induced transformation from austenite to ferrite. The requirements for this appear to be high strain induced by shear in the strip surface layers, and thermal gradients created by heavy quenching of the strip surface by the work rolls. The yield strength was markedly higher than conventionally processed strip, although there was little work hardening even though total elongation of over 20% was achieved.

INTRODUCTION

One of the key goals in the development of steels has been to refine the ferrite grain size, as this leads to an increase in both strength and toughness. Historically this has been achieved by normalising or by cold rolling and annealing to produce fine grained steels. More recently, thermomechanical processing has been used to control both the state of the austenite prior to transformation and the transformation history. To date, the optimum reduction in ferrite grain size is achieved through heavy controlled rolling combined with accelerated cooling. For a simple steel composition in plate rolling this can reduce the ferrite grain size from 10μm for hot rolling and air cooling to 5μm for controlled rolling and water cooling, giving an increase in the yield strength of approximately 80 MPa.

In previous studies related to the austenite to ferrite transformation of controlled rolled steels it was shown that there is a limiting ferrite grain size of approximately 5μm regardless of the level of retained strain introduced into the austenite [1]. Priestner and Hodgson [2] have discussed how this is at least partly due to ferrite coarsening during transformation, associated with growth of ferrite rafts/films from the austenite grain boundary where the grains first nucleate, into the austenite grain interior. This coarsening was particularly pronounced at high levels of retained strain where full ferrite to ferrite impingement along an austenite boundary was observed when the ferrite grains were only 1μm, yet the final average ferrite grain size was 5-7μm. If the 1μm grain size, here called ultrafine ferrite (UFF), could be retained then Hall Petch analysis suggests that this would increase the yield strength by almost 350 MPa compared to a 5μm ferrite microstructure.

At the same time there has been some evidence that strain induced transformation (ie. transformation during rather than after deformation) could lead to much finer ferrite grains than those suggested above. Priestner [3] observed local areas of very fine grains in intercritically rolled steels, while the strain induced ferrite in hot ductility studies also appeared finer than expected [4]. In these cases the regions of fine ferrite were quite small. However, Yada and co-workers at Nippon Steel Company claim to have produced coils of hot rolled strip where the ferrite grain size approached 1μm by rolling close to the transformation temperature [5,6]. They suggested that this refinement was through a combination of strain induced ferrite and dynamic recrystallisation of the ferrite. One of the current authors has also produced uniform ferrite microstructures approaching 1μm in laboratory hot torsion tests [7]. In that work, however, the deformation and cooling conditions were extreme and outside the capability of any realistic bulk metalworking process.

From the above analysis and other literature it appears that there is the potential to markedly refine the ferrite if strain induced transformation can be activated over a significant volume of the austenite. This, in turn, appears to require rolling close to the transformation temperature (ideally as far below the A_{e3} as possible) and high rolling strains. The experimental conditions described below achieve these aims. Firstly, a large austenite grain size was used to suppress conventional pro-eutectoid ferrite formation. Secondly, earlier studies of the recrystallisation of stainless steels [8-11] had used a technique where thin strip was hot rolled under conditions of limited lubrication. This produced a shear layer in the strip which markedly increased the rate of recrystallisation and reduced the recrystallised grain size. Both the microstructural and numerical analysis [10] of the test suggest that the strain in the surface layer is significantly higher than the macroscopic strain given by the reduction in thickness. As shown in the current work this combination of austenite conditioning, deformation conditions and temperature control can lead to the formation of significant volumes of UFF.

EXPERIMENTAL

A series of steel compositions, including low and high carbon and microalloyed steels (Table 1) were rolled in the BHPR 2-high laboratory rolling mill. Steels A and D were experimental laboratory heats and steels B, C and E were commercial plate and strip grades.

Table 1. Chemical composition of the steels investigated (wt%)

Steel	C	Mn	Nb	Ti	B
A	0.06	0.59	-	-	-
B	0.09	1.00	-	-	-
C	0.09	0.71	0.027	0.025	-
D	0.18	1.68	-	0.017	0.0016
E	0.77	0.71	-	-	-

All specimens had an initial thickness of 2mm, and were reheated in stainless steel foil bags to prevent excessive oxidation of the strip. Reheating conditions were 1250°C for 15 minutes, resulting in an austenite grain size of 100-200μm. Strips were removed from the bags and subjected to a single rolling pass of 25-45% reduction, at a roll speed of 24m/min. Rolled samples were simply air cooled or held in a fluidised sand bed at 600°C for 1 hour and then slow cooled between sheets of Kaowool to simulate coiling. Rolling temperatures were monitored by optical pyrometers at the roll gap entry and exit locations. An estimation was made of the A_{r3} of each alloy and this was used as the roll entry temperature. These were typically between 700 and 780°C. The average temperature drop in the roll gap was 70-100°C, resulting in roll exit temperatures of between 620 and 710°C.

Specimens for optical microscopy were prepared by polishing and etching in 2% nital solution. Mechanical properties were determined from full thickness tensile specimens cut from the rolling direction. Testing was performed on a Sintech universal testing machine at a strain rate of 10^{-4} s^{-1}. Further microstructural analysis was performed using SEM and TEM techniques.

RESULTS

Microstructures

The microstructures of all rolled samples were inhomogeneous through the strip thickness (Figure 1). The general feature of all microstructures was the presence of equiaxed UFF grains of about 1μm in a region near the surface (Figure 2). The UFF generally penetrated to a depth of 1/4 to 1/3 of the total thickness, and discrete carbides were also present throughout this region. The centre of the strip consisted of a more conventional microstructure, such as significantly coarser ferrite grains (5-10μm) and pearlite. The microstructures produced by each of the steel types studied are outlined below.

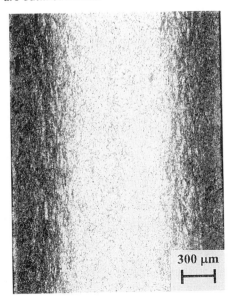

Fig. 1 Typical layered microstructure consisting of UFF surface grains and coarse centre grains

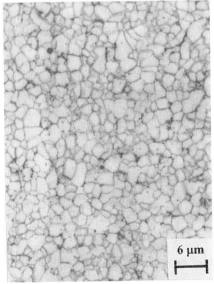

Fig. 2 UFF grains in low carbon grade (Steel A)

Low Carbon (Figure 2)

Steels A and B both displayed microstructures consisting of surface UFF of 1μm grain size and coarse ferrite in the core regions. The surface grains were equiaxed in nature, and some discrete carbides were also visible.

Microalloyed (Figure 3)

Steels C and D displayed surface UFF with less than 1μm grain size, which was almost impossible to resolve by optical microscopy. The core regions consisted of refined angular and Widmanstatten ferrite and a small amount of pearlite.

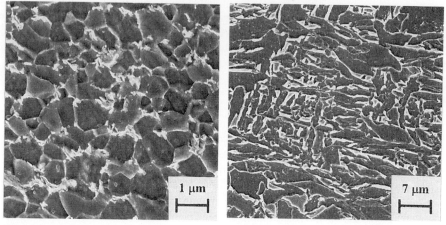

Fig. 3 Surface (left) and core (right) microstructures of microalloyed grade (Steel C)

High Carbon (Figure 4)

The eutectoid steel E displayed a unique microstructure, with the expected lamellar pearlite being completely decomposed into its ferrite and carbide components. As such, the surface regions consisted of UFF grains (less than 1μm in size) with a high volume fraction of discrete carbides. Regular pearlite was present in the core of the samples.

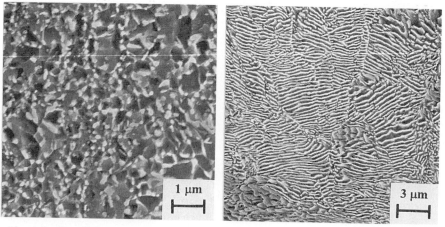

Fig. 4 Surface (left) and core (right) microstructures of high carbon grade (Steel E)

The samples subjected to simulated coiling displayed virtually identical surface and core microstructures to those strips which underwent natural air cooling. This indicates that the UFF grains are extremely stable and do not coarsen during cooling after transformation.

The details of the ferrite-carbide structure were studied using SEM (Figure 5), which revealed the presence of carbides at many of the triple points between the ferrite grains. This indicates that the UFF was formed by transformation, rather than by static or dynamic recrystallisation.

The TEM analysis showed an equiaxed ferrite structure with planar boundaries and little evidence of an internal dislocation structure (Figure 6). Again, grain boundary carbides were present between many of the grains. Similar structures have been noted in dynamically recovered ferrite [12], although the high degree of polygonisation required for this would not be expected at the strains and strain rates used in the current work. In fact, a small amount of proeutectoid ferrite was visible at prior austenite grain boundaries in one sample. These grains must have formed prior to deformation, and show a very low level of polygonisation despite being subjected to the maximum possible strain.

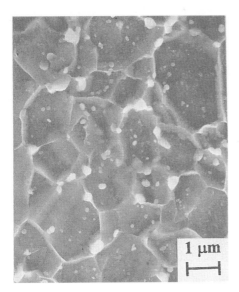
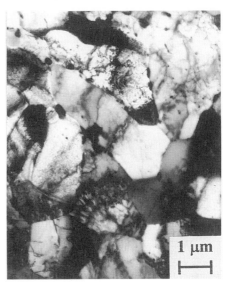

Fig. 5 Scanning electron micrograph showing carbides at ferrite grain triple points (Steel A)

Fig. 6 Transmission electron micrograph of UFF grains showing planar ferrite grain boundaries and low dislocation density (Steel A)

Mechanical Properties

Tensile properties indicate that very high strength levels combined with good ductility are achievable with this type of processing (Table 2). In particular, the increases in yield stress are quite remarkable. It is seen that a plain low carbon steel (Steel A) displayed an LYS of 590 MPa, virtually twice that displayed by the same alloy under conventional processing conditions. Significant increases in LYS and TS are also observed for the microalloyed and higher carbon grades. Ductility remained acceptable in most cases.

Table 2. Mechanical properties of steels with UFF and conventional microstructures

Steel	*Microstructure*	*LYS (MPa)*	*TS (MPa)*	*LYS/TS*	*TE (%)*
A	UFF	490	510	0.96	33
	Conventional	250	360	0.70	30-40
B	UFF	470	530	0.89	28
	Conventional	295	400	0.74	30-40
C	UFF	575	575	1.00	15
	Conventional	380	460	0.83	23-30
D	UFF	490	500	0.98	23
E	UFF	730	1070	0.68	9
	Conventional	560	760	0.74	6-12

One of the more unusual aspects of these results was the nature of the stress-strain curves. A typical example (Figure 7) indicates the absence of significant work hardening after yielding, particularly for the low carbon grades. This is reflected in the ratio LYS/TS, which in many cases was between 0.90 and 1.00.

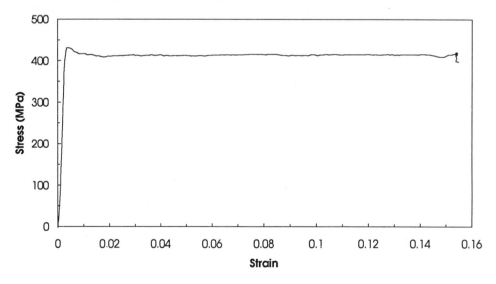

Fig. 7 Stress-strain curve typical of Steel A showing absence of work hardening

DISCUSSION

The remarkable aspects of this work are the simplicity of the processing conditions and its applicability to a wide range of steel chemistries. Conventional rolling processes incorporate significant demands on the control of rolling schedules, temperatures and cooling rates. The current experiments require relatively small reductions and air cooling is sufficient to produce UFF microstructures.

There are potentially three mechanisms which may produce the level of ferrite grain refinement observed in the surface layers. These are:
- strain induced transformation
- transformation from dynamically recrystallised austenite
- dynamic recrystallisation of ferrite

Of the above, strain induced transformation appears to be the only mechanism that is operating here. The other two mechanisms would not be expected to decompose the eutectoid steel to equiaxed ferrite and carbides. Strain induced transformation is the least understood mechanistically, and it had previously appeared that the processing window was rather small. Laboratory studies [3] have shown it is difficult to homogeneously transform the total microstructure to UFF; in most cases only a small volume fraction transformed at a given pass, although Yada and co-workers have stated [5] that they have produced UFF in commercial hot strip rolling.

The experiments to date have shown that the strain induced transformation rolling (SITR) process can be easily reproduced in thin strip using single pass reductions. The presence of a surface layer suggests two factors: strain inhomogeneity and thermal gradients induced by heavy quenching of the strip surface by the work rolls.

It appears that strain inhomogeneity is the dominant factor, although roll cooling also plays an important role. Experiments with austenitic stainless steel rolled under identical conditions have clearly shown a heavy shear layer which has a similar thickness to the UFF layer [10]. Finite element modelling of the rolling geometry has shown that the 'effective' strain at the surface, which accounts for both the compressive and shear strains, can be over 3 times the nominal strain. The quenching by the work rolls is also important as it provides a high level of undercooling which favours rapid, homogeneous transformation, and avoids the formation of conventional proeutectoid ferrite.

Figure 8 shows the strengths that can be theoretically obtained from a typical carbon-manganese steel as a function of ferrite grain size, using the equations developed in [13]. One of the interesting observations is that below 3μm the predicted LYS exceeds the TS. Previous work examining the tensile behaviour of very fine grained steels [14] found increases in yield strength but little work hardening as the ferrite grain size was decreased. Considering this, the flatness of the stress-strain curves obtained from the current steels is not that surprising, although the present levels of ductility are higher than those reported in [14]. To study the deformation behaviour during tensile testing, a tensile specimen with a polished surface was tested and the surface videotaped. It was found that the deformation proceeded by the migration of Luders bands throughout the entire test. It is interesting to note that these bands were developed even though the through the thickness microstructure was highly inhomogeneous.

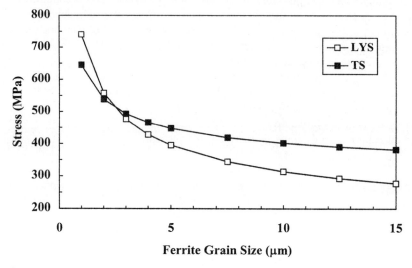

Fig. 8. Relationship between LYS and TS and ferrite grain size according to [12]

Practical Implications
The current work has used an extremely simple thermomechanical process to produce UFF. At the same time, though, this process is greatly removed from current commercial hot rolling operations. Significant work is therefore required to utilise this result on an industrial scale.

There are also a number of issues relating to the processing window and the mechanism for producing UFF which need to be resolved. Amongst those is the potential for increasing the depth of the UFF layer, perhaps even to produce a strip with a homogeneous UFF through the entire thickness. Other factors requiring consideration include austenite condition, level of rolling reduction, strain rate, surface conditions (lubrication and oxidation), chemistry and cooling rate. The mechanical behaviour of these types of microstructures also requires further investigation, particularly to improve the work hardening behaviour. It would appear that if these issues could be addressed then this would be the most promising method for obtaining UFF.

The other two methods of grain refinement referred to above, which both rely on dynamic recrystallisation, have pronounced limitations. In particular, the dynamic recrystallisation of ferrite [15] appears limited to interstitial free steels with very high applied strains, such as those occurring in rod or bar mills. The dynamic recrystallisation of austenite by strain accumulation [16] is also unproven, and relies on the assumption that austenite grain sizes of less than 2μm will transform to 1μm ferrite. This is also only possible for a limited range of conditions. Hence, of the techniques that have been proposed for the manufacture of UFF only the current method appears to work for a wide range of steels.

The potential applications of microstructures consisting of UFF could be grouped into two broad categories:

- Full thickness UFF would most likely display extremely high strength and toughness, but may not be very useful due to the poor yield/tensile ratio. It has been suggested [5] that it may be possible to improve the ratio LYS/TS by controlling the second phase of UFF steels. In that work, accelerated cooling was used to produce a small volume fraction of martensite or bainite.
- Surface UFF with a ferrite/pearlite core could have possible applications in plate or structural products. Such steels would have high strength with improved surface toughness. Nippon Steel has recently announced the development of plate steel with a surface layer of 'super-fine ferrite' [17]. The steel allegedly has improved brittle-crack resistance and is intended for use in large applications such as ships and marine structures.

CONCLUSIONS

1. A new thermomechanical process has been developed to produce ultrafine ($1\mu m$) equiaxed ferrite grains in steel strip. Strips were reheated to produce a coarse austenite grain size, then rolled in a single pass at or just above the A_{r3}. The ultrafine grains formed in the surface regions of the strip, penetrating to a depth of about 1/4 to 1/3 of the total thickness. Coarser ferrite grains (5-10μm) and carbides were produced in the core of the strip.
2. The rolling process was very simple and has been applied to a range of steel chemistries, including plain low and high carbon and microalloyed steels.
3. The observed refinement was due to strain induced transformation from austenite to ferrite. The requirements for this appear to be large strain and high undercooling. In the current method this was achieved by strain inhomogeneity in the surface layers and thermal gradients induced by heavy quenching of the strip surface by the work rolls.
4. The mechanical properties of these steels were very promising. The yield stress was increased by up to 100%, together with improved tensile stress and reasonable ductility. Many samples, however, displayed a lack of work hardening after yielding.

ACKNOWLEDGEMENTS

The authors are grateful to The Broken Hill Proprietary Co. Ltd for support of this work and for permission to publish it. The technical assistance of Mr J. Whale is also acknowledged.

REFERENCES

1. R.K. Gibbs, P.D. Hodgson and B.A. Parker, The modelling of microstructure evolution and final properties for Nb microalloyed steels, Morris E. Fine Symposium, (ed. P.K. Liaw et al), TMS, (1991), p.73
2. R. Priestner and P.D. Hodgson, Ferrite grain coarsening during transformation of thermomechanically processed C-Mn-Nb austenite, Mat. Sci. Tech., 8, (1992), p.849
3. R. Priestner, Strain-induced γ-α transformation in the roll gap in carbon and microalloyed steel, Thermomechanical Processing of Austenite, (ed. A.J. De Ardo et al), TMS-AIME, (1982), p.78

4. B. Mintz, S. Yue and J.J. Jonas, Hot ductility of steels and its relationship to the problem of transverse cracking during continuous casting, Int. Mat. Rev., 36, (1991), p.187

5. H. Yada, Y. Matsumura and K. Nakajima, United States Patent No. 4,466,842 (1984)

6. H. Yada, Y. Matsumura and T. Senuma, A new thermomechanical heat treatment for grain refining in low carbon steels, Thermec '88, (ed. I. Tamura), ISIJ, (1988), p.200

7. J.H. Beynon, R. Gloss and P.D. Hodgson, The production of ultrafine equiaxed ferrite in a low-carbon microalloyed steel by thermomechanical treatment, Materials Forum, 16, (1992), p.37

8. K. Kato, Y. Saito and T. Sakai, Investigation of recovery and recrystallisation during hot rolling of stainless steels with high speed laboratory mill, Trans. ISIJ, 24, (1984), p.1050

9. T. Sakai, Y. Saito, M. Matsuo and K. Kawasaki, Inhomogeneous texture formation in high speed hot rolling of ferritic stainless steel, ISIJ Int., 31, (1991), p.86

10. X.J. Zhang, P.D. Hodgson and P.F. Thomson, The effect of through-thickness strain distribution on the static recrystallisation of hot rolled austenitic stainless steel strip, J. Mat. Proc. Tech., 60, (1996), p.615

11. K. Tsuzaki, K. Young-Soo and T. Maki, Hot rolling microstructure and recrystallisation behaviour of solidified columnar crystals in a 19% Cr ferritic stainless steel, Proc. 7th Int. Symp. on Physical Simulation of Casting, Hot Rolling and Welding, (ed. H.G. Suzuki et al), NRIM, (1997), p.169

12. S. Gohda, K. Watanabe and Y. Hashimoto, Effects of the intercritical rolling on structure and properties of low carbon steel, Trans. ISIJ, 21, (1981), p.7

13. P.D. Hodgson and R.K. Gibbs, A mathematical model to predict the mechanical properties of hot rolled C-Mn and microalloyed steels, ISIJ Int., 32, (1992), p.1329

14. W.B. Morrison and R.L. Miller, The ductility of ultrafine-grain alloys, in Ultrafine-Grain Metals, (ed. J.J. Burke and V. Weiss), Syracuse Univ. Press, (1970), p.183

15. A. Najafi-Zadeh, J.J. Jonas and S. Yue, Effect of dynamic recrystallisation on grain refinement of IF steels, Recrystallisation '92, (ed. M. Fuentes and J. Gil Sevillano), Mat. Sci. Forum, 113-115, (1993), p.441

16. J.J. Jonas, Dynamic recrystallisation in hot strip mills, Recrystallisation '90, (ed. T. Chandra), TMS, (1990), p.27

17. Nippon Steel News, March/April (1997)

Materials Science Forum Vols. 284-286 (1998) pp. 73-82
© 1998 Trans Tech Publications, Switzerland

Modelling Strain Induced Precipitation of Niobium Carbonitride during Hot Rolling of Microalloyed Steel

C.M. Sellars

IMMPETUS, Institute for Microstructural and Mechanical Process Engineering:
The University of Sheffield, Sir Robert Hadfield Building,
Mappin Street, Sheffield, S1 3JD, UK

Keywords: Strain Induced Precipitation, Nb Carbonitride, Modelling, Thermomechanical Processing, Nucleation

Abstract

Because of its industrial importance the behaviour of microalloyed steels during thermomechanical processing has been extensively studied in the laboratory. Semi-empirical models have been developed and successfully applied. Nevertheless there is still not a full understanding, which can be used as a physical basis for a complete model of industrial rolling involving multiple passes in roughing and in finishing. Such a model should define when and where carbonitride precipitates form, and possibly redissolve, their composition and how much micro alloy element is precipitated in the austenite. This paper considers for a simple niobium microalloyed steel

(i)   the effect of particle size distribution during reheating
(ii)   deformation in the roughing temperature range leading to significant precipitation, even though recrystallisation is not retarded, and to acceleration of the kinetics of precipitation during finishing,
(iii)   repeated nucleation and possible reversion of strain induced precipitation between passes in finishing.

It poses a number of questions which have important implications for modelling. Finally some critical experimental observations required to answer the questions are discussed.

Introduction

Despite a great deal of research on thermomechanical processing of microalloyed steels over the last thirty years, a number of questions remain if one attempts to provide a full description of the interaction of the external process variables with the microstructural variables of recrystallisation of austenite and precipitation of microalloy carbonitrides in austenite during industrial processing. Within a full description the particle composition, the size distribution, the spacial distribution and the volume fraction are all important because of their effects on transformation during cooling and on the amount of microalloy addition left in solution for precipitation strengthening. It is the purpose of this paper to pose some of the questions by considering the simple case of steel microalloyed with niobium only and subjected to controlled rolling to plate. Each of the stages of reheating, roughing, and finish rolling is considered in turn to examine the particle interactions. In doing this, it is helpful to consider the behaviour in terms of the thermodynamic stability of small particles and the kinetics governed by the diffusion coefficient of niobium, as illustrated in Fig. 1. For this figure, the solubility equation of Irvine et al [1] has been modified as proposed previously [2] for small particles to give

$$\log [Nb][C + \frac{12}{14} N] = 2.26 - \frac{\left(6770 - ad^{-1}\right)}{T} \qquad (1)$$

where [Nb] [C] and [N] are weight percent, d (m) is the particle diameter. As shown in the Appendix, a = 1.34 x 10^{-6} Km for particles on grain boundaries and 2.01 x 10^{-6} Km for particles in the matrix. The lines in Fig. 1(a) present the stability of grain boundary particles and the sizes must be increased by 50% for matrix particles. For compatibility with the model of Dutta and Sellars [3] the diffusion coefficient for niobium [4] is also taken to calculate distance x (m) over which concentration gradients are halved in time t (s) as

$$x^2 = [1.4 \times 10^{-4} \exp - 270000/RT] \, t \qquad\qquad (2)$$

The resulting values are shown in Fig. 1(b).

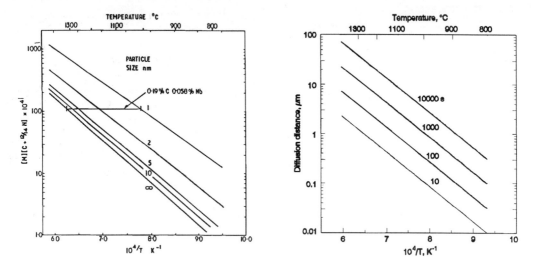

Figure 1 -   Effect of temperature on (a) equilibrium solubility product as a function of particle size [2] (b) diffusion distance as a function of time

The first question must be whether the equations selected are appropriate, because, particularly for the solubility product, many alternatives are available [5]. These would significantly shift the curves on the temperature scale in Fig. 1(a), but the concepts illustrated in Fig. 1 remain valid and provide the essentials to pose the further questions.

Reheating

It is well known that grain coarsening during reheating occurs at a temperature significantly (100-150°C) below the solubility temperature calculated for coarse particles. Recent work [5] using atom probe analysis of steels reheated from room temperature for half and hour at the austenitising temperature shows a discrepancy between the observed amount of Nb in solution and the theoretical amount calculated for Nb(C,N), as shown in Fig. 2. The trend is exactly the same, but less pronounced than found in early work [6,7]. At lower temperatures the larger solubility can simply be understood from Fig. 1(a), if the particle size distribution includes large numbers of small particles in the range 1-10nm as observed in the steels investigated. This then raises the question:

- are the particle size distributions and hence the grain coarsening temperature measured in the laboratory appropriate for industrial processing conditions?

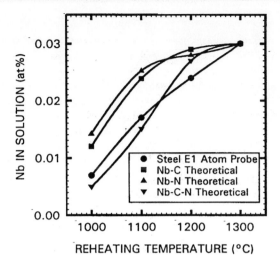

Figure 2 - Comparison of niobium in solution as a function of reheating temperature for an 0.09%C, 0.049%Nb steel measured using an atom probe and calculated from solubility product equations [5].

This is relevant in relation to the top end as well as the bottom end of the particle size range, because solution of the occasional large particle (~ 1μm in size) will produce local enrichment in microalloy content and a spacing between large particles of (say) 50μm requires significant time, even at reheating temperatures, see Fig. 1(b), to homogenise the matrix composition. The presence of a few large particles may thus account for the discrepancies at high reheat temperatures in Fig. 2

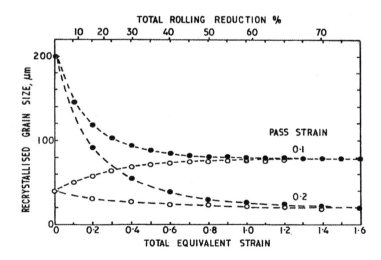

Figure 3 - Computed recrystallised grain size of Nb steel at entry to successive passes in multipass roughing with equivalent pass strains of 0.1 and 0.2 [2]

Any residual segregation after reheating will influence all subsequent precipitation behaviour. Although it is probable, as shown in Fig. 3, that any mixed grain structure will be eliminated by recrystallisation during roughing, a complete model for reheating and its subsequent effects requires detailed inputs about the particles present before reheating. It should therefore be linked to a model for the prior casting and processing conditions.

Roughing Rolling

During roughing, recrystallisation occurs rapidly between passes to give progressive refinement of austenite grain size when the pass strain is sufficiently high, Fig. 3. For lower pass strains fine grains after reheating may actually become coarser, but again convergence from different initial grain sizes is expected.

Early work by Davenport and Dimicco [8] indicated that a significant volume fraction of precipitation occurred during roughing - up to 40% of the total available niobium. The occurrence of such high temperature precipitation was confirmed by Janampa [9], Fig. 4. He also found that there was a rapid burst of precipitation in the shortest time the specimens could be quenched after deformation, then volume fraction increased only slowly after longer holding times. Both these investigations used particle extraction methods to determine volume fraction, so the locations of particles is not known, but extraction replicas indicated that they occur mainly at grain boundaries, as expected because of the higher particle stability at such sites, see Appendix. Particles were generally 20-30μm in size and sufficiently widely spaced for Zener drag effects on recrystallisation to be negligible. For 0.03% Nb steel the radius of matrix completely denuded of niobium in solution to form a particle is ~ 15 times the particle radius i.e. 0.15 to 0.2μm. If this radius is equated to x in Fig. 1(b) it can be seen to be above the limit expected for bulk diffusion in a quench time of ~ 2s after the pass. This diffusion distance also means that only a small volume on either side of a grain boundary should have reduced solute content.

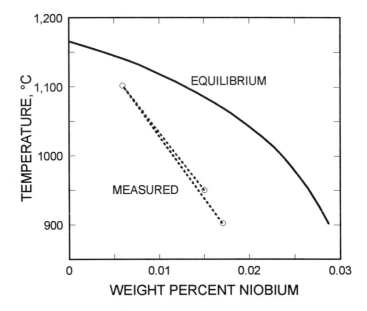

Figure 4 -   Weight percent of niobium expected as precipitates in equilibrium in an 0.08%C, 0.015%N, 0.03%Nb steel, and observed after a single pass experimental rolling by 33% reduction at 1100°C and after two pass rolling, 33% at 1100°C + 25% at 950 or 900°C [9]

These observations raise a number of questions:

- does the burst of precipitation occur because of excess vacancies generated by deformation, or does it occur as a result of grain boundary diffusion as boundaries migrate during recrystallisation?

- is coarsening of stable particles accelerated immediately after each subsequent pass?
- are significant numbers of particles of such a size that they are stable on grain boundaries but not in the matrix, so that reversion occurs when boundaries migrate away as a result of recrystallisation after the next pass?

These issues are important in relation to the total volume fraction precipitated during roughing and to the local concentration gradients created by multi-pass roughing. They could also be significant in relation to the acceleration of precipitation under finish rolling conditions observed as a result of the roughing deformation, Fig. 5. This figure indicates that when a certain pass strain (time) is exceeded in roughing the precipitation after a finishing pass is accelerated by a factor which depends only on the final roughing temperature. These observations have recently been confirmed [11,12] and further support has been given to the concept that the acceleration arises as a result of

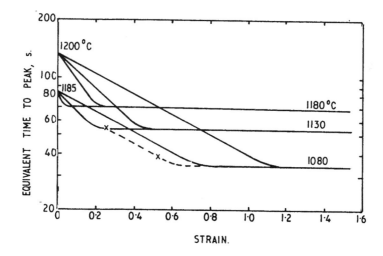

Figure 5 -   Influence of reheating and roughing temperature, and of strain in single roughing pass on time for the onset of strain induced precipitation at 900°C after a finishing pass of 15% reduction at 955°C [10]

clustering of niobium atoms during roughing. However, direct evidence for the occurrence of such clusters is not available, giving rise to the question:

- are clusters formed during roughing and are they the cause of accelerated precipitation during finish rolling?

Finish Rolling

Finish rolling of plate takes place below the recrystallisation stop temperature, which is defined as the temperature at which sufficient strain induced precipitation of fine particles takes place before the onset of recrystallisation to prevent recrystallisation occurring between or after subsequent rolling passes. This can be predicted with reasonable accuracy [3], but the subsequent precipitation kinetics during multi-pass finishing have received little attention, although they determine the amount of microalloy element remaining in solution at the end of rolling.

Detailed study of the particles in microalloyed steels is difficult because fine particles of 2-3nm in diameter appear to be critical, but cannot be observed on extraction replicas, and are difficult to observe in thin foils because of the high dislocation density produced by the martensite transformation during the rapid quench usually used to preserve the particle and grain structure existing at high temperature.

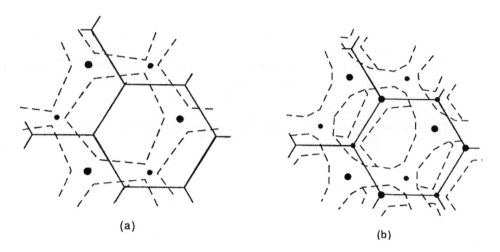

(a)

(b)

Figure 6-   Schematic diagram of the dislocation and particle distribution created (a) by a second finishing deformation and (b) by holding after a second finishing deformation. Dashed lines indicate the limits of the solute depleted regions [13].

Limited observations indicate that a very high initial number density of particles is nucleated, which decreases rapidly with time as precipitation progresses. This led Dutta et al [13] to propose that nucleation occurs at dislocation nodes in the three-dimensional network produced by deformation, and that rapid coarsening then occurs by pipe diffusion along the dislocation links between particles, as shown in Fig. 6(a). As indicated in this figure, and estimated from Fig.1(b), the regions of niobium depletion around particles and dislocations should be relatively small in the times available between passes. Subsequent passes even at constant temperature can lead to additional increments of strengthening, depending on interpass times, shown in Fig. 7, indicating that nucleation of additional particles has occurred. A mechanism for this, indicated in Fig.6(b), is possible if new dislocation nodes produced by each pass are in the non-depleted volumes of matrix. However, by an analogous argument to that used for possible reversion when grain boundaries move away from particles in roughing, reversion of fine particles in finishing may occur when dislocations move away from them in a subsequent pass. Without direct evidence, the mechanisms taking place in multi-pass finish rolling give rise to many questions.

- are dislocation nodes the nucleation sites?
- are the coarsening kinetics consistent with pipe diffusion?
- does new nucleation and reversion take place after successive passes, and what are the net effects on the increase in volume fraction?

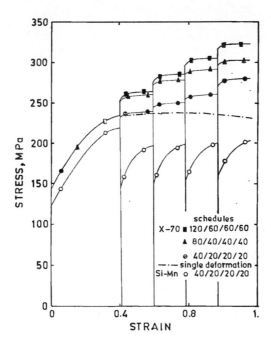

Figure 7 -   Stress-strain curves at 850°C and a strain rate of 5s^{-1} for reduction schedules of
30/15/15/15/15% with various time intervals between deformation for X70 (0.098%C,
0.012%N, 0.03%Nb, 0.05%V) steel [14]

Concluding Remarks

The list of questions posed above is by no means exhaustive, but it illustrates the point that further
basic research is required before a soundly based physical model of the complex interactions in even
the simpler microalloyed steels during multi-pass processing can be developed. There are partial
answers already available to some of the questions, based on indirect evidence within the limitations
of the experimental techniques employed, but to obtain more definitive answers it is necessary to
apply new techniques. In IMMPETUS, current research is investigating microalloyed iron - 30%
nickel as an analogue for normal low carbon microalloyed steels so that the austenite structure can
be retained to room temperature to enable detailed TEM to be carried out to characterise the
niobium carbonitride particle distributions. Local analysis on a nanometer scale is planned using a
newly available FEG TEM instrument. The overall volume fraction of precipitation will be assessed
as a function of thermomechanical history by working in collaboration with British Steel, using new
instrumentation to measure residual carbon and nitrogen in solution. The techniques used previously
on microalloyed steels will also be benchmarked against the new results using both the iron - 30%
nickel and a normal C-Mn microalloyed steel. In this way it is hoped that quantitative answers can
be obtained to some of the questions.

Acknowledgements

The author is grateful to Dr Yao Hou Li for assistance in producing Figs. 1(b) and 4, and to Dr Eric
Palmiere for his comments on an early draft of the paper.

References

1. K J Irvine, F B Pickering and T Gladman, J. Iron Steel Inst., 1967, 205, 161-182
2. C M Sellars, HSLA Steels: Metallurgy and Applications, Ed J M Gray, T Ko, Zhang Shouhua, Wu Baorong, Xie Xishan, ASM International, 1986, 73-81.
3. B Dutta and C M Sellars, Mat. Sci. and Tech., 1987, 3, 197-206.
4. S Kurokawa, J E Ruzzante, A M Hey and F Dymat, 36th Annual Cong. ABM, Recife, Brazil, July 1981, Vol 1, 47-63
5. E J Palmiere, C I Garcia and A J DeArdo, Metall. and Mater. Trans. A, 1994, 25A, 277-286.
6. R Simoneau, G Begin and A H Marquis, Met. Sci., 1978, 12, 381-387.
7. A LeBon, J Roffes Vernis and C Rossard, Met. Sci., 1975, 9, 36-40.
8. A T Davenport and D R Di Micco, Internat. Conf. on Steel Rolling, Iron and Steel Inst. of Japan, Tokyo, 1980, 1237-
9. C J Janampa Ramos, "The Role of Nitrogen in the Hot Working of Niobium Microalloyed Steels" PhD Thesis, University of Sheffield, 1982.
10. E Valdes and C M Sellars, Mat. Sci and Tech., 1991, 7, 622-630
11. E S Siradj, "Strain Induced Precipitation Kinetics of Nb(CN) in Nb-HSLA Steel as a Function of Thermomechanical History" PhD Thesis, University of Sheffield, 1997.
12. E S Siradj, C M Sellars and J A Whiteman, this conference.
13. B Dutta, E Valdes and C M Sellars, Acta. Metall. Mater., 1992, 40, 653-662.
14. B Dutta and C M Sellars, Thermec-88, Internat. Conf., Physical Metallurgy of Thermomechanical Processing of Steels and Other Metals, Ed. I Tamura, Iron Steel Inst., Japan, 1988, Vol 1, 261-268

Appendix - Effect of Particle Size on Solubility

Following the arguments of Dutta and Sellars [3], the change in free energy per mol on forming small particles

$$\Delta G = RT \ln k_s + \gamma V_m . A/V \qquad [A1]$$

where R is the gas constant, T is absolute temperature and from the equation of Irvine et al [1]

$$k_s = [Nb][C + 12N/14]_{soln}/10^{2.26 - 6770/T} \qquad [A2]$$

where [Nb] [C] and [N] are wt% of the elements, interfacial energy per unit area, $\gamma \simeq 0.5$ Jm^{-2}, molar volume, $V_m = 1.28 \times 10^{-5}$ m^3 mol^{-1} and A/V is the interfacial energy per unit volume of particle $= 6d^{-1}$ for spherical particles of diameter d (m) in the matrix.

For equilibrium, $\Delta G = 0$ and

$$\log [Nb][C + 12N/14] = 2.26 - \frac{6770}{T} + \frac{6}{2.303} . \frac{\gamma V_m}{RTd} \qquad [A3]$$

$$= 2.26 - \frac{\left(6770 - 2.01x10^{-6} d^{-1}\right)}{T} \qquad [A4]$$

For particles on grain boundaries, if it is asumed that the grain boundary energy per unit area, $\gamma_{gb} \simeq \gamma$, the particles assume a lens shape with an included angle of 120°, In this case when a particle of diameter, d, forms, the energy of an area of grain boundary $\pi d^2 \gamma_{gb}/4$ is eliminated and the net change in interfacial area per unit volume of particle, $A/V \simeq 4d^{-1}$. For these particles,

$$\log [Nb] [C + 12N/14] = 2.26 - \frac{\left(6770 - 1.34x10^{-6}d^{-1}\right)}{T}$$　　　　[A5]

as calculated previously [2] and as shown in Fig. 1 (a).

Comparison of equations [A4] and [A5] shows that the minimum particle size in equilibrium at a given temperature must be 50% larger for particles in the matrix than for particles on grain boundaries. Thus particles in this size range precipitated on grain boundaries will become unstable and redissolve if the boundary migrates away from them.

Materials Science Forum Vols. 284-286 (1998) pp. 83-94
© *1998 Trans Tech Publications, Switzerland*

Microalloyed Forging Steels

D.J. Naylor

British Steel plc, Swinden Technology Centre, Moorgate,
Rotherham, S60 3AR, UK

Keywords: Vanadium, Forging, Microstructure, Properties, Automotive, Heat Treatment, Air Cooled, Medium Carbon Steel, Bars

Abstract

Microalloyed forging steels have been developed to improve the competitiveness of wrought steel components, especially in the automotive sector, by achieving the desired properties in the as-forged condition, thus eliminating the need to subsequently heat treat, straighten and stress relieve the previously specified low alloy steels. Significant cost reductions are realised by adopting microalloyed steels.

This paper reviews the metallurgical principles on which microalloyed forging steels are based, including the relationships between steel composition, thermomechanical processing, microstructure and the resulting properties, highlighting the various strengthening mechanisms that are invoked.

The properties, characteristics and applications of the initial development grade, 49MnVS3, are described. Research and development then focussed on increasing the strength and/or the toughness of this steel to improve its appeal to the market, especially for safety critical applications. The various metallurgical options are described and discussed.

Attention has also been placed on maximising the machinability of these steels by controlled additions of sulphur, the adoption of inclusion modification techniques and other free machining additives.

The fatigue properties and toughness of microalloyed steel forgings have been demonstrated to be fit for purpose, but compared with heat treated low alloy steels their fracture toughness is lower, albeit still significantly superior to castings.

A wide range of forged automotive applications has been successfully converted to air cooled microalloyed steels over the past 25 years, with a large proportion of crankshaft and connecting rods now being made by this route.

Future challenges have been identified to further extend the attainable properties and to improve the combination of strength and toughness, to broaden the market applications and the product range to include bar and rod. The use of warm near net shape forming processes for microalloyed steel is also anticipated. Greater exploitation of computer aided modelling and design techniques is encouraged to facilitate rapid prototyping, in order to improve further the competitiveness of forged engineering steels.

Introduction

The application of air cooled medium carbon microalloyed steels as a replacement for quenched and tempered low alloy steels is now over 25 years old. The principal motivation behind the development of these steels is to improve the competitive position of steel forgings (compared with castings, powder/sinter forgings and composites) by a reduction in the cost of the wrought component whilst preserving acceptable properties and in-service performance. This paper will review the relevant metallurgical developments and product and market applications over the last quarter of the 20th century and will look forward into the next millennium. The three most appropriate references are the two conferences in Golden, Colorado in 1986 and 1996 and the ECSC Information Day on Microalloyed Steels in Dusseldorf in 1988[1-3].

Steelmakers have made significant progress in reducing the cost of their products through the adoption of modern steelmaking and continuous casting technologies. Typically the value of steel billets and bars doubles after forging and heat treatment and doubles again after subsequent machining to the final shape[4]. The typical thermal cycle (Fig. 1) for conventionally processed alloy steel forgings involves reheating the billet/bar to ~1200°C, forging, finishing at >1000°C, cooling in a bin, reheating to 850-900°C, quenching in oil or polymer, reheating to 500 - 650°C for tempering, straightening (e.g., for crankshafts) to remove the distortion produced by quenching and finally reheating again for stress relieving.

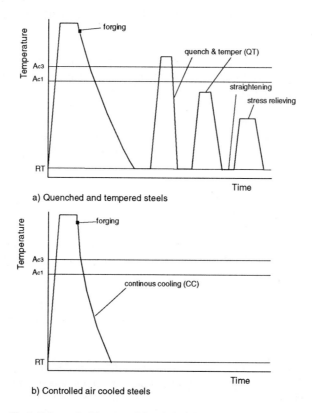

Fig.1 Schematic Diagram of Crankshaft Manufacture

Eliminating the cost of heat treating steel forgings is the main target for the introduction of microalloyed steels, which achieve the desired properties during controlled air cooling after forging. Because the microalloyed steels are not quenched to martensite there is no significant distortion and consequently no straightening or subsequent stress relieving treatment are required. Hence there is less handling of the forgings, shorter delivery schedules, lower manpower requirements and few opportunities for processing variations and errors to occur. Considerable effort has also been made to improve the machinability of engineering steels, by inclusion engineering techniques and these have also been incorporated into microalloyed steel developments.

Automotive engine, steering and suspension components have been the main applications of the air cooled microalloyed steels, initially directed at engine parts such as crankshaft and connecting rods, where the impact toughness requirements are not particularly demanding or critical.

Metallurgical Principles

The underpinning metallurgical understanding on which the development of these steels is based was initially established in the 1950s and early 1960s when the quantitative relationships between microstructure (ferrite grain size, pearlite content and pearlite interlamellar spacing) and properties were determined for low carbon steels, later extended to higher carbon levels [5-8].

During the 1960s extensive industrial development of low carbon microalloyed steels led to extensive studies of the strengthening mechanisms. Additions of vanadium, niobium and titanium were found to give improvements in strength by dispersion strengthening and grain refinement. Transmission electron metallography elucidated the precipitation characteristics of microalloy carbides and nitrides and their formation mechanisms, including interphase precipitation[9]. The strengthening increment was found to relate to the volume fraction and particle size, through the Ashby-Orowan relationship[10]. A detailed understanding of the solubility products of various carbides and nitrides, the associated thermodynamics and consequences for grain growth and hardening were vital to the continued development of microalloyed steels[11]. This research demonstrated that the vanadium is the most suitable microalloying element in medium carbon steels in view of its high solubility as a carbide/nitride in austenite and low solubility in ferrite. The effectiveness of vanadium carbide as a precipitation strengthener in interlamellar ferrite in pearlite has also been confirmed[12].

This physical metallurgy provided the framework for the successful development and exploitation of vanadium treated medium carbon steel for forgings.

Since the properties are created during the forging and cooling process, it is clearly very important not only to select the right composition but also to ensure that the appropriate thermomechanical cycle is chosen and then consistently controlled in an industrial environment. This requires attention to

- the soaking temperature, which must be high enough to dissolve the microalloying constituent such that it is available to reprecipitate during transformation from austenite to ferrite and pearlite[13].

- the forging temperatures and deformation sequence, with finish forging temperature particularly important in controlling the austenite grain size and hence the subsequent ferrite/pearlite structure, with a lower finishing temperature improving the yield to tensile strength ratio and toughness but having a relatively negative effect on tensile strength[13].

- cooling rate after forging, which should be consistent from part to part within and between batches[13]. The faster the cooling rate, the lower the transformation temperature, the finer the precipitate size, the higher the strength and the lower the toughness. If too fast a cooling rate is introduced then lower temperature transformation products and incomplete precipitation may occur with consequential variable properties[1]. Forgers have often found it necessary to introduce cooling conveyors into the post-forging phase of the operation. In some cases quite sophisticated cooling systems have been incorporated involving fans and insulated hoods.

49MnVS3

The first commercial application of medium carbon microalloyed steels was in Germany with the introduction of 49MnVS3 grade[14,15]. The first experiments were conducted on a 0.48% C, 0.75% Mn, 0.06% S with additions of niobium or vanadium. The specification for 49MnVS3 is shown in Table 1. The tensile strength range achievable is quite limited, being restricted to <900 N/mm^2. The yield strength is relatively low, compared with the YS/TS ratio of up to 0.9 for quenched and tempered steels, and the ductility and toughness are lower than those attained in quenched and tempered alloy steels. This is attributed to the coarse grained ferrite-pearlite microstructure in the as-forged and air-cooled condition.

Table 1
Specification for 49MnVS3

Composition, Wt%					
C	Si	Mn	P	S	V
0.44/0.50	0.50 max	0.70/1.00	0.075 max	0.04/0.07	0.08/0.13
YS N/mm^2	TS N/mm^2	Elongation %	R of A %	Charpy V Notch, J
>450	750 - 900	>8	>20	>15

Good success was achieved in utilising this steel for crankshafts and connecting rods in which satisfactory properties and performance were realised with cost savings over conventionally quenched and tempered heat treated engineering steels.

Initial applications tended to be as replacements for heat treated medium carbon steel, where the cost savings in heat treatment etc. were offset by the additional cost of the microalloying addition.

Greater financial benefits and a wider range of applications were envisaged if the strength of the microalloyed steel could be increased such that heat treated low alloy steels were replaced. This would further enable component designers to reduce the weight of vehicles in a cost effective manner, thereby improving fuel economy and performance and reducing emissions.

The lower ductility and toughness of the air cooled microalloyed steel was not perceived to be a problem for these initial applications, as the components were found to be fit-for-purpose.

Crankshafts and connecting rods are not subjected to severe impact loading condition, embodied as they are within the engine and engulfed in oil. The acceptance by some automotive companies of cast iron crankshafts with significantly lower ductility and toughness is another indication of the non-criticality of the toughness of these particular parts.

However, for more safety critical components in the steering and suspension systems of a vehicle a higher level of toughness was seen as desirable.

Hence metallurgical options to increase the yield/tensile strength and/or toughness/ductility of these air cooled microalloyed steels were sought by several steel companies and researchers.

Higher Strength Microalloyed Forging Steels

The obvious metallurgical approaches to increase the strength are additions of carbon, nitrogen, silicon, manganese and microalloying precipitation hardening constituents. Refinement of the microstructure and precipitates and increasing the pearlite content are also beneficial. Apart from microstructural refinement all these strengthening mechanisms are detrimental to toughness. Other approaches involve the use of alternative microstructures, e.g. bainitic structures, or the use of direct quenching technologies.

British Steel[13] introduced a series of four microalloyed forging steels which offered a range of tensile strength levels from 770 up to 1160 N/mm^2, carefully tailored to the standard tensile strength ranges normally specified by steel users for quenched and tempered carbon and alloy steels. The properties of the VANARD series are given in Table 2. The high manganese content provides solid solution hardening, reduces the transformation temperature and increases the pearlite fraction and dilution. The attractions of these steels compared with the original 49MnVS3 steels are a higher yield strength and superior ductility and toughness for the same tensile strength as 49MnVS3 and higher levels of yield and tensile strength. The composition needed to develop the required properties for a given application and forging and cooling conditions is selected from within the range quoted.

Table 2
Properties of the VANARD Steels

C	Si	Mn	P	S	V
0.30/0.50	0.15/0.35	1.0/1.5	0.035 max	0.10 max	0.05/0.20

VANARD Grade	YS N/mm^2	TS N/mm^2	Elongation %	Charpy V Notch J	
850	>540	770 - 930	>18	>20	
925	>600	850 - 1000	>10	>20	
1000	>650	930 - 1080	>12	>15	
1100	>700	1000 - 1160	>8	>10	

The German C38 mod, 27MnSiVS6 and 38MnSiVS6 steels have also become extremely popular and these offer similar properties, as shown in Table 3. The steels are incorporated into an ISO Standard (ISO 11692:1994).

Table 3
Standard Specifications of Microalloyed Steels

Designation	Material Number	Composition, Wt.%						
		C	Si	Mn	P	S	V	N
27MnSiVS6	1.5232	0.25/0.30	0.50/0.80	1.30/1.60	≥0.035	0.03/0.05	0.08/0.13	
38MnSiVS5	1.5231	0.35/0.40	0.50/0.80	1.20/1.50	≥0.035	0.03/0.065	0.08/0.13	
44MnSiVS6	1.5233	0.42/0.47	0.50/0.80	1.30/1.60	≥0.035	0.02/0.035	0.08/0.13	
C38mod		0.36/0.40	0.5/0.65	1.35/1.45	≥0.015	0.03/0.045	0.08/0.12	0.013/0.017

Designation	Material Number	Properties			
		YS N/mm^2	TS N/mm^2	Elongation %	R of A %
27MnSiVS6	1.5232	>500	800 - 950	>14	>30
38MnSiVS5	1.5231	>550	820 - 1000	>12	>25
44MnSiVS6	1.5233	>600	950 - 1100	>10	>20
C38mod		>580	850 - 1000	>12	>25

In common with other steelmakers regression equations and other calculation techniques were established to assist in the alloy design process and to advise forgers of the optimum thermomechanical processing cycle to achieve the specified properties.

Enhanced nitrogen contents up to 0.03% have also been used to advantage to increase the strength of the vanadium microalloyed steels[1]. However, there may be practical limits to the nitrogen level that can be achieved on an industrial scale.

French steelmakers[1] used a combination of niobium and vanadium and a high nitrogen content in their METASAFE 1000 product.

Takada and Koyasu[2] have assessed the properties of 0.24% C, microalloyed steels alloyed with up to 2% Cr, 2.8% Mn, striving to achieve a tensile strength >1000 N/mm^2 with adequate toughness (>50 J/cm^2). The bainitic steels had to be tempered at 300 - 500°C in order to decompose retained austenite and to attain acceptable yield strength and toughness in bainitic microstructure. An alternative approach involves the use of a two stop cooling treatment[2].

This "bainitic" approach to achieve higher strength levels has two economic disadvantages - higher alloy cost and the need for a tempering treatment, both of which significantly reduce the competitiveness of these steels compared with air cooled pearlitic microalloyed steels, provided an acceptable combination of strength and toughness can be realised in the latter. The slow market exploitation of bainitic forging steels bears witness to this conclusion. However, as automotive design engineers seek more cost effective solutions at high strength levels (>1000 N/mm^2) with higher levels of toughness to replace conventionally heat treated low alloy steels, the bainitic approach may yet see wider commercial realisation.

Another method of reducing the cost of heat treated alloy steel forgings is to use a direct quenching operation to utilise the residual heat of the hot forged component as a way of eliminating the post-forging reaustenitising treatment normally required prior to conventional quenching and subsequent tempering. However special quenching facilities and more streamlined operations have to be introduced into the forge to enable the hot forgings to be directly quenched in a reproducible manner on a regular production basis. With directly quenched medium carbon low alloy/boron steels an additional tempering heat treatment is also required. Nevertheless for components such as axle beams, the direct quenching process has proven to be a cost effective way of achieving a good combination of strength and toughness[16]. When the concept is extended to low carbon slightly alloyed steels it is possible to direct quench the forgings and to rely on auto-tempering to realise the desired balance of properties[17].

Improved Toughness

As stated previously the market development potential of air cooled microalloyed steel forgings would be significantly enhanced through an improvement in ductility and toughness, thus making them more attractive to design engineers for a wider range of safety critical components. The most obvious metallurgical approach to improve toughness is through microstructural refinement, e.g., austenite grain size, and the subsequent ferrite grain size, pearlite interlamellar spacing and bainite packet size. Another potentially fruitful approach is to reduce the carbon content and compensate with alternative strengthening additions.

The austenite grain size can be refined by reducing the soaking temperature prior to forging and especially by reducing the finish forging temperature. Unfortunately this incurs cost/operational penalties - higher forging loads/energy, greater die wear and possible delays in the hot deformation sequence to finish colder. Most forgers are reluctant to adopt lower finish forging temperatures, despite the improved toughness of their products which would satisfy their customers' needs.

A refinement of the transformation products can also be achieved through accelerated cooling after forging. Linaza et al[2] demonstrated that a microalloyed steel CV joint which had been accelerated cooled to an essentially acicular ferrite, with some pearlite, had a lower impact transition temperature than a "conventionally" air cooled ferrite-pearlite variant. The important microstructural parameter was the ferrite unit size, comprising ferrite and pearlite with the same ferrite orientation. The finer the ferrite unit size the better the ductility and toughness. The finer the interlamellar spacing the higher the crack initiation energy.

Dong et al[2] showed a toughness trough on increasing the cooling rate coinciding with a thin network of ferrite delineating the prior austenite grain boundaries. At higher cooling rates, lower temperature transformation products are formed with better toughness.

Small austenite grain refining additions of titanium (typically 0.015%) have been used by Japanese and European steelmakers in order to achieve finer grain sizes without recourse to lower finish temperatures. There are many published examples[2,3] where this approach has proved effective, although it is important to keep close control of the Ti:N ratio and the solidification rates (ideally concast billets) in order to avoid coarse titanium nitride particles which themselves are detrimental to toughness, by promoting cleavage fracture. The grain refining effects are only achieved with a uniform dispersion of fine titanium nitrides.

The use of a reduced carbon content with a compensating increase in solid solution strengthening by silicon has been effectively exploited, although again there appears to be an optimum silicon content (0.6 - 0.8%), above which the toughness deteriorates. Manganese and chromium may also play a similar role, but their hardenability effects also have to be considered with consequences on lower temperature transformation products.

There have been made conflicting statements in published research on the effects of bainitic microstructures in as-forged microalloyed steels, some claiming that bainite improves strength and/or toughness and ductility, others demonstrating detrimental effects, especially on toughness. Drobnjak and Koprivica[2] have demonstrated that certain bainitic structures are harmful to toughness, whilst acicular ferrite is beneficial. Improved toughness has been associated with a fine bainitic structure and a large fraction of retained austenite with lower bainite as opposed to upper bainite and an absence of coarse carbides. Reeder et al[2] have shown that high manganese levels can result in some bainite formation in an otherwise ferrite-pearlite structure with a consequential lower strength, attributed to an inhibition of the vanadium carbide precipitation due to the low transformation temperature.

Perhaps perversely sulphur can be beneficial to toughness in these steels since manganese sulphide inclusions act as nuclei for intragranular ferrite nucleation within the prior austenite grains and hence to microstructural refinement. Sulphide stringers can also lead to higher impact energy values due to delamination during fracture.

It has been demonstrated[18] that microalloyed steels can be successfully heat treated in the conventional manner to achieve an improved combination of strength and toughness. This represents some cost saving over traditional heat treated low alloy steel on account of the alloy content, albeit without the cost saving of air cooled steels.

Machinability of Microalloyed Forging Steels

For forged components, the cost of machining is typically 50% of the total cost of the part. Therefore the machinability of the steel, especially when a change of specification, microstructure and properties is being undertaken, is of critical economic as well as technological importance. Despite some initial adverse reaction from the market place about the machinability of microalloyed forging steels it is now recognised that these ferrite-pearlite steels offer superior machining characteristics to quenched and tempered martensitic/bainitic steels at equivalent hardness levels. Many microalloyed steel specifications incorporate slightly enhanced sulphur levels (up to 0.10%) with optional sulphide shape modification by calcium and/or tellurium additions improve transverse ductility and toughness, lead or bismuth free machining additions. As with other engineering steels, machinability deteriorates as the strength of the steel is raised. However the lower ductility of the ferrite-pearlite structure is deemed to be beneficial to machinability. It should be recognised that the optimum machining conditions with microalloyed steels (cutting speed, feed rate and tool type/geometry) may be different from the heat treated steels and therefore machine shops may have to conduct trials in order to maximise the benefits of these steels[1,2].

Performance of Microalloyed Forging Steels in Service

Before a design engineer adopts a new material such as a microalloyed steel for forged components there has to be clear evidence of fitness-for-purpose, especially when not all the properties are identical to the material it is replacing. In this case cognizance has to be taken of the different microstructure, lower yield strength, lower ductility and lower impact toughness of the microalloyed steel for a given tensile strength. Fatigue properties and resistance to impact loading have been the major concerns which have had to be addressed, especially for safety critical applications.

Fatigue tests on machined specimens, notched specimens, strain controlled tests and cyclic loading of components by many workers have demonstrated that microalloyed forging steels have comparable fatigue properties to heat treated low alloy steel. Even the enhanced sulphur and leaded steels with improved machinability have acceptable fatigue characteristics. Fatigue crack initiation is retarded due to the fine precipitates and short interparticle spacing.

Whilst the standard monotonic tensile test reveal a lower flow stress for a microalloyed steel, cyclic stress-strain tests show that these steels cyclically harden whilst a tempered martensite structure cyclically softens. Therefore in a repeatedly stressed situation a microalloy steel exhibits at least as high a flow stress as a conventionally heat treated steel.

The lower Charpy impact energy values evident in microalloyed forging steels must also be put into context. With less sharp notches and slower strain rates, precipitation strengthened ferrite-pearlite structures demonstrate toughness levels approaching those of the heat treated low alloy steel. The fracture toughness values of microalloyed steels are typically 35 - 50% of those of heat treated low alloy steels but can be better than heat treated carbon steels[2]. Through design optimisation it is possible to accommodate the lower tolerance of these steels to defects.

Microalloyed forging steels are satisfactorily surface hardened by induction hardening, nitriding and shot peening to give localised improvements to fatigue strength.

Applications of Microalloyed Forging Steels

There is an ever growing range of applications of components, still mainly in the automotive sector, for these steels. This currently includes crankshafts, connecting rods, wheel hubs, spindles and knuckles, axle beams and shafts, suspension, steering and support arms, yokes and support shafts, bearing blocks, sun wheels, counterweights, shift forks and piston crowns[1,2,3]. There has also been success in applying medium carbon microalloyed steels in the as-rolled condition for various shaft, steering, cylinder and piston rod components and springs. In addition to the well established replacement of quenched and tempered carbon and alloy steels some carburised low alloy components have also been substituted by air cooled microalloyed steels. Whilst initial exploitation of microalloyed forging steels was relatively slow in North America compared with European and Japanese industries, this has increased in recent years.

It is often not easy to get end users to make public declarations of the financial benefits of converting from heat treated low alloy steel to an air cooled forging steel. In some cases the argument is advanced that because heat treatment furnaces still have to be maintained and operated for components that are not converted then the potential cost savings are not fully realised. However the Rover Group[3] stated that at 1982 prices the crankshaft in the A series engine used in the famous Mini car was converted to VANARD 925 at an annual saving of £½ million.

Cost savings with microalloyed steel crankshafts of 13% have been claimed compared with heat treated low alloy steel[19]

Future Challenges

The development and application of microalloyed forging steels have made much progress over the past 25 years. They are now very well established for crankshaft and connecting rods in cars and trucks. There is a continuing need for wrought steels to remain competitive and indeed for some car connecting rods it has proved possible to reduce the cost of air cooled forgings by relying on plain carbon steels, without a microalloying addition. The adoption of fracture splitting techniques to further reduce manufacturing costs is expected to grow.

The drive for low cost, air cooled forging steels capable of consistently attaining tensile strength levels of >1000 N/mm^2 with a high YS/TS, high ductility and toughness (>50 J/cm^2) will continue to be a major challenge for steelmakers and forgers. A demand for even higher strength levels is also anticipated in the search for vehicle weight reduction and improved performance. Close dialogue and co-operation between component designer, forger and steelmaker will be necessary to ensure that the optimum cost effective metallurgical solution is found to satisfy such requirements. This may involve novel cooling regimes, alternative alloying elements and microstructures and appropriate surface hardening treatments.

The application of microalloyed forging steels has been primarily directed at the automotive industry and efforts should be made to promote their benefits to other sectors which currently use heat treated low alloy steels.

Given the success of using microalloyed steels forgings it is surprising that greater conversion of heat treated low alloy steel long products to as-rolled microalloyed steel bars and rods has not occurred. There is clearly considerable potential for more development work in this area and this will lead to greater market development opportunities.

Given the high cost of machining after forging to realise the required component shape and finish, there is a growing interest in (near) net shape forming technologies. These include cold and warm forging. Opportunities are envisaged to develop attractive combinations of properties by warm forging of microalloyed steels, incorporating precipitation hardening from the microalloying addition and the microstructural refinement from the warm working.

Within the past 5 years computer aided modelling techniques have developed rapidly to predict metal deformation behaviour, shape and microstructural evolution. Linked to finite element analysis of component and die design, these tools will be very powerful facilitators of rapid prototyping of wrought steel products, enabling new microalloyed forging steel applications to be taken to market more quickly than previously experienced. This is important as the foundry industry has traditionally been able to manufacture prototypes much cheaper and faster than the forging sector on account of the high cost and time to produce forging dies.

Therefore microalloyed forging steels can be proud of their achievements to date but cannot afford to rest on their laurels.

Acknowledgements

The author wishes to thank Dr. K.N. Melton, Research Director, Swinden Technology Centre, British Steel plc for permission to publish this paper and also his colleagues Mr. P.E. Reynolds and P. Remijn for help with its preparation.

References

[1] Fundamentals of Microalloying Forging Steels, Ed. G. Krauss, S.K. Banerji,
     The Metallurgical Soc. Golden, Colorado. July, 1986.

[2] Fundamentals and Applications of Microalloying Forging Steels Ed. C.J. Van Tyne, G. Krauss,
     D.K. Matlock, The Minerals, Metals and Materials Society, Golden, Colorado, July 1996.

[3] Proceedings of Information Day on Microalloyed Engineering Steels, ECSC.Dusseldorf 1988.

[4] H.K. Tonshoff and H. Winkler, VDI-Z, 198 2, 124, (13), p481

[5] E.O. Hall, Proc. Phys. Soc. 1951, 64B, p747

[6] F.B. Pickering and T. Gladman, Metallurgical Developments in Carbon Steels, ISI, London
     1963, p10.

[7] J.P. Hugo and J.H. Woodhead, JISI, 1957, 186, p174.

[8] T. Gladman, I.D. McIvor and F.B. Pickering, JISI, 1972, 210, p916.

[9] R.W.K. Honeycombe, Met. Trans., 1976, 71A, p915

[10] T. Gladman, D. Dulieu and I.D.McIvor, Microalloying '75, ed. M. Korchynsky, Union Carbide
     Corp, New York, 1977, p32.

[11] K. Narita, Trans. ISIJ, 1975, 15, p145.

[12] N. Ridley, M.T. Lewis and W.B. Morrison, Advances in the Physical Metallurgy and
     Applications of Steels, The Metals Society, London, 1982, p199.

[13] G. Thewlis and D.J. Naylor, Advances in the Physical Metallurgy and Applications of Steels,
     University of Liverpool, The Metals Society, Book 284, 1982, p331.

[14] D. Frodl, A. Randak and K. Vetter, Hartereitechn. Mitt., 1974, 29, 4, p169.

[15] A. Von de Steinen, S. Engineer, E. Horn and G. Preis, Stahl Und Eisen, 1975, 95, (6), p209

[16] C.A. Schemper, M. Heinritz, E. Schreiter and E. Wetter, SAE paper 860130, 1986.

[17] P.E. Reynolds, Private Communication

[18] L. Yang and A. Fatemi, Journal of Engineering Materials and Technology, 1966, 188, p71

[19] J. Bellus, P. Jolly and C. Pichard, HSLA Steels '95, Beijing, 1995.

david_naylor@technology.britishsteel.co.uk

Materials Science Forum Vols. 284-286 (1998) pp. 95-104
© *1998 Trans Tech Publications, Switzerland*

Microstructural Change during the Hot Working of As-Cast Austenite

R. Priestner

University of Manchester / UMIST Materials Science Centre,
Grosvenor Street, Manchester, M1 7HS, UK

Keywords: Direct Rolling, Microalloyed Steel, Thin Slab Casting, Near-Net Shape, Microstructure

Abstract.
The direct hot rolling of the as-cast austenite of thin slab castings requires the microstructural changes within the austenite phase field that are necessary for a satisfactory product to be accommodated within a smaller degree of total reduction, within a smaller number of passes, and within a shorter time scale. Particularly in the case of microalloyed steel, the constitution of the as-cast austenite prior to rolling differs from that of austenite of thick slabs reheated for conventional hot rolling. Consequently, the models developed for conventional rolling employing reheated austenite cannot be relied upon for predicting microstructural change during the rolling of as-cast austenite. The as-cast austenite of microalloyed steels has larger grains, is segregated, and carbonitride precipitates may be present that are not in equilibrium with their solid solution. These factors are discussed in the light of recent research. It is suggested that recrystallisation kinetics should be linearly, rather than quadratically, related to austenite grain size, and that above a limiting carbon content the addition of Nb and Ti results in excess, eutectically precipitated, Nb-rich carbonitride due to segregation during freezing. The former leads to more rapid recrystallisation than would be predicted by the conventional models, and helps to ensure that plain C-Mn steels can be satisfactorily processed. The presence of eutectic carbonitrides means that the supersaturation of the austenite with respect to microalloy constituents is unknown, and recrystallisation-stop temperatures cannot be modelled satisfactorily.

Introduction.

<u>Development of hot direct rolling</u>. In a conventional integrated plant the production of hot rolled plate or strip steel begins with continuously cast slabs approximately 250mm thick. The slabs are allowed to cool to ambient temperature, and are then inspected and their surfaces conditioned for hot rolling to strip or plate. They are then reheated, typically over a period of 6-8 hours to 1200-1250^0C, during which as much as 10% is lost as scale, which must be removed before hot rolling. The reheating conditions the austenite by breaking up the as-cast microstructure and creating a reproducible austenite grain size of about 200 μm, by achieving a degree of homogenisation, and, in the case of microalloyed high strength steels, by bringing precipitated species into equilibrium with their solution in the austenite.

A little over ten years ago some plants, particularly in Japan [1], began to roll slabs without allowing them to cool all the way to room temperature, a variation referred to as hot direct rolling, or HDR. Part of the impetus for HDR was the economic and ecological benefits gained by retaining the heat of the casting within the slabs, but improvements in strength were also reported, for example in Ti-microalloyed steels [2,25]. In 1989 the Institute of Metals (now the Institute of Materials) held a meeting in London entitled "Achieving The Hot Link." [3] Most of the meeting was concerned with the technicalities and the potential economic benefits of removing the reheating stage, either partially or completely, from the hot rolling process. Virtually nothing was said about any differences in metallurgy that might be expected from direct rolling of as-cast austenite, compared with the rolling of reheated austenite. There was one paper [4] on a novel process of compact strip production (CSP), in which 50 mm thick slabs were continuously cast and directly rolled while still hot. The plant described in that paper was for Nucor Steel and was commissioned at Crawfordsville, USA in 1989.

Since then a number of plants have been commissioned around the world, but particularly in the USA, for casting thin or "thinner" slabs for hot direct rolling [5]. The trend has been towards thicker slabs than 50 mm (but still considerably thinner than conventional, thick slab casting), using electric arc furnace steel derived mainly from scrap. Capital costs per tonne are less than half those of a conventional BF-BOF hot strip mill [6] and productivity (with respect to manpower) can be nearly three times as great. It seems certain, therefore that the hot direct rolling of thin slab cast steel will continue to expand. However, in order to do so, it must match the quality produced by conventional plants in higher grades [5], particularly in microalloyed, HSLA steels.

Comparison of conventional and thin-slab direct rolling.

Fig. 1 compares thin slab casting and direct rolling (TSDR) with conventional thick-slab casting and hot rolling.

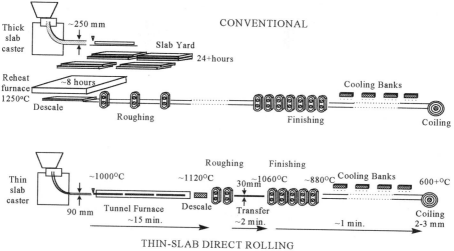

Fig. 1. Comparison of conventional and thin-slab direct rolling.

The data included in the figure for TSDR are characteristic of a recently installed plant in the USA [7]. The most striking characteristic of the TSDR process is its rapidity: the coiled product is removed from the mill within 20 minutes of the slab leaving the caster. There are only seven passes through rolls, two at a "roughing" stage reducing 90 mm slab to 30 mm, and 5 in a finishing train reducing the thickness to 2-3 mm. The "roughing" stage is omitted, and only 5 or 6 passes are applied in mills utilising thinner slabs of only 50 mm thickness. After being cut from the straightened cast strand the slab enters the tunnel furnace at a mean slab temperature of approximately 1000°C, and leaves it with its temperature equilibrated at 1120°C. Thus, the first passes are at a lower temperature than the roughing stage in conventional processing. Further, the microstructural changes within the austenite phase field that are necessary for a satisfactory product have to be accommodated within a smaller degree of total reduction, within a smaller number of passes and within a shorter time scale.

Fig.2. compares the thermal histories of thick [8] and thin [9] slabs during casting and prior to direct rolling.

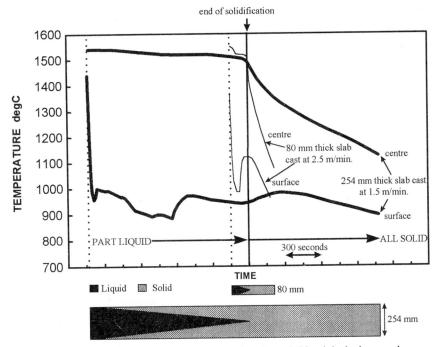

Fig.2. Thermal histories of thick (conventional) and thin slab during casting.

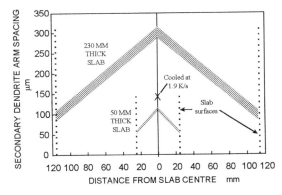

Fig.3. Secondary dendrite arm spacings in thin and thick slab

In Fig.2, the characteristic difference is the much more rapid solidification and the faster cooling rates in the thin slab. The thin slab, in this case 80 mm thick, solidified completely within 2½ minutes and the bulk mean temperature of the slab reached about 1000°C within a further 3 minutes. In thick slab the austenite grain size of low carbon steel soon after solidification is about 3 mm, whereas in the thin slab it is about 1 mm, and may be less if the sulphur content is appreciable (see the paper by Frawley and Priestner in this conference). An additional result of the rapid solidification is illustrated in Fig.3, [10]. The secondary dendrite arm spacing is considerably finer in the thinner slab, and, consequently, so is the scale of segregation of solutes such as manganese, with obvious consequences with respect to homogeneity in the final product. The data point labelled "cooled at 1.9 K/s" is from our own measurements of secondary dendrite arm spacing in a laboratory casting cooled to simulate thin slabs, in which the segregation ratio of Mn was measured to be approximately 1.3, similar to that in conventional casting, but on a scale determined by the finer dendritic structure.

Laboratory simulation of HDR and TSDR.

The methodology of laboratory-based metallurgical research into conventional reheating and rolling processes is relatively straightforward and well established. A single plate of homogeneous composition and initial microstructure can be divided into many samples, and reproducible results obtained for many variations of experimental parameters. For example, identical samples may be reheated to different temperatures and rolled to different strains, either by real rolling or by simulation in a plane strain or torsion testing machine, and realistic temperature histories can be mimicked. For research into hot direct rolling of as-cast austenite, however, a newly melted and cast slab is needed for every experiment and each parametral variation. Early work in this area [24,25] used large cast ingots, and demonstrated that low carbon steels could be successfully hot direct rolled [24], but that microalloying resulted in a degree of grain size heterogeneity in the final product. In the case of Ti-microalloyed steel [2,25] containing small additions of Ti it was shown that strength was increased by direct rolling, probably as a result of a higher supersaturation of Ti in solution at the rolling temperature in austenite cooled from the solidification temperature than in reheated austenite. However, experiments with large castings are expensive in the use of material, and it is difficult to quench large samples efficiently enough for investigation of sequential microstructural changes throughout processing.

Fig.4 illustrates the laboratory simulation of TSDR developed at the Manchester Materials Science Centre, starting in 1991 and described in more detail elsewhere [11,12]

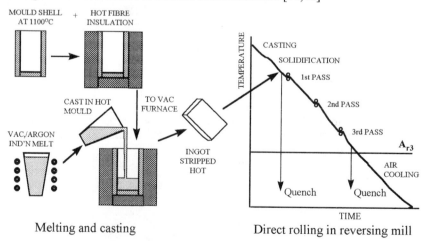

Fig.4. Laboratory simulation of thin slab direct rolling.

Essentially, a 1 kg melt of steel is made to the required chemistry under low pressure argon and cast into a hot ceramic mould insulated with hot ceramic fibre. A thermocouple incorporated on the centreline of the ingot is used to measure the cooling rate (1.9 K/s between 1400-1200°C) and to record temperature throughout processing. The ingot is 15 mm thick and 75 mm wide, small enough to be quenched efficiently at any stage of processing. The ingot is stripped from the mould, which is re-usable, and rolled, typically, to 3 mm thickness in three passes. The main limitation of the mill is the inter-pass time interval, which is 25 seconds minimum.

Other notable simulations of hot direct rolling have been conducted at CANMET [13] using a similar method, but with a larger sample size, and by Kaspar at the Max-Planck Institute fur Eisenforschung, Dusseldorf using a hot plane strain simulator of rolling. Kaspar in particular has conducted systematic research showing that direct rolling can yield a product comparable to conventionally processed strip [14,15].

Segregation of microalloying elements in as-cast austenite.

Because of the finer microstructure of thin slab castings relative to thick slab castings, segregation of manganese is not likely to be a particular problem in TSDR strip, despite the absence of the significant degree of homogenisation that occurs in conventionally reheated austenite. However, segregation of microalloying elements may be a problem in microalloyed steels. Using the melting and casting scheme outlined in Fig.4, Park [16] made a series of steels with a base composition of 1.4 wt%Mn, 0.25 wt%Si, 0.003 wt%N, P and S < 0.008 wt%, and with varying C and Nb contents. After solidification the ingots were stripped from the mould and quenched from approximately 1400°C and examined metallographically for the presence of a NbC eutectic. The results are shown in Fig.5.

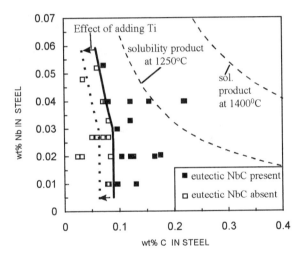

Fig.5. Incidence of NbC eutectic in as-cast austenite at 1400°C.

The figure is separated into two well defined areas by the solid line: to the left of the line, at lower carbon contents, no NbC eutectic precipitates were observed in the as-cast austenite at 1400°C; at higher carbon contents, to the right of the line, eutectic NbC precipitates were present, whose typical morphology is illustrated in Fig.6(a) and (b). The NbC precipitate generally had a herringbone appearance made up of parallel, thin, lath-like rods or lamellae. The NbC was usually associated with MnS, which was often present as large particles on which the NbC formed, and which often coated the rods and formed nodules on their ends. These complexes of particles were nearly always associated with microporosity and located at austenite grain corners and boundaries, indicating that they formed in the final stage of solidification.

The limiting carbon content was constant at about 0.09 wt%C up to 0.03 wt%Nb, and then decreased slightly at higher Nb contents. Thus, for carbon contents higher than the values indicated by the line, the addition of Nb resulted in the effective loss of some Nb in the form of precipitated NbC in the as-cast austenite. Solubility isotherms for NbC are superimposed on the figure for the temperatures of 1400°C and 1250°C, using the solubility product recommended by Irvine *et al* [17]. These dashed curves represent the equilibrium solubility of NbC at these temperatures. Quite clearly, the constitution of the as-cast austenite departs strongly from equilibrium, since the NbC precipitate should be completely dissolved at all compositions to the left of the 1400°C isotherm. Further, the concentrations of Nb and C in solution at the start of rolling will be less than

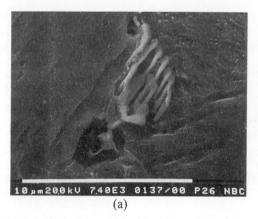
(a)

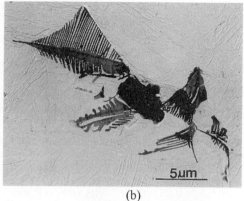
(b)

Fig.6. Morphology of eutectic precipitates;

(a) SEM micrograph, 0.1 wt%C, 0.048 wt%Nb

(b) Extraction replica, 0.12 wt%C, 0.033 wt%Nb

(c) Extraction replica,

0.082 wt%C, 0.01 wt%Nb, 0.015 wt%Ti

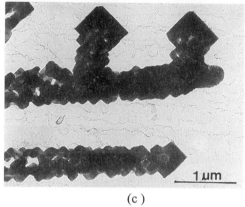
(c)

would be expected in austenite reheated to 1250°C, for compositions to the right of the solid line and below the 1250°C isotherm in Fig.5.

The effect of adding Ti to Nb-bearing steel on the composition and incidence of eutectic precipitates has been studied previously [12,18], and of adding Ti to Nb-V-bearing steel by P.H. Li in a paper in this conference. The effect of adding Ti was to restrict the composition range over which the as-cast austenite was free of eutectic carbonitride precipitate, but the precipitate remained Nb rich. Adding sufficient Ti resulted in the presence TiN cubes which had formed in the liquid state. The results of Park's systematic studies of adding 0.012-0.02 wt% Ti to Nb steels of the Nb and C combinations in Fig. 5 are shown by the dotted line in that figure, and an example of the morphology of the eutectic precipitates in a steel containing 0.082 wt%C, 0.01 wt%Nb and 0.015 wt%Ti is shown in Fig.6(c). The effect of adding Ti was to further limit the composition range in which eutectic carbonitrides appear. Fig.6(c) suggests that at low Nb/Ti ratio this may be due to segregation of Ti to the final volume of liquid before freezing is complete, and the initial formation of TiN cubes, on which NbC may more readily precipitate. At higher Nb/Ti ratios the cubes of TiN did not appear and the eutectic precipitate was a Nb-rich carbonitride (see P.H. Li's paper in this conference).

Nb and Ti are added to conventionally reheated and controlled rolled steel, in part to control austenite grain growth on reheating, but, importantly, also to control recrystallisation of the austenite and to precipitation strengthen ferrite. Rolling austenite below its nil-recrystallisation temperature leads to refinement of the ferrite grain size and increases the strength and toughness of the product. The importance of the results in Fig.5 is that the effectiveness of Nb additions to thin-slab cast steel that is to be directly rolled is likely to be compromised for carbon contents above a limit, which is about 0.09 wt%C for up to 0.03 wt%Nb and decreases to about 0.06 wt%C at 0.06

wt%Nb. The addition of Ti lowers the carbon limit above which this reduction in the effectiveness of Nb additions is likely to occur.

Modelling the recrystallisation of austenite during TSDR.

The constitution of the austenite at the end of hot working determines, together with the cooling rate, the grain size and strength of the ferrite/pearlite transformation product. The constitution of the austenite at any stage during processing is determined by the kinetics of recrystallisation, the dependence of recrystallised grain size on prior history, and on the kinetics of grain growth. Considerable effort has been expended in many research laboratories on modelling the kinetics of change in austenite microstructure during controlled hot rolling, and the resulting models have been applied with considerable success to predict and control the microstructure and properties of reheated and controlled rolled plain carbon and Nb-microalloyed, low carbon steels.

The kinetics of recrystallisation can be represented by an Avrami-type equation:

$$X = 1 - \exp\left\{ \left[\ln(1.-f) \right] \cdot \left(\frac{t}{t_f} \right)^n \right\}$$ Eq.1

where X is the fraction recrystallised in time t, t_f is the time for some specified fraction of recrystallisation, f, and n is the Avrami exponent, usually a value near to 1. For low carbon, manganese steels, Sellars proposed [19]:

$$t_{0.5} = 2.5x10^{-19} \cdot d_0^2 \cdot \varepsilon^{-4} \cdot \exp\left(\frac{Q_{Rx}}{RT} \right)$$ Eq.2

where d_0 is the austenite grain size in μm before deformation to the strain ε, and Q_{Rx} is the activation energy for recrystallisation (300 kJ/mol.). To account for the effect of solute drag due to the concentration [Nb] of Nb in solution in the austenite, Dutta and Sellars [20] proposed that

$$t_{0.05} = 6.75x10^{-20} \cdot d_0^2 \cdot \varepsilon^{-4} \cdot \exp\left(\frac{Q_{Rx}}{RT} \right) \cdot \exp\left\{ \left[\left(\frac{275000}{T} \right) - 185 \right][Nb] \right\}$$ Eq.3

The recrystallised grain size is given by [19]

$$d_\gamma = D \left(\frac{d_{\gamma 0}}{\varepsilon} \right)^{\frac{2}{3}}$$ Eq.4

where $d_{\gamma 0}$ is the prior austenite grain size in μm and D is about 1. Using these equations and referring to the information in Fig.1 for TSDR, it may be predicted that full recrystallisation of a plain, low carbon austenite after roughing would take less than a second, assuming an initial austenite grain size of 1000 μm. The grain size after recrystallisation would be about 95 μm, thus increasing the rate of recrystallisation in the first stages of finishing (recrystallisation would be complete in <1s at 1000°C after a strain of 0.4). It seems probable that static or dynamic recrystallisation would occur during the first part of the finish rolling, refining the grain size further, but that recrystallisation would be somewhat retarded at the end of the finish rolling taking about 5s at 880°C. Because the first passes do recrystallise the austenite to a grain size comparable to that of reheated and conventionally rolled austenite after roughing it is not surprising that TSDR can produce microstructures in low carbon steels comparable with those produced conventionally [24].

However, the situation with respect to microalloyed steel is complicated by solute drag and precipitation.

The presence of microalloying elements in solid solution reduces the kinetics of recrystallisation by a solute drag effect. For the case of Nb solute drag was taken into account by Dutta and Sellars in Eq.3. However, Zhou *et al* [21] and Khan [22] found that the quadratic dependence of the time to start of recrystallisation on austenite grain size in Eq.3 did not extrapolate satisfactorily, from the limited, lower, grain size range over which Dutta and Sellars established their equation, to the large grain size of as-cast austenite. They found that as-cast austenite recrystallised faster than predicted by Eq.3 and proposed that the dependence of the start of recrystallisation on grain size should be linear instead of quadratic:

$$t_{0.05} = 1.2x10^{-17}.d_0.\varepsilon^{-4}.\exp\left(\frac{Q_{Rx}}{RT}\right)\exp\left\{\left[\left(\frac{275000}{T}\right)-185\right][Nb]\right\}$$ Eq.5

Patel *et al* [23] also concluded that a linear dependence of recrystallisation kinetics on grain size was more appropriate in as-cast austenite of a V-microalloyed steel, and that the effect of solute drag was less for V than for Nb. They proposed the following equation for the time to 5% recrystallisation for V austenite:

$$t_{0.05} = 1.74x10^{-17}.d_0.\varepsilon^{-4}.\exp\left(\frac{Q_{Rx}}{RT}\right)\exp\left\{\left[\left(\frac{275000}{T}\right)-185\right][V]/10\right\}$$ Eq.6

These equations predict faster recrystallisation of the coarse grained, just cast austenite than the earlier equations would, but make little difference in the 50-150 µm range.

Strain induced precipitation.

In microalloyed steels which have been reheated before hot rolling the microalloy carbonitride is brought to equilibrium with respect to its solution in austenite. The concentrations of, for example, Nb and C in solution are fixed by the solubility product of NbC and a mass balance calculation at temperatures below its solvus, or by their concentrations in the steel at temperatures above the solvus. The supersaturation increases as the temperature is then decreased below the solvus during working. The natural precipitation kinetics of NbC are quite slow. However, the application of strain and the consequential introduction of dislocations considerably enhances the kinetics of precipitation, and a competition then exists between the onset of strain-induced precipitation and recrystallisation. If strain-induced precipitation occurs before recrystallisation starts it can completely inhibit recrystallisation. Since recrystallisation becomes slower as the temperature falls, a temperature may exist, the recrystallisation-stop temperature, below which recrystallisation cannot occur. Dutta and Sellars [20] proposed that if 5% of the potential amount of precipitate was induced to precipitate before 5% of recrystallisation occurred, then recrystallisation would be inhibited at all lower temperatures. They deduced that the time to the start of strain-induced precipitation was given for Nb steel austenite by the following equation:

$$t_{0.05P} = \frac{3x10^{-6}}{[Nb].\varepsilon.\sqrt{Z}}.\exp\left(\frac{270000}{RT}\right).\exp\left(\frac{2.5x10^{10}}{T^3.(\ln k_s)^2}\right)$$ Eq.7

where ε is the strain in the austenite, Z is the Zener-Hollomon parameter relating to the prior deformation, and k_s is the supersaturation ratio for NbC at the temperature, T. Eq.7 together with Eq.3 and a known solubility product for NbC are then sufficient to predict the recrystallisation-stop temperature in Nb-microalloyed austenite.

However, Zhou *et al* [21] found it necessary to use the modified recrystallisation equation, Eq.5, and slightly faster strain-induced recrystallisation kinetics in order to explain the recrystallisation-stop temperatures in as-cast Nb-austenite during single-pass TSDR experiments. In retrospect, the faster precipitation kinetics may have been due to the segregation of Nb that it is now known would lead to precipitation of eutectic NbC; faster strain-induced precipitation in the Nb-enriched periphery of dendrites may have been capable of suppressing general recrystallisation.

Patel *et al* [23] deduced an equation similar to Eq.7 for the strain induced precipitation of VN in as-cast V-microalloyed austenite,

$$t_{0.05P}(forVN) = \frac{1.15x10^{-6}}{[Nb].\varepsilon.\sqrt{Z}}.\exp\left(\frac{293000}{RT}\right).\exp\left(\frac{1.64x10^{10}}{T^3.(\ln k_s)^2}\right)$$ Eq.7

Together with Eq.6 and the solubility product for VN (to calculate k_s), Eq.7 predicted the recrystallisation-stop temperature during single-pass TSDR experiments on as-cast V-austenite.

When segregation of Nb and Ti (V segregates less than either) during freezing leads to the precipitation of eutectic carbonitrides during casting, then the concentrations of microalloying elements in the as-cast austenite are unknown. Thus, the supersaturation of the austenite during TSDR at lower temperatures is unknown, and the recrystallisation-stop temperature becomes unpredictable. Even for carbon contents less than the limit at which eutectic precipitates appear, segregation in the as-cast microstructure may influence the recrystallisation-stop temperature. Zhou and Priestner [12] quenched as-cast austenite of Nb-Ti-bearing steels, after solidification and cooling to different temperatures in the austenite phase field, and estimated the amounts of (Nb+Ti) in solution at the quench temperature by tempering the quenched, martensitic microstructure. The degree of secondary hardening found on tempering was taken to be a measure of the (Nb+Ti) in solution at the quench temperature. It was found that at 1400°C the austenite was considerably under saturated with respect to (NbTi)(CN) due to an excess of solute being tied up in the eutectic precipitate. The amount of (Nb+Ti) in solution in the as-cast austenite was about that which would have been expected in equilibrium at 1200°C. This limited the degree of supersaturation at lower processing temperatures.

Summary and conclusions.

Because the thin slab casting and direct rolling process is compressed in duration and more efficient than the controlled rolling of reheated thick slabs, it introduces significant differences in the metallurgy of hot working.

The as-cast austenite that is hot worked in TSDR has a finer dendritic structure and smaller grain size than thick slabs, which is generally beneficial with respect to the segregation of solutes. In the case of un-microalloyed, C-Mn steels recrystallisation kinetics are fast enough at the lower start-rolling temperatures of TSDR to ensure that hot rolled products are comparable with conventionally rolled products, despite the starting austenite grain size being about five times larger than reheated austenite grains.

In microalloyed steels, segregation of Nb and Ti during freezing can lead to the non-equilibrium presence of eutectic carbonitrides in the as-cast austenite. This can lower the supersaturation of the austenite with respect to carbonitrides, by an amount that is as yet unquantified, so that the degree of supersaturation at lower rolling temperatures is unknown and recrystallisation-stop temperatures cannot be effectively modelled. Even for the low carbon contents at which eutectic carbonitrides do not appear, segregation could still lead to unpredictable recrystallisation-stop temperatures. Importantly, these effects of segregation need to be understood in more detail in order to ensure the efficient use of microalloys in TSDR.

Acknowledgements.

This research is supported by EPSRC Grants GR/J 80191 and GR/L 06997. The author is also grateful to J.S.Park for contributing some of his original PhD research results.

References.

1. S. Uchida, T. Masaoka and T. Mori, Nippon Kokan Tech. Rep. 113, July, 1986, pp.1-8.

2. K. Kunishige and N. Nagao, ISI Japan, vol.25, (1985), p.315.

3. Proc. Conf. "Achieving The Hot Link", 11-12 May, 1959, London, The Inst. of Metals.

4. H-F Marten, *ibid* ref.3, pp.2.1-2.11.

5. J.K. Brimacombe and I.V. Samarasekera, Proc. Int'l Symposium "Near Net Shape Casting in the Minimills", Vancouver, 19-23 Aug., 1995, Ed. Brimacombe and Samarasekera, The Mat. Soc. of CIM, pp.33-53.

6. J. Edington, 43rd Hatfield Memorial Lecture, Ironmaking and Steelmaking, 1997, vol.24, p23.

7. D.A. Dunholter, "Design and Start-up of the North Star BHP Steel Minimill", Iron and Steel Engineer, Dec. 1997, pp.49-52.

8. R.A. Carr, E.C. Hewitt and J.H. Watters, *ibid* ref.3, pp.1.1-1.23.

9. K. Doring, H. Weisinger, A. Eberle, A. Flick, F. Hirschmanner, F. Wallner, G. Giedenbacher, D. Gabel, P. Kohle and J. Loose, Metall. Plant and Tech., 1990, vol.13, pp16-29.

10. G. Flemming, P. Kappes, W. Rohde and L. Vogtmann, Metall. Plant and Tech., 1988, vol.11, pp.16-35

11. R. Priestner and C. Zhou, Ironmaking and Steelmaking, 1995, vol.22, pp.326-332.

12. C. Zhou and R. Priestner, ISIJ Int'l, 1996, vol.36, pp.1397-1405.

13. E. Essadiqi, L.E. Collins and G. E. Ruddle, *ibid* ref 5, pp.341-352.

14. N. Zentara and R. Kaspar, Mater. Sci. and Tech., 1994, vol.10, pp.370-376.

15. R. Kaspar, N. Zentara and J.C. Herman, Steel Res., 1994, vol.65, pp.279-283.

16. J.S. Park, Private communication: current PhD research at Manchester Materials Science Centre.

17. K.J. Irvine, F.B. Pickering and T Gladman, JISI, 1967, vol.205, p.161.

18. R. Priestner, C. Zhou and A.K. Ibraheem, "Carbonitride Precipitation in Cast Low-C Steel Microalloyed With Ti and Nb", in Titanium Technology in Microalloyed Steels, Ed. T.N. Baker, publ. The Inst. of Materials, London, 1997, pp.150-168.

19. C.M. Sellars, Hot Working and Forming Processes, Ed. C.M. Sellars and G.J. Davies, publ. The Met. Soc., London, 1980, p.3.

20. B. Dutta and C.M. Sellars, Mat. Sci. and Tech., 1987, vol.3, p.197.

21. C. Zhou, A.A. Khan and R. Priestner, Proc. Int'l. Conf. "Modelling of Metal Rolling Processes", 21-23 Sept. 1993, publ. The Inst. of Materials, pp.212-223.

22. A.A. Khan, "Multipass Hot Direct Rolling of Thin Slab Cast Nb-microalloyed Steel", PhD Thesis, University of Manchester, 1997.

23. P. Patel, C. Zhou and R. Priestner, "The Deformation and Recrystallisation Behaviour of As-cast V Microalloyed Austenite During Hot Direct Rolling", in ReX '96, 3rd Int'l Conf. On Recrystallisation and Related Phenomena, 22-24 Oct., 1996, Monterey, Ca.

24. T. Wada, H. Tsukamoto and M. Suga, ISI Japan, 1988, vol.74, pp.1438-1445.

25. R.K. Gibbs, R. Peterson and B.A. Parker, Proc. Int'l Conf. On Processing, Microstructure and Properties of Microalloyed and Other HSLA Steels, ISS-AIME, 1992, pp.201-207.

Correspondence: e-mail; ronald.priestner@umist.ac.uk or Fax R Priestner at (UK) 0161 200 3586

Materials Science Forum Vols. 284-286 (1998) pp. 105-116
© *1998 Trans Tech Publications, Switzerland*

Modelling of Microstructure and Texture during Annealing of Titanium Alloyed Interstitial Free Steels

D. Artymowicz[1], W.B. Hutchinson[1], P.J. Evans[2] and G.J. Spurr[2]

[1] Swedish Institute for Metals Research,
Drottning Kristinas v. 48, S-114-28 Stockholm, Sweden

[2] British Steel Strip Products, Welsh Technology Centre,
Port Talbot, West Glamorgan, SA13 2, UK

Keywords: Recrystallisation, Ti-Alloyed Steels, IF Steels, Texture Evolution, Grain Growth

Abstract:
This presentation describes the principles of a computer model which has been developed to aid in the processing of Ti-alloyed IF steels with special reference to continuous annealing conditions for both uncoated and metallised sheets. The aim of the model is to predict reliably the progress of recrystallisation, grain growth and texture evolution throughout the annealing cycle. The criteria employed in the model are a mixture of physical principles (wherever possible) and enlightened empiricism.

1 Introduction

Attempting to model the evolution of microstructure and texture during annealing of cold rolled steel is truly a daunting task; so many variables are known to play subtle but important roles, not least the interstitial atoms carbon and nitrogen [1,2]. Furthermore there exist no physically based relationships which can be used to provide a trustworthy quantitative basis for predictions. The present approach acknowledges all these limitations. It seeks its legitimacy by dividing up the complex process into distinct stages involving the specific phenomena. As much physical insight as possible is incorporated into each stage and this is supplemented by empirical rationalisations and simplified parametric descriptions which are then tuned using experimental measurements. The method resembles in character that which was successfully pioneered by Sellars and Whiteman for modelling grain structures during hot working [3]. The aim has not been physical rigor but rather usefulness - to be able to predict the effect of changes in steel chemistry or processing conditions on the microstructure and hence the properties of the annealed sheets.

Even acknowledging the many simplifications involved, it was felt that Ti - alloyed IF steels represent the only category which can reasonably be handled at the present time. Since these are fully stabilised with respect to carbon and nitrogen by formation of TiC and TiN the complications caused by interstitial elements are avoided. The particles are sufficiently coarse to have little influence on the recrystallisation behaviour [4] unlike the case of Nb-alloyed IF steels. However the material cannot be regarded as simply single phase since the TiC particles have a strongly inhibiting action on grain growth. It is therefore essential that their evolution by Ostwald ripening is taken into account.

Following the reasoning given previously [5] it is considered that the two factors which principally control the recrystallisation texture are the grain size prior to cold rolling (hot band grain size) and the degree of cold rolling. As discussed recently [6] these two factors may be combined into a single parameter, the specific area (Sv) of grain boundary existing in the cold rolled state. The fact that nucleation of {111} oriented recrystallised grains is observed to take place along prior grain boundaries [7] is in good conformity with this viewpoint. It has also been observed that {111} grains tend to nucleate preferentially at shorter times then do other {hkl} orientations [8, 9, 10], which may be due to the favourable conditions existing in grain boundary regions. It is also supposed [11,12] that {111} nuclei may develop inside deformed grains and this possibility is included in the model although such intra-granularly nucleated {111} grains are considered to develop in the same way as for the random {hkl} grains.

Growth of all grains during primary recrystallisation is assumed to follow the same law. Both the nucleation and growth rates decay as recrystallisation proceeds as a result of parallel recovery and also the fact that stored energy of deformation is not homogeneously distributed [11]. Nucleation and growth rates depend on temperature and have the same activation energies. This later assumption is justified by observations which show that as-recrystallised grain size in steel is essentially independent of annealing temperature over a very wide range [13]. Increase in cold rolling reduction accelerates both nucleation and growth due to the increase in stored energy of deformation. Growth rate is assumed to vary linearly with true strain, as does dislocation density to a first approximation [14]. Nucleation rate must vary more strongly with strain to account for the well known refinement in recrystallised grain size as the degree of cold rolling increases.

At present the model includes no effect of steel chemistry on nucleation or growth rates during recrystallisation. There is evidence that higher levels of titanium in solid solution may decrease the rate of recrystallisation [15, 16] and this can be included relatively easily in the future.

The preferential nucleation of {111} grains at grain boundaries means that these have on average a longer period for growth and so their size distributions tend to be skewed to larger sizes than {hkl} grains at the end of primary recrystallisation. Subsequent grain growth on prolonged annealing occurs by the growth of large grains and the shrinkage of the small ones [17]. There will therefore be a tendency for the {111} texture to strengthen further during grain growth, which is well established experimentally [18, 19]. However, the extent of the grain growth must be limited by the Zener pinning force of the second phases [4], in the present case mainly TiC. It is therefore the kinetics of particle coarsening of TiC which sets limits to grain growth and the associated texture evolution. The model calculates distribution of sizes for {111} and {hkl} grains at the end of recrystallisation and then allows these to evolve further during grain growth according to Hillert's equation for growth and shrinkage [17].

A summary of input parameters in the model is presented in Table 1.

Table 1 List of input and output parameters of the model

INPUT	OUTPUT
1. Ti content	1. Temperature range of recrystallisation
2. C content	2. Average grain size at the end of recrystallisation
3. S content	3. Volume fraction of {111} at the end of
4. N content	recrystallisation
5. Hot rolled band grain size	4. Increase in average grain size
6. Coiling temperature	5. Growth of volume fraction of {111}
7. Cooling rate during coiling	
8. Cold rolling reduction	
9. Maximum temperature of annealing Tpeak	
10. Heating rate during continuous annealing between 500^{0}C and Tpeak	
11. Time of isothermal holding	

2 Physical bases of the model

The discussed model describes changes of three elements of microstructure: precipitation and coalescence of TiC, recrystallisation of cold rolled grains and growth of recrystallised grains. An example of thermomechanical cycle for the IF steel is given in Fig. 1. A possible position of the temperature for complete recrystallisation, T_{rec}, is indicated and the modelled processes are marked relative to the cycle elements.

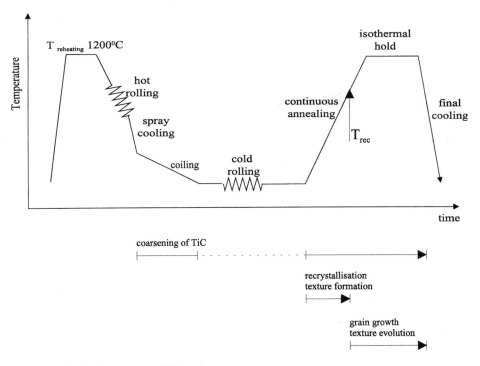

Fig. 1 Thermomechanical treatment of IF steels

2.1 Precipitation and coalescence of TiC

The input steel chemistry parameters are concentrations of C, N, Ti and S. The elements N and S are considered to be combined as TiN and TiS respectively. These particles precipitate during casting or shortly afterwards and do not dissolve upon reheating of the slab. They are large enough not to influence recrystallisation and have only a minor effect on grain growth. Their main effect is to lower the effective content of Ti. The phase $Ti_4C_2S_2$ is not included in the present treatment which is aimed at normal steelworks practice with a high soaking temperatures before hot rolling. However, the program can be adapted to include this phase if this becomes appropriate. The remaining Ti (effective Ti) is able to form TiC. TiC dissolves upon slab reheating and precipitates again during coiling. The volume fraction of TiC depends on the current equilibrium concentration of Ti and C in ferrite given by the solubility product of TiC in ferrite [20]:

$$\log[Ti_s]\cdot[C_s] = -\frac{A}{T} + B \qquad\qquad \textbf{Eq. 1}$$

where

$[Ti_s]$ is the Ti in solution in weight %
$[C_s]$ is the C in solution in weight %
A, B are numerical constants
and T is absolute temperature.
Equilibrium concentration of Ti and C in ferrite is assumed throughout coiling and the continuous annealing cycle.

The increase in average size of TiC particles is calculated according to Wagner's coalescence equation:

$$r^3 - r_0^3 = \frac{8}{9} \cdot \frac{\gamma \cdot D \cdot [Ti_s] \cdot V_m}{R \cdot T} \cdot t \qquad \text{Eq. 2}$$

where :
r is the mean particle radius at the time t
r_0 is the initial mean particle radius at the time t_o
γ is the interface surface energy
D is the diffusivity of Ti in ferrite
$[Ti_s]$ is the fraction of Ti dissolved in ferrite
V_m is the molar volume of TiC
R is the gas constant

The equation is integrated over all preceding stages where the temperature profile is approximated by a series of isothermal steps 1°C apart. Coalescence of TiC takes place during cooling of the coiled hot band and continues during continuous annealing and isothermal hold according to the above description.

2.2 Recrystallisation of cold rolled ferrite during continuous annealing.

In the current model three types of recrystallisation nuclei are considered: $\{111\}_m$ and randomly oriented $\{hkl\}_m$ nuclei that nucleate generally throughout the structure (matrix) and $\{111\}_{gb}$ nuclei that form at prior grain boundaries. Nucleation rates for all three types of nuclei occur at a rate decaying in time (fraction recrystallised) and are a function of the cold rolling reduction . Additionally, $\{111\}_{gb}$ nucleation rate is proportional to the specific grain boundary area S_v . Nuclei forming in the matrix, $\{111\}_m$ and $\{hkl\}_m$, have an incubation period and start forming only after the fraction recrystallised reached a small threshold. As a result of the above assumptions, the ratio of the numbers of $\{111\}$ to $\{hkl\}$ nuclei decreases as recrystallisation proceeds.

Nucleation rate equations [number /unit time /unit volume of unrecrystallised matrix] are assumed to have the following form:

$$\dot{N}_{matrix} = \begin{cases} 0 & \text{for } x < 5.e-5 \\[2mm] A_{matrix}^n \cdot \varepsilon^4 \cdot \exp\left(\frac{-Q}{RT}\right) \cdot (1-x) & \text{for } x \geq 5.e-5 \end{cases} \qquad \text{Eq. 3}$$

for nucleation throughout the matrix, where
A_{matrix}^n is a preexponential constant, different for $\{111\}_m$ and $\{hkl\}_m$ nuclei, found by fitting
ε is the true strain, with the exponent 4 deduced by fitting to measurements of the as-recrystallised grain size,
Q is the activation energy for nucleation
R is the gas constant
T is the absolute temperature
x is the fraction recrystallised, and

$$\dot{N}_{111_{gb}} = A^n_{111\_gb} \cdot \varepsilon^4 \cdot \exp\left(\frac{-Q}{RT}\right) \cdot (1-x) \cdot S_v \cdot \gamma(\varepsilon) \qquad \text{Eq. 4}$$

for nucleation of {111} oriented grains on grain boundaries, {111}$_{gb}$, where:

$A^n_{111\_gb}$ is a preexponential constant found by fitting,

S_v is the specific area of grain boundary in the hot rolled band and equals $2/\bar{d}$ where $\bar{d}$ is the mean intercept length of the grains,

$\gamma(\varepsilon) = 1 + 0.12965 \cdot \varepsilon + 0.23545 \cdot \varepsilon^2 + 0.199131 \cdot \varepsilon^3$ which represents the relative area of grain boundary in the cold rolled structure. This is a very close approximation to the elliptical integration solution [21]. Other symbols are as in Eq. 3.

Growth rates of {hkl} and both types of {111} nuclei are the same and decay with increasing time (fraction recrystallised). A 'mean field' assumption is made that all grains have at the same instant the same fraction of their extended surface areas 'free to grow' i.e. unimpinged.

Growth rate is modelled with the following equation:

$$G = A^g \cdot \varepsilon \cdot \exp\left(-\frac{Q}{RT}\right) \cdot (1-x) \qquad \text{Eq. 5}$$

where:

A^g is the growth constant and other symbols are as described before.

At the completion of primary recrystallisation (normally taken as x=0.95) the as-recrystallised texture is defined in terms of the volumetric proportions of the {111} and {hkl} components.

It is assumed that particles of TiC do not influence the kinetics of recrystallisation (neither nucleation nor growth).

The model based on Eq. 3 - Eq. 5 makes predictions about how the volume fraction of recrystallised microstructure increases during heating or during the isothermal holding period of the continuous annealing cycle. It also predicts the volume fractions of {111} and {hkl} texture components as the summation of the respective grain volumes. Since the grains nucleate after different periods of time and accordingly are allowed different lengths of time for growth, they constitute a distribution of sizes at the end of recrystallisation. These size distributions. both for the {111} and {hkl} families, are also output of this part of the model.

2.3 Grain growth after recrystallisation

After the completion of primary recrystallisation a process of grain growth starts. The evolution of the grain size distribution depends on average radius and volume fraction of the continuously coarsening TiC particles. Grain growth occurs up to a mean grain size determined by the Zener limit based on the TiC particles and other particles present in the material. The Zener limited grain diameter is calculated according to the formula below:

$$\frac{1}{D_{zener}} = 0.5 \cdot \frac{f}{r} + \frac{1}{30} \qquad \text{Eq. 6}$$

Here f is the volume fraction of TiC and r is the average particle radius. The first term on the right hand side represents pinning from TiC and the second from all other particles presented in the steel. The implication of this is that the Zener drag due to other particles will prevent grains from growing to more than 30 μm even though the effect of the TiC may disappear. The factor 0.5 is chosen in agreement with theoretical and experimental evaluations of Zener drag [22].

To achieve the necessary grain growth, grains with radii smaller than the mean radius of the current distribution shrink and their volumes are redistributed to the larger grains (these with radii larger than the mean radius). The speed of shrinking and growing is related the radius if the grain: the smallest grains shrink the fastest and the biggest grains grow the fastest. In this way the {111} texture also strengthens during grain growth since the {111} grains are on average larger than {hkl} due to their faster initial rate of nucleation.

2.4 Textures

In the present simplified model, grain orientations are considered to be either close to {111} or other, randomly scattered, {hkl} values. Experimental data were obtained from electron back-scattering patterns and from x-ray diffraction ODF measurements. When determining the volume fraction of {111}, the {111} grains were defined as having their sheet plane normal within 15° of <111>. These volume fractions were then used to obtain best estimates of the coefficients in equations Eq. 3 - Eq. 5. A significant feature of the present model is that texture is not a primary parameter in the calculations. Rather, the texture evolution is a by-product of the processes of recrystallisation and grain growth, where the volume fractions of the texture components are controlled according to the densities of their nucleation sites and their respective incubation times.

3 Results

A set of experimental structures was collected for comparison with simulations. All the steels were melted, cast and hot rolled in full scale industrial production. Some cold rolling was also carried out in the plant while some was done on a laboratory mill. Table 2 lists ranges in which chemical compositions and thermomechanical parameters varied for the these steels. The average hot rolled band ferrite grain diameters varied between 13 and 23 μm.

Table 2 Range of compositions and thermomechanical parameters of laboratory simulated samples.

Chemical composition wt %			
C	Ti	N	S
0.0018 - 0.0074	0.04 - 0.065	0.0022 - 0.0031	0.01 - 0.014

thermomechanical parameters				
$T_{coiling}$ [°C]	CR reduction [%]	heating rate [°C/s]	T_{peak} [°C]	isothermal hold [s]
625 - 725	53 - 86	3 - 30	750 - 850	0 - 180

Comparisons of numerically simulated and experimentally obtained recrystallisation kinetics data are shown in Fig. 2. The influence cold rolling reduction and heating rate are modelled with an excellent agreement.

The influence of the cold rolling reduction on the recrystallisation temperature, grain size and volume fraction of {111} in the as recrystallised material is summarised in Table 3. Recrystallisation temperatures and grain sizes are predicted within the range of experimental error. However the volume fraction of {111} is not sensitive enough to the extent of cold deformation. Although it could be corrected , this led to increasing errors in other predictions.

An example of the grain size distribution in the as-recrystallised material is shown in Fig. 3. It is important to notice the tail of large grains in the {111}family which is due to early nucleated grains at prior grain boundary sites. The increase in grain frequency attributable to the onset of intra-granular nucleation is abrupt but is not a mathematical step function.

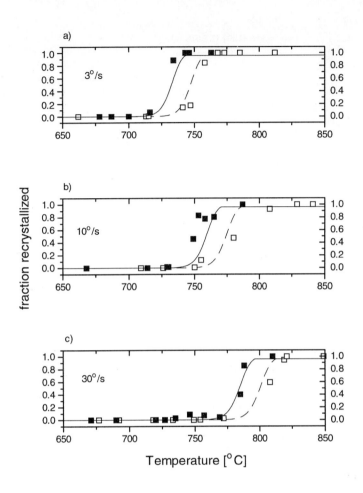

Fig. 2 Simulated and experimental recrystallisation kinetics data for steel coiled at 650°C and cold rolled 65% (solid line and solid squares) and 78% (dashed line and white squares). Heating rates during continuous annealing were: a) 3°/s, b) 10°/s, c) 30°/s.

Table 3 Microstructural data of the as-recrystallised samples for different rolling reductions

cold rolling reduction	Trec(x=0.95) [°C]		grain radius [μm]		volume fraction of {111}	
[%]	experimental	simulated	experimental	simulated	experimental	simulated
53	810	826	7	6.6	.48	.59
65	790	811	5	5.2	.55	.61
75	780	799	4.3	4.2	.62	.63
86	750	783	3.6	3.1	.72	.69

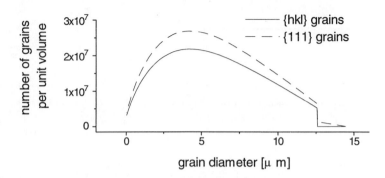

Fig. 3 Examples of grain size distributions in the as-recrystallised state.

Simulation of the evolution of microstructure during isothermal holding is shown in Fig. 4.   The microstructure evolves so as to make the average grain size equal the maximum allowed value which is defined by TiC and other particles.   The grain distribution is allowed to change only if the difference between the Zener limited grain diameter and current average grain diameter is bigger than $1\mu m$.   Volume fraction of {111} changes in tact with the grain distribution evolution.

Results of simulations are compared with the experimental data in Fig. 5.   In general simulations predict a somewhat too small radius in the as - recrystallised material and too much growth with the progress of the continuous annealing.   Nevertheless the agreement between the simulated and experimental data is rather good.

Simulation results for a hypothetical sample are shown in Fig. 6.   Corresponding chemical composition and processing input parameters are listed in Table 4.   In this example, the low peak temperature inhibits the progress of recrystallisation,   which is complete only after the onset of isothermal hold.   The large volume fraction of TiC (only little dissolution and coarsening at $700^{\circ}C$) gives rise to strong pinning of grain boundaries.   As a result, there is no grain growth or increase in volume fraction of {111} oriented grains after completion of recrystallisation.

Experience of operating the model has shown that it can predict a wide range of behaviours.   For example, recrystallisation may occur either during heating or during isothermal holding (compare Fig. 5 and Fig. 6) or only partially or not at all, depending on the deformation level and temperature profile.   Grain growth commences immediately after primary recrystallisation if the Zener pinning force is sufficiently low.   This depends mainly on the steel chemistry but also on the coiling temperature.   Otherwise there is a period where the grain size remains stable until sufficient coarsening /dissolution of the TiC particles has occurred to reduce the Zener pinning to a level where grain growth can commence.   The strength of the {111} texture depends directly on the initial grain size and cold rolling reduction.   It is, however , also sensitive to the steel composition and thermal cycle to the extent that these permit grain growth to proceed. The model has also been applied to conditions of simulated batch annealing (slow heating) and transverse-flux induction heating (very fast).   Although not tested against experimental data, the results were in agreement with expectations and revealed no instabilities produced by such extreme input parameters.

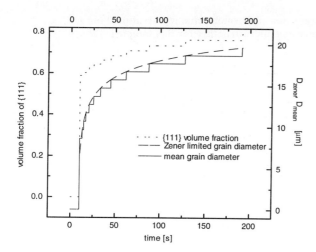

Fig. 4 Simulated microstructural changes in sample coiled at 725°C, cold rolled 75%, annealed with 30°/s rate and isothermally held at 850°C for 180s.

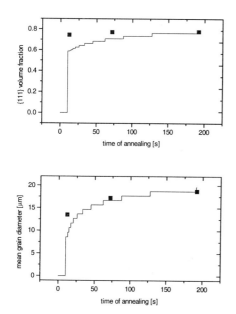

Fig. 5 Comparison of simulated and experimental isothermal annealing data for sample H159_1. The continuous lines represent simulated values, points correspond to experimental data.

Table 4 Chemical composition and parameters of thermomechanical treatment for a hypothetical sample. Evolution of texture and microstructure for this sample is shown in Fig. 6.

Ti content [w/%]	0.045
C content [w/%]	0.002
S content [w/%]	0.01
N content [w/%]	0.0031
Coiling temperature [°C]	700
cooling rate during coiling [deg/h]	10
HRB ferrite grain size [µm]	20
cold rolling reduction [%]	80
maximum temperature of annealing (Tpeak) [°C]	700
heating rate between 500°C and Tpeak [deg/s]	30
time of isothermal hold [s]	200

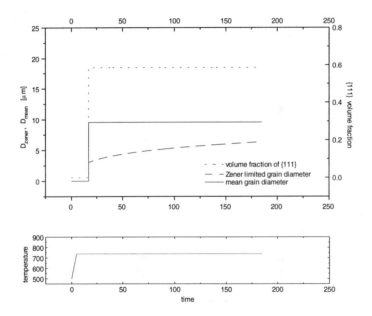

Fig. 6 Texture and microstructure evolution during annealing of a hypothetical sample. Chemical composition and thermomechanical treatment are described in Table 4.

4 Conclusions

The recrystallisation, grain growth and accompanying texture evolution in selected Ti alloyed IF steels were successfully described with a numerical model. The model is based on empirical and physical relations and addresses the continuous annealing process by approximation of a temperature profile with a set of isothermal steps. Well known processing parameters together with steel chemistry and hot rolled band ferrite grain size are used as input parameters. Recrystallisation kinetics as well as changes of grain size distribution, average grain size and volume fraction of the {111} texture component throughout grain growth are calculated as outputs. Recrystallisation kinetics are modelled with very good agreement with the experimental data; average grain size and volume fraction of {111} are predicted with somewhat less accuracy.   Further refinements are desirable to improve   the agreement between predictions and observations, especially with regard to more realistic grain size distributions and to accommodate the influence of dissolved Ti on recrystallisation kinetics.

Acknowledgements:

The authors thank Dr. B. J. Hewitt, Director Technology for the British Steel Strip Products for permission to publish. The work was carried out as part of the ECSC research programme under contract 7210.EC/808.

References:

1 H. Abe, Nishiyama Memotial Lecture, Iron and Steel Institute of Japan (1983) p.3.

2 W. B. Hutchinson, Int. Metall. Reviews, 29 (1984) p.25.

3 C. M. Sellars and J.A. Whiteman, Metal. Sci. 13 (1979) p.187

4 C. M. Suzuki, Y. Ishii, A. Atami, K. Ushioda, N. Yoshiniga and M. Tezuka, Mater. Sci. Forum 204-206 (1996) p.673

5 W. B. Hutchinson, Mater. Sci. Forum, 157-162 (1994) p.1917

6 W. B. Hutchinson and L. Ryde, Thermomechanical Processing in Theory, Modelling and Practice, Stockholm (1997) ASM. p.145

7 W. B. Hutchinson, Acta Met., 37 (1989) p.1047

8 I. L. Dillamore and W. B. Hutchinson, Trans, ISIJ, 11 (1974) p.877

9 Y. Meyzaud, P. Parniere, B. J. Thomas and R. Tixier, Proc. ICOTOM 5, Aachen (1978) p.243

10 E. Lindh, W. B. Hutchinson and P. Bate, Mater. Sci Forum, 157-126 (1994) p.997

11 I. L. Dillamore, C. J. E. Smith and T. W. Watson, Metals Sci. J., 1 (1967), p.49

12 D. Vanderschueren, N. Yoshinaga and K. Koyama, Proc ICOTOM 11, China (1996) p.1400

13 G. R. Speich and R. M. Fisher, Recrystallisation, Grain Growth and Textures, ASM, Ohio (1965) p.563

14 M. B. Bauer, D. L. Holt and A. L. Titchener, Prog. in Mater. Sci., 17 (1973) p.1

15 H. Hayakawa, Y. Furuno, M. Shibata and N. Takahashi, Trans. ISIJ 23 (1983) p.B 434

16 R. Yoda, CAMP - ISIJ 5 (1992) p.831

17 M. Hillert, Acta Met., 13 (1965) p.227

18 W. B. Hutchinson, T. W. Watson and I. L. Dillamore, JISI, 207 (1969) p.1479

19 D. A. Karlyn, R. W. Veith and J. L. Forand, Metal Working and Steel Processing VII AIME (1969) p.127

20 S. Akamatsu et. al., *ISIJ International*, (1994), p.9

21 P. Bates, private communication

22 S. P. Ringer, W. B. Li, and K. E. Easterling, *Acta metall.* 37,(1989) p.831

HOT WORKING

Materials Science Forum Vols. 284-286 (1998) pp. 119-126
© *1998 Trans Tech Publications, Switzerland*

Recrystallization Kinetics of Microalloyed Steels Determined by Two Mechanical Testing Techniques

K. Airaksinen[1], L.P. Karjalainen[1], D. Porter[2] and J. Perttula[1]

[1] University of Oulu, Department of Mechanical Engineering,
P.O. Box 444, FIN-90571 Oulu, Finland

[2] Rautaruukki Oy, Research Centre, P.O. Box 93, FIN-92101 Raahe, Finland

Keywords: Recrystallization Rate, Static Recrystallization, Metadynamic Recrystallization, Stress Relaxation Method, Double-Compression Test, Recrystallization Controlled Rolling

Abstract

Data on the static and post-dynamic recrystallization have been determined in five Ti-microalloyed steels. Both the stress relaxation and interrupted deformation techniques have been employed. The effect of a strain rate change on the flow stress and the subsequent softening kinetics was also investigated. A reasonable agreement is obtained between the results of both the stress relaxation and double-compression methods, which further confirms the reliability of the stress relaxation technique. The results indicate that steels with plain Ti or with Ti-Ni-V or Ti-Ni-Cu alloying recrystallize at temperatures above 900°C (pass reduction ≥0.15) for interpass times characteristic of plate rolling, but Nb (ca. 0.03%) retards the recrystallization rate so that the final rolling temperature should be about 1000°C for full recrystallization between passes. The characteristics of static and metadynamic recrystallization are distinctly different. Softening becomes independent of strain and highly dependent on the strain rate even at strains leading to a small fraction of dynamic recrystallization. Nb has only a small retarding effect in metadynamic recrystallization. The flow stress level and softening kinetics are independent of the strain rate history only being dependent on the final strain rate.

1. Introduction

Recrystallization controlled rolling followed by accelerated cooling (RCR+ACC) is able to provide a good combination of strength, fracture toughness and weldability in HSLA steels. This process is characterized by a high reduction and a small number of rolling passes. From an economical point of view, the productivity of thermomechanical rolling tends to be poor due to the delay while cooling to below the austenite recrystallization temperature range for finishing rolling. RCR is a simpler and more productive alternative [1, 2].

The basic philosophy of RCR is to achieve a fine as-reheated austenite grain size, grain size refinement by repeated static recrystallization events and subsequently, a fine final ferrite grain size by utilizing controlled accelerated cooling through the austenite to ferrite transformation region. A prerequisite for successful processing is that grain growth at high reheating temperatures and between passes is inhibited by finely dispersed TiN particles. In order to optimize the design of RCR schedules, the influence of microalloying additions on the recrystallization rate and grain growth must be well understood.

In the current work the static as well as post-dynamic recrystallization behaviour of five Ti microalloyed steels have been investigated. Two of the steels were simply C-Mn-Si based (355 MPa grades) and three had additions of Ni-Nb, Ni-Cu, or Ni-V (460 MPa grades). The stress relaxation method, recently developed at the University of Oulu [3-6], was applied to measure the recrystallization kinetics in hot deformed austenite. The second objective of this study was to compare the data on recrystallization kinetics determined by the stress relaxation method with the results from double-compression tests. In rolling, the deformation rate is not constant but changes in the roll gap in the course of a deformation pass. More refined data are needed for the accurate modelling of recrystallization. Therefore, the effects of a rapid strain rate change on flow stress behaviour and subsequent recrystallization kinetics were also briefly investigated, utilizing the stress relaxation technique.

2. Experimental procedure

The investigation was carried out on five Ti microalloyed steels from the charges produced by Rautaruukki Raahe Steel. The compositions are given in Table 1. Rectangular bars of 15 x 15 x 220 mm were cut from production rolled plates, reheated at 1200°C for 45 minutes and water quenched. Cylindrical specimens (ø10 x 12 mm) for the relaxation and double-compression tests were machined from these bars.

The tests were performed using a Gleeble 1500 thermomechanical simulator. For the stress relaxation tests, the schedule was as follows: the specimen was heated at 10°C/s to 1200°C and soaked for 4 min prior to cooling at 5°C/s to the deformation temperature of 900°C, 1000°C, 1100°C or 1200°C, held for 15 s, and subsequently compressed at a true strain rate of 1 s^{-1} to a strain of 0.1, 0.2, 0.3 or 0.4. The strain was held constant after the deformation and the declining compressive force was recorded as a function of time at the data acquisition rate of 30 Hz. A second compression cycle was executed after a relaxation time of 200 s to obtain the stress-strain curve for checking the degree of softening. The analysis of relaxation curves is described elsewhere [3, 5]. The conventional interrupted deformation technique, the double-compression test, was employed to measure the fractional recrystallization using the 2% offset, which excludes the recovery effects better than the 0.2% offset method [7].

For metallographic examination, a few specimens were quenched by a water spray from the deformation temperature before the onset of deformation. Specimens were halved in the axial direction along the central plane and the prior austenite grain boundaries were revealed by etching in a saturated aqueous picric acid solution at 70°C, and the austenite grain size was determined using the linear intercept method.

Table 1. Chemical compositions of the steels (mass %).

Grade	C	Si	Mn	Al	Ni	Cu	Nb	Ti	V	N (ppm)
Ti 1	0.08	0.16	1.42	0.03	0.03	-	-	0.011	-	46
Ti 2	0.06	0.18	1.57	0.05	0.03	-	-	0.019	-	60
Ti-Ni-Nb	0.08	0.23	1.54	0.03	0.34	-	0.034	0.017	-	63
Ti-Ni-Cu	0.10	0.20	1.57	0.04	0.33	0.37	-	0.018	-	53
Ti-Ni-V	0.10	0.21	1.54	0.03	0.34	-	-	0.018	0.087	52

3. Results and discussion

3.1. Effect of microalloying elements on recrystallization

The initial austenite grain size is an important factor affecting the static recrystallization (SRX) kinetics. The austenite grain size after reheating at 1200°C and cooling to the deformation temperature is given in Table 2. It is relatively fine in all steels as a result from Ti microalloying which prevents the grain growth.

Table 2. Austenite grain size before deformation (1200°C, 3 min).

Grade	d_0 (µm)
Ti 1	45
Ti 2	52
Ti-Ni-Nb	50
Ti-Ni-Cu	40
Ti-Ni-V	56

Examples of the recrystallization data obtained by the relaxation technique are presented in Figs. 1a and 1b, which display the time for 50% recrystallization, t_{50}, as a function of the true strain at three temperatures. Data for a C-Mn steel are also plotted in Fig. 1b for comparison. The recrystallization kinetics of the steels are almost identical, except for the Ti-Nb steel, which exhibits more sluggish softening (see Fig. 1a).

There seems to be a small difference between the softening of the plain Ti grades Ti 1 and Ti 2, i.e. the steels containing 0.011% and 0.019% Ti, respectively. However, in order to discern the effect of Ti, the effect of their different grain sizes must be taken into account. If the recrystallization time is proportional to grain size squared, the value commonly proposed [8, 9], then the ratio of recrystalli-zation times (t_{50}) for steels Ti 1 and Ti 2 should be $(45/52)^2$, i.e. 0.75. This can only account for a part of the difference between the two, however, as the measured ratio is ca. 0.5, especially at lower temperatures, such as 900°C (Fig. 1b). Hence, the higher Ti content seems to create a significant retarding effect. This is consistent with the results of Siwecki [9] for Ti-V-N steels which had slightly slower SRX kinetics compared to the C-Mn steels. In high Ti steels ($\geq$0.07% Ti), Ti exerts a pro-nounced retarding effect, especially at lower temperatures, which is presumably associated with the precipitation of TiC [4, 10].

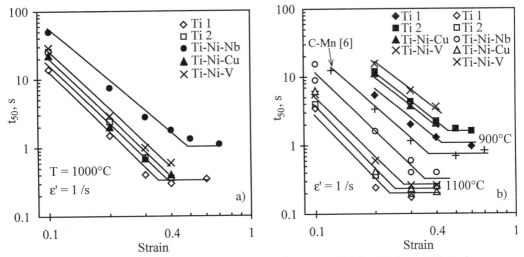

Fig. 1. t_{50} values from relaxation tests a) for the tested steels and b) for C-Mn and Ti steels.

Comparison of the recrystallization times (t_{50}) of the steels Ti 2 and Ti-Ni-V shows that the differ-ences between the two steels are small (Fig. 1) and most of this difference can be accounted for by the small difference in austenite grain size (Table 2). Therefore, at the levels used in the present steels Ni and V seem to have practically negligible effect on the rate of SRX. Medina et al. [11] have also shown that V in solution hardly influences SRX kinetics.

A finer grain size of the Ti-Ni-Cu grade compared to Ti 2 is the main reason for slightly faster soften-ing (see Fig. 1). The role of Cu is not often been investigated. Abe et al. [12] and Devaraj et al. [13] did, however, find that Cu retards the austenite recrystallization, presumably due to a solute drag effect, for Cu does not precipitate in austenite.

For predicting the fractional softening using the t_{50} times, the Avrami exponent is also needed. The fraction of recrystallized austenite can be calculated from the recorded relaxation curves and the Avrami equation fitted with the data to give the exponent [5, 6]. The present data, in agreement with previous studies on C-Mn and Nb bearing steels, suggest that 1.5 - 1.7 is a reasonable value for the exponent [5, 6].

Times for full recrystallization (t_{95}) were calculated using the t_{50} values, obtained by fitting with the measured data together with the Avrami exponent of 1.7. The strain rate exponent (k) and strain exponent as well as the apparent activation energy (Q) are discussed in section 3.2. Some data and the results of predictions are given in Table 3. For the practical design of the RCR schedules the pre-dictions indicate that the steels with plain Ti or with Ti-Ni-V or Ti-Ni-Cu can recrystallize at temper-atures above 900°C at pass strains $\geq$0.15 in reasonable interpass times in a finishing mill, where a finer austenite grain size and a slightly higher strain rate during the final passes accelerate SRX. However, in the Ti-Ni-Nb steel precipitation prevents recrystallization at 950°C so that the final roll-ing temperature has to be at least around 1000°C. The minimum reduction for the occurrence of SRX seems to be about 0.1, but then SRX is quite slow even at 1000°C. Siwecki [9] reports a critical strain of 0.058 for recrystallization in Ti-V-N steels, apparently due to the drag force of TiN particles.

Table 3. Measured and calculated t_{95} values from relaxation tests.

Grade	T (°C)	ε	t_{95} (s), meas. ($\dot{\varepsilon}=1\ s^{-1}$)	A	t_{95} (s), calc. ($\dot{\varepsilon}=1\ s^{-1}$)	t_{95} (s), calc. ($\dot{\varepsilon}=10\ s^{-1}, \varepsilon=0.15, d_0=20\ \mu m$)
Ti 1	900	0.2	13	1.7	17	6
Ti 2	900	0.2	29	3.3	33	8
Ti-Ni-Cu	900	0.2	26	15	32	14
Ti-Ni-V	900	0.2	37	4.3	50	11
Ti-Ni-Nb	950	0.3	incomplete soft.	-	-	-
Ti-Ni-Nb	1000	0.2	17	0.65	22	5

$t_{50} = A \cdot 10^{-15}\dot{\varepsilon}^{k}\varepsilon^{-2.8}d_0^2 \exp(Q/RT)$, k=-0.23 and -0.12 for the Ti-Ni-Nb and other steels, respectively. The Avrami exponent 1.7.

3.2. Characteristics of static and post-dynamic recrystallization

As illustrated in Fig. 1b, at low strains t_{50} decreases with increasing strain, behaviour typical of SRX, but at high strains it becomes independent of strain, which is typical of post-dynamic softening (PDS), commonly also called metadynamic recrystallization (MDRX) [5, 6, 8, 14, 15]. The power of strain on the SRX kinetics was found to be in a range of -2.5...-3.0, in agreement with the earlier measurements for C-Mn and Ti-Nb steels [5, 6].

In order to analyse the relationship between dynamic recrystallization (DRX) and MDRX, the peak strains (ε_p) were determined from the compression curves (see Table 4). Some DRX (≈ 5 - 25%, depending on the deformation conditions) has already occurred at ε_p [16]. Hence, the critical strain (ε_c) for the onset of DRX is slightly lower than ε_p, and commonly taken as given by $\varepsilon_c = 0.8\varepsilon_p$ [8]. These values are also given in Table 4, along with the strains where static changes into metadynamic recrystallization (ε_{S-M}). It should be noted that the recrystallization data are fitted by two straight lines in Fig. 1, even though there may be a short transition stage. ε_{S-M} is given by the intersection of these lines. Nevertheless, on the basis of the values it is quite apparent that the change takes place at a strain which is even smaller than ε_p. In agreement, Hodgson [14] suggests that the rate of recrystallization becomes strain independent near the peak strain. This means that even a small fraction of DRX is capable of producing PDS which has a rate independent of the amount of previous deformation. These observations are also consistent with earlier observations on a Ti-Nb steel in single-pass compression [6], but this is different from the behaviour in multipass torsion tests, where a significant pass strain dependence exists even at the cumulative strains above the peak strain [17].

Table 4. ε_p, ε_c and ε_{S-M} in tested steels ($\dot{\varepsilon}=1\ s^{-1}$).

Grade	d_0 (μm)	T (°C)	ε_p	$0.8\varepsilon_p$	ε_{S-M}	Ref
C-Mn	17	900	0.42	0.34	0.36	6
Ti 1	45	900	0.58	0.46	0.42	
Ti 2	52	900	0.60	0.48	0.44	
Ti 1	45	1000	0.44	0.35	0.32	
Ti-Ni-Nb	50	1000	0.62	0.50	0.49	
Ti 1	45	1100	0.26	0.20	0.21	
Ti 2	52	1100	0.30	0.24	0.24	
Ti-Ni-Nb	50	1100	0.38	0.30	0.31	
Ti-Ni-Cu	40	1100	0.30	0.24	0.24	
Ti-Ni-V	56	1100	0.32	0.26	0.26	

Data on the effect of temperature on t_{50} can be found in Fig. 1. The temperature dependence of t_{50} is also shown in Fig. 2 at the strain of 0.4. In this figure, it can be seen that the slope, which gives the apparent activation energy of the process, changes from a small value at high temperatures to a much higher value at lower temperatures. The transition temperature is dependent on the chemical composition of the steel and the strain rate, as demonstrated in Fig. 2b. Values of the apparent activation energy for SRX and PDS/MDRX were determined from the measured data and they are listed in Table 5. The data for SRX (valid for strains 0.1 - 0.4) are in reasonable agreement with the values reported in the literature. The apparent activation energy of MDRX is much lower, about 60 kJ/mol, for both the Ti 2 and Ti-Ni-Nb steels. Values from 0 to 153 kJ/mol are reported for carbon and microalloyed steels [14,19], and the present values are in the same range.

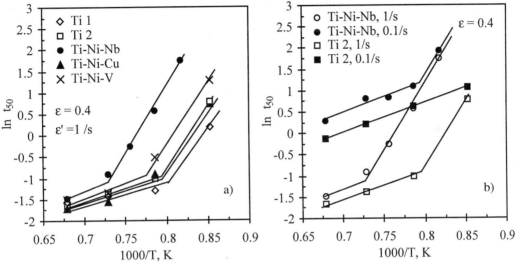

Fig. 2. Temperature dependence of the time for 50% recrystallization a) for the tested steels at strain of 0.4 and b) for Ti and Ti-Ni-Nb steels at different strain rates.

Furthermore, Fig. 2b indicates that the strain rate has a pronounced effect on the MDRX rate, but hardly anything on the SRX rate. The strain rate exponent of -0.12 and -0.23 for SRX was measured for the Ti 2 and Nb-Ni-Ti steels, respectively. For MDRX, the corresponding values were -0.77 and -0.83, respectively, which values are close to -0.8, suggested by Hodgson [14] as an universal strain rate exponent for MDRX. It is interesting to note that Nb significantly retards the SRX rate, but its effect is much smaller on MDRX, especially at high temperatures ($\geq$1100°C). This suggests that solute drag affects more the nucleation rate than the growth rate of statically recrystallized nuclei. Of course, MDRX has no nucleation stage as the nuclei are formed during DRX.

Table 5. Apparent activation energies of recrystallization for some steels.

Steel	Apparent activation energy		Ref.
	SRX (kJ/mol)	MDRX (kJ/mol)	
Ti 1	210	-	
Ti 2	230	58	
Ti-Ni-Nb	265	61	
Ti-Ni-Cu	220	-	
Ti-Ni-V	230	-	
Ti-V-N	280	-	9
C-Mn-(Ti-V)	230	-	15
Plain carbon steel	248-263	127-133	18
C-Mn	-	50	6
Ti-Nb	-	90	6
Nb / Ti	-	153 / 125	19
Nb	-	0	14

3.3. Comparison of softening determined by the stress relaxation and double-compression techniques

In the current work, the fraction of recrystallized austenite was mainly determined using the efficient stress relaxation technique. However, it is interest to know whether the considerable stress present during the relaxation test accelerates recrystallization. Comparisons between results from these two mechanical tests are shown in Figs. 3a and 3b for the Ti 2 and Ti-Ni-V steels, respectively. As can be seen, the accelerating effect of stress is very small and quite insignificant in practice. It has been shown elsewhere that the stress may shorten the recrystallization time by over 50%, if the grain size or strain is very small, but in most instances it has no effect at all or it only slightly enhances the recrystallization rate [7].

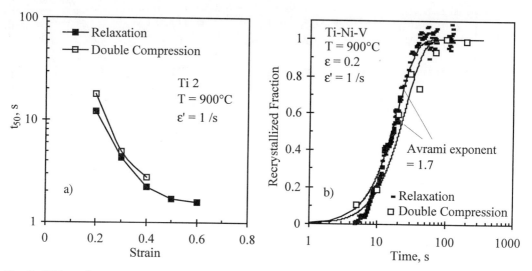

Fig. 3. Effect of stress on recrystallization kinetics in a) Ti 2 and b) Ti-Ni-V steels.

3.4. Effect of strain rate change on flow stress and recrystallization

In order to reveal the effect of strain rate history on deformation resistance and the subsequent softening rate, the strain rate was changed instantly after a certain deformation. The flow stress curves are shown in Figs. 4a and 4b. In Fig. 4a the strain rate change was made at the strain of 0.15, i.e., in the dynamic recovery (DRV) regime for the Ti-Ni-Nb steel. As seen, the Gleeble machine starts to reduce it in advance. For Ti 2 (Fig. 4b) the change was made at $\varepsilon = 0.3$ near the peak strain, i.e., after some DRX. In the DRV regime, the flow stress changes within a short transient strain to a new level, which is identical to the flow stress level in the constant strain rate test at this strain rate. Hence, the flow stress is controlled by the current strain rate. This conclusion is consistent with the observations of Urcola and Sellars [20]. The same holds true in the MDRX regime, although the change creates some peculiar features as also observed by Sakai and Jonas [21].

The softening kinetics after deformation at a constant strain rate or in the case of a strain rate change are shown in Figs. 5a and 5b in the instances of the change in the DRV or DRX regimes, respectively. The effect of strain rate is much higher for the MDRX rate than for the SRX rate, as shown earlier. However, it is quite evident that the recrystallization rate is independent of the strain rate history so that it is determined by the final strain rate in both instances. In the present tests a new steady state has been reached after the strain rate change. However, during a rolling pass the strain rate changes almost continuously so that a stable state is hardly reached. Therefore, the effect of very short deformation stages should be investigated in future.

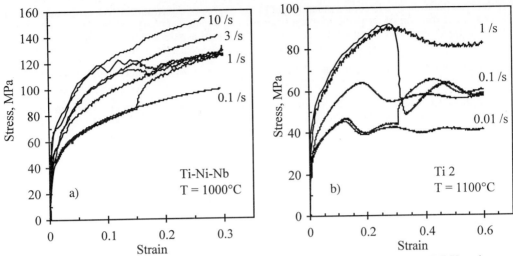

Fig. 4. Effect of changing strain rate on stress-strain behaviour in a) SRX and b) MDRX regime.

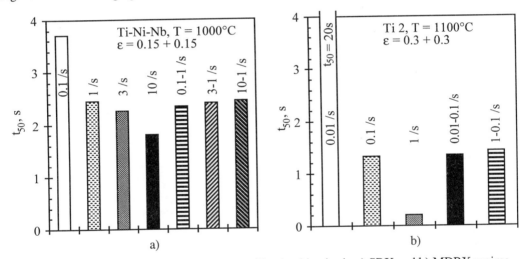

Fig. 5. Effect of changing strain rate on recrystallization kinetics in a) SRX and b) MDRX regime.

4. Conclusions

The static (SRX) and metadynamic (MDRX) recrystallization behaviour of five Ti microalloyed HSLA steels have been determined by the stress relaxation method and, for comparison, by the conventional double-compression tests applying the 2% offset method. The results show that:

1. Ti has a small retarding influence on the rate of SRX, but Nb (0.03%) has a considerably larger effect. V and Cu seem to have only very small influence. The apparent activation energy of SRX is about 225 kJ/mol for the steels tested, except for the Nb-bearing grade for which it is 265 kJ/mol.

2. The minimum strain leading to SRX is about 0.1. In the rolling of Nb-free grades SRX can be expected to occur at strains ≥0.15 at temperatures above 900°C in finishing rolling. For 0.03%Nb-bearing grades, however, temperatures of about 1000°C are needed.

3. Recrystallization rate becomes independent of the degree of deformation even at strains resulting in a small fraction of dynamic recrystallization. The activation energy of the MDRX process is about 60 kJ/mol. Increased strain rate has a pronounced accelerating effect on MDRX, but the retarding influence of Nb on the rate of MDRX is much weaker than it is on the rate of SRX.

4. The stress relaxation test provides data in fair agreement with the double-compression test so that it can be employed with good reliability.

5. The flow stress level as well as both the SRX and MDRX rates subsequent to a rapid strain rate change are determined by the final strain rate after a short transient period. However, during a rolling pass the strain rate changes almost continuously so that a stable state is not reached. Such effects require further investigation.

5. Acknowledgements

The financial support from the Technology Development Centre of Finland is acknowledged with gratitude. The experimental materials were supplied by Rautaruukki Oy Raahe Steel. Rautaruukki Oy is also thanked for the permission to publish these results.

6. References

[1]   T. Siwecki, B. Hutchinson and S. Zajac, Microalloying '95, Conf. Proc., ed. by M. Korchynsky et al., Iron and Steel Society, Warrendale, USA, (1995), p. 197.

[2]   A. Tamminen, R. Laitinen, L. Myllykoski, D. Porter and P. Sandvik, THERMEC '97, Int. Conf. on Thermomechanical Processing of Steels & Other Materials, (1997), University of Wollongong, Wollongong, Australia (in printing).

[3]   L. P. Karjalainen, Mater. Sci. Technol. 11 (1995), p. 557.

[4]   L. P. Karjalainen, HSLA Steels '95, The Third Int. Conf. on HSLA Steels, ed. by G. Liu et al., The Chinese Society of Metals, China Science and Technology Press, Beijing, China, (1995), p. 179.

[5]   L. P. Karjalainen and J. S. Perttula, ISIJ Intern. 36 (1996), p. 729.

[6]   L. P. Karjalainen and J. S. Perttula, ReX '96, The Third Int. Conf. on Recrystallization and Related Phenomena, ed. by T. R. McNelley, MIAS, Monterey, USA, (1997), p. 413.

[7]   J. S. Perttula and L. P. Karjalainen, to be published in Mater. Sci. Technol.

[8]   C. M. Sellars, Hot Working and Forming Processes, Int. Conf. on Hot Working and Forming Processes, ed. by C. M. Sellars and C. J. Davis, The Metals Society, London, (1980), p. 3.

[9]   T. Siwecki, ISIJ Intern. 32 (1992), p. 368.

[10] D. P. Dunne, T. Chandra and S. Misra, HSLA Steels: Metallurgy and Applications, Int. Conf. on HSLA Steels '85, ed. by J. M. Cray et al. ASM International, (1986), p. 207.

[11] S. F. Medina, J. E. Mancilla and C. A. Hernández, ISIJ Intern. 34 (1994), p. 689.

[12] T. Abe, M. Kurihara, H. Tagawa and K. Tsukada, ISIJ Intern. 27 (1987), p. 478.

[13] S. Devaraj, Y. Dake and T. Chandra, Recrystallization '90, Int. Conf. on Recrystallization in Metallic Materials, ed. by T. Chandra, TMS Publ., Warrendale, USA, (1990), p. 237.

[14] P. D. Hodgson, THERMEC '97, Int. Conf. on Thermomechanical Processing of Steels & Other Materials, (1997), University of Wollongong, Wollongong, Australia (in printing).

[15] P. D. Hodgson and R. K. Gibbs, ISIJ Intern. 32 (1992), p. 1329.

[16] A. Fabreque, Advances in Hot Deformation Textures and Microstructures, ed. by J. J. Jonas et al., The Minerals, Metals & Materials Society, USA, (1994), p. 75.

[17] L. P. Karjalainen, P. Kantanen, T. M. Maccagno and J. J. Jonas, THERMEC '97, Int. Conf. on Thermomechanical Processing of Steels & Other Materials, (1997), University of Wollongong, Wollongong, Australia (in printing).

[18] W. P. Sun and E. B. Hawbolt, ISIJ Intern. 37 (1997), p. 1000.

[19] C. Roucoules, S. Yue, and J. J. Jonas, Metall. Mater. Trans. A 26A (1995), p. 181.

[20] J. Urcola and M. C. Sellars, Acta Metall. 35 (1987), p. 2637.

[21] T. Sakai and J. J. Jonas, Acta Metall. 32 (1984), p. 189.

Materials Science Forum Vols. 284-286 (1998) pp. 127-134

Effect of the Chemical Composition on the Peak and Steady Stresses of Plain Carbon and Microalloyed Steels Deformed under Hot Working Conditions

J.M. Cabrera[1], J.J. Jonas[2] and J.M. Prado[1]

[1] Dept. of Ciencia de Materiales e Ingeniería Metalúrgica, Universidad Politécnica de Catalunya, Av. Diagonal 647, E-08028 Barcelona, Spain

[2] Dept. de Metallurgical Engineering, McGill University, 3610 University Street, Montreal, H3A-2B2, Canada

Keywords: Constitutive Equations, Hot Working, Flow Behaviour

ABSTRACT

Two different behaviours are classically observed during the high temperature deformation of metals: i) power law creep and ii) exponential law creep. The first is observed at relatively low stresses and is considered as a deformation process controlled by diffusion. At higher stresses the above behaviour is converted into an exponential one, i.e. the power law breaks down. Both phenomena can be described by a single expression of the form:

$$\dot{\varepsilon} = A(sinh\alpha\sigma)^n \cdot \exp(-Q/RT)$$

Here the parameters A, n, α and Q depend on the material being considered, and are usually referred to as apparent values because no account is generally taken of the internal microstructural state. In the particular case of microalloyed steels, a broad range of values have been reported in the literature for the latter constants, and clear trends have not always been evident. In recent work, it has been shown that the high temperature behaviour of medium carbon microalloyed steels can be accurately described by the classical hyperbolic sine relation provided the stresses are normalised by Young's modulus $E(T)$ and the strain rates by the self-diffusion coefficient $D(T)$. According to this formulation, only two parameters need to be determined to characterise the hot flow behaviour: A and α (n can be set equal to 5 for carbon steels).

In the present work, the latter expression is extended to plain carbon and low carbon microalloyed steels, and applied to the peak and steady stresses of the flow curve. To attain this goal, experimental results corresponding to several different steels reported by many authors are employed. The effect of chemical composition on the above constants is derived statistically.

INTRODUCTION

Although microalloyed steels were discovered during the 40's, they have undergone continuous development since that time and, according to the topic of this meeting, further growth is expected in the forthcoming years. The subject of microalloyed steels interacts significantly with many aspects of metallurgy, such as strengthening mechanisms, toughness, ductility, hot and cold working, recrystallization, inclusions, grain refinement, phase transformations and weldability. An excellent recent review of all these aspects can be found in reference [1].

One of the main purposes of employing microalloyed steels is to improve on the mechanical properties achieved by C-Mn-Si steels, or even alloyed steels, while offering economical savings in relation to conventional steels and processes. To attain this purpose, an accurate design of the thermomechanical process is required. In order to produce improved properties, the parameters of the forming process must be optimised. This is especially important when the final microstructure is produced directly by the forming process, and no subsequent heat treatment is carried out.

Appropriate understanding of the microstructural behaviour of the steel under consideration is therefore required, together with the constitutive relation describing material flow. The latter can then be used to carry out computer simulations by means of which the final microstructural state can be improved and the design of the forming process can be optimised.

The hot flow behaviour of steels involves two dynamic softening processes, namely, dynamic recovery and dynamic recrystallization. The view is taken here that the initial part of the flow curve can be described in terms of dynamic recovery parameters, while the region beyond the peak stress (see Fig. 1) is characterised by dynamic recrystallization parameters.

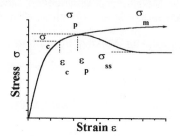

Fig. 1. Plastic flow behaviour of γ-Fe under conditions of high temperature deformation.

Various models have been proposed to describe the behaviour of metals undergoing dynamic recovery. Here the models proposed by Estrin and Mecking [2] and Bergström [3] are used to predict the observed flow curve σ-ε. The resulting relation is given by:

$$\sigma^2 = \left[\sigma_s^2 + \left(\sigma_o^2 - \sigma_s^2\right) \cdot e^{-\Omega\varepsilon}\right]$$

(1)

where

$$\sigma_o = \alpha'\mu b\sqrt{\rho_o}$$

(2)

and

$$\sigma_s = \alpha'\mu b\sqrt{U/\Omega}$$

(3)

Here α' is a geometric constant, μ the shear modulus, b the burgers vector, σ_o the yield stress for the initial dislocation density ρ ($\rho=\rho_o$, $\varepsilon=0$), U and Ω are the characteristic parameters describing the work hardening and dynamic recovery behaviours, respectively, and σ_s is the extrapolated saturation stress of the flow curve in the absence of dynamic recrystallization. Modelling of the latter softening process can be accomplished by treating it as a solid state transformation. In this case, the kinetics of dynamic recrystallization can be represented by an Avrami equation. For this purpose, it is assumed that the mechanical softening is directly proportional to the recrystallized volume fraction. In this way, the constitutive equation that applies after the initiation of dynamic recrystallization is:

$$\sigma = \sigma_s - \left(\sigma_s - \sigma_{ss}\right) \cdot X$$

(4)

where σ_{ss} is the steady state stress at large strains and X is the recrystallized volume fraction. Eqs. (1) to (4) can be used to represent the flow behaviour of the material under consideration, although some rate equations must also be provided. These are the ones that specify the saturation and steady state stresses, the recrystallized volume fraction and the work hardening and recovery terms. When these are available, the flow stress can be expressed as an explicit function of the temperature, strain rate and strain. In this work, the study will be focused on the rate equations for σ_s and σ_{ss}. The effects of chemical composition will also be included.

RATE EQUATIONS FOR THE PEAK AND STEADY STRESSES

Two different behaviours are classically observed during the high temperature deformation of metals [4-7]: i) power law creep and ii) exponential law creep. The first is observed at relatively low stresses (or low strain rates and high temperatures), and is considered as a deformation process controlled by diffusion. Under these conditions, the glide and climb of dislocations take place simultaneously, and although the glide step is mainly responsible for the strain, the average dislocation velocity is determined by the frequency of climb. In other words, the rate-controlling mechanism is diffusion to or from the climbing dislocation rather than the thermally activated glide of the dislocation itself. Such behaviour can be satisfactorily described by the following power law:

$$\dot{\varepsilon} = A' \cdot \sigma^n \cdot \exp(-Q/RT) \qquad (5)$$

where A' and n are constants, Q is the activation energy for hot working, $\dot{\varepsilon}$ the strain rate involved, T the absolute temperature and R the universal gas constant.

At higher stresses (or lower temperatures and higher strain rates) the above behaviour is converted into an exponential one, i.e. the power law breaks down. This phenomenon is an indication of a transition from a climb-controlled mechanism to a glide-controlled process. The latter can be represented by an equation of the form:

$$\dot{\varepsilon} = A'' \cdot (\exp \beta\sigma) \cdot \exp(-Q/RT) \qquad (6)$$

A'' and β being two constants. Eqs. (5) and (6) can be combined [4-7] into a single expression of the form of eq. (7):

$$\dot{\varepsilon} = A''' (sinh\alpha \ \sigma)^n \cdot \exp(-Q/RT) \qquad (7)$$

where α is the inverse of the stress associated with power-law breakdown and the parameters α, β and n are linked by the relation $\beta = \alpha \cdot n$. The terms A''', n, α and Q depend on the material being considered, and are usually referred to as apparent values because no account is generally taken of the internal microstructural state. In the particular case of microalloyed steels, a broad range of values have been reported in the literature for the latter constants. For example, the activation energy is reported to vary from 270 kJ/mol to 400 or higher values, the creep exponent falls between 4 and 6, and the inverse stress is usually considered to be close to 0.012 MPa^{-1}. Clear trends associated with the chemical composition have not always been evident. A notable exception can be found in the work of Medina et al. [8-11]. These authors proposed the following expression for the activation energy as a function of the chemical composition (in weight %):

$$Q(kJ/mol) = 267 - 2.5(\%C) + 1.0(\%Mn) + 33.6(\%Si) + 35.6(\%Mo) + 70.7(\%Nb)^{0.56} + 93.7(\%Ti)^{0.59} + 31.6(\%V) \quad (8)$$

In recent investigations [12,13], the present authors have shown that the high temperature behaviour of medium carbon microalloyed steels can be accurately described by the classical hyperbolic sine relation provided the stresses are normalised by Young's modulus and the strain rates by the self-diffusion coefficient:

$$(\dot{\varepsilon}/D(T)) = A \cdot (sinh\alpha \cdot (\sigma/E(T)))^n \qquad (9)$$

According to this formulation, only two experimental parameters need to be determined to characterise the flow behaviour of the material under consideration: i.e. A and α. (The creep exponent n can be set equal to 5 when the substructure is independent of the strain [14]; this is the most frequently reported value for carbon steels.)

According to this new approach, the activation energy is considered to be constant and equal to the self-diffusion one (270 kJ/mol), and therefore relatively independent of the chemical composition. The latter is consistent with the expression of Medina and co-workers if typical amounts of alloying and microalloying elements are considered. In other words, the effect of the alloying and microalloying elements on the γ-Fe self-diffusion activation energy is, at most, relatively small.

A final remark must be made regarding eqs. [5] to [9]. Although these rate expressions are intended for steady state stresses, it has been shown [7,15] that they remain valid when other types of stress are being considered, such as σ_s. Due to the difficulties involved in determining the saturation stress, in the present work, σ_s will be replaced by the peak stress σ_p. The effect of chemical composition on σ_p has already been discussed by the present authors [16]. The current study extends the latter treatment to σ_s and considers differences between the two stresses.

MATERIALS STUDIED AND RESULTS

Various results reported in the literature concerning the hot flow behaviour of low and medium carbon microalloyed steels and C-Mn-Si steels were employed (see the chemical compositions listed in Table I). Only data corresponding to tests carried out under or close to isothermal conditions (i.e. at strain rates lower than 10 s^{-1}) were used. A further condition was that the initial grain size had to be sufficiently large ($d_o > 30$ μm) to avoid strengthening by grain size refinement [12,13,16]. Furthermore, tests or studies carried out under conditions where dynamic precipitation was occurring were not considered (these usually involved relatively low temperatures or low strain rates). The precipitation of carbides, nitrides and carbonitrides of the microalloying elements strengthen the matrix so that an additional stress term must then be included in the constitutive equation (9). The latter would complicate the aim of this work, which was to determine the effect of the chemical elements in solid solution on the coefficients of the latter equation. In all the steels selected, the austenitization temperature prior to the test was sufficiently high for the microalloying elements, especially Al, Mo, Nb and V, to be put into solution. However, in the case of Ti, which has high stability in the form of TiN particles, the reheating temperature was generally not high enough to dissolve the TiN precipitates.

Applying a least squares method to eq. (7), the parameters $A^{1/5}$ and α associated with σ_p and σ_{ss} were derived for each steel. Young's modulus and the self-diffusion coefficient for γ-Fe were taken from reference [7]. The estimates of $A^{1/5}$ and α, listed in Table I, were then fitted by statistical regression methods to the chemical compositions of the corresponding steels. For this purpose, the concentrations of C, Mn and Si, and those of the microalloying elements (Al, Mo, Nb, Ti and V) were considered. The effects of P and S were not analyzed because of the low levels of these elements in most of the steels. In order to better quantify the effect of each element, and because the physical interpretations of A and α involve atomistic processes, the compositions quoted in weight percent were converted into atom fractions, and the latter values were employed in the statistical analysis.

Several types of linear and non-linear regression equations were tested. The best results were obtained with the following linear equations (expressed in atomic %):

σ_p

$$A^{1/5} = 3.4 \cdot \%Fe + 293 \cdot \%C - 88 \cdot \%Mn + 17 \cdot \%Si + 5375 \cdot \%Al - 2391 \cdot \%Mo - 1920 \cdot \%Nb - 770 \cdot \%Ti - 838 \cdot \%V$$

$$\alpha = 17.1 \cdot \%Fe - 234 \cdot \%C + 201 \cdot \%Mn - 90 \cdot \%Si - 5149 \cdot \%Al + 2755 \cdot \%Mo + 1073 \cdot \%Nb - 694 \cdot \%Ti + 1410 \cdot \%V$$

σ_{ss}

$$A^{1/5} = 0.92 \cdot \%Fe + 535 \cdot \%C + 5.2 \cdot \%Mn - 433 \cdot \%Si + 13314 \cdot \%Al - 4742 \cdot \%Mo + 1034 \cdot \%Nb - 1440 \cdot \%Ti + 222 \cdot \%V$$

$$\alpha = 43.4 \cdot \%Fe - 1442 \cdot \%C - 234 \cdot \%Mn + 1184 \cdot \%Si - 44695 \cdot \%Al + 8117 \cdot \%Mo - 23211 \cdot \%Nb + 3365 \cdot \%Ti - 383 \cdot \%V$$

The relatively good agreement between the functions fitted and the experimental results is displayed in Fig. 2. It is also apparent that there is a clear relation between $A^{1/5}$ and α, as shown in Fig. 3. The larger the $A^{1/5}$ value, the lower the α. This means that once the self-diffusion activation energy is employed, and the deformation mechanism is controlled by the glide and climb of dislocations ($n=5$), only one parameter is necessary to describe the flow behaviour of the material (A or α). Although it is widely recognized [17] that the flow behaviour can be affected by the type of test employed, i.e. compression or torsion, such a trend was not observed in the present investigation.

DISCUSSION AND CONCLUSIONS

According to some models [6,32], the constant A is a type of structure factor and is linked to the burgers vector, atomic vibration frequency, average distance between jogs, etc. By contrast, the value of $\alpha/E(T)$, which is associated with the activation volume [6] (burgers vector times area swept by the dislocation during deformation), corresponds to the inverse of the stress associated with power-law breakdown. This inverse stress is temperature dependent through the temperature dependence of Young's modulus, an effect that is usually neglected in many treatments. Because the two parameters are related through the structure, the good experimental correlation displayed in Fig. 3 is not surprising.

The following conclusions can now be derived by analysing the regression equations:

* The characteristic parameters A and α associated with σ_p attain values of 340 m^{-2} and 1710 for pure iron, respectively. In the case of σ_{ss}, these values are 92 m^{-2} and 4340.

* In general, an element that increases A decreases α and conversely. The latter generalization apparently fails with Ti in the expression for σ_p.

* The effect of the microalloying elements on A and α in the expression for σ_p is stronger than that of C, Mn and Si. This behaviour changes in the equation for σ_{ss}. Under steady state conditions, the amounts of C and Si have more effect than additions of V, while the rest of the microalloying elements also have stronger effects than the alloying elements.

* In the case of the peak stress, an increase in the amounts of C and Si, and of microalloying elements such as Al and Ti, leads to a decrease in the value of α, and therefore extends the range of power-law validity to higher stresses. In other words, the climb of dislocations remains controlling at higher strain rates. The opposite trend applies to Mn and the microalloying elements Mo, Nb and V. With regard to the steady state stress, an inverse effect (opposite sign of the coefficients) of Mn, Si, V and Nb is observed. Accordingly, an increase in the concentration of C, Mn, Al, Mo, Nb and V promotes a rate equation of the power-law type. Apparently, only additions of Si and Ti are expected to induce an exponential relation between the steady state stress and the strain rate.

It is worth noting the effect of C on A and α. Its influence is similar (i.e. of the same sign) on both expressions, with a stronger effect on the steady state conditions than on the peak stress. It must also be mentioned that the statistical expressions systematically failed when high carbon steels were included in the analysis. This observation is consistent with recent results reported by Collinson, Hodgson and co-workers [18,19]. These authors detected that increasing the C content reduces the flow stress at low values of the Zener-Hollomon parameter, but increases it dramatically under high Z conditions. The reason for these changes in the expected influence of C is unclear at this moment.

With regard to the opposite roles played by the Mn and Si additions, it must be noted that Mn increases the activity of C, while Si tends to diminish it. Furthermore, Mn is an austenite stabilizer, while Si is a ferrite stabilizer. This would explain the same signs of the Mn and C terms in the expressions for σ_{ss} and the different signs of the Si terms. However, in the σ_p equations, the C terms follow the signs of the Si terms.

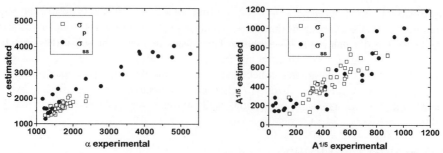

Fig. 2. Comparison between the experimental and fitted values of the parameters $A^{1/5}$ and α for σ_p and σ_{ss}.

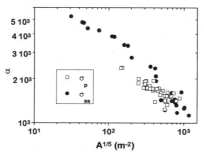

Fig. 3. Experimental relation between the structure factor A and the power-law breakdown coefficient α

It is generally accepted that the effect of the alloying and microalloying elements on retarding the onset of dynamic recrystallization is related to the atomic size difference between γ-Fe and Mn, Si, V, Mo, Ti and Nb, which increases in the order listed. However, in the present case, the above correlation does not apply to the parameters A and α. As the stresses considered here are associated with the balance between work hardening and dynamic softening (mainly recovery in σ_p and recrystallization in σ_{ss}), the lack of a clear correlation indicates that the alloying elements have different effects on hardening and softening. This can also explain why the effect of a given element changes when σ_{ss} is being considered; in this case, hardening effects are further compensated by the initiation of an additional and powerful softening mechanism (dynamic recrystallization). Modulus and electronic differences between γ-Fe and the element under consideration can also play a role here and, although they are usually neglected, according to the present results they can have a significant influence.

Finally it must be noted that the effect of Ti was not clearly discerned because, in most of the cases, it was precipitated in the form of TiN, although for analysis purposes it was considered here as being in solution. This simplification was responsible for introducing some scatter in the regression results.

ACKNOWLEDGEMENTS

The authors express their thanks for the kind co-operation of Dr. S.F. Medina in providing flow stress data from his extensive study of references [8] to [11]. Financial support was received from the Comisión Interministerial de Ciencia y Tecnología (CICYT) of Spain through the project MAT 97-0827-C02-01.

REFERENCES

[1] T. Gladman in "The Physical Metallurgy of Microalloyed Steels", The Institute of Materials, UK (1997).

[2] Y. Estrin and H. Mecking, Acta Metall., 32 (1984), p. 57.

[3] Y. Bergström, Mater. Sci. Engineering, 5 (1969-1970), p. 193.

[4] C.M. Sellars and W.J.McG. Tegart, Memoires Scientifiques de la Revue de Métallurgie, Vol. LXIII, n°9 (1966), p. 731.

[5] F. Garofalo, Trans. AIME, 227 (1963), p. 351.

[6] J.J. Jonas, C.M. Sellars and W.J.McG. Tegart, Metall. Reviews, 14 (1969), p. 1.

[7] H.J. Frost and M.F. Ashby in "Deformation-Mechanism Maps", Pergamon Press, Oxford, (1982).

[8] S.F. Medina and C.A. Hernández, Acta Metall. et Mater, 44, (1996), p. 137.

[9] S.F. Medina and C.A. Hernández, ibid: p. 149.

[10] C.A. Hernández, S.F. Medina and J. Ruiz, ibid: p. 155.

[11] S.F. Medina and C.A. Hernández, ibid: p. 164.

[12] J.M. Cabrera, J.J. Jonas and J.M. Prado, Mat. Sci. and Technology, 12 (1996), p. 579.

[13] J.M. Cabrera, A. Al Omar, J.J. Jonas and J.M. Prado, Metall. and Mater. Trans., 28A (1997), p. 2233.

[14] K.T. Park, E.J. Lavernia and F.A. Mohamed, Acta Metall. et Mater, 42 (1994), p. 667.

[15] W. Roberts, in "Deformation, Processing and Structure", Ed. G. Krauss, ASM, Ohio (1982), p. 109.

[16] J.M. Cabrera, A. Al Omar, J.J Jonas and J.M. Prado, in "Proceedings of ReX'96. The Third International Conference on Recrystallization and Related Phenomena". Ed. Terry R. McNelley, Monterey Institute of Advanced Studies, Monterey, California (1997), p. 373.

[17] I. Weiss, T. Sakai and J.J. Jonas, Metal Science, 18 (1984), p.77.

[18] J. Jaipal, C.H.J. Davies, B.P. Wynne, D.C. Collinson, A. Brownrigg and P.D. Hodgson and, Proceedings of the International Conference on Thermomechanical Processing of Steels and Other Metals, THERMEC'97, July 7-11 (1997), Wollongong, Australia. In press.

[19] D.C. Collinson, P.D. Hodgson and C.H.J. Davies, Proceedings of the Australasia-Pacific Forum on Intelligent Processing and Manufacturing of Materials, IPMM'97, July 14-17 (1997), Gold Coast, Australia. In press.

[20] J.M. Cabrera, A. Al Omar, J.J. Jonas, J.M. Prado, in "Fundamentals and Applications of Microalloying Forging Steels", Eds. C.J. Van Tyne, G. Krauss and D.K. Matlock, TMS, Warrendale (1996), p. 225.

[21] A. Laasraoui and J.J. Jonas, Metall. Trans, 22A (1991), p. 1545.

[22] B. Bacroix, These de Docteur-Ingenieur de l'INPL, Nancy, (1982).

[23] E. Doremus, J. Ondin, J.P. Bricout and Y. Ravalard, Journal of Materials Processing Technology, 26 (1991), p. 257.

[24] G.T. Velarde, C.J. Van Tyne and Y.W. Cheng in "Fundamentals and Applications of Microalloying Forging Steels", Eds. C.J. Van Tyne, G. Krauss and D.K. Matlock, TMS, Warrendale (1996), p. 209.

[25] C. Ouchi and T. Okita, Transactions ISIJ, 22 (1982), p. 543.

[26] Y. Xu, D. Hou, D. Wang, W. Xu, T. Huang, D. Lou and D. Zhong in "Advances in Hot Deformation Textures and Microstructures", (Eds. J.J. Jonas, T.R. Bieler and K.J. Bowman), TMS, Warrendale (1994), p. 183.

[27] J.G. Anderson and R.W. Evans, Ironmaking and Steelmaking, 23, (1996), p. 130.

[28] K.P. Rao, E.B. Hawbolt, H.J. McQueen and D. Baragar, Canadian Metallurgical Quarterly, 32, (1993), p. 165.

[29] C. Roucoules, Ph.D. Thesis, McGill University, Montreal (1992).

[30] C.M. Sellars, International Conference on Hot Working and Forming Processes, Sheffield (1979), p. 3.

[31] J. Sankar, D. Hawkins and H.J. McQueen, Metals Technology (1979), p. 325.

[32] A. K. Mukherjee, J. E. Bird and J. E. Dorn. Trans. of the ASM, 62 (1969), p. 155.

Correspondance with readers: cabrera@cmem.upc.es

Ref	Test	C	Mn	Si	P	S	V	Ti	Al	Nb	N	Cr	Mo	Cu	Ni	B	d_o	$A(\sigma_p)$	$\alpha(\sigma_p)$	$A(\sigma_{ss})$	$\alpha(\sigma_{ss})$
12	C	0,34	1,52	0,72	0,025	0,025	0,083	0,018	0,0145	0	0,0114	0	0	0	0	0	90	603	1338	813	1168
20	C	0,29	1,19	0,19	0,012	0,025	0,09	0,002	0,0114	0	0,0131	0	0	0	0	0	120	573	1481	798	1318
21	C	0,03	1,54	0,19	0,008	0,005	0	0,02	0,02	0,055	0,0048	0	0	0	0	0	0	396	1936	691	1592
21	C	0,026	1,42	0,16	0,007	0,007	0	0,02	0,02	0,058	0,0063	0	0	0	0	0,003	0	319	1709	497	1430
21	C	0,026	1,38	0,18	0,007	0,006	0	0,017	0,019	0	0,006	0	0	2,03	0	0	0	413	1388	560	1217
22	C	0,06	1,43	0,24	0,006	0,012	0	0	0,025	0	0,006	0,04	0	0	0	0	110	663	1503	-	-
23	C	0,15	0,45	0,35	0,04	0,05	0	0	0,007	0	0	0	0	0	0	0	0	570	1386	429	1928
24	C	0,22	1,53	0,52	0,011	0,021	0,13	0,01	0,007	0	0,0122	0,13	0,23	0,262	0,11	0	0	300	1905	-	-
24	C	0,24	1,52	0,54	0,012	0,016	0,12	0	0,01	0,087	0,0097	0,12	0,23	0,25	0,11	0	0	152	2376	-	-
24	C	0,23	1,54	0,53	0,013	0,019	0	0	0,008	0	0,009	0,12	0,23	0,25	0,11	0	0	146	2355	-	-
24	C	0,25	1,54	0,55	0,01	0,02	0,13	0	0,012	0	0,0102	0,12	0,24	0,24	0,11	0	0	249	1861	-	-
25	C	0,1	1,5	0,22	0,015	0,01	0	0	0,023	0	0,0069	0	0	0	0	0	360	728	1594	-	-
25	C	0,09	1,54	0,22	0,016	0,01	0	0	0,027	0,032	0,0069	0	0	0	0	0	270	607	1612	-	-
25	C	0,09	1,51	0,27	0,015	0,01	0	0	0,034	0,072	0,0051	0	0	0	0	0	230	528	1658	-	-
25	C	0,09	1,55	0,26	0,015	0,008	0	0	0,029	0,124	0,0061	0	0	0	0	0	195	500	1603	-	-
26	C	0,157	1,4	0,45	0,019	0,016	0	0	0	0	0	0	0	0	0	0	556	1527	1003	1248	
27	C	0,044	0,23	0,002	0,009	0,017	0	0,001	0,031	0	0,002	0,04	0,016	0,01	0	0	680	1514	691	1407	
27	C	0,122	1,27	0,03	0,002	0,001	0	0,01	0,005	0	0,0077	0,02	0,01	0,011	0	0	467	1719	-	-	
28	C	0,34	0,69	0,013	0,005	0,01	0	0	0,031	0	0	0,023	0,002	0,019	0,008	0	0	878	1469	1172	1122
18	T	0,15	1	0,14	0,013	0,009	0	0	0	0	0	0,016	0	0,019	0,022	0	0	593	1523	929	1620
28	T	0,34	0,69	0,013	0,005	0,01	0	0	0,031	0	0	1,023	0	0,019	0,008	0	0	878	1469	424	2081
29	T	0,063	1,2	0,22	0,011	0,008	0	0,029	0,029	0	0,0027	0	0,18	0	0	0	50	390	1656	693	1600
29	T	0,061	1,2	0,2	0,009	0,008	0	0,032	0,032	0,039	0,0036	0	0	0	0	0	50	561	1419	765	1374
29	T	0,055	1,3	0,24	0,009	0,008	0	0,16	0,036	0	0,0036	0	0	0	0	0	50	600	1323	1020	1270
30	T	0,42	0,94	0,25	0	0,03	0,02	0,02	0	0,02	0,008	0,88	0	0	0,16	0	110	793	1409	46	4753
8	T	0,15	0,74	0,21	0	0	0	0	0	0	0	0	0	0	0	0	187	407,4	1689,3	44	4802
8	T	0,11	0,55	0,26	0	0	0	0	0	0	0	0	0,26	0	0	0	143	340	1774	-	-
8	T	0,11	0,47	1,65	0	0	0	0	0	0	0	0	0,38	0	0	0	104	559,3	1233,6	-	-
8	T	0,11	0,68	0,26	0	0	0	0	0	0	0	0	0,18	0	0	0	212	360,8	1723	-	-
8	T	0,11	1,55	0,25	0	0	0	0	0	0	0	0	0	0	0	0	192	349	1790	110	3867
8	T	0,44	0,79	0,24	0	0	0	0,055	0	0	0,01	0	0	0	0	0	180	453,5	1659,7	160	3377
8	T	0,44	0,79	0,23	0	0	0	0,075	0	0	0,0102	0	0	0	0	0	205	373,8	1708,8	-	-
8	T	0,42	0,79	0,27	0	0	0	0	0	0	0	0	0,18	0	0	0	193	437,4	1704,2	415	2353
8	T	0,15	1,25	0,27	0	0	0	0,021	0	0	0,0105	0	0	0	0	0	95	385,5	1708,3	74	4227
8	T	0,15	1,1	0,26	0	0	0	0	0	0	0,0105	0	0	0	0	0	90	325	1732,5	-	-
8	T	0,15	1,12	0,24	0	0	0,043	0	0	0	0,0123	0	0	0	0	0	29	354,6	1712,5	52	4377
8	T	0,11	1,1	0,24	0	0	0,06	0	0	0	0,0144	0	0	0	0	0	172	316,4	1927,8	118	3843
8	T	0,12	1,1	0,24	0	0	0,093	0	0	0	0,0112	0	0	0	0	0	167	309	1917,6	31	5258
8	T	0,11	1	0,24	0	0	0	0	0	0,041	0,0119	0	0	0	0	0	165	251,4	1997,3	358	2412
8	T	0,11	1,23	0,24	0	0	0	0	0	0,093	0	0	0	0	0	0	122	303,3	1833,8	180	3337
8	T	0,11	1,32	0,24	0	0	0	0	0	0	0	0	0	0	0	0	116	292,4	1733,8	202	2775
31	T	0,12	0,94	0,007	0,003	0,016	0,007		0,005	0,05	0,0045	0,04	0,007	0,09	0,04	0		485	1549	761	1445

Table I. Chemical compositions of the steels studied. The test method is indicated in the column "Test" (C: compression, T: torsion). The initial grain size d_o, as well as the experimental values of the parameters $A^{(1/5)}$ (m^{-2}) and α for each stress considered are also indicated.

Materials Science Forum Vols. 284-286 (1998) pp. 135-142
© 1998 Trans Tech Publications, Switzerland

Effect of Coarse γ Grain Size on the Dynamic and Static Recrystallisation during Hot Working in Microalloyed Nb and Nb-Ti Steels

A.I. Fernández, R. Abad, B. López and J.M. Rodríguez-Ibabe

CEIT and ESII, Univ of Navarra,
P° Manuel de Lardizábal 15, E-20009 San Sebastián, Basque Country, Spain

Keywords: Hot Direct Rolling, Dynamic Recrystallisation, Static Recrystallisation, Nb Steels, Nb-Ti Steels

ABSTRACT

The effect of coarse austenite grain size on the dynamic and static recrystallisation kinetics of two microalloyed Nb and Nb-Ti steels has been investigated in the present work. To characterize the dynamic recrystallisation behaviour of the austenite, continuous-torsion tests were carried out after the reheating of the specimen at different temperatures in the range 1000-1420°C. It has been observed that the occurrence of dynamic recrystallisation is dependent on the initial grain size and the deformation conditions (temperature and strain-rate). Decreasing values of the Zener-Hollomon parameter (Z) and grain size promotes dynamic recrystallisation. However for the coarser grain sizes no peaks appear on the flow curves above a determined value of Z. This value seems to decrease with increasing the grain size. An equation to predict the ε_p peak strain for a wide range of grain sizes has been obtained for both steels. The effect of strain on the static recrystallisation of the austenite, having a large grain size, has been also studied. Interrupted-torsion tests were performed to determined the fractional softening. A quadratic dependence of $t_{0.5}$ on strain has been observed, denoting a less dependence of recrystallisation on strain than proposed previously by other authors in the range of lower grain sizes.

INTRODUCTION

In recent years, the trend in steel processing technology is to integrate the rolling process with the continuous casting process, which has attendant economic advantages [1-4]. This allows a consistent reduction of the energy consumption compared with the conventional rolling processes (CCR). Two alternatives are possible: to roll the casting directly as its temperature falls to a suitable rolling temperature, or to charge it into a high temperature soaking furnace when its temperature has dropped to some lower level. These processes are known as hot direct rolling (HDR) and hot charge rolling (HCR) respectively.

However, the as-cast austenite microstructure at the start of HDR differs in several ways from the reheated austenite at the start of CCR. The as-cast austenite is characterised by a very large austenite grain size [1]. In contrast the as-reheated austenite is significantly refined by the γ/α and α/γ transformations on cooling and reheating respectively which also result in an homogenisation of the composition. It is not directly known if the interactions between deformation, precipitation and recrystallisation in as-cast austenite during HDR are different from those in reheated austenite during CCR. However, it has been found that the final microstructure and the mechanical properties of the microalloyed steels processed by hot charge rolling and hot direct rolling can differ significantly from those obtained by the conventional cold charge rolling [3, 4]. The different thermal conditions can influence such microstructural changes as dynamic and static recrystallisation, and also precipitation characteristics during hot working.

In the present work the effects of HDR/HCR processing on the dynamic and static recrystallisation during hot working of two microalloyed steels have been investigated and compared with the case of CCR processing. Hot torsion tests, applying a very high soak temperature of 1400°C-1420°C, have been used to simulate the hot direct or charge rolling. Lower reheating temperatures have been used in the case of simulations of the conventional CCR process.

EXPERIMENTAL

Two low carbon steels have been used in the present work: a Nb microalloyed steel and a Nb-Ti microalloyed steel whose base compositions are listed in Table 1. To analyse the influence of the austenite grain size on the dynamic recrystallisation behaviour, continuous isothermal torsion tests were performed after soaking the specimen at different temperatures in the range 1000°C-1420°C for 15 min. After reheating and cooling ($\approx$1°C/s) to the deformation temperature (5 min. for stabilisation) the specimens were deformed at temperatures of 1000, 1100 and 1200°C and strain rates in the range 0.02 -5s^{-1}.

Table 1: Chemical composition of the steels (wt%)

Steel	C	Mn	Si	S	P	Nb	Ti	N
Nb-Steel	0.1	1.42	0.31	0.008	0.018	0.035	-	0.0053
Nb-Ti Steel	0.07	0.62	0.012	0.006	0.011	0.034	0.067	0.0043

The effect of hot direct rolling (large austenite grain size) on the static recrystallisation behaviour has been investigated in the case of the Nb microalloyed steel. Two-pass torsion tests were carried out under isothermal conditions to examine the softening behaviour of the austenite between intervals of hot working. After reheating at the high soak temperature of 1400°C for 15 minutes and cooling ($\approx$1°C/s) to the deformation temperature (1100°C), a given specimen is prestrained at a constant strain rate, unloaded, and held for increasing times. After the interruption, the specimen is reloaded at the same strain rate and temperature. The fractional softening has been determined by the 2% offset method [5]. Strain per pass between ε=0.1 and 0.5 and a constant strain rate of $\dot{\varepsilon}$ =1 s^{-1} have been used in these tests.

RESULTS AND DISCUSSION

Figure 1 shows the evolution of the austenite grain size with the reheating temperature for the steels analysed in the present work. It can be observed that in the case of the Nb microalloyed steel grain coarsening occurs at lower temperatures than for the Nb-Ti steel. The different behaviour observed between the Nb-steel and the Nb-Ti steel can be related with the nature of the particles present in each steel. In the Nb-steel the dissolution of the particles which can pin the grain boundaries and prevent grain coarsening (Nb(C,N)) occurs at lower temperatures than in the Nb-Ti microalloyed steel. In the Ti containing steel it is obvious that very stable TiN particles can withstand dissolution at higher temperatures than Nb(C,N) so that a fine grain austenite microstructure prevails at higher temperatures.

Figure 2(a) shows the stress-strain curves corresponding to the Nb steel, preheated at 1400°C (Do=806 μm) and deformed at different temperatures (1000-1200°C), and strain-rates (0.1-5s^{-1}). From the figure it can be observed that for the higher deformation temperatures of 1100°C and

1200°C and all the strain-rates used, the curves exhibit a peak flow stress followed by a work-softening and a steady-state region, denoting that dynamic recrystallisation is taking place. However for the lowest temperature of 1000°C peak occurrence is not observed. This behaviour is also dependent on the initial austenite grain size, as can be observed in Fig.2(b), where the effect of initial grain size on the flow curve is shown for the case of a deformation temperature of 1000°C and $\dot{\varepsilon}=1$ s^{-1}. From this figure it is clearly evident that for the same deformation conditions, the peak occurrence is more tangible for the lower grain sizes. These results are in accordance with that observed by other authors, which have demostrate the existence of a limiting value of the Zener-Hollomon parameter, Z_{lim}, above which a peak stress no longer appears in the flow curves. They have also found that this limiting value tend to decrease with increasing the initial austenite grain size [6, 7]. The previous results denote that in industrial operations involving coarse austenite grain sizes, as for example "thin slab" or "hot direct rolling", dynamic recrystallisation will be more difficult to occur during the first passes than in the case of conventional controlled rolling processes.

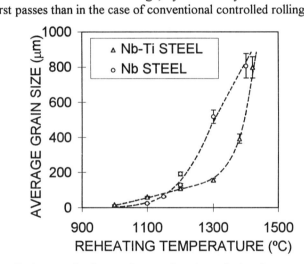

Fig.1 Grain growth of austenite as a function of reheating temperature.

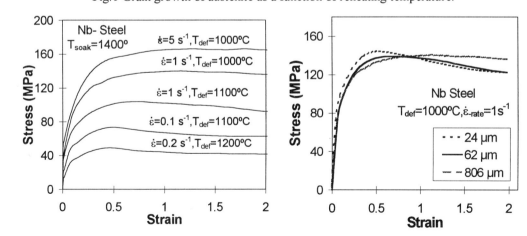

Fig. 2 :(a) Effect of temperature and strain-rate, (b) effect of initial austenite grain size, on the dynamic recrystallisation flow curves in the Nb steel.

The ε_p peak strains worked out from the flow curves are plotted in Fig.3 (on a logarithmic scale) as a function of the Zener-Hollomon parameter (Z), where Z is calculated as $Z = \dot\varepsilon \exp(Q_{def}/RT)$, using the corresponding values of the activation energy, Q_{def}=341 kJ/mol and 311kJ/mol determined for the Nb and the Nb-Ti steel respectively [8]. From the figure it can be observed that the initial grain size is an important variable for the onset of dynamic recrystallisation. Usually the peak strain is related to the initial grain size (D_o) and the Zener-Hollomon parameter (Z) by an equation of the type [9]:

$$\varepsilon_p = BD_o^m Z^p \tag{1}$$

where the coeficient B and the exponents m and p are dependent on the material.

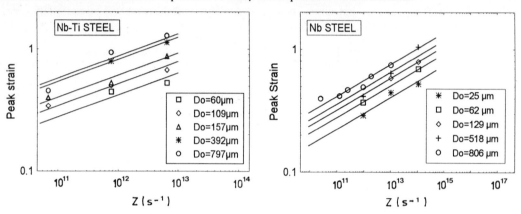

Fig.3 Dependence of peak strain ε_p on Zener-Hollomon parameter Z:(a) Nb-Ti steel; (b) Nb steel

The relationship between ε_p and the initial grain size D_o can be found by plotting the corresponding pairs of values (D_o, ε_p) obtained from the plots shown in Fig.3, for a given level of the Z, as shown in Fig.4. From the figure it is observed that the points can be fitted by straight lines giving the following equations:

$$\varepsilon_p \propto D_o^{0.16} \qquad \text{for the Nb steel} \tag{2}$$

$$\varepsilon_p \propto D_o^{0.28} \qquad \text{for the Nb-Ti steel} \tag{3}$$

The exponents in equations (2) and (3) are lower than the values of 0.3 and 0.5, suggested by Sellars for plain carbon and Nb microalloyed steels [9, 10]. However, the present results are in good agreement with other studies where a lower dependence of ε_p on grain size has been observed for plain carbon steels [6, 11, 12]. A subsequent logarithmic representation of $\varepsilon_p D_o^{-m}$ against Z yields the plots of Fig.5, using the corresponding values of the exponent m (Fig.4) for the Nb and the Nb-Ti steel. Fitting straight lines through the data leads to the relationships:

$$\varepsilon_p D_o^{-0.16} = 5.18 \times 10^{-3} Z^{0.13} \qquad \text{for the Nb steel} \tag{4}$$

$$\varepsilon_p D_o^{-0.28} = 9.8 \times 10^{-4} Z^{0.18} \qquad \text{for the Nb-Ti steel} \tag{5}$$

The values of the exponent p=0.13 and 0.18 are within the range 0.12 to 0.23 reported for steels [9-14].

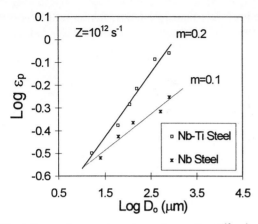
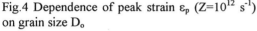

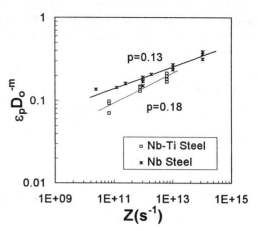

Fig.4 Dependence of peak strain ε_p ($Z=10^{12}$ s^{-1}) on grain size D_o

Fig.5 Dependence of peak strain compensated for grain size on Zener-Hollomon parameter.

The influence of having a large austenite grain size on the static recrystallisation characteristics following hot torsion has been analysed on the Nb-steel. The initial austenite grain size ($D_o=806\mu m$), the deformation temperature and strain rate were held constant for all the tests, 1100°C and 1s^{-1} respectively. The effect of pass strain on the recrystallisation behaviour was investigated in this case, using pass-strains in the range 0.1-0.5. Figure 6 shows the variation of recrystallised fraction as a function of time. In this figure the 2%-total strain offset method was used to determine the fractional softening, which excludes the effect of recovery better than the 0.2% offset [5].

For each strain the recrystallised fraction increases with time following closely an Avrami type equation with an exponent value $n=1$. However, it can be observed that a lower value of n would give a better fit to the data for a strain of 0.1. The values of the Avrami exponent n have been reported to be between 1 and 2, but it has been observed that n tends to decrease from 2 to 1 when the grain size is increased from 140 to 530μm [15]. Otherwise values of n close to 1 have been also reported for lower grain sizes [16, 17]. Sellars suggested that such variations in n could be due to differences in the grain size distribution [18].

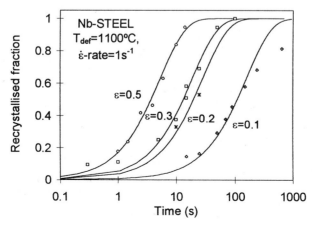

Fig.6 Effect of strain on the static recrystallisation curves for an initial grain size of 806 μm

The times for 50% recrystallised fraction ($t_{0.5}$) have been determined from the fraction *vs* time curves shown in Figure 6 and are plotted in Fig.7 as a function of the strain. The results in Fig.7 indicate a dependence of $t_{0.5}$ on strain:

$$t_{0.5} \propto \varepsilon^{-2} \tag{6}$$

Similar dependence has been also observed by other authors for Nb microalloyed steels [17, 19].

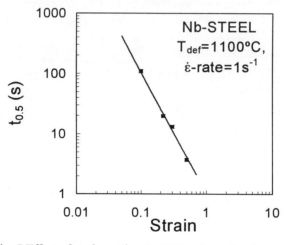

Fig. 7 Effect of strain on time to 50% volume fraction recrystallised

Dutta and Sellars proposed the following equation to describe the recrystallisation kinetics (time for 5% recrystallisation) for Nb microalloyed steels including the effect of the amount of Nb present in solution in the austenite before deformation is applied [20]:

$$t_{0.05X} = CD_o^a \varepsilon^{-b} \exp\left(\frac{300000}{RT}\right) \exp\left[\left(\frac{275000}{T} - 185\right)[Nb]\right] \tag{7}$$

where D_o is the initial grain size, ε is the true strain, $[Nb]$ is Nb concentration in solution in wt%, Z is the Zener-Hollomon parameter, T the temperature in Kelvin, and R is the gas constant. Dutta and Sellars found the following values for the constants: $C=6.75 \times 10^{-20}$, $a=2$, $b=4$.

Priestner et al. modified equation (7) to extend the application of this equation to the large grain size range used in their work [1, 21]. These authors found that the quadratic dependence of $t_{0.05X}$ on D_o ($a=2$) was inappropiate to the large grain size range applied in their work and that a linear dependence was more appropiate ($a=1$), together with a change in the pre-exponential constant C to 1.231×10^{-17}. Equation (7) has been again modified, taking into account the quadratic dependence of $t_{0.5}$ on ε observed in the present work, i.e. an exponent $b=2$ is considered instead of $b=4$, and the modified dependence of $t_{0.5}$ on D_o reported by Priestner et al [1, 21]. From the above considerations equation (7) can be rewritten for the present Nb steel as:

$$t_{0.5X} = 13.58 \times 2.21 \times 10^{-16} D_o \varepsilon^{-2} \exp\left(\frac{300000}{RT}\right) \exp\left[\left(\frac{275000}{T} - 185\right)[Nb]\right] \tag{8}$$

where the ratio $t_{0.5}/t_{0.05}$ was taken as 13.58 , which was calculated from the Avrami equation using the exponent $n=1$.

Some data on recrystallisation for Nb microalloyed austenite have been taken from the literature [17, 19, 22] and compared with the values calculated using the equation (8) in Fig.8. The reported experimental values correspond to austenite compositions ranged from 0.01-0.06 wt% Nb, 0.08-0.18 wt%C, 0.004-0.0112 wt%N and initial grain sizes between 25 and 122 μm. From Fig.8 it is seen that the predictions of equation (8), which has been developed for a large austenite grain size (Do=806 μm), are in reasonable agreement with the data for the times of 50% recrystallised fraction in the range of lower grain sizes.

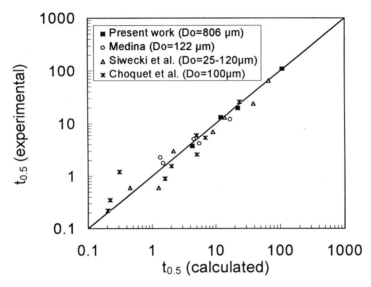

Fig.8 Relationship between model (equation (8)) and experimental $t_{0.5X}$.

CONCLUSIONS

-During hot deformation, dynamic recrystallisation, as indicated by a peak in the flow curve, takes place on the Nb and Nb-Ti steels analysed in the present work. Decreasing values of the Zener-Hollomon parameter (Z) and grain size promotes dynamic recrystallisation and therefore decreases the value of the peak strain (ε_p). However for the coarser grain sizes no peaks appear on the flow curves above a determined value of Z. A limiting value of the Zener-Hollomon parameter, above which dynamic recrystallisation does not occur, can be defined. This limiting value tends to decrease with increasing grain size. Some implications can derive from these results for industrial operations involving coarse austenite grain sizes, as it is the case of "thin slab casting" or "hot direct rolling". In these processes dynamic recrystallisation can be difficult to occur during the first passes if the value of the Zener-Hollomon parameter is greater than the limiting value.

-In the Nb steel used, the austenite having a large initial grain size exhibits a quadratic dependence of $t_{0.5}$ on strain. This suggests that over the range of grain size obtained here (≈800 μm), recrystallisation kinetics may be less powerfully dependant on strain than proposed by Dutta and Sellars [20].

ACKNOWLEDGEMENTS

Part of this research work has been performed under an ECSC Project. The authors thank the European Coal and Steel Community for partial funding of the research, the project leaders at the partner institutes and their colleagues. The authors thank also J.I. Astiazaran, for the help in performing the experimental tests. The authors wish to dedicate this article to the memory of Prof. J.J.Urcola, who was directly involved in this project before his untimely death.

REFERENCES

[1] C.Zhou, A.A. Khan, R.Priestner, 1st Int.Conf. of Modelling of Metal Rolling Processes, (1993), p. 212-218.

[2] R.Priestner and C. Zhou, Ironmaking and Steelmaking, vol.22, n°4, (1995), p. 326-332.

[3] R.K.Gibbs, R.Peterson and B.A. Parker, Proc. Int. Conf. on Processing, Microstructure and Properties of Microalloyed and Other Modern High Strength Low Alloy Steels, Pittsburgh, PA, ISS-AIME, (1992), p. 201-207.

[4] V.Leroy and J.C. Herman, Microalloying '95, M. Korchynsky, A.J. DeArdo, P. Repas, G. Tither eds, ISS, Pittsburgh, (1995), p. 213-223.

[5] G. Li et al., ISIJ Intern., 36 (12), (1996), p. 1479-1485.

[6] W.P. Sun and E.B. Hawbolt, ISIJ Int., vol. 37, (1997), no. 10, p. 1000-1009.

[7] C.A. Muojekwu, D.Q. Jin, V.H.Hernandez, I.V. Samarasekera and J.K. Brimacombe, 38TH Conf. Proc., ISS, Vol. XXXIV, (1997), pp.351-366.

[8] A. Fernández, Ph. D. Thesis in course.

[9] C.M.Sellars, Hot Working and Forming Processes, ed. C.M. Sellars and G.J. Davies, Metals Soc., London, (1980),p. 3-15.

[10] C. M. Sellars, Mater. Sci. Technol. 6, (1990), p. 1072.

[11]E. Anelli, ISIJ Int., vol. 32, (1992), no. 3, p. 440-449.

[12] S. F. Medina and C.A. Hernandez, Acta Metall. vol. 44, (1996), p. 149-154.

[13] L.N. Pussegoda, P.D. Hodgson and J.J. Jonas, Mater. Sci. Technol. 8, (1992), p. 63.

[14] C.Roucoules, S.Yue and J.J. Jonas, 1st Int.Conf. of Modelling of Metal Rolling Processes, (1993), p. 165-179.

[15] D.R.Barraclough and C.M. Sellars, Met. Sci., (1979),p. 257-267.

[16] A. Laasroui and J.J. Jonas, Metall. Trans., vol.22A, (1991), p. 151-60.

[17] S. F. Medina, Scripta Metallurgica et Materialia, vol.32, No.1, (1995), p. 43-48.

[18] C.M. Sellars, Proc. 7th Symp.on Metallurgy and Materials Science, Risø, N.hansen, D.Juul Jensen, T.Leffers and B. Ralph, eds., (1986), p. 167-87.

[19] T. Siwecki, S. Zajac and G. Engberg, Proc. 37TH MWSP Conf., 23, ISS, (1996), p. 721-734.

[20] B.Dutta and C.M.Sellars, Mat.Sci. and Technology, vol.3, (1987), 197-206.

[21] P.Patel, C.Zhou and R. Priestner, Proc.Third Int. Conf. on Recrystallisation and Related Phenomena, Rex 96, Terry R.McNelley ed., (1997), p. 421-428.

[22] P. Choquet, A. Le Bon and Ch. Perdrix, Proc.Conf. "Strength of metals and alloys", H.J.McQueen et al. eds., Oxfford Pergamon Press, (1985), p. 1025-1030.

B.López
E-mail: blopez@ceit.es
Fax: (34)43 21 30 76

Materials Science Forum Vols. 284-286 (1998) pp. 143-150

The Influence of Roughing Strain and Temperature on Precipitation in Niobium Microalloyed Steels after a Finishing Deformation at 900°C

E.S. Siradj[1], C.M. Sellars[2] and J.A. Whiteman[2]

[1] Department of Metallurgy, Engineering Faculty, University of Indonesia, Kampus Baru, Depok, Indonesia

[2] Department of Engineering Materials, The University of Sheffield, Mappin Street, Sheffield S1 3JD, UK

Keywords: Microalloyed Steels, Stress Relaxation, Strain Induced Precipitation, Nb Carbonitride, Roughing, Strain and Temperature Effects

Abstract

This work demonstrates that the technique of stress relaxation in plane strain compression can be used to investigate the effect of process variables on precipitation of micro alloy carbides in austenite. In particular it has been used to study the influence of roughing deformation on precipitation after working at 900C.

It has been demonstrated that increase in roughing strain decreases the time to the onset of precipitation after a constant strain at 900C in a range of steels with different niobium and carbon levels, but that beyond a certain value of roughing strain the time for the onset of precipitation remains constant. The flow stress in the austenite at 900C is increased with increase in roughing strain and decrease in roughing temperature.

The stress relaxation technique has allowed a wider range of roughing conditions to be examined as it uses only one specimen to obtian information that would have required at least six specimens with the techniques previously used for measuring the kinetics of strain induced precipitation.

Introduction

The kinetics of strain induced precipitation in low carbon steels with microalloying additions of Nb has been widely investigated as it is the basis of controlled rolling operations. This is due to the influence of Nb on the retardation of the kinetics of the recrystallisation of austenite in such steels. The influence of the roughing deformation on this reaction has had limited investigation. Dutta [1] showed as a result of rolling an X70 steel at successively lower temperatures, that the maximum stress measured in the third pass depended critically on the holding time at 900C after the second deformation, which was given at about 950C. The most comprehensive work was done by Valdes and Sellars [2] who used deformation at three successively lower temperatures in rolling and plane strain compression on several niobium containing steels to study the effect of strain induced precipitation after the second deformation on the maximum strength found in the third deformation. The main findings were that the time to peak strength (measured as an equivalent time at 900C) decreased as the roughing strain increased up to a limiting value, which was higher as the roughing temperature decreased. The limiting time to peak strength decreased as the roughing temperature was decreased, Figure 1.

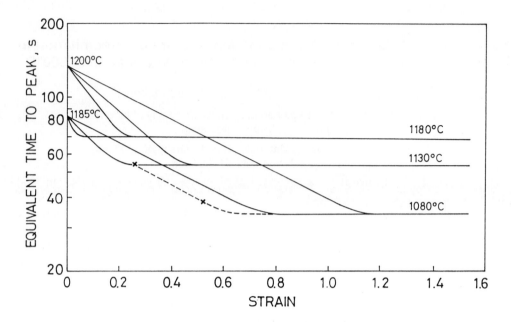

Figure 1   Summary of the influence of roughing/reheating temperature and strain in roughing
                  deformation on equivalent isothermal holding time at 900°C, after 15% reduction at
                  950/956°C, to attain peak strengthening in final finishing deformation [2]

A more limited study by Choi [3] was undertaken using stress relaxation experiments in plane
strain compression, this generally confirmed the earlier work, and showed the potential of the
technique. It did however lead to shorter times for the onset of precipitation than those obtained by
the peak stress measurements of Valdes and Sellars[2]. Dutta, Valdes and Sellars [4] developed a
model of strain induced precipitation of niobium carbide and its effect on the flow stress of niobium
containing steels. The model assumed that precipitation took place on dislocation nodes and that
bulk diffusion of niobium controlled the growth and pipe diffusion controlled the coarsening of
these precipitates. This model was based on observations of particle size distibutions.

Experimental Techniques
Stress relaxation experiments were performed on three carbon niobium steels, two of the steels had
compositions close to the stoichiometric line for NbC whilst the third had an excess of carbon, see
Table 1.

Steel	C	Si	Mn	P	S	Al	N	Nb
1	0.012	0.45	1.30	0.005	0.003	0.020	0.0024	0.10
2	0.013	0.47	1.21	0.005	0.004	0.012	0.0032	0.09
3	0.10	0.37	1.35	0.020	0.019	0.041	0.0042	0.03

Table 1   Compositions of steels used in current investigation. Wt %

Steels 1 and 2, which were vacuum melted into 75mm diameter ingots in the Department of Engineering Materials, were extruded into bar. Steel 3 was conventional hot rolled plate. Plane strain compression specimens of 60mm by 30mm by 10mm were obtained from all the steels.

To limit scaling and decarburisation the specimens were plated with a layer of chromium about 15 microns thick. To take all the niobium carbonitride into solution the specimens were austenitised at temperatures of 1150C for steels 1 and 2 and 1200C for steel 3. These temperatures were calculated to be sufficient using the Irvine et al [5] solubility relationship.

Specimens were held at temperature for 15 minutes before air-cooling to a roughing temperature between 1150 and 1000C. For Steel 1 after a roughing strain of 0.1,0.3,0.5 or 0.7 the specimen was furnace cooled to a test furnace temperature of 900C (this took some 300 seconds). Some specimens of Steel 2 were treated in the same way but others were air cooled after roughing and held for 60 seconds in the test furnace before the finishing deformation. A finishing strain of 0.5 was given and the displacement was then fixed to follow stress relaxation for 400 seconds before quenching. The strain rate for all these deformations was 1 sec^{-1}.

Steel 3 had a more complex series of strains which included two roughing strains and a range of finishing strains at three strain rates between 10sec^{-1} and 1 sec^{-1} all specimens were air cooled between roughing and finishing.

The temperature of each specimen was measured by a thermocouple inserted into the centre of the specimen half way along the specimen length. All the strains, strain rates and temperatures were logged by use of a data logger and processed by the computer used to control the plane strain compression machine.

The effects of temperature and reduction on the recrystallised and unrecrystallised austenite grain size were studied on the deformed specimens after quenching. To reveal the austenite boundaries aqueous picric acid containing a small amount of HCl and teepol as a wetting agent was used. For Steel 3 cupric chloride was added to this etchant. Grain size measurements were made using the linear intercept method, at least 200 grains were counted in both longitudinal and transverse directions, this gave an error of about +/- 7%.

Results

The austenite grain size in the three steels investigated increased with increase in austenitising temperature. The grain size in steel 3 was smaller than that in the other steels and grain growth was retarded to higher temperatures. The data obtained are similar in characterstics to those obtained by previous workers [3,7], see Fig 2. The roughing deformation produces fully recrystallised austenite grains in all three steels. The grain size is largest (~ 65μm) in the specimens that have been furnace cooled after roughing. There is only a slight influence of roughing temperature on the observed grain size after air cooling (35 - 45 μm). Small roughing strains also result in an increase in the austenite grain size. Siradj [6] calculated the grain sizes expected after the roughing deformation and cooling to the finishing temperature and showed that recrystallisation followed by grain growth occurs. The predicted grain sizes were close to those found experimentally.

The stress relaxation technique relies on the fact that the reduction in stress with log time is linear in the regime controlled by recovery and the recovery process is retarded when precipitation starts leading to a reduction in the slope of the curve. The time for initiation of precipitation is taken as the point where the two curves of different slope intersect. An example of the type of stress relaxation curve obtained is shown in Figure 3.

The effect of different roughing strains before a final deformation to a strain of 0.45 at 900C and a strain rate of 1sec-1 is shown for steel 1 in Figure 4.

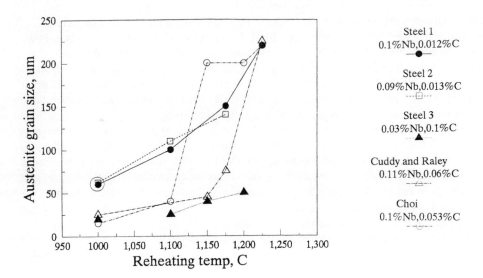

Figure 2   The austenite grain growth behaviour of the experimental steels compared with results obtained by other investigators.

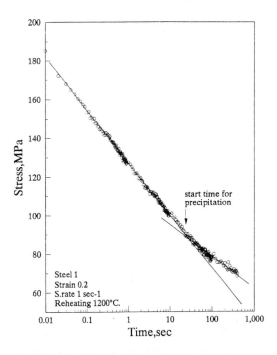

Figure 3 An example of a stress relaxation curve at 900°C showing how the precipitation start time is determined

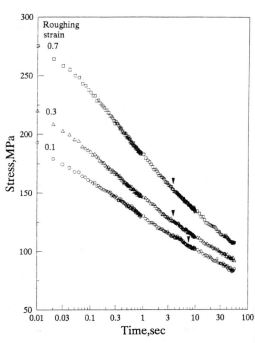

Fig 4 Stress relaxation curves Steel 1 showing the effect of roughing strain at 1140°C after a finishing strain of 0.45 at 900°C.

It is clear that the greater the roughing strain, up to 0.3 the shorter is the time for the onset of precipitation, then the time is constant. Similar results were obtained for all three steels after roughing deformation at several different temperaturesThe influence of roughing strain on the stress strain curves on finishing at 900C for steel 2 is shown in Figure 5. Increase in roughing strain increases the flow stress at all strains in finishing, although the increase at low values of roughing strain is limited and the curves become closer at higher roughing strains of 0.3 and 0.47. The grain sizes after roughing are large and cannot contribute significantly to this increased flow stress. The results thus indicate that roughing has some effect additional to that of reducing grain size

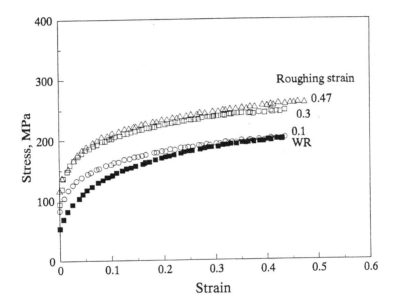

Figure 5   Stress strain curves showing the effect of roughing strain at 1025°C on finishing deformation at 900°C, for Steel 2.

In order to compare the results of the present work with previous work the time to the reduction in the stress relaxation rate, assumed to be the onset of precipitation is plotted as a function of roughing strain for steels 1 and 2 in Figures 6a and 6b.The roughing strain has been calculated from the reduction in thickness i.e. the nominal equivalent strain in Fig 6a, a correction for slip line field effects from the work of Colas and Sellars [8] has been made in Fig 6b. These curves show that the time for the onset of precipitation is reduced as the roughing deformation is increased up to a limiting value. This value is between 0.3 and 0.4 strain for most of the conditions investigated. The correction for slip line field strain gives slightly higher limiting strains and shows that the limiting strain increases with decrease in roughing temperature. Higher strains do not produce any further reduction in the time for the onset of precipitation. The absolute value of the minimum time is dependent on the roughing temperature with lower roughing temperatures leading to shorter limiting times for onset of precipitation.

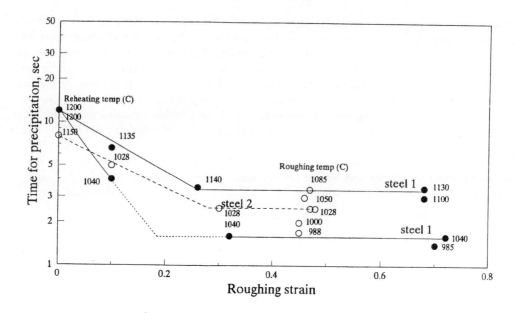

Figure 6a   The effect of roughing strain and roughing temperature on precipitation kinetics of Steels 1 and 2 (nominal equivalent strain)

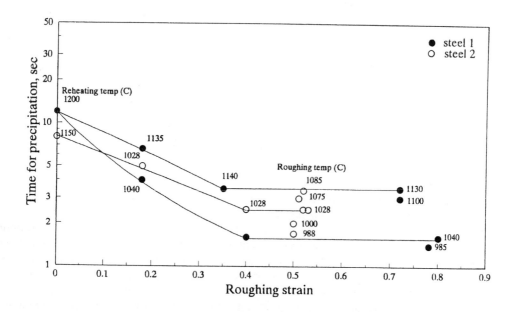

Figure 6b   The effect of roughing strain and roughing temperature on precipitation kinetics of Steels 1 and 2 (strain corrected to that in the active slip line field)

Discussion

The aim of this work was to determine whether the technique of stress relaxation is a viable technique for the investigation of strain induced precipitation of niobium carbonitrides with consequent effects on the flow stress and recrystallisation of austenite during further deformation. This can first be assessed by comparing the current results with those of previous investigations [2]. Comparison of figures 1 and 6 shows that the same characteristics are found in both investigations. Increase in roughing strain reduces the time for the attainment of peak stress in the previous work and the onset of precipitation in the current work. In both investigations there is a limit to the effect of the roughing strain in that increase in strain does not further reduce the time beyond a limiting value. The differences between the two investigations relate to the limiting times. The time to peak stress as measured by Valdes and Sellars [2] is about 5 times longer than that for the onset of precipitation as measured in the present work. The temperature dependence of the two investigations is however the same, as is shown in Figure 7.

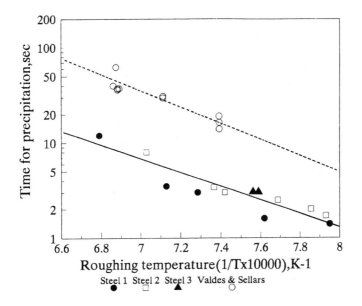

Figure 7   The effect of roughing temperature and time for the start of precipitation compared with results of Valdes and Sellars [2]

The strains at which the minimum time for precipitation or peak stress is found is also different in each investigation. The limit being found at larger strains in the previous work. In part this is due to the inhomogeneity of deformation in the plane strain compression test. In plane strain compression the deformation is concentrated in slip line fields and as the strain increases the number of such fields and their position in the specimen changes. Strains are thus localised when there is a low overall strain as has been reported in several previous investigations and is a characteristic of this investigation. This localised strain in plane strain compression has been characterised by Colas and Sellars [8] and their data have been used to obtain corrected strains. It is also likely that the maximum stress as measured by Valdes and Sellars was associated with a larger fraction of precipitation than is measured at the onset of precipitation in the present work.

A calculation by Siradj [6] using a modified Johnson Mehl equation reported by Herman et al [9] suggests that the stress relaxation technique detects the start of precipitation when 2-3% of the volume fraction is precipitated. A calculation by Dutta et al [4] indicates that the peak strenth is associated with about 6% of the volume fraction precipitated. Such a change would produce a change in precipitation time of a factor of 2.5. A change from 2% to 10% would give a factor of 5, which is similar to the difference found between this investigation and that of Valdes and Sellars [2] and seems the probable reason for the different times obtained from the different techniques.

Conclusions

This work has demonstrated that the stress relaxation technique can be used in plane strain compression to define the onset of strain induced precipitation of niobium carbonitride in niobium microalloyed steels. Although the characteristic times for the onset of precipitation are shorter than those to the peak stress in successive deformation tests used previously, the temperature dependence of the reaction is the same.

The main advantage of this technique is that only one specimen is required, in place of a series of specimens held after an intermediate deformation for different lengths of time, to define the maximum flow stress in the final deformation. This is therefore a considerably more efficient way of obtaining information about the strain induced precipitation reaction.

Increase in the roughing strain reduces the time for the onset of strain induced precipitation up to a limiting value of strain beyond which the reaction time remains constant. Decrease in the roughing temperature reduces the time at which the onset of precipitation becomes constant.

The limiting strain that produces the mimimun precipitation time is greater that the typical strain in roughing. It is likely that several roughing strains may be accumulated to reach the limiting strain. Strains up to this limiting value produce considerable strengthening in the austenite.

Acknowledgements

EJS wishes to thank the Research Grant Programme of the Faculty of Engineering in the University of Indonesia for financial support during this work.

References

[1]   B Dutta, PhD thesis, 1985,University of Sheffield
[2]   E Valdes and C M Sellars, Mat Sci and Tech , 7 ,1991, p 622
[3]   D Y Choi, M Phil thesis, 1992, University of Sheffield
[4]   B Dutta, E Valdes and C M Sellars, Acta Met, 40, 1992, p653
[5]   K J Irvine, F B Pickering and T Gladman, J Iron Steel Inst.,205, 1967, p161
[6]   E S Siradj, PhD thesis, 1997, University of Sheffield
[7]   L J Cuddy and J C Raley, Met Trans A, 14A, 1989, p1989
[8]   R Colas and C M Sellars, J Testing and Evaluation, 15, 1987, p342
[9]   J C Herman, B Donnay and V Leroy, ISIJ Int, 32, 1992, p779

Materials Science Forum Vols. 284-286 (1998) pp. 151-158
© *1998 Trans Tech Publications, Switzerland*

Influence of Processing Conditions and Alloy Chemistry on the Static Recrystallisation of Microalloyed Austenite

E.J. Palmiere[1], C.I. Garcia[2] and A.J. DeArdo[2]

[1] The University of Sheffield, Department of Engineering Materials, Sir Robert Hadfield Building, Mappin Street, Sheffield, S1 3JD, UK

[2] University of Pittsburgh, Department of Materials Science and Engineering, 848 Benedum Hall, Pittsburgh, PA 15261, USA

Keywords: Fractional Softening, Precipitate Pinning Force, Static Recrystallisation, Solute Supersaturation, Thermomechanical Processing

ABSTRACT

This paper draws from both current and previous research to address the manner in which microalloying elements such as Nb influence the recrystallisation-stop temperature of austenite (T_{RXN}). Results from quantitative microstructural characterisation, microchemical analysis and hot deformation simulations are presented, and show how the T_{RXN} increases with solute supersaturation.

INTRODUCTION

The thermomechanical processing of microalloyed steel has been employed for some time in the production of plate and sheet material in order to optimise properties such as strength and impact toughness [1-7]. This technology was subsequently applied to the bar and forging industry with considerable success, considering the more complicated geometry, thicker sections and localised states of strain [8-10]. The central feature of thermomechanically processed steel austenite is the fine grain size in the final transformation product. This fine grain size is known to cause both high strength and high resistance to brittle fracture by cleavage. While achieving high strength in structural steel is rather straight forward and well-understood, achieving a fine, *and uniform*, grain size is a more complex task related to the synergy which exists between the composition of the steel and its processing.

The range in behaviour of austenite during hot deformation is exhibited schematically in Figure 1. This figure shows the influence of both deformation temperature (T_ϵ) and amount of strain on the microstructure of statically recrystallised austenite [11]. It can be observed from Figure 1 that for constant deformation variables such as amount of strain per pass, strain rate and interpass holding time, the austenite microstructure will be completely recrystallised at high deformation temperatures (i.e., when $T_\epsilon \geq T_{95\%}$). Multi-pass hot deformation sequences which end in this regime, and where there is a pre-existing pinning force to suppress grain coarsening, are entitled recrystallisation controlled rolling (RCR) practices. Hence, a distinguishing feature of RCR processing is that a pre-existing pinning force (precipitate or solute) be present which is small enough to allow for static recrystallisation to occur but large enough to suppress grain coarsening.

At the other extreme regarding deformation temperature, a completely unrecrystallised microstructure is present when deformation occurs below the recrystallisation-stop temperature of austenite, T_{RXN} (i.e., when $T_\epsilon \leq T_{5\%}$). Multi-pass hot deformation sequences which occur largely in this regime are entitled conventional controlled rolling (CCR) practices. For a fixed rolling schedule which includes a specific number of roughing and finishing passes, the higher the T_{RXN}, the larger will be the amount of rolling strain imparted in the non-recrystallisation region. Earlier work has shown that the density of the near-planar crystalline defects (i.e., grain boundaries, deformation bands, twin boundaries), labeled S_v, increases with increasing deformation in the non-recrystallisation region [3,12]. Since these defects act as nucleation sites for pro-eutectoid ferrite during subsequent cooling, there is a strong relationship between the final ferrite grain size and S_v [13,14]. The addition of microalloying elements such as Nb, Ti and V is known to increase T_{RXN}, with Nb having the strongest effect per atomic percent addition [13,15]. Hence, Nb bearing steels with high T_{RXN} and high S_v values are known to have a very fine ferrite grain size.

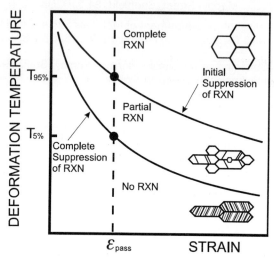

Figure 1. **Schematic illustration of austenite microstructures resulting from various deformation temperatures (T_e) at a constant level of strain [11]**

If, however, the deformation temperature is greater than T_{RXN}, but less than the temperature where the austenite grains would be fully recrystallised (e.g., $T_{95\%} > T_e > T_{5\%}$), a partially recrystallised microstructure is observed. This microstructure is often referred to as being duplex because of a non-uniform austenite grain size. Upon transformation to the low temperature product, this non-uniform grain size would be inherited, and the result would be a steel product exhibiting highly variable, and usually poor mechanical properties. Hence, deformation within this region should be avoided at all costs. It is therefore surprising that very little attention has been given to this area, and in particular, quantifying how the extent of this microstructural region can be effected by processing parameters.

It is important to recognise that the temperatures $T_{95\%}$ and $T_{5\%}$ (henceforth referred to as T_{RXN}) will vary with deformation processing conditions. In general, those factors which accelerate recrystallisation (high strain per pass, high strain rate) reduce both of these temperatures, whereas those factors which retard recrystallisation (low strain per pass, low strain rate) increase both of these temperatures. However, at present it is not well understood if both $T_{95\%}$ and T_{RXN} change by the same magnitude with changes in processing parameters. Hence, there are deficiencies in our fundamental understanding of those factors which are important in controlling the recrystallisation behaviour of austenite. Previous research had concentrated on the influence of Nb supersaturation on T_{RXN}, and constitutes the majority of this paper. Current research is in progress to investigate how both $T_{95\%}$ and T_{RXN} vary with processing parameters.

EXPERIMENTAL PROCEDURE

The approach involved a series of low carbon, Si-killed steels having compositions as shown in Table 1. Details regarding the choice of steel composition and primary processing are described elsewhere [15]. Four grades of steel (E1-E4) microalloyed with Nb were developed based on the reference steel E0. Three of the compositions (E1, E3 and E4) contained similar N levels while the Nb concentrations were varied. The fourth steel (E2) exhibited a Nb level similar to that of steel E1 but contained three times the N level. Hence, upon reheating these steels prior to deformation, large differences between the amount of dissolved Nb in austenite should be realised. Since the supersaturation is directly related to the driving force for precipitation, these steels were expected to yield evidence as to the role of solute supersaturation on the suppression of recrystallisation.

Element	E0	E1	E2	E3	E4
C	0.090	0.090	0.080	0.080	0.080
Mn	1.490	1.490	1.470	1.440	1.430
P	0.009	0.009	0.009	0.008	0.010
S	0.006	0.012	0.006	0.006	0.006
Si	0.410	0.410	0.410	0.400	0.290
Nb		0.049	0.048	0.020	0.090
N	0.008	0.008	0.024	0.008	0.008

Table 1.   **Steel compositions in wt%**

Static softening studies were conducted using an MTS unit modified for compression under constant true strain rate conditions. To minimise any adverse frictional effects, a water-based glass lubricant was employed for axisymmetric compression testing. Testing was such that the fractional softening of austenite was determined by an interrupted compression technique. The testing parameters conformed to those currently employed in plate rolling practice. Hence, deformation occurred at a strain rate of 10 sec^{-1}. The temperature was monitored and controlled throughout each test by means of a thermocouple inserted at the mid-height of the specimen. Each specimen was given an intermediate deformation ($\epsilon = 0.3$) at a specified temperature (ranging between 900-1100°C) and held at temperature for a period of 10 seconds. After this 10 second delay, specimens experienced another deformation ($\epsilon = 0.3$) followed by quenching in an iced brine bath.

Following deformation, specimens were prepared for quantitative microstructural analysis using light microscopy, TEM and APFIM.

RESULTS AND DISCUSSION

The percent fractional softening for all steels is shown in Figure 2. It is apparent that for each steel, the amount of fractional softening observed within the 10 second holding time decreased with decreasing temperature. In fact, all steels except E0 and E3 exhibited fractional hardening at the lowest deformation temperature of 900°C. It is important to note that since the amount of fractional softening was measured from mechanical testing data, this softening corresponds to the total net softening of austenite during the 10 second delay. Hence, the total softening would be comprised of all static events:

softening due to recovery and recrystallisation plus hardening due to precipitation. However, since the stacking fault energy of austenite is relatively low (75 mJ/m^2), softening attributed to recovery processes would be limited because of the difficulty for dislocations to cross-slip or climb [16]. Therefore, the overall fractional softening, as depicted in Figure 2, would largely be expected to reflect the softening due to static recrystallisation. In fact, previous work [17-19] has shown that static recovery processes may comprise anywhere from 15-20% of the overall softening behaviour of austenite.

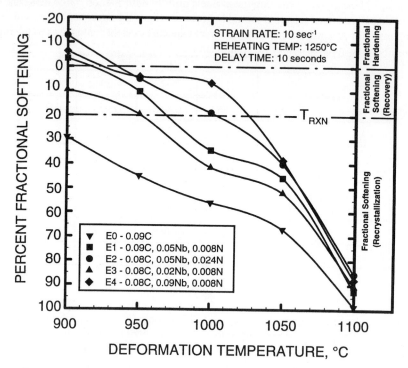

Figure 2.   **Percent fractional softening of austenite as determined from interrupted compression testing [11]**

Employing the criterion that 20% of the overall softening is due to recovery, the T_{RXN} for each steel can readily be obtained from Figure 2. All softening greater than 20% can be attributed to static recrystallisation. As the deformation temperature increases and particles begin to coarsen and possibly go into solution, accelerated recrystallisation kinetics are observed. Therefore, the convergence of the curves in Figure 2 above 85% fractional softening is indicative of complete recrystallisation for the delay time of 10 seconds. Hence, recrystallisation-stop temperatures (temperatures corresponding to $T_{5\%}$, Figure 1) of 942, 971, 999 and 1030°C were obtained for steels E3, E1, E2 and E4. The results shown in Figure 2 are unique in that the fractional softening of austenite is measured as a function of deformation temperature at a constant delay time. The delay time of 10 seconds was chosen to represent plate processing conditions. However, delay times of 1 or 100 seconds could have also been selected to represent a strip rolling or open-die forging operation, respectively.

The fractional softening data shown in Figure 2 were further complimented using quantitative metallography. This was necessary to verify the correspondence between 20% fractional softening and T_{RXN}. All steels showed good agreement between microstructure and mechanical softening data. At temperatures below the respective T_{RXN} (corresponding to the temperature where 20% fractional softening was measured after a 10 second holding time) for each steel, prior-austenite grains are completely unrecrystallised. These grains were elongated in a direction which was perpendicular to the axis of compression.

In a related investigation [15], the amount of Nb in solution in austenite was experimentally determined using atom probe analysis. Using these data, it was possible to determine the Nb supersaturation in austenite. When combined with the results from the fractional softening studies, some interesting behaviour can be noted. These results are summarised in Figure 3, which depicts the Nb supersaturation in austenite at the respective T_{RXN} of each steel. Figure 3 shows that the Nb supersaturation at the T_{RXN} varies with each steel and is dependent on the initial steel composition.

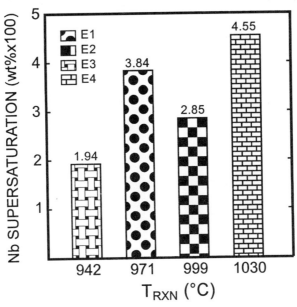

Figure 3. **Nb supersaturation in austenite at the respective recrystallisation-stop temperatures for steels E1 through E4 [11]**

The trend noted from Figure 3 is that increasing recrystallisation-stop temperatures are associated with increasing Nb supersaturation in austenite. The exception to this trend is found in steel E2. This steel had the same initial Nb concentration as steel E1 but had three times the nitrogen concentration. This increased N concentration provided for higher precipitate stability [15]. Therefore, at the reheating temperature of 1250°C, steel E2 had significantly less Nb in solution in austenite than steel E1. Hence, for the same reheating and deformation temperature, steel E2 would have a smaller Nb supersaturation than steel E1. The reason that steel E2 exhibits a higher T_{RXN} than steel E1 is most likely due to an increased amount of undissolved particles in addition to the strain-induced precipitation of Nb(CN).

It is also important to note that the *measured* supersaturations shown in Figure 3 for the three low N steels are smaller than supersaturations *calculated* using published solubility products. Hence, it was found that the thermodynamic stability of the precipitate was higher than what had originally been reported, due in part to the interaction between elements such as Si and Mn [15]. This was particularly true for temperatures greater than 1000°C (e.g., within the temperature range of industrial plate rolling). The significance of this is shown in Figure 4, which illustrates how the T_{RXN} varies with the amount of Nb in solution in austenite. In this figure, the measured results from the fractional softening studies [11] coupled with atom probe results [15] are compared with data calculated from published solubility products [13]. At low solute levels, corresponding to relatively low temperatures, there is very good agreement between the two curves. This is consistent with the results from another investigation when extensive precipitation was observed to have taken place along subgrain boundaries [20]. However, at high levels of Nb in solution, corresponding to relatively high deformation temperatures, the curves shown in Figure 4 begin to deviate from one another. The curve which incorporated the solubility product (which assumed an activity coefficient of unity) exhibits a larger gradient, and predicts a higher T_{RXN} at a given level of solute than does the curve which incorporates the measured solubility of Nb in austenite. Again, this shows the importance that alloy interactions (which cause deviations in the activity coefficient from unity) can have on precipitate stability. Hence, as the precipitate stability becomes greater, the solute supersaturation becomes smaller for a given reheat and deformation temperature. This in turn yields a lower T_{RXN} at any given level of solute.

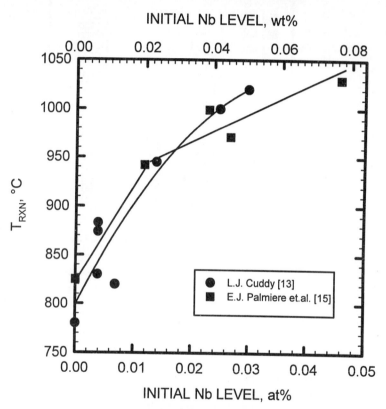

Figure 4.   **Variation in T_{RXN} with soluble Nb in austenite illustrating the importance of the thermodynamic stability of precipitate**

CONCLUSIONS

The following conclusions can be made regarding the factors important to the static recrystallisation of microalloyed austenite.

1. Isothermal interrupted compression testing proved to be a relatively easy and reliable method for determining the T_{RXN}. The T_{RXN} was associated with 20% of the total fractional softening of austenite and was verified by quantitative microscopy. A T_{RXN} of 942, 971, 999 and 1030°C was

measured for steels E3, E1, E2 and E4, respectively.

2   The coupling of the results from atom probe analysis and the results from interrupted compression testing indicate that the T_{RXN} increases with increasing Nb supersaturation in austenite. This level of Nb supersaturation was not constant from steel to steel.

3.   The thermodynamic stability of particles is important in governing the level of solute supersaturation at a given temperature. This is important because the level of solute supersaturation is directly related to the precipitation potential, which in turn dictates T_{RXN}.

REFERENCES

[1]   K.J. Irvine, F.B. Pickering and T. Gladman, T., JISI, (1967), p. 161.

[2]   K.J. Irvine, Low Alloy High Strength Steels, (Düsseldorf, BRD: The Metallurg Companies, 1970), p. 1.

[3]   I. Kozasu and T. Osuka, Processing and Properties of Low Carbon Steel, J.M. Gray, Ed., (New York, NY: TMS-AIME, 1973), p. 47.

[4]   F.B. Pickering, Microalloying 75, M. Korchynsky, Ed., (New York, NY: Union Carbide Corporation, 1977), p. 9.

[5]   W.J. McG. Tegart and A. Gittins, The Hot Deformation of Austenite, J.B. Ballance, Ed., (Warrendale, PA: TMS-AIME, 1977), p. 1.

[6]   P.K. Amin and F.B. Pickering, Thermomechanical Processing of Microalloyed Austenite, A.J. DeArdo, G.A. Ratz and P.J. Wray, Eds., (Warrendale, PA: TMS-AIME, 1982), p. 1.

[7]   A.J. DeArdo, Microalloying 95, (Warrendale, PA: ISS-AIME, 1995), p. 15.

[8]   J.H. Woodhead, Fundamentals of Microalloying Forging Steels, G. Krauss and S.K. Banerji, Eds., (Warrendale, PA: TMS-AIME, 1987), p. 3.

[9]   C.I. Garcia, E.J. Palmiere and A.J. DeArdo, Mechanical Working and Steel Processing Proceedings, (Warrendale, PA: ISS-AIME, 1989), p. 59.

[10]   E.J. Palmiere, C.I. Garcia and A.J. DeArdo, Fortieth Sagamore Army Materials Research Conference on Metallic Materials for Lightweight Applications, M.G.H. Wells, E.B. Kula and J.H. Beatty, Eds.,1993, p. 479.

[11]   E.J. Palmiere, C.I. Garcia and A.J. DeArdo, Metall. Trans., 27, (1996), p. 951.

[12]   I. Kozasu, C. Ouchi, T. Sampei and T. Okita, Microalloying 75, M. Korchynsky, Ed., (New York, NY: Union Carbide Corporation, 1977), p. 120.

[13]   L.J. Cuddy, Metall. Trans., 12, (1981), p. 1313.

[14]   T. Gladman, The Physical Metallurgy of Microalloyed Steels, (London: Institute of Materials, 1997), p. 251.

[15]   E.J. Palmiere, C.I. Garcia and A.J. DeArdo, Metall. Trans., 25, (1994), p. 277.

[16]   W. Charnock and J. Nutting, J. Met. Sci., 1, (1967), p. 23.

[17]   W. Roberts and B. Ahlblom, Acta Metall., 26, (1978), p. 801.

[18]   R.A.P. Djaic and J.J. Jonas, Metall. Trans., 4, (1973), p. 621.

[19]   R.A. Petkovic, M.J. Luton and J.J. Jonas, Can. Metall. Quart., 14, (1975), p. 137.

[20]   B. Dutta and C.M. Sellars, Mat. Sci. and Tech., 3, (1987), p. 197.

Materials Science Forum Vols. 284-286 (1998) pp. 159-166
© *1998 Trans Tech Publications, Switzerland*

Comparison of the Deformation Characteristics of a Ni-30wt% Fe Alloy and Plain Carbon Steel

P.J. Hurley[1], B.C. Muddle[1], P.D. Hodgson[2], C.H.J. Davies[1], B.P. Wynne[1], P. Cizek[1] and M.R. Hickson[3]

[1] Department of Materials Engineering, Monash University, Clayton, Vic., 3168, Australia

[2] School of Engineering and Technology, Deakin University, Waurn Ponds, Vic., 3217, Australia

[3] BHP Research, Melbourne Laboratories, 245 Wellington Rd, Mulgrave, Vic., 3167, Australia

Keywords: Austenite Modelling, Deformation-Induced Ferrite, Microbands, Hot Torsion, Hot Compression, Nickel Alloy

ABSTRACT

A major barrier confronting researchers studying the hot deformation of plain and low carbon steels is the inability to directly observe the deformation microstructures of hot worked austenite due to the unavoidable transformation to martensite on quenching to room temperature. Various model materials, such as austenitic stainless steels have been used to overcome this difficulty. However, these materials have markedly different stacking fault energies from plain and low carbon steels and this will affect the evolving deformation structure. In this work, a model austenitic Ni-30wt%Fe alloy, calculated to have a stacking fault energy similar to that of low carbon steel, has been tested using hot compression. Stress-strain curves obtained during hot deformation show characteristics similar to those generated during identical tests on a 0.15wt%C steel. This suggests that the two materials behave similarly during deformation under similar experimental conditions.

An application of the Ni-Fe alloy in the study of microstructural changes in austenite during hot deformation is demonstrated. A series of hot torsion experiments on a 0.11wt%C steel have been found to produce deformation-induced, intragranular nucleation of ferrite from austenite when a single deformation pulse is applied at 675°C. A similar set of experiments have also been performed on the Ni-Fe alloy at 750°C. Transmission electron microscopy carried out on the Ni-Fe alloy torsion specimens has revealed that likely preferred sites for intragranular ferrite nucleation appear to be microbands produced in the austenite during deformation.

INTRODUCTION

An important area of research over the years has been the study of work-hardening, recovery and recrystallisation mechanisms occurring in austenite during hot deformation. It is not possible to examine directly the austenite deformation microstructures developed when low alloy steels are hot worked, due to the martensitic transformation which occurs on quenching of these steels. As a result, knowledge of deformation mechanisms occurring in austenite has been built up using indirect methods such as flow curve analysis or the study of ferrite textures. Another approach has been to study the deformation characteristics of austenitic stainless steels and to directly relate these to low alloy steels [1]. However, the usefulness of such model materials is expected to be limited given the fact that the stacking fault energy of austenitic stainless steel is significantly lower than that of low alloy steels [2]. Stacking fault energy has been shown to have a strong influence on the behaviour of metals and alloys during deformation and, provided that other factors which affect deformation behaviour such as grain size and alloying content remain constant, materials with similar stacking fault energies will exhibit similar deformation characteristics [3].
Nickel alloys, predicted to have stacking fault energies similar to that of austenite at high temperatures in plain and low carbon steels, have been used to model austenite textures [4]. The

present work employs an austenitic Ni-30wt%Fe alloy as a material for examining and modelling austenite deformation at elevated temperatures. This alloy has been calculated to have a stacking-fault energy similar to that of austenite in pure iron at 1100°C [5], and it would thus seem to be an ideal model material for studying such things as deformation microstructures and texture development in hot worked austenite. One particular application of interest is the investigation of potential intragranular nucleation sites for deformation-induced ferrite. There has been much speculation recently as to what deformation features inside austenite are responsible for the nucleation of ferrite during or following deformation [6-10]. This paper illustrates the usefulness of the Ni-Fe alloy for providing insight into such matters.

EXPERIMENTAL METHODS

Materials

The compositions of the two vacuum cast materials used in this work are set out in Table I. The as-cast 75 kg billets were reduced in thickness to 20mm by hot rolling at temperatures between 1200°C and 1000°C. Cylindrical compression specimens, with diameters of 10mm and lengths of 15mm were machined from the 20mm plate with the axis of compression parallel to the rolling direction of the plate. Torsion specimens, each with a gauge length of 20mm and a gauge diameter of 6.7mm, were machined from the 20mm plate with the axis of the sample parallel to the rolling direction.

Table I: Chemical compositions (wt %) of the alloys.

Alloy	Fe	C	P	Mn	Si	S	Ni	Cr	Mo	Al
Ni-30Fe	bal.	0.001	-	0.0031	-	0.001	71.5	0.0022	0.011	0.011
Steel A	bal.	0.111	0.022	1.68	0.20	0.002	0.013	0.019	0.023	0.027

Hot Compression Experiments

The hot compression testing involved deformation of the Ni-Fe alloy using similar experimental conditions to those used on a 0.13wt%C steel, outlined in [11]. Strain rates ranging from $0.07s^{-1}$ to $70s^{-1}$, and temperatures ranging from 800°C to 1100°C, were used on an MTS TestStar high speed machine. Samples were initially heated to 1000°C and held for 2 minutes before being air cooled to the test temperature. Once the test temperature was reached, a delay of 15 sec. was allowed for temperature stabilization. The specimens tested at 1100°C were heated to this temperature, held for 2 min. and then deformed. Boron nitride lubrication paste was applied between the contact surfaces of the specimen and the anvils to minimise friction.

Hot Torsion Experiments

The torsion machine located at BHP Research- Melbourne Laboratories was used. Samples of both steel A and the Ni-30Fe alloy were induction heated to 1250°C for 300 sec., air cooled to the testing temperature and given a single deformation pulse before being water quenched to room temperature. In the case of the steel, the deformation temperature was 675°C and von Mises strains of 0.8, 1.3, 1.6 and 2.0 were used. The Ni-30Fe alloy was tested at 750°C and strains of 0.35, 0.70 and 1.0 (failure) were used. A temperature of 750°C was used rather than 675°C because it enabled higher strains to be achieved as a result of increased ductility at the slightly higher temperature, while still remaining within the processing window in which intragranular ferrite is observed to form in the steel with the application of strain. Intragranular ferrite nucleation has been found to occur at temperatures as high as 750°C in this steel [12]. Metallographic examination was performed on sections cut through the centre of the specimens parallel to the long axis. The steel specimens were etched using 2% nital and the Ni-30Fe specimens were etched using a solution containing the following: HNO$_3$ (100ml), HCl (500ml), CuCl (25g), FeCl$_3$ (25g), distilled water (500ml).

Foils for transmission electron microscopy were prepared from tangential sections cut from the surface of the torsion samples of the Ni-30Fe alloy. Foils were thinned using a Tenupol-3 Twin-Jet Electropolisher with a solution of 5% perchloric acid in methanol at -30°C, and a voltage of 35V (current = 0.17A). A Philips CM20 transmission electron microscope operating at 200kV was used

RESULTS

Hot Compression Testing

Two sets of results for the Ni-30Fe alloy, selected to be representative of the behaviour of this alloy over a range of strain rates and temperatures, are plotted in Fig. 1(a) and Fig. 2(a). To allow direct comparison of the hot deformation behaviour of this alloy and that of the plain carbon steel, results obtained using the same test conditions on the 0.15wt% carbon steel have been reproduced from [11], and are shown in Fig. 1(b) and Fig. 2(b). It is evident that the flow stress-true strain curves for each material are very similar under similar deformation conditions. For example, for results obtained at a strain rate of 0.7s^{-1}, Fig. 1(a), neither material recrystallises at 800°C. Above this temperature, both materials are found to recrystallise at similar levels of strain. As a second example, the Ni-30Fe alloy deformed at 900°C exhibits a peak strain of 0.38±0.03, while the steel begins to recrystallise at the same temperature at a strain of 0.33±0.3, Fig. 1(b). Furthermore, the flow curves generated at 900°C indicate that the both materials do not recrystallise at strain rates above 0.7s^{-1}, Fig. 2(a)-(b). It is important to realise that the actual strength levels observed for the Ni-30Fe alloy are higher than for the steel. This is most likely due to the fact that the addition of 30wt% Fe to Ni gives rise to solid solution strengthening.

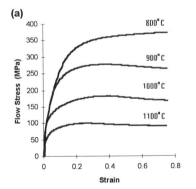 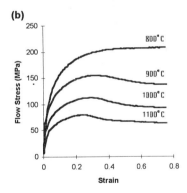

Figure 1: Flow stress-true strain curves for (a) Ni-30wt%Fe and (b) a 0.13wt%C steel at a strain rate of 0.7s^{-1} for varying temperatures.

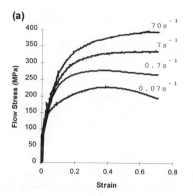 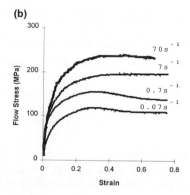

Figure 2: Flow stress-true strain curves for (a) Ni-30wt%Fe and (b) a 0.13wt%C steel at 900°C for varying strain rates.

Given the similarity of the stress-strain curves in Fig. 1 and Fig. 2, it would seem reasonable to assume that the two materials behave similarly during hot deformation. As a result, several applications for this alloy may be envisaged, such as modelling of austenite texture after hot deformation and analysis of deformation microstructures produced during hot deformation of austenite.

An Application of the Model Ni-30Fe Alloy - Observation of Sites for Intragranular Nucleation of Ferrite During Torsion Testing

The set of torsion experiments performed on steel A gave rise to the nucleation of deformation-induced ferrite within the austenite during testing. Interruption of the deformation stage at various levels of applied strain allowed the progression of this intragranular ferrite nucleation to be observed. At a strain of 0.8, Fig.3 (a), fine ferrite grains had begun to nucleate inside a few of the austenite grains. The ferrite grains form within the austenite grains in what appear to be parallel and closely-spaced linear arrays (in two-dimensional cross-section) traversing the austenite grains, Fig. 3(a). Close observation of the microstructure developed by a strain of 0.8, reveals that the ferrite formed as discrete grains along each particular row making up the arrays. As the strain was increased to 1.3, Fig. 3(b), the extent of intragranular ferrite nucleation had increased, and the number of austenite grains containing intragranular ferrite nuclei had also increased substantially. As the strain was further increased to 1.6, Fig. 3(c), and 2.0, Fig. 3(d), the number of ferrite nuclei also increased giving rise to almost complete transformation of some austenite grains to ferrite. Throughout this process, the existence of closely-spaced, parallel linear arrays of fine ferrite grains were observed within transformed austenite. This is a characteristic of the ferrite nucleated using the torsion schedules outlined in this work. Ferrite grains formed early in the process have grown along individual rows, eventually linking up to form raft-like bands across individual austenite grains, Fig. 3(c)-(d). Further analysis indicated that the spacing between the rows of ferrite grains within individual austenite grains decreased with increasing strain. For example, the rows of ferrite developed by a strain of 0.8 have an average spacing of between 3 and 5 μm, while at a strain of 1.6 the spacing between the rows, Fig. 3(c) is less than 1 μm. Clearly, the deformation process is responsible for the enhanced ferrite nucleation. Figures 3(a)-3(d) imply that the ferrite nucleates on planar defects, developed in the austenite as strain is increased.

In an attempt to determine what deformation events are occurring in the austenite during the torsion experiments described above, a similar set of experiments were performed on the model Ni-

30Fe alloy. The maximum strain achievable in this alloy at 750°C was 1.0. This is adequate for modelling the early stages of the torsion tests performed on the steel, during which ferrite first begins to nucleate intragranularly as a result of the applied strain. In order to elucidate why ferrite nuclei do not form at strains below about 0.8, strains of 0.35, 0.7 and 1.0 were imparted to the Ni-30Fe alloy. Water-quenching of the specimens immediately following the tests was used to effectively freeze the deformation microstructure, allowing the dislocation arrangements to be observed using transmission electron microscopy.

Figure 4 illustrates the resulting microstructures. At a strain of 0.35, Fig. 4a, a network of equiaxed dislocation cell structures delineated by dense dislocation walls (DDWs) is observed (see [13] for a description of the terminology used in this work). There is also some evidence of early microband formation. This network of DDWs and undeveloped microbands manifests itself metallographically as a pattern of closely-spaced ripples in the surface following etching, Fig.4(b). At a strain of 0.7, a series of well-developed microbands has replaced the equiaxed cell network observed at lower strains, Fig. 4(c). Two families of intersecting microbands are evident. Figure 4(d) shows that the ripple-like pattern which was evident in Fig. 4(b) after 0.35 strain has developed into a more clearly-defined slip pattern. By a strain of 1.0, Fig. 4(e), one series of microbands has developed to become the predominant feature of the microstructure. In accordance with the observations of Hughes and Nix [14], it appears likely that, of the two varients of microbanding observed by a strain of 0.7, Fig. 4(c), that which is more closely aligned parallel to the macroscopic shear direction has developed to become the most prominent feature of the microstructure by a strain of 1.0. The spacing of these microbands is approximately 0.5 to 1 μm. Figure 4(f) shows the corresponding microstructure using optical microscopy. The microbands are revealed as closely-spaced parellel striations crossing individual grains.

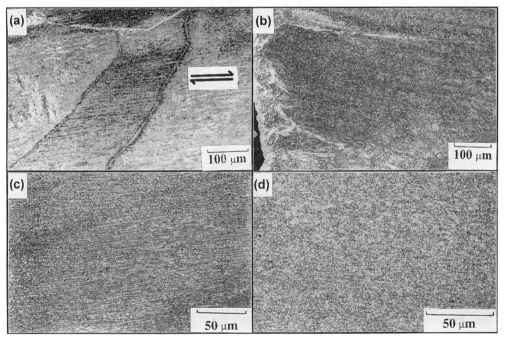

Figure 3: Reflected light micrographs of samples of steel A deformed in torsion to applied strains of: (a) 0.8; (b) 1.3; (c) 1.6; (d) 2.0. Double arrays shown in (a) indicate direction of applied shear stress for all micrographs.

DISCUSSION

An extensive survey failed to reveal any previous reports in the literature of intragranular ferrite nucleation in the characteristic linear arrays that have been observed in this work, and initial uncertainty thus existed as to the origin of the nucleation sites in the austenite. No nucleation of intragranular ferrite is observed when strain levels are below 0.8. Moreover, when the strain level is increased above 0.8, the extent of intragranular ferrite nucleation is increased significantly. Therefore, it seems quite clear that ferrite nucleation is directly associated with deformation of the austenite. It is implied that deformation produces an array of planar-like defects in the austenite which traverse entire individual grains. At strain levels around 0.8, these defects become effective as nucleation sites for ferrite within the austenite at 675°C, where the driving force for ferrite nucleation is quite high given the large undercooling at this temperature. Increasing the strain level further enhances the potency of these defects as ferrite nucleation sites.

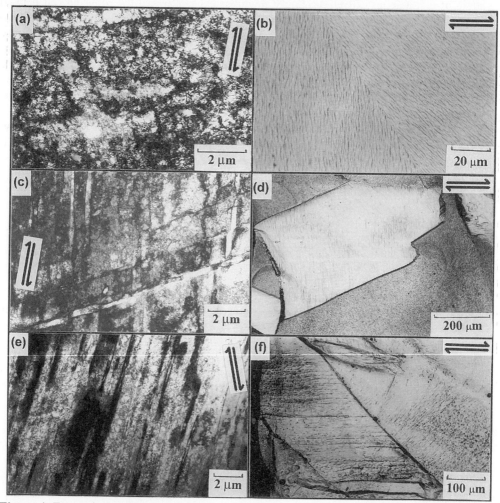

Figure 4: Transmission electron micrographs (left column) and reflected light micrographs (right column) of the microstructures of Ni-30Fe alloy deformed in torsion to strains of: (a,b) 0.35; (c,d) 0.70; (e,f) 1.0. Double arrays indicate direction of applied shear stress.

The results of compression testing presented above indicate that an alloy of Ni-30wt%Fe may provide a suitable model material for study of austenite deformation representative of the behaviour of plain carbon steels at elevated temperatures.   Based on this evidence, it is plausible to assume that the deformation microstructures produced during the torsion testing of the model alloy would be similar to those generated in steel A subjected to a similar series of tests.   Results obtained from the transmission electron microscopy shown in Fig. 4 reveal that, during the torsion tests employed, a substructure of microbands develops with increasing strain.   By a strain of 0.7, two intersecting networks are commonly observed, Fig. 4(c), and, with further strain, a single parallel array of microbands aligned near to the macroscopic shear direction dominate the deformation microstructure, Fig. 4(e).   In the case of steel A, the form of the ferrite microstructures suggested that ferrite grains nucleate along some kind of closely-spaced planar defects in the austenite, Fig. 3(a)-(d). The morphology and separation of the microbands observed in the model alloy correspond well with the expected nucleation sites for ferrite in the steel.   Direct comparison of Fig. 3(a) with that of Fig. 4(f), provides further evidence for the notion that such microbands provide potential sites for intragranular ferrite nucleation.   Clearly, the rows of ferrite grains in Fig. 3(a) form a pattern consistent with the array of microbands in Fig. 4(f).

Korbel et al. [15] have suggested that a significant amount of shear strain may be accommodated by such microbands based on the fact that they cause steps to form in grain boundaries during deformation.   The stored energy resulting from this shear strain, as well as the presence of the interface provided by the planar microband, may enable ferrite to nucleate along the microband interfaces.   Unfortunately, the maximum strain achievable for the present model alloy at 750°C was 1.0, and it was not possible to pursue the affect of strain on the microband substructure. However, work on Ni alloys by Hughes and Nix [14] suggests that the laminar microband structures continue to dominate at higher strains than were achieved in this work.   Furthermore, the average spacings between sub-boundaries in the laminar substructures was observed to decrease with increased strain.   Thus, it is plausible to conclude that increasing the applied strain will have the effect of further developing the laminar microbanding in the steel to provide further sites for ferrite nucleation.

CONCLUSIONS

1. A Ni-30wt%Fe alloy, which has been found to have a stacking fault energy similar to that of austenite in pure iron, was tested in hot compression.   The stress-strain curves generated during these tests are closely comparable with those obtained for a plain carbon steel using a similar series of tests.   This suggests that the two materials exhibit similar hot working characteristics, and it is concluded that the Ni-30wt%Fe alloy is a potentially useful model material for simulating hot deformation processes in steel.

2. Torsion testing was used to produce deformation-induced, intragranular ferrite nucleation in a 0.11wt% carbon steel at 675°C.   The ferrite is first observed at a strain level of 0.8.   With application of further strain, the ferrite nucleation process is enhanced so that the number of ferrite nuclei is increased with increased strain.   The ferrite is observed to nucleate on linear arrays of closely-spaced defects within individual austenite grains.   These defects appear to be the result of the deformation process.

3. An application for the Ni-30wt%Fe alloy as a model material for plain low carbon austenite is demonstrated.   The torsion tests used to produced deformation-induced ferrite in the steel were repeated on the model Ni-Fe alloy.   At the appropriate strain levels, transmission electron microscopy revealed a pattern of closely-spaced deformation microbands which developed during torsion testing and which had a form and spacing consistent with the arrays of ferrite grains produced in the hot-worked steel.

ACKNOWLEDGMENTS The authors would like to thank BHP Research for permission to publish this research and acknowledge the support of staff at BHP Research-ML.   One of the authors (P.J.H) would like to thank BHP Research and the Australian Government for financial support (APA).

REFERENCES

[1]   Brown, E. L. and DeArdo, A. J., Hot Working and Forming Processes, ed. Sellars, C. M. and Davies, G. J., The University of Sheffield, July 17-20, 1979, The Society, (1980), p.21-26.

[2]   McQueen, H. J. and Bourell, D. L, Formability and Metallurgical Structure, ed. Sachdev, A. K. and Embury, J. D., Orlando, Florida, October 5-9, 1986, The Metallurgical Society, Inc, (1987), p.341-368.

[3]   McQueen, H. J. and Jonas, J. J., Treatise on Materials Science and Technology, 6, (1975), p.393-493.

[4]   Ray, R. K. and Jonas, J. J., International Materials Reviews, 35, 1, (1990), p.1-36.

[5]   Charnock, W. and Nutting, J., Metal Science Journal, 1, (1967), p.123-127.

[6]   Priestner, R., Thermomechanical Processing of Austenite , ed. DeArdo, A. J., Ratz, G. A., and Wray, P. J., Pittsburgh, Pennsylvania, August 17-19, 1981, The Metallurgical Society of AIME, (1982), p.455-466.

[7]   Priestner, R. and Ali, L., Materials Science and Technology, 9, (1993), p.135-141.

[8]   Yada, H., Matsumura, Y., and Senuma, T., Proceedings of The International Conference on Martensitic Transformations (1986), p.515-520.

[9]   Matsumura, Y. and Yada, H., Transactions ISIJ, 27, (1987), p.492-498.

[10] Yada, H., Chunming Li , Yamagata, H., and Tanaka, K., Thermec-97: International Conference on Thermomechanical Processing of Steels and Other Materials, Wollongong, Australia, 7-11 July, 1997, (in press).

[11] Jaipal, J., Davies, C. H. J, Wynne, B. P., Collinson, D. C., Brownrigg, A., and Hodgson, P. D., ipid.

[12] Hurley, P. J., Muddle, B. C., Hodgson, P. D., and Davies, C. H. J., (to be published).

[13] Bay, B., Hansen, N., Hughes, D. A., and Kuhlmann-Wilsdorf, D., Acta Metallurgica et Materialia, 40, 2, (1992), p.205-219.

[14] Hughes, D. A. and Nix, W. D., Materials Science and Engineering A, A122, (1989), p.153-172.

[15] Korbel, A., Embury, J. D., Hatherly, M., Martin, P. L., and Erbsloh, H. W., Acta Metallurgica et Materialia, 34, 10, (1986), p.1999-2009.

Correspondence should be addressed to: Peter.Hurley@eng.monash.edu.au (e-mail)
                                              or (61-3)-99054940 (Fax).

Materials Science Forum Vols. 284-286 (1998) pp. 167-174

Combined Effect of Nb and Ti on the Recrystallisation Behaviour of Some HSLA Steels

R. Abad, B. López and I. Gutierrez

CEIT and ESII, Univ of Navarra,
P° Manuel de Lardizábal 15, E-20009 San Sebastián, Basque Country, Spain

Keywords: No-Recrystallisation Temperature (T_{nr}), Hot Multipass Torsion Test, Strain Induced Precipitation

ABSTRACT

The recrystallisation behaviour of several Nb-Ti bearing high-strength low-alloy steels was investigated during multipass deformation by torsion under continuous cooling conditions. The effect of the strain on the no-recrystallisation temperature (T_{nr}), the temperature at which recrystallisation is no longer complete, was determined. It was observed that the T_{nr} decreases with increasing strain. It has been observed that for similar Nb concentrations, increasing the Ti content increases the T_{nr}, but this effect seems to saturate at the higher Ti-concentrations and over a given Ti content there is no additional increase of the T_{nr}. This effect can be related to the presence in the austenite of an increased amount of undissolved (Ti,Nb)(C,N) particles when the Ti/N ratio is increased by the addition of titanium. Because of this, the supersaturation level is reduced leading to a low driving force for precipitation during deformation. Additionally these undissolved particles may favour heterogeneous precipitation of carbonitrides and in consequence the precipitation on dislocations, which is the responsible for the retardation of recrystallisation, can be seriously impaired. Both effects can significantly reduce the effectiveness of the titanium addition in increasing the T_{nr}.

INTRODUCTION

The main objetive of the thermomechanical processing of steels is to refine the ferrite grain size, which is well known to be the only parameter which can simultaneously improve the strength and toughness in steels. One method that can lead to a significant refinement of the ferrite grain size is to accumulate as much strain as possible in the austenite before transformation occurs. This is the philosophy of the process known as conventional controlled rolling (CCR). The strain retained during finishing rolling operations causes the austenite grains to be flattened and elongated and introduces intragranular defects, such as deformation bands and twin boundaries. The strained austenite is thus effectively refined, in the sense that its grain boundary area, including intragranular defects, is significantly increased and it transforms to a much finer ferrite grain structure [1-4].

Accumulation of deformation takes place when the austenite is deformed at temperatures low enough to produce strain induced precipitation of carbonitrides during deformation which inhibit recrystallisation [5-9]. The temperature below which this occurs is known as the no-recrystallisation temperature (T_{nr}). Increasing this temperature allows the finishing rolling operations to be performed at higher temperatures which translates into lower efforts to be applied by the mills. Some data found in the literature point out the beneficial effect of the addition of titanium to the niobium microalloyed steels in increasing the value of the T_{nr} [10].

In the present work the effect of the combination of Ti and Nb on the no-recrystallisation temperature of some HSLA steels has been studied. It has been observed that the addition of titanium to Nb-bearing steels can influence the kinetics of the Nb(C,N) precipitation and thus the kinetics of recrystallisation. The undissolved TiN particles may favour heteregoneous precipitation of carbonitrides and also reduce the amount of solute available for precipitation during deformation which can result in a decrease of the T_{nr} for a given composition.

EXPERIMENTAL

Four microalloyed low carbon steels with different contents of Nb and Ti were used in the present work. Additionally two Nb-bearing low carbon steels were included in the study for comparison. Their chemical composition is listed in Table 1.

Table 1 Chemical composition of the steels

Steel	C	Mn	Si	P	S	Cu	Al	Ti	Nb	Nppm
A	0.044	1.39	>0.02	0.013	0.001	>0.02	0.027	0.066	0.062	52
B	0.06	1.47	>0.02	0.012	0.001	>0.02	0.028	0.025	0.060	75
C	0.132	1.44	0.23	-	-	-	0.03	0.012	0.043	56
D	0.07	0.62	0.012	0.011	0.006	0.026	0.053	0.067	0.034	43
E	0.11	1.14	0.51	0.005	0.005	-	0.029	-	0.049	60
F	0.1	1.4	0.3	0.018	0.008	<0.02	0.039	-	0.035	53

Multipass torsion tests have been carried out to determine the no-recrystallisation temperature (T_{nr}). Before deformation the samples were preheated for 15-20 min at 1200°C. After reheating the specimens were deformed under multipass torsion tests at decreasing temperature at a rate of 1°C/s in the range 1180°C-700°C. The tests were carried out using different pass-strains in the range ε=0.1-0.7, pass strain-rate of 1 and 2 s^{-1} and interpass times of 30s. In particular tests, the strain per pass, strain-rate and interpass time were held constant. After the tests the specimens were water quenched. The dispersion of precipitates in the deformed region of the torsion specimens was analysed using carbon extraction replicas, which were examined in a Philips CM12 STEM fitted with an EDAX PV 9900 X-ray microanalyser.

RESULTS AND DISCUSSION

Determination of Tnr

Figure 1 shows a typical stress-strain curve obtained in a 17-pass torsion test corresponding to the steel A deformed at decreasing temperature and using a strain per pass of 0.2 and strain-rate of 2 s^{-1}. The stress-strain curve can be divided in three zones. The first one is the zone where full recrystallisation takes place between passes and the stress increase is only due to the temperature decrease. In the second zone the material is not able to recrystallise completely between passes and begins to accumulate deformation. This is the zone where partial or no recrystallisation at all occurs. In this zone, the stress increases more rapidly due to both, the temperature decrease and the strain accumulation. At the end, in the third zone, a decrease in the flow stress is clearly observed. The stress reduction results from the start of the transformation of the austenite to the softer ferrite phase (A_{r3} temperature). When the transformation is more or less completed the stress increases again as a result of the strain hardening of the ferrite-pearlite structure and the decrease in temperature.

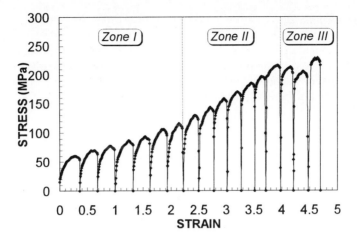

Fig. 1. Stress-strain curves for sample deformed using a 17-pass schedule
A steel: T_{soak} = 1200°C; ipt = 30 s, ε = 0.2, $\dot{\varepsilon}$ ≈ 2 s^{-1}

In Fig.2 the dependence of the mean flow stress (MFS) on inverse pass temperature has been plotted. The mean flow stress corresponding to each pass has been calculated from the flow stress curves in Fig.1 using the equation [11]:

$$\overline{\sigma} = \frac{1}{\varepsilon_b - \varepsilon_a} \int_{\varepsilon_a}^{\varepsilon_b} \sigma \cdot d\varepsilon \qquad (1)$$

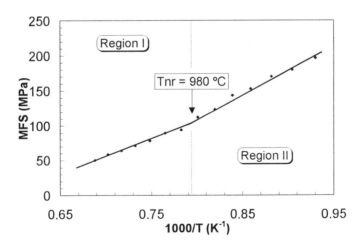

Fig. 2. Dependence of the MFS on inverse pass temperature during the multipass torsion testing of A steel

The slope change divides the temperature range in two regions. In region I, full recrystallisation between passes takes place. The region II corresponds to deformations below the T_{nr}, thus there is only partial recrystallisation or no recrystallisation at all. As Fig.2 shows, the no-recrystallisation temperature can be obtained from the intersection between both straight lines.

Effect of pass-strain on the T_{nr}

Figure 3 shows the influence of the pass-strain on the no-recrystallisation temperature for the Nb-Ti microalloyed steels analysed in the present work. From the figure it is clearly evident that the T_{nr} decreases with increasing the strain per pass. The dependence of the T_{nr} on pass strain can be described empirically by the following relationship:

$$T_{nr} = \beta \, \varepsilon^{-0.072} \qquad\qquad (2)$$

where β is the same for steels A and B (877°C) and slightly lower (835°C) for steel C. These results are consistent with those previously reported for Nb-microalloyed steels by Bai et al. [11].

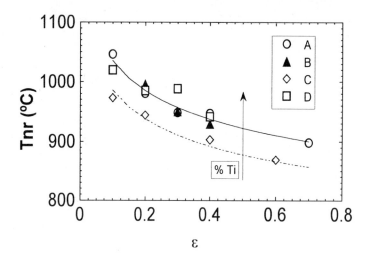

Fig. 3. Dependence of the T_{nr} on the pass strain

Another feature that can be observed in Figure 3 is the influence of the Ti content on the T_{nr} value. It is observed that for steels having similar contents of Nb, increasing the Ti content from the value of 0.012 corresponding to steel C to that of 0.067 which is the content of the steel D, the T_{nr} increases about 40°C for all the strain range. But it is also clearly evident that the increase in the Ti content from 0.025 in steel B to 0.066 in steel A does seem not to produce any additional increment in the T_{nr}. From the previous results it can be deduced that when the Nb content is maintained nearly constant, the addition of Ti leads to higher values of the T_{nr}. However this effect seems to saturate for the higher Ti concentrations used in the present work. This means that over a given Ti content there is no additional increase of the T_{nr}.

Precipitation

The study of the precipitation present in the deformed zone of the torsion specimens has been carried out for steels B and A. This precipitation has been observed in two steps. Optical microscopy has been used to observe coarse precipitates (>0.5μm) and extraction replicas have been prepared to analyse the finest ones by transmision electron microscopy.

A wide range of sizes for the precipitates has been observed. For both steels the presence of coarse precipitates, >0.5 μm, is clearly evident, but these coarse particles were found at a much higher frequency in steel A than in steel B, as has been reported elsewhere [12]. The size of the precipitates can be related to the composition, more exactly, with the different Ti/N ratios. In the present steels, which have a relatively low content of nitrogen, the increase of the Ti content, over the stiochiometric Ti/N=3.42 ratio, would favour precipitation starting in the liquid prior to solidification [13]. These particles formed at high temperature have time to coarsen and are very stable. This could be the reason why steel A, which has a very high content of titanium (Ti/N = 12.69), exhibits precipitates coarser than those observed in steel B, which has a lower titanium concentration (Ti/N =3.33, close to the stiochiometric composition).

Figure 4 shows some TEM micrographs of the typical precipitation observed in these steels. From the figure, several distinct morphologies of precipitate can be observed, *i.e.* plate, cuboid, and spheroid. The larger particles have been observed to be mainly rich in Ti and to present a plate-cuboid shape. On the contrary, the smaller ones (<25 nm) are rich in Nb and predominantly spherical, and, in certain cases, elongated (Figs.4(c) and (d)). Figure 5 shows the size distribution of the precipitates observed on the carbon replicas (< 0.5μm) for both steels. From the figure some differences in the size of the precipitates, especially for the coarser particles, can be observed between both steels. In the case of the steel B a normal distribution is observed, with the major fraction of particles lower than about 75 nm. For steel A a bimodal distribution is observed. About 40% of particles are less than 25 nm, the precipitates between 25 and 75 nm amount to less than 20%, whilst those in the range 75-100 nm are again more numerous and constitute a 25%. However, for both steels about 30% of the precipitates observed by TEM are less than 10 nm, and are in the range of those reported to be able to inhibit recrystallisation [14].

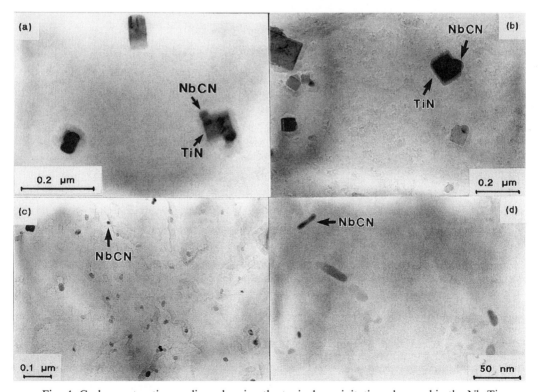

Fig. 4. Carbon extraction replicas showing the typical precipitation observed in the Nb-Ti microalloyed steels

As pointed out above, the proportion of Ti and Nb in the different particles changes according to their size. Ti concentrates, mainly in coarser ones, while the smaller size particles (<25nm) are mainly rich in Nb. This is in agreement with some thermodynamic calculations for the complex precipitation of carbonitrides of the type $Ti_xNb_{1-x}C_yN_{1-y}$, carried out appliying a regular solution model [12]. From this study it was deduced that the particles formed at high temperatures were rich in Ti and N, of the type TiN, but with decreasing temperature of precipitation, the content of Nb and C present in the newly formed precipitates increased. From the previous, it is deduced that the Ti-rich cuboids and plates observed in Figure 4 could correspond to titanium nitrides precipitated at high temperatures (probably some of them in the liquid) and which have not been dissolved during reheating. On the other hand the Nb-rich precipitates would be formed at lower temperatures and of lower size consequently (see Fig. 4(d)).

The analysis carried out by Diffraction patterns shows that both Ti and Nb richs phases agree with a f.c.c. type of structure having lattice parameters close to 0.43 nm, which is consistent with titanium and niobium nitrides. Additionally, Ti rich large particles act frequently as nucleation sites for Nb-rich precipitates (see Fig.4). It has been observed, in this case that the latters form coherently with the Ti rich substrate.

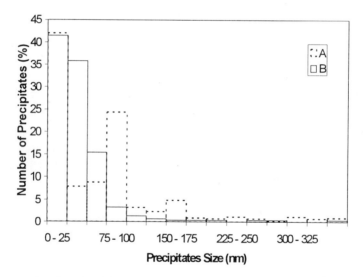

Fig. 5. A histogram showing the difference in the size distribution of the precipitates between high Ti and low Ti steels.

The no-recrystallisation temperature is mainly related to the precipitation of carbides/nitrides of the microalloyed elements which takes place on dislocations (strain induced precipitation), preventing the formation and growth of recrystallised nuclei [15]. The occurrence of precipitation will depend on the supersaturation level reached when the temperature decreases during deformation after reheating. High levels of supersaturation will promote the precipitation starting at higher temperatures, thus the values of T_{nr} will be increased in this way. Dissolution of precipitates during reheating is required to provide high supersaturation which gives the driving force for precipitation at lower temperature. At the relatively low reheating temperature of 1200°C, many particles will remain undissolved, both in steel A and in steel B. However, in steel A coarser particles are present in the starting material, which are more difficult to dissolve, and in consequence the amount of solute available to precipitate later, at lower temperatures, will be probably less than would be expected for the high titanium content of this steel. In addition these undissolved particles constitute preferential nucleation sites for the precipitates formed later at decreasing temperature, as they provide low energy interfaces favourable for precipitation (see Fig. 4). As a consequence of this, the precipitation on dislocations, which is responsible for the retardation of recrystallisation, can be seriously impaired. Both effects could contribute to the saturation of the T_{nr} observed at the higher Ti contents in Fig.3.

Comparison between Nb-Ti microalloyed steels and Nb-microalloyed steels

In Figure 6 the values of the T_{nr} obtained for the steels C and A are compared to those obtained for the Nb-microalloyed steels analysed in the present work. From the figure it can be observed that for similar contents of niobium, the T_{nr} of the Nb-Ti steel is significantly lower than that observed for both Nb-steels at all strains. Similar T_{nr} tempertures are only obtained if a large amount of Ti and Nb are used in Nb-Ti steel (steels A and B). This behaviour can be explained as follows. In the case of the only Nb-bearing steels, the equilibium solution temperatures for Nb(C,N) can be estimated from the equation given by Irvine et al. [16] and the obtained values are 1168 and 1228°C for steels F and E respectively. However in the presence of titanium and niobium it has been observed that they are mutually soluble and the precipitation will start at very high temperatures as (Ti,Nb)N because a part of the Nb will be tied up with the Ti (Ti-rich particles observed in Fig.4). These particles formed at high temperature are very stable and high reheating temperatures would be required to dissolve them. In the present tests a reheating temperature of 1200°C has been used. This temperature would be enough to dissolve all the niobium in steel F, maximising in this way the effect of niobium in retarding recrystallisation by strain induced precipitation during deformation. However this temperature is not high enough to dissolve the (Ti,Nb)N particles, and a significant fraction of the added niobium may remain undissolved in the austenite. Consequently, the supersaturation becomes lower and results in the necessity of going to lower temperatures for the precipitation of Nb carbonitrides, able to pin the dislocations and halt the recrystallisation process (decrease of T_{nr}). Additionally, part of the Nb effect is lost by the coherent precipitation of Nb-carbonitrides on the undissolved high temperature stable Ti-Nb carbonitrides.

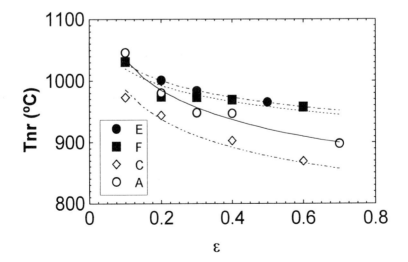

Fig. 6. Dependence of the T_{nr} on pass strain for Nb-Ti and Nb-bearing steels with similar base composition.

CONCLUSION

The presence of Ti in addition to Nb, instead of increasing the supersaturation due to the presence of Ti, can produce a decrease in the supersaturation due, firstly to the decrease in Nb solution and, secondly to the preferential precipitation of Nb rich carbonitrides on Ti rich particles which remain undissolved during reheating. This lower supersaturation results in a reduction of the no-recrystallisation temperature of the Nb-Ti steels as compared with the only Nb-bearing steels having similar base compositions.

ACKNOWLEDGEMENTS

Part of this research work has been carried out under an ECSC Mega-Project No 7210.EC/934. The authors thank the European Coal and Steel Community for partial funding of the research, the project leaders at the partner institutes and their colleagues. The authors wish to dedicate this article to the memory of Prof. J.J.Urcola, who was directly involved in this project before his untimely death.

REFERENCES

[1] I. Tamura, Int. Conf. on Physical Metallurgy and Thermomechanical Processing- Thermec 88, The ISI of Japan, (1988),p. 1-10.
[2] C. Ouchi, T. Sampei and I.Kozasu, Trans. ISIJ, vol. 22, (1982),p. 215-222
[3] T. Tanaka, "Microalloying 95", ISS, Pittsburgh, M. Korchynsky, A.J. DeArdo, P. Repas, G. Tither eds, (1995), p. 165.
[4] A. J. DeArdo, "Microalloying 95", ISS, Pittsburgh, M. Korchynsky, A.J. DeArdo, P. Repas, G. Tither eds., (1995), p. 15.
[5] L.J.Cuddy, "Thermomechanical Processing of Microalloyed Austenite", A.J.DeArdo and al. eds., AIME, Warrendale PA, (1982), p. 129.
[6] Sun, W.P., Liu, W.J. and Jonas, J.J., Met. Trans.,Vol.20A, (1989), p. 2707.
[7] J.G. Speer, J.R. Michael, and S.S. Hansen, Metall. Trans. A, vol. 18A, (1987), p. 211-22.
[8] J.G. Speer, and S.S. Hansen, Metall.Trans. A, vol. 20A, (1989), p. 25-38.
[9] O.Kwon and A.J. DeArdo, Acta Metall, vol.39, (1991), p. 529.
[10] S.V. Subramanian and H.Zou, Proc. of the Int. Conf. on Processing, Microstructure and Properties of Microalloyed and Other Modern High Strength Low Alloy Steels, A.J.DeArdo ed., ISS, Pittsburgh, (1992), p. 23.
[11] D.Q. Bai, S. Yue, W.P. Sun and J.J. Jonas: Met. Trans, Vol. 24A, (1993), p. 2151-2159.
[12] B.López et al. "New development in Thermomechanical Treatments of Nb-Ti Steels. Optimisation and Modelisation", Final Report, ECSC Contract 7210.EC/933-934-935-936, CEIT, (1996).
[13] S.Zajac, R.Lagneborg and T. Siwecki, Proc. of Conf. "Microalloying 95", (1995), p. 321-338.
[14] S.S. Hansen, J.B. Vander Sande and M. Cohen, Met. Trans. vol 11, (1980), p. 387-402.
[15] B. Dutta and C.M. Sellars, Mater. Sci. Technol., vol. 22A, (1987), p. 1511-23.
[16] K.J. Irvine, F.B. Pickering, and T.Gladman, J. Iron Steel Inst., vol. 205, (1967), p. 161-82.

B.López
E-mail: blopez@ceit.es
Fax: (34)43 21 30 76

PHASE TRANSFORMATION

Materials Science Forum Vols. 284-286 (1998) pp. 177-184
© *1998 Trans Tech Publications, Switzerland*

Effect of Composition and Prior Austenite Deformation on the Transformation Characteristics of Ultralow-Carbon Microalloyed Steels

P. Cizek[1], B.P. Wynne[1], C.H.J. Davies[1], B.C. Muddle[1], P.D. Hodgson[2] and A. Brownrigg[3]

[1] Department of Materials Engineering, Monash University, Clayton, Vic., 3168, Australia

[2] School of Engineering and Technology, Deakin University, Waurn Ponds, Vic., 3217, Australia

[3] BHP Research, Melbourne Laboratories, 245 Wellington Rd, Mulgrave, Vic., 3170, Australia

Keywords: Ultralow-Carbon Microalloyed Steel, Deformation Dilatometry, Hardenability

Abstract

Deformation dilatometry has been used to simulate controlled hot rolling followed by controlled cooling of a group of ultralow-carbon microalloyed steels containing additions of B and/or Mo to enhance hardenability. Each alloy was subjected to simulated recrystallisation and non-recrystallisation rolling schedules, followed by controlled cooling at rates from 0.1°C/s to about 100°C/s. The resulting transformation products were analysed using optical microscopy in conjunction with hardness measurements which, together with the dilatometry data, facilitated the construction of CCT diagrams. The resultant microstructures ranged from polygonal ferrite (PF) for combinations of slow cooling rates and low alloying element content, through to lath bainitic ferrite accompanied by martensite for fast cooling rates and high concentrations of alloying elements. The combined addition of B and Mo was found to be most effective in increasing steel hardenability, while B was significantly more effective than Mo as a single addition. Large plastic deformation of the prior austenite markedly enhanced PF formation, both in the base and Mo-containing steels, thus decreasing their hardenability significantly. In contrast, the steels containing B displayed only very small decreases in hardenability, resulting from the lack of sensitivity to strain in the austenite of the non-equilibrium microstructure constituents forming in the absence of PF.

Introduction

The continually increasing requirement for steels to exhibit enhanced strength in good balance with high toughness and excellent weldability has led to recent intensive examination of the transformation products obtained by continuous cooling of low- and ultralow-carbon microalloyed steels [1]. These products often display non-equilibrium, non-equiaxed ferrite microstructures as a result of suitable microalloying combined with accelerated cooling. Although formed in the temperature range typical for classical bainite in medium-carbon steels, these microstructures do not contain cementite and possess some unique morphological features. For these reasons, a new terminology has recently been proposed [2,3] in an attempt to describe all possible ferrite morphologies formed by the decomposition of austenite in this new family of steels. Apart from martensite (M), this terminology recognises five separate forms of ferrite: (i) polygonal ferrite (PF), characterised by equiaxed grains with smooth boundaries, containing a low dislocation density and no substructure; (ii) Widmanstätten ferrite (WF), defined by elongated crystals of ferrite with minimal presence of dislocation substructure; (iii) quasipolygonal ferrite (QF), characterised by grains with irregular boundaries containing dislocation substructure and occasionally also martensite-austenite (M/A) microconstituent; (iv) granular bainitic ferrite (GBF), which consists of a matrix of either equiaxed or elongated ferrite crystals with low misorientations and a high dislocation density, containing equiaxed islands of M/A microconstituent; and (v) lath bainitic ferrite (LBF), which consists of fine elongated ferrite crystals with a high dislocation density and low-angle boundaries, but in contrast to GBF the M/A microconstituent retained between the ferrite laths has an acicular morphology.

Recently, work has commenced [4-7] to characterise the transformation behaviour and microstructures developed in low-carbon microalloyed steels processed to simulate controlled rolling in conjunction with continuous cooling. This work has indicated that the transformation characteristics depend significantly on chemical composition as well as processing variables such as strain in the parent austenite. The aim of the present investigation was to further elucidate this dependence in a group of ultralow-carbon microalloyed steels containing additions of boron and/or molybdenum to enhance hardenability.

Experimental Procedures

The chemical compositions of the steels studied are given in Table I. These steels were vacuum melted as 60 kg ingots and hot rolled to a thickness of 6 mm at BHP Research-Melbourne Laboratories (BHPR-ML).

Table I. Chemical Compositions of Steels (wt%)

Steel	C	Si	Mn	P	S	Mo	Nb	Ti	B (ppm)
A	0.005	0.189	1.68	0.020	0.010	0.027	0.023	0.014	<1
B	0.007	0.212	1.67	0.021	0.011	0.309	0.024	0.016	<1
C	0.007	0.200	1.64	0.017	0.011	0.008	0.022	0.015	20
D	0.009	0.205	1.70	0.024	0.011	0.273	0.030	0.016	35

Deformation dilatometry was performed at BHPR-ML using a computerised MMC-High Speed Quenching and Deformation Dilatometer. The dilatometer specimens had a diameter of 4 mm and a length of 8 mm. Each steel was subjected to simulated recrystallisation and non-recrystallisation rolling schedules followed by controlled cooling at rates from 0.1°C/s to about 100°C/s (Fig. 1). The cooling rates were defined by the time interval required for the specimens to cool from 800°C to 500°C.

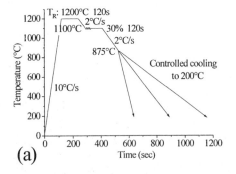
(a)

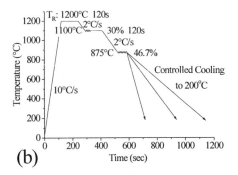
(b)

Figure 1 - Schematic of simulated rolling schedules: (a) recrystallisation schedule;
(b) non-recrystallisation schedule involving a reduction of 47% at 875°C.

Metallography specimens were prepared using standard procedures and etched using 2% nital. All specimens were Vickers hardness tested using a 5 kg load and analysed by optical microscopy. The information obtained facilitated the construction of continuous cooling transformation (CCT) diagrams. In order to evaluate the impact of different variables on the transformation temperatures of a given microstructural constituent, the comparisons between these temperatures were always made at a similar microstructure (a roughly equivalent position on a CCT digram) rather than at a similar coling rate.

Results

1. Effect of Composition

The CCT diagrams of the steels studied are presented in Fig. 2. It is evident that the addition of Mo (steel B) to the base composition (steel A) did not cause a significant change in the transformation characteristics. The transformation temperatures of steel B did not seem to differ noticeably from those of steel A. The addition of B alone (steel C) had a more significant effect on the change in transformation characteristics of the base steel than that of Mo. In steel C, the formation of PF was completely suppressed throughout the range of cooling rates studied, even in the case of transformation from the deformed austenite. As a result, the formation of a mix of QF and GBF extended over the whole range of low to intermediate cooling rates and the corresponding transformation temperatures became approximately constant throughout this range. The transformation field corresponding to the formation of LBF was shifted towards the slower cooling rates compared to steel A. Consequently, some M was introduced to the microstructure at the highest cooling rates. Nevertheless, no significant differences in the transformation temperatures between steels A and C were detected when comparing transformation products with similar microstructural characteristics. The addition of both B and Mo (steel D) to the base composition did not alter the appearance of the CCT diagrams significantly compared to steel C. However, in the case of recrystallised austenite, both the transformation start and transformation finish temperatures in steel D were mostly markedly lower than those in steel C. As a result, the microstructure containing predominantly GBF accompanied by only a tiny amount of QF was formed in steel D in the range of low to intermediate cooling rates, instead of the mix of QF and GBF observed in steel C. In the case of deformed, unrecrystallised austenite, the transformation start temperatures in steel D were largely similar to those in steel C, whereas the transformation finish temperatures were mostly significantly lower.

2. Effect of Austenite Deformation

In steel A, deformation of the prior austenite brought about a significant expansion of the PF transformation field in the CCT diagram, the highest cooling rate for PF formation being shifted from about 1°C/s to about 15°C/s, compare Figs. 2a and 2b. The corresponding transformation temperatures remained either approximately unchanged or underwent only a small increase, as illustrated by the small deviations of the transformation points (marked by solid circles) from the superimposed open triangles in Fig. 2b. The transformation temperatures represented by the triangles were derived from the corresponding CCT diagram of the recrystallised austenite (Fig. 2a) at comparable microstructures. Thus, such a comparison (also illustrated in Figs. 2d, 2f and 2h) shows the effect of the austenite deformation on the transformation characteristics of a given microstructural constituent rather than the same effect at a given cooling rate, that can simply be evaluated using a direct overlap of the respective CCT diagrams. The effect of the parent austenite deformation on the transformation behaviour of steel B was found to be fundamentally similar to steel A, compare Figs. 2c and 2d. Steel C appeared to be the least sensitive to austenite deformation of all the steels studied. The corresponding transformation characteristics remained basically unchanged, except for a small reduction of the M transformation region, compare Figs. 2e and 2f. Steel D was found to be somewhat more sensitive to the prior austenite deformation than steel C. As a result of the deformation, the transformation start temperatures for the GBF formation increased,

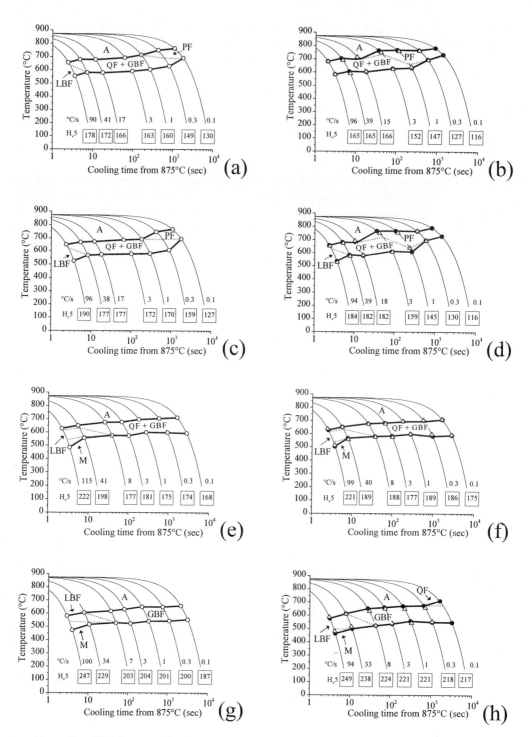

Figure 2 - CCT diagrams for the recrystallised (a,c,e,g) and deformed (b,d,f,h) austenite:
Steel A (a,b); Steel B (c,d); Steel C (e,f); Steel D (g,h).

while the transformation finish temperatures remained approximately unchanged and, thus, the corresponding temperature interval somewhat expanded, compare Figs. 2g and 2h. The transformation start temperatures of the mix of LBF and M remained roughly the same, whereas the transformation finish temperatures were slightly lowered, perhaps in accord with the slightly larger M volume fraction observed. Moreover, deformation of the austenite introduced an increased volume fraction of QF to the microstructure at the lowest cooling rate (0.1°C/s).

3. Transformation Microstructures

The microstructures of steels A and B, obtained by very slow cooling from the recrystallised austenite, consisted of coarse PF grains characterised by mostly fairly straight boundaries and absence of substructure. Some WF sideplates, extending from the polygonal grains, were also occasionally observed. Microstructures composed of a mix of QF and GBF were observed at the medium cooling rates in steels A and B and at the slow to medium cooling rates in steel C, Figs. 3a and 3d. As both QF and GBF contained a pronounced substructure and some very fine M/A microconstituents, it was often difficult to distinguish between them. The ragged QF grains frequently crossed the prior austenite grain boundaries and, thus, the locations of these boundaries were only partially conserved. In steel D, the microstructure formed at the slow to medium cooling rates was composed almost entirely of GBF, accompanied by only a tiny amount of QF, Figs. 3b and 3e. Consequently, the locations of the original austenite grain boundaries remained largely preserved. At the high cooling rates, the microstructure of all the steels studied was dominated by LBF, Figs. 3c and 3f. There was also some M accompanying LBF in steels C and D. The locations of the prior austenite grain boundaries were fully conserved.

The microstructures obtained after transformation from the deformed austenite differed detectably from those transformed from the recrystallised austenite. The microstructures containing PF and Qf became finer and both GBF and LBF became more fragmented.

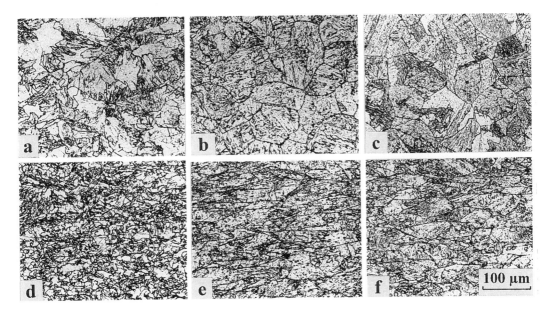

Figure 3 - Optical micrographs of the microstructures transformed from the recrystallised (a,b,c) and deformed (d,e,f) austenite: (a) steel A, 41°C/s, QF+GBF; (b,e) steel D, 1°C/s, GBF; (c) steel D, 34°C/s, LBF; (d) steel A, 39°C/s, QF+GBF; (f) steel D, 33°C/s, LBF.

Discussion

1. Effect of Composition

The base steel A contains Nb that is known to enhance hardenability [1,7,8]. Thus, the observed significant improvement in hardenability arising from both an individual addition B (steel C) and a combined addition of Mo plus B (steel D) is, in fact, the result of a coupled effect of these elements in conjunction with Nb. Hardenability mechanisms that operate in low-carbon microalloyed steels appear to be rather complex and are not fully understood at the present time. It has been suggested [1,8] that microalloying elements such as Nb, Mo and B segregate at the prior austenite grain boundaries, and also at the boundaries introduced by deformation, which reduces the surface energy of these boundaries and thus decreases their effectiveness as ferrite nucleation sites. Alternatively, it has been proposed [9] that the same effect could be brought about by extremely fine carbonitride or carboboride particles precipitating continuously at the austenite grain boundaries with a specific orientation relationship and covering the incoherent boundary facets with coherent precipitate/austenite interfaces having a comparatively lower surface energy. The presence of Nb and Mo in solid solution appears to decrease both the nucleation and growth of ferrite through reduction in carbon diffusivity in austenite and also by a partitioning within the product phases that imposes a strong diffusive drag on the phase transformation [8,10]. It has been proposed that the microalloying elements could either segregate [10] or create fine carbonitride precipitates [11] at the austenite/ferrite interface thus hindering its motion by solute drag or pinning effects respectively.

In the present study, the addition of Mo alone did not seem to enhance noticeably the hardenability of the base steel. The individual addition of B to the base steel composition was found significantly more effective in promoting hardenability, completely suppressing the formation of PF throughout the entire range of cooling rates studied. Several possible mechanisms have been proposed to account for such a prominent synergistic effect of the combined addition of B and Nb, which has been reported to enhance hardenability significantly more than an individual addition of B in the absence of Nb [8]. As already mentioned above, Nb can reduce the diffusivity of carbon in austenite [8]. Therefore, Nb dissolved in austenite might protect boron from forming carboborides thus maintaining B in solid solution and increasing its availability for segregation at all possible PF nucleation sites. Moreover, the addition of Nb brings about precipitation of Nb carbonitrides, which limits the supply of carbon required for the precipitation of carboborides. It has been shown [8] that this effect could be particularly important in the case of transformation from deformed, unrecrystallised austenite and appears to intensify with decreasing carbon content. It has also been postulated [1,8] that the addition of Nb could be very effective in both retarding the austenite grain boundary motion at high reheating temperatures through a solute drag effect and suppressing recrystallisation of the deformed austenite at lower temperatures via the combination of solute drag and pinning by fine strain-induced carbonitrides. As a consequence, B would have sufficient time to diffuse to the austenite boundaries, which enhances its segregation capacity and increases austenite hardenability. The results of the present study indicate that an especially large increase in hardenability could be achieved by the combined addition of Mo and B to the base steel containing Nb, which is also in good correspondence with the published data [1]. This phenomenon could possibly be attributed to the contribution of solute Mo to the deceleration of carbon diffusion [8] and also to the stabilisation of austenite grain boundaries by a solute drag effect [1], which further enhances the opportunity for B segregation.

2. Effect of Austenite Deformation

The large deformation of the austenite, carried out at a non-recrystallisation temperature of 875°C in the present study, quite markedly enhanced the formation of PF in steels A and B, significantly extending its presence in the CCT diagrams to the higher cooling rates. A pronounced shift of the PF transformation field towards higher cooling rates due to the prior austenite deformation has frequently been reported in the literature [4,7]. It has often been claimed that this shift is also accompanied by a significant increase in the corresponding transformation start temperatures [7]. This phenomenon, however, appears to be mainly a consequence of comparing these temperatures at a similar cooling rate rather than at a similar microstructure (a roughly equivalent position on a CCT diagram) as has been done in the present study. The present results indicate only small increases in the PF transformation temperatures due to prior austenite deformation. The decrease in hardenability due to the deformation-enhanced PF formation, described above, may be attributed to both an increase in the austenite stored energy due to deformation and a rise in the density of ferrite nucleation sites, as well as to the lowering of the amount of Nb in solid solution due to strain-induced precipitation of Nb carbonitrides [4,7].

In the present study, the prior austenite deformation appeared to promote QF formation through a shift of the corresponding transformation field in the CCT diagram towards comparatively higher cooling rates, as evidenced by the observed introduction of some fraction of this constituent to the microstructure of steel D at the lowest cooling rates. Such a behaviour seems to be principally similar to that of PF discussed above, which is also in good correspondence with the observations reported in the literature [6]. The experimental results obtained for steels A, B and C suggest that the transformation start temperatures corresponding to QF formation might be quite insensitive to austenite deformation. In steel D, GBF was observed to be formed almost entirely without the interference of QF formation, which allowed the effect of the parent austenite deformation on the GBF transformation characteristics to be evaluated. It appears that GBF nucleation might be enhanced due to the deformation, as indicated by the noticeable increase in the transformation start temperatures, while the progress of the transformation seems to have been retarded. As far as the LBF transformation characteristics were concerned, they appeared rather insensitive to the parent austenite deformation in the present investigation.

The formation mechanisms of QF, GBF and LBF still remain to be clarified. It is widely believed that QF might be a product of massive transformation, accomplished by short-range diffusion across transformation interfaces and there is some evidence indicating that GBF might grow by a diffusion-controlled ledge mechanism [2,3]. There have been suggestions that LBF might be formed through displacive growth events followed by partitioning of the excess carbon into the residual austenite, a mechanism that has often been proposed for the formation of classical bainite [5]. In principle, prior austenite deformation might influence the transformation characteristics qualitatively in a similar manner, irrespective of the transformation mechanism involved. Both the austenite free energy and the density of nucleation sites will generally increase as a result of deformation, thus raising the nucleation rate. In addition, the progress of the transformation may be similarly retarded when either a diffusional or displacive mechanism is operative. A displacive transformation could be susceptible to mechanical stabilisation, implying that the overall progress of the transformation might be suppressed because the deformation-induced dislocation substructure hinders the growth of ferrite crystals [12]. Conversely, the progress of a diffusion-controlled transformation might be retarded by strain-induced precipitation of fine carbonitrides and also by other mechanisms [7]. Moreover, both the present results and those published in the literature [6,8] indicate that changes in the transformation characteristics of the non-equilibrium ferrite constituents, despite being induced by large strains in the prior austenite, might often be quite subtle. Thus, it would be somewhat speculative to attempt to contribute to the discussion on the transformation mechanisms of QF, GBF and LBF on the basis of the results obtained in the present study.

Conclusions

Effects of composition and prior austenite deformation on the transformation characteristics of a group of four ultralow-carbon microalloyed steels, containing hardenability-enhancing additions of B and/or Mo, have been studied using deformation dilatometry. The combined addition of B and Mo was found to be especially effective in improving hardenability, while the individual addition of B was significantly more effective than that of Mo. In both the base and Mo-containing steels, the austenite deformation brought about a significant decrease in hardenability due to enhanced PF formation. Conversely, the steels containing B displayed only a very small decrease in hardenability, resulting from lack of sensitivity to deformation of the non-equilibrium microstructure constituents forming in the absence of PF.

Acknowledgement

The authors acknowledge gratefully the support of an Australian Research Council Collaborative Research Grant in conjunction with BHP Research-Melbourne Laboratories.

References

[1] C.I. Garcia, Proc. Int. Conf. Microalloying '95, Pittsburgh, USA, June 1995, ed. M. Korchynsky et al. (Iron and Steel Society, 1995), p. 365.
[2] T. Araki, M. Enomoto and K. Shibata, Mater. Trans. JIM 32 (1991), p. 729.
[3] G. Krauss and S.W. Thompson, ISIJ Int. 35 (1995), p. 937.
[4] J.R. Yang, C.Y. Huang and S.C. Wang, Proc. Int. Symp. on Low-Carbon Steels for the 90's, Pittsburgh, USA, October 1993, ed. R. Asfahani and G. Tither (TMS, 1993), p. 293.
[5] J.R. Yang, C.Y. Huang and C.S. Chiou, ISIJ Int. 35 (1995), p. 1013.
[6] S. Yamamoto et al., ISIJ Int. 35 (1995), p. 1020.
[7] P.A. Manohar, T. Chandra and C.R. Killmore, ISIJ Int. 36 (1996), p. 1486.
[8] H. Tamehiro, M. Murata and R. Habu, Proc. Int. Conf. on HSLA Steels '85, Beijing, November 1985, ed. J.M. Gray et al. (ASM International, 1986), p. 325.
[9] Y.C. Jung et al., ISIJ Int. 35 (1995), p. 1001.
[10] J.S. Kirkaldy, B.A. Thomson and E.A. Baganis, Hardenability Concepts with Applications to Steel, ed. D.V. Doane and J.S. Kirkaldy (TMS-AIME, Warrendale, USA, 1977), p. 82.
[11] M.H. Thomas and G.H. Michal, Solid-Solid Phase Transformations, ed. H.I. Aaronson et al. (TMS-AIME, Warrendale, USA, 1981), p. 469.
[12] P.H. Shipway and H.K.D.H. Bhadesia, Mater. Sci. Technol. 11 (1995), p. 1116.

Correspondence should be addressed to: Pavel.Cizek@eng.monash.edu.au (e-mail)
                                           or (61-3)-99054940 (Fax)

Materials Science Forum Vols. 284-286 (1998) pp. 185-192

Nb and the Transformation from Deformed Austenite

M. Onink, Th.M. Hoogendoorn and J. Colijn

Hoogovens Research and Development, Metallurgy of Hot Rolled Steel,
PO Box 10.000, NL-1970 CA Ijmuiden, The Netherlands

Keywords: Transformation, Grain Refinement, Widmanstätten Ferrite, Dilatometry

ABSTRACT

The influence of Nb on the transformation behaviour depends on the thermal and mechanical history of the steel. In this paper experiments are described which were performed to study the relation between deformation and transformation behaviour in a typical HSLA steel. It was found that Nb in solution raises the start temperature of transformation but delays the transformation kinetics. Retained strain also raises the start temperature and has a small reducing effect on the temperature range of transformation. The competition of ferrite formation and Widmanstätten formation is influenced by the amount of Nb in solution. More Nb in solution leads to a larger fraction of Widmanstätten ferrite.

INTRODUCTION

Grain refinement is a very powerful mechanism in strengthening steels. This can be achieved either by transformation from a fine-grained or from a heavily deformed austenite. Microalloying with Nb is often used to obtain a finer ferrite grain. Nb in solid solution influences recrystallisation of austenite, the phase equilibrium between austenite and ferrite and the kinetics of the austenite to ferrite transformation. Although it is well established that deformation in Nb bearing steels has an influence on the subsequent transformation behaviour quantification of this phenomenon started only recently [1]. In this paper the Hoogovens R&D contribution to an ECSC project "The Improvement of Hot Rolled Products by Physical and Mathematical Modelling" is described [2]. Parts of this work have been presented previously [3]. The influence of deformation in austenite on the transformation behaviour of a Nb bearing steel is analysed. The recrystallisation behaviour in austenite has been studied using double compression testing with varying annealing times between the deformations. During annealing of the deformed specimens in austenite, softening occurs due to the elimination of stresses, but at the same time hardening can occur due to precipitation or grain refinement. The progress of the precipitation process has been monitored by chemical analysis and compared to the equilibrium calculations using the thermodynamic package MTData [4].

EXPERIMENTAL SET-UP

The experiments were carried out using a Bähr 805A/D deformation dilatometer. The cylindrical samples have a diameter of 5 mm and a length of 10 mm. Graphite foil (0.25 mm) has been used as a lubricant. Some samples were chromium-plated, others coated with BN to avoid carbon diffusion from the graphite into the sample. The tests were performed on a commercial HSLA steel (1.2 wt% Mn; 0.136 wt% C; 0.02 wt% Nb; 0.11 wt% other elements; balance Fe).

The experimental set-up can be divided into three main parts:

1. Pre-conditioning of the austenite: all the samples are heated (4 K/s) to 1473 K. A soaking period of 15 min. has been used to dissolve the pre-existing Nb(C,N) precipitates. After cooling to 1323 K at a rate of 2.5 K/s, a 30% deformation at a rate of 5 s^{-1} has been applied to obtain a smaller starting austenite grain size. The sample has been held at 1323 K for one minute, followed by further cooling at 2.5 K/s to the test temperature.

2. Conditioning of the austenite : at the test temperatures of 1223 and 1123 K a 30% deformation at a rate of 5 s^{-1} has been given. Holding times of respectively 1, 5, 15 and 60 s were used at 1223 K, at 1123 K the holding times were 1, 5, 15 and 60 min. to allow for recrystallisation of the matrix and precipitation to occur. To investigate the softening a second deformation of 5 % at a rate of 5 s^{-1} was applied. To study the austenite structure after the isothermal annealing samples were quenched to room temperature to determine the austenite structure. The amount of Nb in solution was determined by ICP OES after electrochemically dissolving the deformed samples and filtering the solvent applying a 0.1 μm filter.

3. Transformation from austenite to ferrite : after part 2 several cooling patterns were applied to the samples :
   a) cooling at 1 K/s to Ar3 and then He-quenched.
   b) cooling at 1 K/s to Ar1 and then He-quenched.
   c) cooling at 1 K/s to room temperature.

The microstructures of above samples were analysed using visible light microscopy and the amount of Nb was determined at the various stages.

RESULTS AND DISCUSSION

AUSTENITE STRUCTURE BEFORE TRANSFORMATION

The austenite grain size found after quenching at Ar3 was roughly similar for all experiments. The values are listed in Table 1. In earlier experiments [3] a large difference in austenite grain size was observed with varying annealing time, however in this work no such variation was observed again.

Temp (K)	Time (s)	Austenite (μm)	Ferrite (μm)
1223	1	37	13
	5	31	9
	15	37	9
	60	44	11
1123	60	37	9
	300	38	9
	900	47	9
	3600	34	10

Table 1 Grain sizes just before transformation (austenite) and after cooling to room temperature at a rate of 1 K/s(ferrite).

The amount of Nb in solution was determined after the isothermal holding and at the temperature just before the transformation (Ar3). Little difference was observed between the amount of Nb in solution at both temperatures. Therefore, the values determined at Ar3 are used for interpretation with respect to the influence of precipitation on the transformation behaviour. The amount of Nb in solution after the different annealing treatments are represented in Fig. 1. The upper dashed line represents the observed effect of the pre-conditioning on the amount of Nb in solution. This was determined by quenching samples to room temperature directly after the pre-conditioning of the austenite and after subsequent cooling at 2.5 K/s to 1123 K and 1223 K. The two lower dashed lines represent the equilibrium amount of Nb in solution at 1223 and 1123 K, respectively calculated using the thermodynamic package MTData [3].The triangles represent the data obtained after deformation and annealing at 1223 K, the squares represent the data for 1123 K. In view of the accuracy of the experimental determination of the amount of Nb in solution it is seen that in the time range investigated little precipitation occurs at 1223 K with respect to the pre-conditioning, whereas a significant decrease in the amount of Nb in solution is observed at 1123 K.

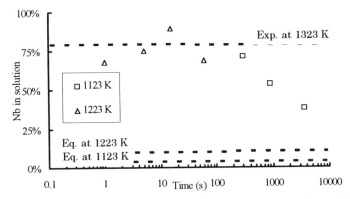

Fig. 1 The amount of Nb in solution after combined deformation and annealing at 1123 and 1223 K. The lower dashed lines represent the equilibrium amounts of Nb in solution. The upper dashed line represents the influence of the preconditioning.

SOFTENING BEHAVIOUR

The recrystallisation behaviour has been characterised by calculating the softening factor from the stress-strain curve before and after the isothermal annealing [5]. The results according to this approximate measure are shown in Fig. 2. It must be reminded that hardening effects due to precipitation and / or grain refinement in the second stress-strain curve will lead to an underestimation of the actual fraction recrystallised, because these effects are not included in the analysis. Beside the experimental points in the figure the calculated recrystallisation behaviour is presented as well. These data were obtained applying a recrystallisation model tuned on hot strip data for both CMn steels and HSLA steels [6]. The calculated recrystallisation behaviour is shown at both temperatures in absence and presence of the Nb. It can be seen that according to the model Nb delays the recrystallisation with respect to similar CMn steels. At 1223 K the experimental data are well represented by the calculated line, whereas for the lower temperature a large deviation is observed. From Fig. 2 and Fig. 1 it can be seen that precipitated Nb retards recrystallisation to a larger extent compared to Nb in solution. Apparently, the applied recrystallisation model works quite well for situation were Nb mainly remains in solution and leads to an considerable underestimation in case of significant precipitation of Nb.

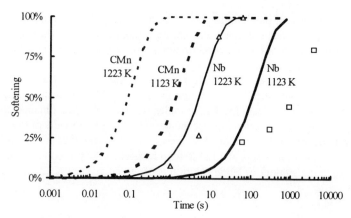

Fig. 2 Observed recrystallisation behaviour of HSLA and CMn steels at 1123K (squares) and 1223 K (triangles) compared to calculated recrystallisation behaviour in absence (dashed lines) and presence (solid lines) of Nb for the low temperature (thick lines) and high temperatures (thin lines).

INFLUENCE OF RETAINED STRAIN ON THE START OF TRANSFORMATION

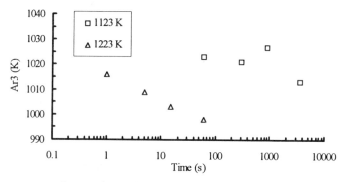

Fig. 3a Temperature of start of transformation vs. annealing time at different temperatures.

In Fig. 3a the starting temperatures for transformation (Ar3) have been plotted vs. annealing time. It can be seen that Ar3 are shifted towards lower values for the longer annealing times at the higher temperature. The effect of annealing time on Ar3 also appears stronger for the higher annealing temperature. The differences between the two sets of data are the amount of retained strain in the austenite before transformation and the amount of Nb in solution.

The lines drawn in the following Figs served to guide the eye rather than represent a fit on the experimental data. The effect of softening on Ar3 is depicted in Fig. 3b. The observed decrease in Ar3 for the data obtained at 1223 K can be solely attributed to softening, whereas the data obtained at 1123 K represent a different amount of Nb in solution combined with a varying amount of softening. From Fig. 3a it can be seen that retained strain will increase Ar3. The difference in starting temperature between fully deformed austenite and fully recrystallised austenite at 1223 K is almost 20 K, which is calculated with MTData to represent an extra driving force of about 30 J/mole. For a larger strain of 0.40 (at a lower strain rate) an additional contribution of 50 J/mole has been reported [1]. The observed decrease of Ar3 with increasing softening at 1123 K is comparable to 1223 K, but is shifted to a higher temperature level. This leads to the conclusion that precipitation of Nb has an opposite effect on Ar3 compared to softening.

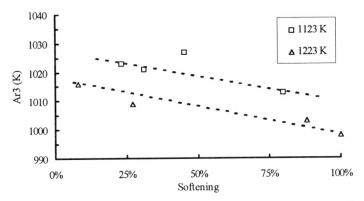

Fig. 3b Temperature of start of transformation vs. softening of the austenite for different annealing treatments..

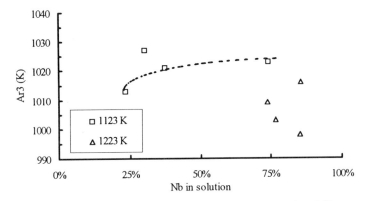

Fig. 3c Temperature of start of transformation vs. Nb in solution for different annealing treatments.

The influence of Nb in solution is represented in Fig. 3c. The data for 1223 K show the influence of the softening but do not allow a conclusion about the relation between Ar3 and the amount of Nb in solution at this temperature. The data obtained at 1123 K show that more Nb in solution will favour ferrite formation, which is in agreement with the expected influence of Nb on the phase stability of austenite in respect to ferrite. Precipitation will suppress the formation of ferrite to a lower temperature. The difference between 25% and 75 % Nb in solution is calculated to correspond to a change in the Gibbs free energy of austenite of 10 J/mole.

INFLUENCE OF RETAINED STRAIN ON TRANSFORMATION TRAJECTORY
Apart from the start of the transformation, another important factor determining the final ferrite grain size is the transformation kinetics and especially the balance between the nucleation and growth of

the ferrite grains. As a measure for the transformation kinetics the difference between Ar1 and Ar3 was used. As was seen earlier (Fig. 3b), retained strain in the austenite raises Ar3 whereas precipitation of Nb decreases Ar3 (Fig. 3c). In Fig 4a the size of the temperature range is depicted for annealing treatments at two different temperatures. A small reduction is observed for the higher temperature whereas the effect at the lower annealing temperature is larger.

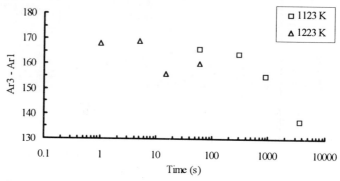

Fig. 4a Transformation trajectory vs. annealing time at different temperatures.

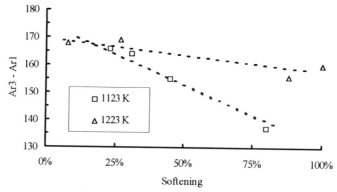

Fig. 4b Transformation trajectory vs. softening of the austenite for different annealing treatments.

The experimental data plotted against softening, in Fig. 4b, apparently show that a higher degree of softening lead to a faster overall transformation behaviour. Thus was also concluded in [1]. From the data at 1223 K it can be seen that the effect of recrystallisation alone is small if even significant. At the lower temperature a much larger influence is observed. In Fig. 4c. the data are represented as a function of the amount of Nb in solution. A lower amount of Nb in solution is seen to be very effective in enhancing the transformation rate. This can be explained by either a reduced solute drag effect during the transformation from austenite to ferrite or a higher nucleation rate compared to undeformed austenite due to the lower start temperatures for the transformation. If the nucleation rate is influenced by the start temperature of the transformation a larger number of ferrite grains per austenite grain should be found in the final structure. However, no significant increase was found (see Table 1). This suggest that the absence of solute drag of Nb at the transformation interface was the most important factor for the enhanced transformation rate.

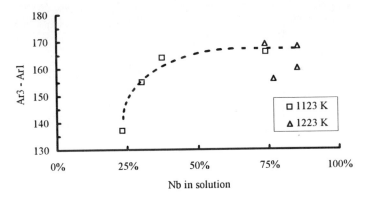

Fig. 4c Transformation trajectory vs. Nb in solution for different annealing treatments.

INFLUENCE OF DEFORMATION ON FINAL MICROSTRUCTURE
Cooling at a rate of 1 K/s after deformation and annealing leads to a mixed microstructure of ferrite, pearlite and Widmanstätten ferrite. Because the amount of pearlite in the microstructure was approximately constant at all annealing times at both annealing temperatures (15 % at 1223 K, 12 % at 1123 K), these data were omitted from Figs 5a and 5b.

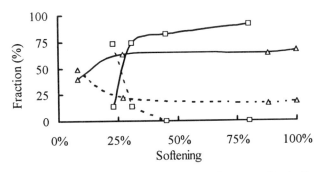

Fig. 5a The composition of the final microstructure, without pearlite indicated (ferrite - solid lines; Widmanstätten ferrite - dashed lines) vs. the softening at 1123 K (squares) and 1223 K (triangles).

The composition of the final microstructure was influenced by deformation of austenite. Fig.5a shows that a large amount of retained strain leads to a significant amount of Widmanstätten ferrite in the microstructure. After annealing at 1223 K a small portion of Widmanstätten ferrite is observed even after complete softening. At the lower temperature of 1123 K Widmanstätten ferrite disappears from the microstructure after about 30 % softening. Combining this with the amount of Nb in solution, as was done in Fig. 5b, shows that Nb in solution increases the amount of Widmanstätten ferrite. The competition of ferrite formation with Widmanstätten ferrite formation appears to be influenced by the solute drag effect, at least for the range of austenite grain sizes investigated. A strong solute drag effect will favour Widmanstätten formation, a weak solute drag will favour ferrite formation.

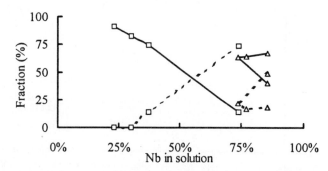

Fig. 5b Final microstructure components (without pearlite) against the amount of Nb in solution at 1123 K (squares) and 1223 K (triangles) (Ferrite - solid lines; Widmanstätten ferrite - dashed lines)

CONCLUSIONS

Deformation of Nb-bearing steels in the austenite phase facilitates the transformation from austenite to ferrite in several ways.

1. Retained strain raises the start temperature of the transformation by generating an additional contribution of about 30 J/mole (after a strain of 0.35) to the driving force for the transformation. This is in rough agreement with data presented in literature.
2. Retained strain has little effect on the transformation rate.
3. Nb in solution increases the start temperature of transformation but delays the transformation kinetics by solute drag. Precipitation of Nb leads to a higher transformation rate by reducing the solute drag effect.
4. The formation of ferrite competes with the formation of Widmanstätten ferrite. Nb in solution will favour the formation of Widmanstätten ferrite by reducing the transformation rate of ferrite formation through solute drag.

REFERENCES

1. X. Liu, P. Karjalainen, J. Perttula, "Thermo-mechanical processing in Theory, Modelling and Practice", ed. Hutchinson et al., 4-6 September 1996, Stockholm, Sweden, 240-248.
2. A.A.Howe, ECSC Contract 7210.EC (D3.02.94), "The improvement of hot rolled products by physical and mathemathical modelling", 1994.
3. J. Colijn, Th.M. Hoogendoorn, M. Onink, Proc. ECSC Workshop, ed. R. Tomellini, 22 January 1997, Bruxelles, 47-57.
4. R.H. Davies, A.T. Dinsdale, J.A. Gisby, S.M. Hodson, R.G.J. Ball, Proc. Conf ASM/TMS Fall Meeting, Rosemont, 2-6 October 1994, Il., USA.
5. H.L.Andrade, M.G. Akben, J.J. Jonas, Metall. Trans A, 14A, (1983), 1967-1976.
6. M. Hartwig, A.J. van den Hoogen, "Numerical predictions of deformation processes and the behaviour of real materials" ed. S.I. Andersen et al., Proceedings of the 15th Risø conference on Materials Science, 5-9 September1994, 335-341.

Materials Science Forum Vols. 284-286 (1998) pp. 193-200
© *1998 Trans Tech Publications, Switzerland*

The Effect of the Austenitisation Temperature on the Transformation Kinetics of an Nb-Containing Steel

Th.A. Kop[1], P.G.W. Remijn[1,2], J. Sietsma and S. van der Zwaag[1]

[1] Laboratory of Materials Science, Delft University of Technology,
Rotterdamseweg 137, NL-2628 AL Delft, The Netherlands

[2] Present Address: British Steel, Swinden Technology Centre, Moorgate, Rotherham S60 3AR, UK

Keywords: Microalloyed Steels, Austenite Microstructure, Phase Transformations, Precipitation

Abstract

An experimental study on the effect of the austenitisation temperature on the austenite decomposition for an Nb-microalloyed steel is performed by means of Differential Thermal Analysis. The start temperature for the formation of allotriomorphic ferrite in a cooling experiment depends strongly on the austenitisation temperature through the austenite grain size. The austenite grain size in its turn depends on the austenitisation temperature and is strongly restricted by the presence of NbC-precipitates. Also the transformation rates depend on the presence and/or formation of NbC-precipitates. Precipitation of NbC during the phase transformation, likely to take place at the austenite/ferrite interface, reduces the effective transformation rate, in some cases leading to the formation of Widmannstätten ferrite and bainite.

1   Introduction

A number of elements, in particular niobium, titanium and vanadium, have been found to have a very strong strength-enhancing effect on steels at alloying levels of only a few hundredths of a percent, provided an appropriate heat treatment is applied [1]. Most of the work in this field, including the present paper, has been performed on niobium-containing steels. It has been established that these microalloying elements owe their strength-enhancing effect primarily to a strong reduction of the average grain size of the ferrite that forms during the austenite-to-ferrite phase transformation. The predominant effect appears to be that Nb(C,N)-precipitates have a crucial grain-refining effect during the austenitisation treatment. In addition, several suggestions for the grain-refining mechanism during the austenite-to-ferrite transformation have been proposed, for instance by Lee *et al.* [2], who suggest that niobium atoms segregate at austenite/ferrite interfaces, thus causing a reduced grain-growth velocity due to solute-drag effects. The reduced growth rate causes a shift in the balance of nucleation versus growth rate. An alternative interpretation has been adopted by Manohar *et al.* [3], who suppose the A_3-temperature to decrease strongly under the influence of niobium dissolved in the iron lattice. Contrary to these suggestions based on the presence of solute niobium, a number of authors [4] ascribe the grain-refining effect of niobium to the formation of Nb(C,N)-precipitates. These precipitates can either work as nucleation sites for the ferrite formation, or have a pinning effect on the moving austenite/ferrite interface. Both mechanisms have a grain-refining effect.

In the present paper the phase-transformation behaviour of a niobium-microalloyed steel is studied by means of Differential Thermal Analysis after different austenitisation treatments. The relation between the austenitisation temperature and the kinetics of the formation of the different phases, as measured by Differential Thermal Analysis, is clarified by relating the transformation temperatures to the austenite grain size. Also, a micrographic analysis of the resulting microstructures has been performed.

2   Transformation and precipitation behaviour

The manufacturing process of a steel brings the metal to temperatures at which the fcc phase, austenite, is the stable phase. In low-alloy steels the austenite forms a single homogeneous phase, the microstructure of which develops during the austenitisation treatment. Like most solid-state phase transformations, the formation of austenite evolves by means of nucleation and growth. It is well-known that the austenite microstructure gets coarser with increasing austenitisation temperature. The presence of precipitates is likely to influence the development of the austenite microstructure. This influence is particularly pronounced if the austenite is subjected to a deformation process, since the presence of precipitates strongly retards the recrystallisation process.

During cooling from the austenitisation temperature, initially the bcc phase, ferrite, is formed in a nucleation-and-growth controlled phase transformation. Nucleation is heterogeneous, and mainly takes place at grain corners, edges or boundaries. Both experimentally and theoretically it is found that grain corners are the most likely nucleation sites [5]. A fine austenite microstructure therefore leads to a high ferrite nucleation rate. The growth rate for the ferrite formation during the (usually fast) cooling also has a strong influence on the eventual microstructure. The growth rate can be envisioned to be governed by the actual velocity by which the interface moves through the structure. Pinning effects of precipitates, present or forming at the interface, can lead to a relatively low growth rate. On the other hand, also solute niobium can cause a reduction of the growth rate. The energetically favourable situation of niobium atoms at the interface can only be maintained if the interface moves with a velocity that is low enough to allow niobium diffusion. This effect is called solute drag, and has been shown to have a significant retardation effect on the interface movement [6]. The growth rate during the decomposition of austenite is not only important for the grain size that develops, but also for the constituting phases in the microstructure. When the growth rate is low, relatively deep undercoolings can be achieved, at which Widmannstätten ferrite is formed rather than allotriomorphic ferrite. At even lower temperatures the formation of bainite can become more favourable than the formation of pearlite, if low growth rates occur.

A key property in the effect of niobium on the formation of both the austenite and the ferrite microstructure is the solubility product of niobium-(nitro)carbide [7], which has a Arrhenius-like temperature dependence. In the present study we will concentrate on NbC as the phase that precipitates. The solubility product k for NbC is defined by

$$k = [\text{Nb}][\text{C}], \tag{1}$$

in which [A] denotes the concentration of element A in solution in the iron lattice. Most commonly, [A] is expressed in weight percentages. The solubility product is strongly temperature dependent, which is expressed by the equation

$$k = k_0 \exp\left(-\frac{Q}{RT}\right), \qquad (2)$$

with k_0 denoting a pre-exponential factor, Q an activation energy, R the gas constant and T the temperature. It is readily derived from eq. (1) that c_{NbC}, the mass percentage of Nb present as NbC-precipitate, is given by

$$c_{NbC} = \frac{1}{2}\left(c_{Nb}^0 + c_C^0 - \sqrt{\left(c_{Nb}^0 + c_C^0\right)^2 - 4\left(c_{Nb}^0 c_C^0 - k\right)}\right), \qquad (3)$$

in which c_{Nb}^0 and c_{Nb}^0 denote the overall contents of Nb and C, respectively. Note that eq. (3) is valid only for $0 \le c_{NbC} \le c_{Nb}^0$.

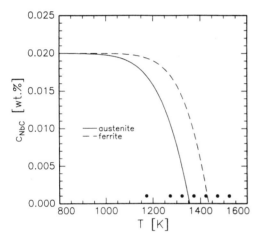

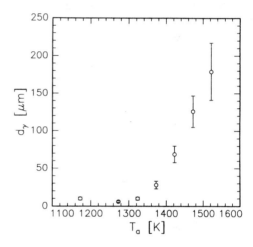

Figure 1: Solubility product for austenite and for ferrite as a function of temperature. The symbols indicate the 7 austenitisation temperatures used in this study.

Figure 2: The austenite grain size as a function of the austenitisation temperature.

Several sets of these parameters for NbC in ferrite and in austenite have been proposed in the literature, as summarised recently by Gladman [1]. Whereas the actual values for k_0 and for Q show considerable differences, the resulting values for k as a function of T show a spread in the austenite temperature range (1000 K $< T <$ 1300 K) that is limited to $\pm 30\%$. For ferrite (800 K $< T <$ 1200 K) the differences increase up to $\pm 60\%$ in the high-temperature range. In this work we will adopt the values $k_0 = 1380$ (wt.%)^2 and $Q = 148$ kJ/mol for the austenite phase, which were obtained with the thermodynamical database programme MT-DATA. The parameters for the solubility product of NbC in ferrite have the values $k_0 = 31600$ (wt.%)^2 and $Q = 194$ kJ/mol. Figure 1 shows c_{NbC} in the austenite and in the ferrite phase as a function of temperature, calculated from the solubility products for the HSLA-steel investigated in this study. It can be seen that for $T > 1350$ K all niobium dissolves in the austenite phase. When the temperature is lowered NbC-precipitates start to form. Note that the lines in figure 1 are valid in equilibrium; during a cooling treatment the kinetics of precipitation are possibly too slow to follow the equilibrium amount.

3   Experimental

The study that is presented in this paper has been performed on an HSLA-steel provided by Hoogovens NV. The main components of the alloy are given in table 1. Microstructural analysis revealed that the material suffered from rolling-induced segregation.

Table 1: Composition of the HSLA-steel under study

element	content [wt.%]	content [at.%]
C	0.136	0.629
Nb	0.020	0.012
Mn	1.20	1.21
Si	0.017	0.034
N	0.0036	0.014
Ni	0.027	0.026
Al	0.053	0.109

The composition shows that the formation of niobium-nitrides will not play an important role in the processing of this steel grade, due to the surplus in carbon. The transformation behaviour has been studied by means of Differential Thermal Analysis (DTA) on a Perkin-Elmer DTA-7. Cylindrical samples with a diameter of 3.0 mm and a height of approximately 2 mm (mass $90-100$ mg) were cut from a 6 mm hot-rolled slab. In order to ensure a good thermal contact between the platinum sample cups and the sample, Al_2O_3-powder is inserted in the sample cup and in the reference cup before introducing the sample. All experiments are carried out under a dynamic argon gas flow (25 ml/min). Temperatures are measured by means of platinum/platinum-rhodium thermocouples. After the austenitisation treatment, for which the conditions are varied, each sample is cooled to room temperature at a constant cooling rate of 20 K/min. Baseline corrections have been applied using separate runs on empty sample cups, and the DTA-signal, being the temperature difference between sample cup and reference cup, has been translated to the apparent specific heat by means of runs on the reference material sapphire [8]. No correction for broadening due to thermal-lag effects has been applied, but all conditions determining the thermal lag that occurs have been kept constant throughout the experiment series.

Microstructural analysis has been performed by light microscopy on the DTA-samples. The austenite microstructure is visible due to thermal etching that takes place at the sample surface positioned on the Al_2O_3-powder. The austenite grain size is determined using the linear-intercept method at amplifications ranging from 30 to 500. Besides the austenite microstructure, also the microstructure resulting after the phase transformation has been determined, after etching the samples in 2% Nital.

4   Results

In a set of 7 experiments the austenitisation temperature T_a was varied between $T_a = 1173$ K and $T_a = 1523$ K, for a fixed annealing time of $t_a = 1200$ s. The temperature dependence of the austenite grain size d_γ is shown in figure 2. At high austenitisation temperatures a strong increase of d_γ with temperature occurs.

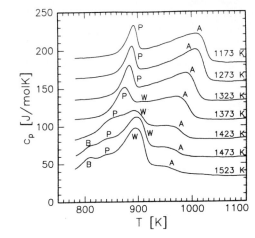
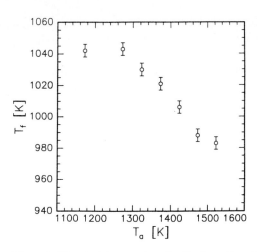

Figure 3: Differential Thermal Analysis results after austenitisation at the temperatures indicated. The peaks are labelled according to the phase that is formed: "A" for allotriomorphic ferrite, "W" for Widmannstätten ferrite, "P" for pearlite, "B" for bainite. Each curve is shifted over 25 J/molK with respect to the previous one.

Figure 4: The start temperature for the formation of allotriomorphic ferrite as a function of the austenitisation temperature.

The transformation behaviour of the HSLA-steel after different austenitisation treatments is visible in the DTA-traces, giving the apparent specific heat c_p as a function of temperature during continuous cooling (figure 3). In the curves of figure 3 the peaks indicate the decomposition of austenite into several phases, dependent on the austenitisation temperature T_a. These curves concern the same samples for which the austenite grain size has been determined (figure 2). Four different transformation stages can be distinguished, indicated by different letters in figure 3. After austenitisation at relatively low temperatures allotriomorphic ferrite readily forms, starting the transformation at $T \approx 1030-1040$ K, slightly dependent on T_a. At $T \approx 900$ K this transformation is followed by the formation of pearlite. After austenitisation at higher temperatures a distinct retardation of the austenite decomposition occurs. The temperature T_f, at which the formation of allotriomorphic ferrite during a DTA-experiment starts, is given as a function of T_a in figure 4. Finally, the DTA-traces in figure 3 show that variation of T_a can also affect the phases that are being formed. The coarse austenite microstructure that develops at a high austenitisation temperatures gives rise to a large undercooling; large enough for the formation of Widmannstätten ferrite and bainite. Figure 5 shows this phenomenon, expressed by the fractions of the four phases allotriomorphic ferrite, Widmannstätten ferrite, pearlite and bainite in the eventual microstructure that were estimated from micrographic analysis.

5   Discussion

An important feature of the solubility of NbC is, as shown in figure 1, that for temperatures above 1350 K all Nb in the presently investigated HSLA steel is in solution. In this temperature range the austenite grain size is found to strongly increase with T_a (figure 2). Although not investigated for this particular steel, it has been reported in the literature [9] that without the presence of niobium the austenite grain size regularly increases with temperature over the entire temperature range shown in figure 2. For $T_a > 1350$ K solute-drag effects of niobium apparently have a much smaller influence on d_γ than the precipitates in the temperature range below 1350 K. Accompanying the austenite grain size increase, the start temperature for the formation of allotriomorphic ferrite, T_f, decreases steadily with T_a (figure 4). This increased undercooling does even lead to the enhanced formation of Widmannstätten ferrite and bainite (figure 5), specifically after austenitisation in the temperature range above 1350 K.

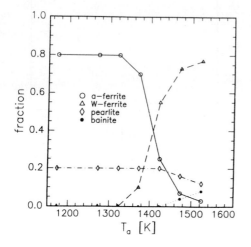
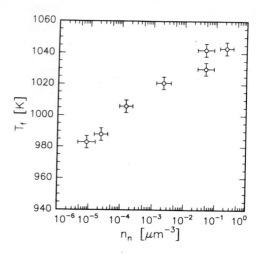

Figure 5: The relative occurrence of the four different phases after austenitisation at different temperatures.

Figure 6: The relation between the density of grain corners n_n and the start temperature for the formation of allotriomorphic ferrite.

It seems a just proposition to assume that the transformation start temperature is predominantly determined by the abundance of nucleation sites connected to the austenite microstructure. Since grain corners can be shown to be the most advantageous nucleation sites [5], in figure 6 the transformation start temperature is given as a function of the density of grain corners n_n, calculated from the average grain size d_γ through

$$n_n = \frac{K}{d_\gamma^3}, \tag{4}$$

in which the proportionality parameter K is taken as $K = 48$, according to the tetrakaidecahedron geometry for the austenite microstructure [1] [10]. Based on the logarithmic relation between T_f and n_n, one can argue that the start of the transformation is marked by a critical nucleation rate J^* in the system. This assumption stems from the well-known expression

$$J = \omega n_n \exp\left(-\frac{\Delta G^*}{RT}\right) \tag{5}$$

for the nucleation rate J at a specific type of nucleation sites, in which ω is a frequency factor and ΔG^* is the activation energy for nucleation. Using $J = J^* = 10^{15}$ m^{-3}s^{-1} at $T = T_f$ and $\omega = 10^{13}$ s^{-1}, the activation energy for nucleation can be calculated to be $\Delta G^* \approx 300$ kJ/mol at $T = 1043$ K, decreasing linearly to $\Delta G^* \approx 205$ kJ/mol at $T = 983$ K. Other choices for the ratio J^*/ω shift these values by approximately 50 kJ/mol per factor 10^3 in J^*/ω.

Figure 6 also indicates that the nucleation behaviour itself is not strongly affected by the presence of NbC-precipitates in the austenite. No change in tendency is observed around the austenitisation temperature above which all Nb is in solution (1350 K). Nevertheless, the growth behaviour can be affected by the precipitation behaviour. Although in equilibrium practically all precipitation has already taken place at the start of the transformation (figure 1), the cooling rate can be high enough for continued precipitation occurring in the temperature range around 1000 K. Precipitation at the γ/α interface will have a retarding effect on the interface mobility. It can indeed be seen in figure 3 that the formation of allotriomorphic ferrite occurs at a lower rate (the height of a peak in the DTA-traces indicates the maximum transformation rate) after austenitisation at a higher temperature. In fact, a transition in the maximum formation rate of allotriomorphic ferrite, as derived from the "A"-peak maximum, is visible at $T_a \approx 1350$ K (figure 7). Moreover, Widmannstätten ferrite is formed only in samples that were austenitised at $T_a > 1350$ K. It must be borne in mind, however, that this effect is blurred by transformation-temperature effects.

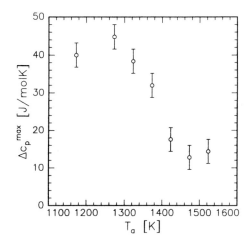

Figure 7: The height of the DTA-peak for allotrio-morphic ferrite as a function of the austenitisation temperature.

6   Conclusions

The austenitisation temperature has a very strong influence on the transformation behaviour of the Nb-microalloyed HSLA-steel that is the subject of this study. A strong increase in the austenite grain size with increasing temperature is found in the temperature range where all Nb is in solution. This indicates that the pinning effect of precipitates on the development of the austenite microstructure is much stronger than the solute-drag effect of niobium in solution. An explicit correlation between the transformation start temperature and the density of grain corners, assumed to be the most important nucleation sites, is found. The experimental evidence indicates that precipitation of NbC during the transformation is the cause of the transformation rate decreasing with increasing austenitisation temperature.

Acknowledgments

This research is financially supported by the Ministry of Economic Affairs through their *IOP-Metalen* programme.

References

[1] T. Gladman, *The physical metallurgy of microalloyed steels*, Institute of Metals, London, 1996

[2] K.J. Lee, K.B. Kang, J.K. Lee, O. Kwon and R.W. Chang, Proc. Int. Conf. on Mathematical Modelling of Hot Rolling of Steel, ed. S. Yue, Hamilton (1990), p. 435

[3] P.A. Manohar, T. Chandra and C.R. Killmore, ISIJ Int. 36 (1996) 1486

[4] V.K. Heikkinnen, Scand. J. Met. 3 (1974) 41

[5] W. Zuang and M. Hillert, Metall. Mater. Trans. 27A (1996) 480

[6] L. Meyer, C. Strassburger and C. Schneider, Proc. Int. Conf. on HSLA Steels: Metallurgy and Applications, ed. J.M. Gray *et al.*, Beijing (1985), p. 29

[7] H. Nordberg and B. Aronsson, J. Iron Steel Inst. 206 (1968) 1263

[8] T.A. Kop, J. Sietsma and S. van der Zwaag, Proc. of the Int. Symposium on Accelerated Cooling and Direct Forge Quenching of Steels, Indianapolis IN, September 15–18, 1997, ed. G.M. Davidson, ASM International, Materials Park, Ohio (1997), p. 159

[9] E.J. Palmiere, C.I. Garcia and A.J. DeArdo, Metall. Mater. Trans. 25A (1994) 277

[10] Y. van Leeuwen, S. Vooijs, J. Sietsma and S. van der Zwaag, to be published (1998)

Materials Science Forum Vols. 284-286 (1998) pp. 201-208
© *1998 Trans Tech Publications, Switzerland*

Effect of Retained Strain on the Microstructural Evolution during the Austenite to Ferrite Transformation

R. Bengochea, B. López and I. Gutierrez

CEIT and ESII, Univ of Navarra,
P° Manuel de Lardizábal 15, E-20009 San Sebastián, Basque Country, Spain

Keywords: Recrystallised Austenite, Deformed Austenite, Phase Transformation, Ferrite Grain Coarsening

ABSTRACT

In the present work the transformation behaviour of a C-Mn-Nb steel deformed by torsion at temperatures both above and below the non recrystallisation temperature for this steel has been studied in detail. After deformation, specimens were cooled at a constant cooling rate of 1°C/s, and interrupted quenching from different temperatures was used to observe different stages of transformation. In the case of transformation from deformed austenite, in addition to the austenite grain boundaries, the deformation bands and twin boundaries act also as preferential nucleation sites for the ferrite. It has been observed that in all the cases when impingement becomes important, the number of grains per unit volume decreases significantly during transformation denoting that ferrite coarsening is taking place.

INTRODUCTION

One of the most important objectives of controlling the thermomechanical processing of low carbon and microalloyed steels is to refine the ferrite grain size. There are two different approaches in hot working to produce ferrite grain size refinement. They are known as conventional controlled rolling (CCR) and recrystallisation controlled rolling (RCR). Accelerated cooling after controlled rolling will enhance refinement [1, 2]. In the recrystallisation controlled rolling the refinement of the ferrite is obtained by refining the austenite by succesive recrystallisations [3]. In the conventional controlled rolling of microalloyed steel the austenite is deformed at temperatures low enough to produce strain induced precipitation of carbonitrides, during deformation, to inhibit recrystallisation [4]. As a result the strain is accumulated in the austenite grains, which is well known to increase the ferrite grain nucleation density [5, 6]. In addition, planar defects as deformation bands and twin boundaries which appear in the austenite grains if deformation exceeds a critical amount [7], act also as nucleation sites for the ferrite [8].

In the present work, a detailed analysis of the kinetics of the ferrite formation and of the microstructure obtained during the course of the transformation of both deformed and recrystallised austenite has been carried out.

EXPERIMENTAL

The composition of the steel used was, as expressed in weight per cent: 0.082%C, 0.36%Si, 1.5%Mn, 0.012%P, 0.005%S, 0.051%Nb, 0.08%V, 0.0082%N. Torsion tests followed by continuous cooling were used to study the evolution of the microstructure during post deformation transformation to ferrite. Specimens, after reheating at 1200°C for 15 min, were deformed using multipass torsion tests at decreasing temperature in the range of 1180°C-800°C. In all cases a strain per pass of $\varepsilon=0.2$, a strain rate of $\dot{\varepsilon}=1s^{-1}$ and interpass times of 30s were used. The post deformation controlled cooling rate was 1°C/s. Two kinds of test were performed:

    A - 6 passes applied at temperatures above the non recrystallisation temperature (T_{nr})
    B - 7 passes applied at temperatures below the T_{nr}

After deformation, the cooling was interrupted at different temperatures in the range of transformation and the specimens were quenched. After quenching, the microstructure was observed at the effective radius $(r^*=0.724\ r)$, where r is the outer radius of the specimen [9]. Etching with nital provided a distinction between the ferrite grains (polygonal ferrite) present at the moment of quench and the remaining austenite which transformed subsequently to martensite. The transformed ferrite volume fraction and the ferrite grain size, d_α, were determined using quantitative metallographic techniques. The recrystallised austenite was analysed in terms of the recrystallised grain size and the deformed austenite in terms of the specific grain boundary area (S_{vgb}).

The migrating grain boundary area per unit volume, A_v was calculated as has been described in more detail elsewhere [10]. The number of grains per unit area, N_A, has been measured at the different stages of transformation. It has been found that this value is sensitive to the magnification, and thus all the quantitative metallography (determination of $f_{v\alpha}$, d_α, A and N_A) has been carried out under the same conditions [10].

RESULTS AND DISCUSSION

The non recrystallisation temperature of this steel was previously determined using the procedure of Bai et al. [11] resulting in a value of $T_{nr} = 1000°C$ for the deformation conditions used in the present work. When deforming the austenite under the B type of tests, the strain is accumulated to reach a value in the plane of measurement of $\varepsilon_{acc}^*=0.724\varepsilon_{acc}\approx1$, where ε_{acc} is 1.4. The initial austenite grain size obtained after reheating at 1200°C for 15 min has been measured, giving a value close to $D_{yo}=125\mu m$. The microstructural parameters referring to the austenite stage given by the different test conditions are shown in Table 1.

Table 1.-Microstructural parameters of the austenite

Test	Recrystallised (equiaxed austenite)		Unrecrystallised (elongated austenite)	
	$d_{rex}(\mu m)$	$S_{vgb}(mm^{-1})$	$d_{bf}(\mu m)$	$S_{vgb}(mm^{-1})$
A	43	46.5	-	-
B	-	-	125	39

*d_{bf} is the austenite grain size present before finishing is applied at $T<T_{nr}$
* S_{vgb} represents the specific grain boundary area prior to transformation

The volume fractions of polygonal ferrite present in specimens quenched from a series of temperatures in the transformation range are shown plotted against the temperature in Fig.1, for specimens deformed following the different schedules. From the figure, it can be observed that the transformed fraction increases with decreasing temperature, until the final volume fraction is reached. Additionally it can be observed that when transformation takes place from a recrystallised austenite, a lower polygonal ferrite volume fraction is obtained, $f_{v\alpha}=0.61$ as compared to the value of $f_{v\alpha}=0.85$ when the austenite is deformed in the non recrystallisation region. A lower transformation starting temperature is also observed for the former, $T_s=765°C$, compared to $T_s=800°C$ for the latter. This is in accordance with the effect of the retained strain raising the starting temperature of transformation and reducing the occurrence of non-polygonal transformation products [12, 13].

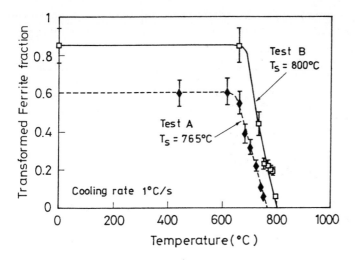

Fig.1 Volume fraction of proeutectoid ferrite as a function of the temperature before quenching.

Figure 2 shows several micrographs of the microstructure observed after quenching from different temperatures during the transformation and corresponding to aproximately the same transformed fractions of about 0.2 (a1, b1) and 0.4 (a2, b2), together with the resulting final microstructures (a3, b3). In Figs. 2(a1) and (a2), which correspond to the transformation from an equiaxed austenite, it can be observed that nearly all the austenite grain boundaries are covered with ferrite grains which coarsen with increasing transformed ferrite volume fraction until the final microstructure is reached (Fig.2(a3)). At this final stage $f_{v\alpha}=0.61$ and $d_\alpha=13$ μm. The observed second phases are principally a mixture of an acicular microstructure (acicular ferrite, bainite) and pearlite.

In Fig. 2(b1) which corresponds to an early stage ($f_{v\alpha}=0.2$) of the transformation from the austenite deformed by schedule B, it is observed that very fine ferrite grains ($d_\alpha=2.6$ μm) have been nucleated on the austenite grain boundaries (allotriomorphic ferrite) and also inside the grains on deformation bands and twin boundaries. Later, when the fraction transformed reaches a value of 0.44 (see Fig. 2(b2)), a significant increase in the ferrite grain size can be observed ($d_\alpha=5.6$ μm) together with some intragranular nucleation within the largest austenite grains. Then, both ferrite volume fraction and grain size increase continuously until the final microstructure is obtained (see Fig.2 (b3)), for which $f_{v\alpha}=0.85$ and $d_\alpha=7.4$μm. In this case, the observed second phase is pearlite. From Fig.2(b3) it is clearly apparent that, even if the final grain size is signifincantly lower than when recrystallised austenite transforms, the ferrite grain size is far from homogeneous, very coarse grains coexisting with much finer ones.

Ferrite grain size is plotted in Fig.3(a) as a function of the fraction transformed. Experimental values are compared with those obtained using the following equation, based on the assumptions of uniform grain sizes and that all the grains which nucleated at the beginning of the transformation remain in the final ferrite microstructure:

$$d_{\alpha i} = d_\alpha \left(\frac{f_{v\alpha i}}{f_{v\alpha}} \right)^{\frac{1}{3}} \qquad (1)$$

where d_α is the mean ferrite grain size after complete transformation, $f_{v\alpha}$ is the volume fraction of ferrite then present (the remainder being pearlite or acicular microstructures depending on the case), and $d_{\alpha i}$ and $f_{v\alpha i}$ represent the corresponding values of these variables at any instant during the incomplete transformation. From Figure 3(a) it is clearly evident that the curves calculated using equation (1) do not fit the experimental points, which means that the aforementioned assumptions are not fulfilled during transformation. From Fig. 3(a), it is to be noted that in both analysed cases, two different steps, concerning the ferrite grain size evolution as the transformation progresses, are observed. Normal growth takes place during the early stages followed by a sudden increase of the ferrite grain size to reach, at $f_{v\alpha} \approx 0.4$, a new step where the ferrite grain size is between 3 and 4 times higher than before. The effect is more pronounced in the case of the recrystallised austenite. Afterwards, the transformation progresses to completion with the grain size approximately following equation (1).

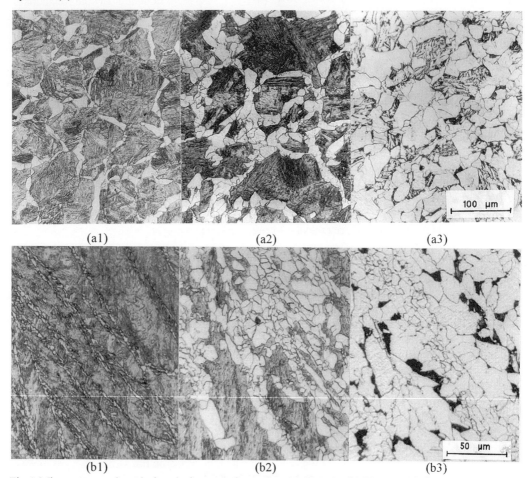

Fig.2 Microstructural evolution during transformation:(a) Test A, (b) Test B; (1) $f_{v\alpha}$=0.2, (2) $f_{v\alpha}$=0.4, (3) final microstructure.

The migrating boundary area per unit volume, A_v, between ferrite and austenite has been measured and the corresponding values are shown, as a function of the fraction transformed, in Fig.3(b). It can be seen that, in the case of transformation from a deformed austenite (Test B), A_v increases initially to reach a peak corresponding to a ferrite volume fraction, $f_{v\alpha} \cong 0.2$, and then decreases due probably to impingement. The maximum value of the migrating grain boundary area per unit volume measured in this transformation is about 220 mm^{-1}, which is nearly five times the value of $S_{vgb} = 39$mm^{-1} determined in this case (see Table 1).This confirms the contribution of the ferrite nucleated inside the

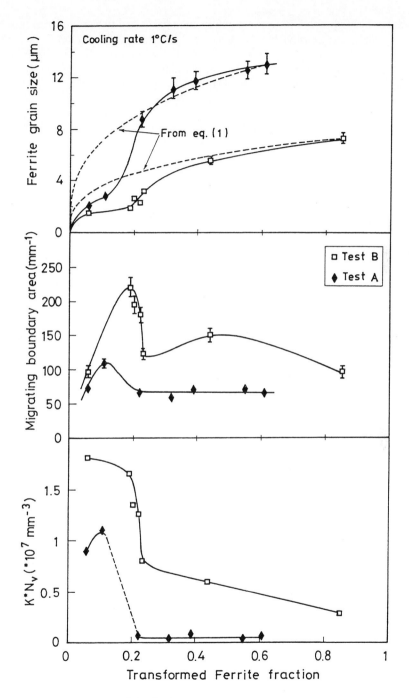

Fig.3: (a) Experimental values of the ferrite grain size at each stage of the transformation plotted together with the prediction from equation (1); (b) Migrating proeutectoid ferrite/austenite interphase boundary area per unit volume as a function of the transformed ferrite fraction and (c) Evolution of the number of grains per unit volume with the transformed ferrite fraction.

grains which has been observed to form on deformation bands and twin boundaries, see Fig.2(b1), in increasing the nucleation density. It is to be noted that from about $f_{v\alpha}=0.25$, A_v remains nearly constant, although it is also observed that it increases slightly for a transformed fraction of about 0.5. This could be attributed to an increase of the interphase boundary area caused by the observed fresh nucleation inside the grains, probably associated with dislocations or subgrain boundaries (see Fig.2(b2)).

In the case of the transformation from equiaxed austenite, the first value of A_v, which corresponds to a small transformed fraction (about 0.06), denotes that, at this stage, not all austenite grain boundaries are covered by ferrite ($A_v < 2S_{vgb}$). For the next stage, $f_{v\alpha} = 0.11$, a value of $A_v \approx 2S_{vgb}$ (see Table 1) is obtained, which denotes that site saturation at grain boundaries is reached at this stage in accordance with the microstructural observations. After that, the value of A_v decreases significantly, due probably to some impingement taking place. Afterwards the value of A_v remains nearly constant until the transformation is concluded.

The number of grains per unit area, N_A, has been determined at different stages of the transformation. The number of grains per unit volume, N_V, can be related to N_A in the following way:

$$N_A = k \, N_V \, d_\alpha \qquad (2)$$

with k a shape factor relating the mean linear intercept, d_α, to the average tangent diameter [14]. Equation (2) is strictly true only when all the grains are of constant shape, independent of their size. Assuming that the ferrite grain shape remains unchanged during the overall transformation, the values ($k \, N_V$) have been calculated from the N_A measurements. The results are plotted against the transformed fraction in Fig.3(c). From the figure it is clearly evident that the number of grains present at the final stage is significantly lower than that measured at the initial stages of the transformation. In the case of transformation from the deformed austenite (Test B), it is observed that even if the number of grains per unit volume is decreasing during the overall transformation, the most important reduction takes place within the early stages in the 0.19-0.23 $f_{v\alpha}$ range. A consequence of this drastic reduction in N_V is that the number of ferrite grains per unit volume is about 6 times lower in the final microstructure as compared with those observed at earlier stages of transformation.

In the case of transformation from equiaxed austenite (Test A) it is also observed that the number of grains per unit volume decreases significantly after reaching a maximum value for $f_{v\alpha} = 0.11$ where site saturation occurs at grains boundaries (maximum value of A_v in Fig. 4). Afterwards, N_V remains nearly constant until the transformation is concluded in agreement with no nucleation taking place inside the austenite grains. The number of grains present in the final microstructure is about 15 times lower than the maximum value reached at $f_{v\alpha} = 0.11$.

The previous results denote that ferrite coarsening is taking place during the course of transformation and this can explain why equation (1) does not fit to the experimental data in Fig.3. Priestner and Hodgson [15] have also observed the occurrence of coalescence of ferrite grains at very early stages of transformation in the presence of high levels of retained strain. However, according to the results plotted in Fig.3, it seems that a similar effect takes place when recrystallised austenite transforms. The main differences between both cases seem to come from the lower nucleation density sites when recrystallised austenite is transforming. Another feature, which is clearly apparent from the data in Fig.3, is that this coarsening effect is very strong when the ferrite grains formed at the early stages of the transformation impinge. This takes place at a ferrite volume fraction of about 0.2 in both cases and the significant drop of the migrating boundary area clearly indicates the occurrence of impingement.

Several authors have proposed equations based on nucleation and growth theories to predict the final ferrite grain size [16, 17]. These expressions are based on the same principles as equation (1) and relate the ferrite grain size, d_α, in the final stage with the total number of ferrite grains nucleated, throughout transformation, per unit volume of austenite, N_V. These equations assume that each nucleus formed will grow uniformily until transformation completion. However, from the results in the present work, it can clearly be seen that these expressions cannot predict the real final grain size, see Fig. 3(a). As pointed out before, for the transformation after Test A, the measured value of N_V at the final stage is 15 times lower than that measured at the early stages. In this case, if the value of N_V measured before impingement is used, the calculated ferrite grain size would be 2.5 times lower than that obtained by using the value of N_V measured at the final stage. In the case of transformation from deformed austenite, a ferrite grain size 1.8 times lower would be obtained if the value of N_v measured at the initial stages of transformation is used instead of the final value. This suggests that the use of expressions which only consider the nucleation at initial stages followed by uniform growth and do not account for the possibility of the occurrence of coarsening during transformation, could lead to significant errors in predicting the final ferrite grain size.

Another feature that has to be considered, in the case of transformation from deformed austenite, is that some intragranular nucleation seems to take place at the latter stages of transformation, i.e. at high values of $f_{v\alpha}$. New nucleation inside the deformed and larger grains is substantiated by the metallographic observations and by the results in Fig.3. The migrating boundary area slightly increases to reach a new relative peak at $f_{v\alpha} \approx 0.5$. This is only possible if new γ/α interfaces are produced. At the end, impingement is dominant and A_V decreases again. Consequently the continuous decrease of N_V as the transformation progresses (see Fig.3(c)) is only possible if the elimination of ferrite grains by coarsening compensates for the formation of new nuclei in the centre of the austenite grains. When recrystallised austenite transforms, no intragranular nucleation is observed and, after the ferrite coarsening taking place after the grains along initial grain boundaries impinge, A_V and N_V remain constant to the end of the transformation.

The actual results substantiate coarsening enhanced by impingement between ferrite grains and suggest that two different stages of ferrite coarsening would be expected, for low ferrite volume fractions ($f_{v\alpha} \leq 0.2$), when impingement takes place parallel to the austenite grain boundary plane and/or twin boundaries and deformation bands, and for high $f_{v\alpha}$, when impingement occurs perpendicular to the former due to additional intragranular nucleation in the case of transformation from deformed austenite. Additional work is required to investigate how impingement activates a very powerful mechanism of selection of the ferrite grain to be present in the final microstructure.

CONCLUSIONS

-Coarsening of ferrite grains occurs during transformation from both recrystallised and deformed austenite at very early stages of transformation which significantly reduce the number of grains. For deformed austenite at the latest stages of transformation coalescence of ferrite grains seems to be accompanied by new nucleation inside the largest austenite grains. As a consequence the final grain size seems to be determined, not directly by the nucleation density, but, by a competition between the transformation rate and ferrite grain coarsening and probably by the number of the nucleation sites being activated in the largest austenite grain interiors.

-The retained strain in austenite enhances ferrite nucleation, leading to a finer final mean grain size, even if the heterogeneity is higher than in the case of a transformation from recrystallised austenite.

ACKNOWLEDGEMENTS

Part of this research work has been carried out under an ECSC Mega-Project No 7210.EC/938. The authors thank the European Coal and Steel Community for partial funding of the research, the project leaders at the partner institutes and their colleagues. The authors wish to dedicate this article to the memory of Prof. J.J.Urcola, who was directly involved in this project before his untimely death.

REFERENCES

[1] M. Umemoto, Z.H.Guo, I. Tamura, Mat. Sci. and Tech., vol 3, (1987), p.249-255.

[2] S.Zajac, T. Siwecki, B. Hutchinson, M. Attlegard, Met. Trans. vol 22A, (1991), p. 2681-2694.

[3] T.Siwecki, B. Hutchinson, S.Zajac, Proc. Conf. Microalloying 95, M. Korchynsky, A.J. DeArdo, P. Repas, G. Tither eds, (1995), p. 197-211.

[4] B. Dutta, E. Valdés, and C.M. Sellars: Acta. Metall., Vol 40, No 4, (1992), p. 653-662.

[5] I. Tamura, Int. Conf. on "Physical Metallurgy and Thermomechanical Processing, Thermec 88", The ISI of Japan, (1988), p. 1-10.

[6] R. Priestner and L.Ali, Mater. Sci. Tech., vol.9, (1993), p. 135-141.

[7] C. Ouchi, T. Sampei and I.Kozasu, Trans. ISIJ, vol. 22, (1982), p. 215-222.

[8] T. Tanaka, "Microalloying 95", ISS, Pittsburgh, M. Korchynsky, A.J. DeArdo, P. Repas, G. Tither eds, (1995), p. 165.

[9] D.R. Barraglough, H.J. Whittaker, K.D. Nair and C.M. Sellars: J.Test Eval., 1, (1973), p. 220-226.

[10] R. Bengochea, B. López and I. Gutierrez, Met. Trans., Vol.29A, (1998), p.417-426.

[11] D.Q. Bai, S. Yue, W.P. Sun and J.J. Jonas: Met. Trans, Vol. 24A, (1993), p. 2151-2159.

[12] I.Kozasu, C. Ouchi, T. Sampei and T. Okita, Proc. Conf. Microalloying 75, M. Korchinsky ed.,ISS, Pittsburgh, (1975), p. 120-134.

[13] B. Ouchi, T. Sampei and I. Kozasu, Trans. ISIJ, vol. 22, (1982), p. 214-222.

[14] R. T. DeHoff, "Quantitative Microscopy", eds. R. T. DeHoff and F. N. Rhines, McGraw-Hill Series in Mat. Sci. and Eng., (1968), p. 129.

[15] R. Priestner and P.D. Hodgson: Materials Science and Technology, vol. 8, (1992), p. 849-854.

[16] I. Tamura, Int. Conf. on "Physical Metallurgy and Thermomechanical Processing, THERMEC 88", The ISI of Japan, (1988), p. 1-10.

[17] M. Umemoto, H. Ohtsuka, I. Tamura, Proc. Conf. on HSLA steels, T.Chandra and D.P. Dunne eds, (1984), p. 96-100.

B.López
E-mail: blopez@ceit.es
Fax: (34)43 21 30 76

Materials Science Forum Vols. 284-286 (1998) pp. 209-216
© *1998 Trans Tech Publications, Switzerland*

Transformation Behaviour and Related Mechanical Properties of B-Added ULC- and IF Steels

G. Krielaart[1], M. Baetens[1], S. Claessens[2], D. Vanderschueren[2] and Y. Houbaert[1]

[1] Laboratory for Iron and Steelmaking, University of Ghent, Technologiepark 9, B-9052 Zwijnaarde, Belgium

[2] OCAS N.V. - Research Centre of the SIDMAR Group, John Kennedylaan 3, B-9060 Zelzate, Belgium

Keywords: Boron, Ultra-Low Carbon Steel, IF Steels, Dilatometry, Transformation, Mechanical Properties

Abstract

Boron may be added to modern ULC- and IF steels for various reasons. It is well known that the presence of this element has a strong effect on the kinetics of the γ/α transformation, which in its turn has a major effect on the resulting microstructures. However, the origin and the nature of the resulting microstructures in ULC-and IF steels is poorly understood.

In principle, the addition of boron offers the possibility of obtaining an excellent combination of strength and ductility in deep-drawable ULC-steels. In P-bearing high strength IF-steels, B is mainly added to avoid Cold-Work-Embrittlement. In this case, the effect of B on the transformation behaviour and the resulting microstructure may easily become detrimental to the mechanical properties and further processing through cold-rolling. For both ULC-steels and IF-steels, the optimisation of properties requires a trade-off in steel composition and processing conditions. This requires knowledge of the prevailing transformation mechanism.

The transformation behaviour of various B-added ULC- and IF steels was studied by dilatometric and microstructural analysis, combined with hardness testing of hot-rolled material. Identification of the various constituents of the microstructure was accomplished by comparing the results of microstructural analysis and hardness tests with literature data. Tuning of the mechanical properties also requires knowledge of the recovery and recrystallisation behaviour, which is investigated in combination with the relevant mechanical properties.

1   Introduction

In the development of modern IF- and ULC steels, much effort is put into improving the balance between properties like strength and formability. Good formability and resistance to ageing requires, among others, that the amounts of free interstitially solved elements are kept sufficiently low. Strength improvement is achieved by methods as an appropriate thermomechanical treatment, resulting in grain refinement and precipitation strengthening. Also solid solution strengthening may be employed.

It is well known that the combination of extremely low amounts of free interstitially dissolved elements and high cooling rates may easily give rise to the occurrence of microstructures like massive ferrite and bainitic ferrite [1-3]. An important feature of these microstructures is their relatively high dislocation density (typically $5 \ 10^{10} \ cm^{-2}$ in Fe-Mn alloys [4]), which give rise to increased strength compared to the polygonal ferrite. Obviously, the increased dislocation density is also detrimental for ductility.

Due to their high strength and relatively poor formability, such constituents of the microstructure are troublesome in cold-processing. The difficulties are particularly clear in P-bearing IF-steels [5,6]. Although P is very effective and economical in improving strength by solid solution strengthening, it is also a source of cold work embrittlement. A solution is to add B to P-bearing IF steels. However, B has a very strong hardening effect in steels [7]. This also holds when the amounts free carbon and nitrogen are extremely low. In practice, B-additions to formable steels are kept limited to the range of 0.0005 to 0.002 mass%.

Commonly, the occurrence of ferritic microstructures different from polygonal ferrite in steel grades with extremely low carbon contents is considered as the result of a change of the transformation behaviour during ferrite formation from austenite. Polygonal ferrite is obtained when the cooling rates are low and the transformations proceeds at low undercoolings. Transformation of iron and steels with sufficiently low carbon content at higher undercoolings, e.g. by applying higher cooling rates, results in a microstructure with ragged ferrite grain boundaries. This ferritic microstructure is known as quasi-polygonal or massive ferrite. The raggedness of the grain boundaries increases with increasing undercooling. According to the ISIJ-bainite committee, there is a gradual transition towards so-called granular ferritic bainite with increasing cooling rate [8]. Literature data on phase transformations in high-purity iron indicates that the formation of massive ferrite occurs mainly in the vicinity of 700-750 °C [9].

Increasingly higher undercoolings result in the formation of bainitic ferrite (occasionally called acicular ferrite), lath-martensite (also called dislocated cube martensite or massive martensite) and twinned martensite respectively. Former austenite grain boundaries usually are clearly visible in these type of microstructures.

In earlier work on the transformation behaviour of high-purity iron [10], extremely high cooling rates had to be used to invoke the formation of transformation products different from polygonal ferrite. Massive ferrite in high-purity iron is reported from cooling rates in the order of magnitude of 1000 K/s and more. In ultra-low carbon steels, massive ferrite is found at similar cooling rates [11]. The use of such high cooling rates is necessitated by the requirement to prevent nucleation and growth near equilibrium conditions, which would result in polygonal ferrite. In the study reported in this work, some amounts of B are added with the purpose to prevent ferrite nucleation near equilibrium conditions and thus to stimulate the change of transformation behavior at moderate cooling rates, more in agreement to industrial practice. This effect can be made even more pronounced by adding significant amounts of Mn, which strongly promotes austenite stability. The result of these additions is that fairly large undercoolings can be realised even at moderate cooling rates.

The dependency of the dislocation density on the applied cooling rate already indicates that the mechanical properties will also depend strongly on the thermal history after completion of the phase transformation. Maintaining high cooling rates until ambient temperatures will result in both interrupted recovery and recrystallisation. Therefore, it can be expected that strength will decrease and formability will be improved if the material is kept at higher tempratures for longer times. In practical conditions, it may therefore be possible to improve formability in cold processing through the selection of a suitable coiling temperature.

2   Experimentals

B-bearing ULC steels (grades 1 and 2) and IF steels (grades 3 and 4) where lab-produced from vacuum degassed ULC-steel with approximately 0.002mass% C and 0.002mass%N, molten under an Ar-flow. Their respective alloy compositions are given in Table 1. The ingots where cut into slices with a thickness of 25 mm. These slices where hot-rolled to a thickness of 6 mm for dilatometric analysis or 2 mm thickness for use with a continuous annealing simulator. The

Table 1:   Steel compositions (10^{-4} mass %).

Grade	C	N	Ti	Nb	S	Mn	Al	B	Si	P
1	18	14	7	4	46	10822	315	68	59	122
2	20	19	8	7	51	14460	218	65	96	130
3	14	18	556	29	48	10814	568	4	83	784
4	16	16	556	207	47	10884	484	11	79	788

materials for use on the continuous annealing simulator further received a cold reduction to 1 mm thickness.

Dilatometric experiments where carried out on a Theta Dilatronic dilatometer using specimens with a length of 5 mm and a diameter of 3 mm, which where machined from the plates with 6 mm thickness. Specimens of grades 1 and 2 have been subjected to a temperature cycle according to Figures 1 and 2. Austenitising took place at 1250 °C for 1 minute, followed by cooling to 980 °C at a rate of 10°C/s. When this temperature was reached, the specimens where cooled to ambient temperature with cooling rates programmed from 0.1 °C/S to 125 °C/s. The IF-grades were austenitised at 950 °C and further cooled to ambient temperature with cooling rates varying from 1 °C/s to 90 °C/s.

Identification of the resulting microstructures was accomplished by comparing the results of metallography and hardness tests with literature data. The results of the dilatometric analysis, the microstructural analysis, as well as the results of the hardness measurements, are summarised in the CCT-diagrams in Fig. 2. In the case of grades 1 and 2, the time axis is related to the part of the temperature program at which the various cooling rates are used (i.e. the time at which 980 °C is reached during cooling from the annealing temperature is set to zero). The shaded areas in Fig. 2 correspond to the equilibrium α/γ two-phase region. Region boundaries not entirely certain are indicated by dashed lines.

To investigate the possibility of changes of the microstructure and the mechanical properties, specimens of the cold rolled materials where subjected to the temperature cycle indicated in Fig. 3. In this case, the cooling was interrupted at the temperature where, according to the CCT-diagrams, the transformation was expected to be completed. The samples where then annealed isothermally at that temperature for respectively 1 and 10 minutes. The samples thus obtained where tested for strength (limit of elasticity R_e 0.2% and tensile strength R_m) and formability (A_{50}). The results are summarised in Fig. 4. Note that the temperature used for isothermal hold decreases with the applied cooling rate, according to the CCT-diagrams in Fig 2. Optical micrographs corresponding to various representative cooling rates in Fig. 2 are given in Figs. 5-8.

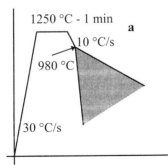
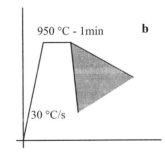

Fig. 1:   Temperature cycle for grades 1 and 2 (1a) and grades 3 and 4 (1b) during the dilatometric experiments. The gray shaded area corresponds to the part where the various cooling rates have been used.

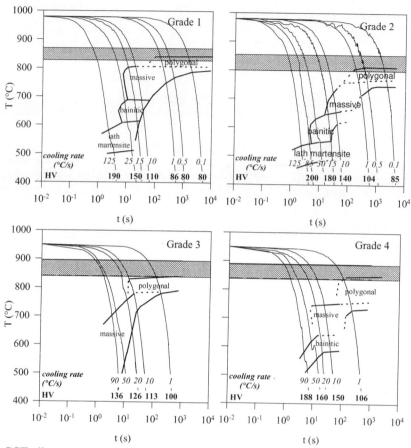

Fig. 2: CCT diagrams summarising the results of dilatometric analysis, hardness tests and microstructural analysis of B-added ULC steels (grades 1 and 2) and IF-steels (grades 3 and 4).

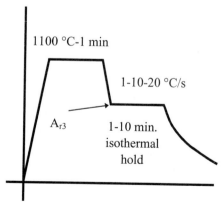

Fig. 3: Schematic temperature profile with accelerated cooling interrupted after complete transformation.

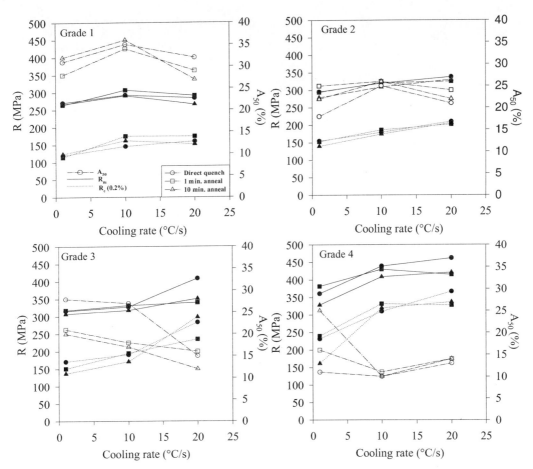

Fig. 4: Limit of elasticity $R_{e\ 0.2\%}$, tensile strength R_m and elongation A_{50} for various cooling rates and isothermal hold times.

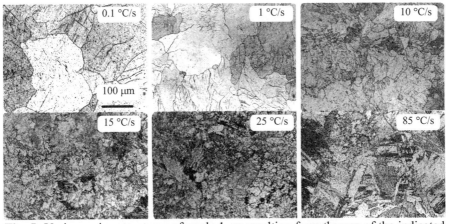

Fig. 5: Various microstructures of grade 1, as resulting from the use of the indicated cooling rate.

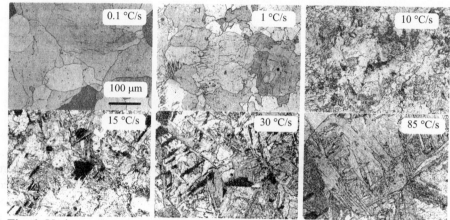

Fig. 6: Various microstructures of grade 2, as resulting from the use of the indicated cooling rate .

Fig. 7: Various microstructures of grade 3, as resulting from the use of the indicated cooling rate.

Fig. 8: Various microstructures of grade 4, as resulting from the use of the indicated cooling rate.

3   Discussion

3.1   Transformation Behaviour

The CCT diagrams given in Fig. 2, indicate that the formation of polygonal ferrite in ULC steels results from transformations at higher temperatures. The undercoolings are low compared to the other constituents of the microstructure. The start of the transformation process is visible in the two-phase region only at low cooling rates. Nevertheless, the transformation proceeds almost entirely in the ferrite one-phase region.

Massive ferrite is the result from the γ/α-transformation at higher undercoolings. The temperatures at which this occurs, depends somewhat on the composition, but generally temperatures well below 750 °C are required. Compared with non-B-bearing ULC type [11], the required cooling rates are low. Microstructural evidence for presence of high dislocation density, in the form of a subgrain structure, can already be found at the lowest cooling rates in grades 1, 2 and

4. These subgrain structures are the result of the occurrence of recovery [12]. The driving force for recovery and recrystallisation is provided for by the presence of dislocations. As the material has not been deformed during the experiment, it seems natural to assume that the dislocations are generated during the phase transformation.

Massive ferrite seemingly has its own C-shaped curve in the CCT-diagram. This probably indicates a change in the transformation mechanism. Either or both the nucleation kinetics and the growth kinetics may change. The irregularity of the ferrite grain boundaries increases strongly with the cooling rates; the amount of massive ferrite increases at the expense of polygonal ferrite.

At deeper undercoolings, resulting from the use of higher cooling rates, bainitic ferrite becomes apparent in the boron bearing grades. The loci of the start temperatures for the formation of massive ferrite and bainitic ferrite appear to have their own distinct curves in the CCT-diagram.

In grades 1 and 2, containing 1% and 1.5% Mn respectively, the transformation temperatures at cooling rates exceeding 15 °C/s are so depressed that the formation of lath martensite becomes possible. In both cases the microstructure consists completely of lath martensite at a cooling rate of 125 °C/s.

3.2 Mechanical Properties

According to Fig. 2, the hardness increases with the cooling rate. This is in agreement with the increasing fraction of massive ferrite and the increasing dislocation density the material acquires with increasing cooling rate. It may be argued that increasing raggedness of the ferrite grains and the increasing presence of subgrain structure with increasing cooling rate is caused by the increasingly slower kinetics of recovery and recrystallisation. With the increasing of the cooling rate, the undercooling at which the transformation proceeds increases and the temperature drops, which obviously causes recovery and recrystallisation to proceed at a lower rate.

With the exception of grade 1, the strength of the material increases monotoneously with increasing cooling rate (see Fig. 4). This is also in agreement with the considerations given above. For the ULC grades, the isothermal hold does not have a large effect. This particularly holds at the lowest cooling rates. The transformation proceeds at high temperatures in this case, the specimen already kept at high temperaures for a relatively long time, and most of the dislocations are already removed before isothermal hold. The susceptibility to isothermal annealing increases somewhat when the transformation proceeds at lower temperatures by applying higher cooling rates. Combined with the observed increasing presence of subgrain structure, it may be concluded that these grades are still fairly susceptible to recovery but much less to recrystallisation at these temperatures. Further research is required to confirm this.

The IF grades have a larger response to the isothermal hold. This is most probably caused by the presence of precipitates, which will decrease the mobility of dislocations and hence decrease the rate of recovery and recrystallisation. By keeping the specimens at a higher temperature for a longer time, by means of the isothermal hold, recovery and recrystallisation can still proceed to a considerable extent. At lower temperatures, the IF-grades are susceptible mainly to recovery, as was the case with the ULC grades. The response however, is stronger.

4 Conclusions

In literature, it is often suggested that massive transformations involve the growth of ferrite across the original austenite grain boundaries [1,2] during the transformation. The results presented here, suggest that the typical microstructure of massive ferrite is to a large extent the result of the interrupted recrystallisation. One of the processes active during recrystallisation is the movement of high angle grain boundaries, as in strain induced boundary migration [12]. Interruption of this process by rapid cooling to low temperatures would result in ragged grain boundaries, as are present in massive ferrite. The driving force for recovery and recrystallisation originates from the high dislocation density which apparently is caused by the phase transformation. The (apparent) crossing of austenite grain boundaries probably is caused by this.

As the formation of massive ferrite proceeds at temperatures where the rate of recrystallisation is slow, adjustment of the mechanical properties either requires that the transformation proceeds at a higher temperature (i.e. cooling rates should be lower) or a suitable annealing treatment at higher temperatures subsequent to phase transformation.

Acknowledgements

The authors are much indebted to dr. ir. Frans Leysen for performing the dilatometric experiments and to ing. Geert Steyaert for performing the experiments with the continuous annealing simulator.

References

1.   T. Maki; in: International Forum for the Physical Metallurgy of IF Steels IF-IFS-94, The Iron and Steel Institute of Japan, (1994), 183.
2.   E. A. Wilson; ISIJ Int., 34, (1994), 615.
3.   G. Krauss, S.W. Thompson; ISIJ Int., 35, (1995), 937.
4.   M. Roberts; Met. Trans., 1, (1970), 3287
5.   L. Meyer, W. Bleck, W. Müschenborn; in: International Forum for the Physical Metallurgy of IF Steels IF-IFS-94, The Iron and Steel Institute of Japan, (1994), 203.
6.   J. Dilewijns, D. Vanderschueren, M. Baetens, K. Mols, S. Claessens; in: International Conference on Thermomechanical Processing of Steels & Other Materials (Thermec '97), (1997), The University of Wollongong, Australia,
7.   J. E. Morral, T.B. Cameron; in: Boron in Steel; S.K. Banjeri, J.E. Morral (eds.), The Met. Soc. AIME, (1980), 19.
8.   T. Araki et. al.; Atlas for Bainitic Microstructures vol. 1, Bainite committee of Iron and Steel Institute of Japan, (1992), 156.
9.   E. A. Wilson, Met. Science, 18, (1984), 471.
10. M.J. Bibby, J.G. Parr; J. Iron Steel Inst., 202, (1964),100.
11. K. Shibata, K. Asakura; ISIJ Int., 35, (1995), 982.
12. F.J. Humphreys, M. Hatherly; Recrystallisation and Related Annealing Phenomena, Elsevier Science Ltd., (1995).

Contact adress :
Prof. dr. ir. Y. Houbaert, Laboratory for Iron and Steelmaking, University of Ghent, Technologiepark 9, B-9052 Zwijnaarde (Belgium), e-mail:Yvan.Houbaert@rug.ac.be.

Materials Science Forum Vols. 284-286 (1998) pp. 217-224
© *1998 Trans Tech Publications, Switzerland*

The Influence of Aluminium and Silicon on Transformation Kinetics in Low Alloy Steels

M.F. Eldridge and R.C. Cochrane

Department of Materials, University of Leeds, Woodhouse Lane, Leeds, LS2 9JT, England

Keywords: Aluminium, Interface, Widmanstätten Ferrite, Acicular Ferrite, Kinetics, Grain Size

1. Abstract

The use of aluminium to deoxidise steel is well known and has been the key to significant improvements in steel cleanness for some 100 years [1,2]. It is common practice to add more than sufficient aluminium to combine with the oxygen in the steel at the end of steel making to ensure current levels of cleanliness. Usually, therefore, steels made with current Basic Oxygen Steelmaking or electric arc practices contain excess "free" or soluble aluminium which can be exploited to an advantage by combining with nitrogen to form aluminium nitride, which then acts to refine grain size during subsequent processing. For most steels therefore, the aluminium additions made play several metallurgical roles; for example, the soluble aluminium to nitrogen ratio affects both the ferrite grain size and the strain ageing behaviour [3]. Many steel specifications therefore, pay particular attention to this aspect and Al/N ratio greater than 2 is normally considered to provide a fine ferrite grain size and freedom from the harmful effects of strain ageing on mechanical properties. As a direct consequence of these attributes, aluminium is usually in excess of the stoichiometric ratio of 2:1 and thus in many steels some aluminium is present in solid solution in the ferrite. Furthermore, the aluminium, invariably present in most modern steels is generally not considered to have an important or significant effect on microstructure. However, interest in the role of aluminium during microstructural evolution has been increased from work based on the formation of steel microstructures based on tough acicular ferrite in the weld heat affected zone (HAZ), thought to be nucleated by TiO compounds [4,5].

One approach which has been taken in the formation of such microstructures has been to adopt compositions which favour the formation of titanium rich inclusions, thought to be responsible for the nucleation of acicular ferrite [6]. Such steels are commonly aluminium free, or very nearly so, since a prerequisite for titanium oxide to form as inclusions is the presence of uncombined oxygen at the end of steel making. Observations made from studies into the microstructural development in weld metals appear to show aluminium having a deleterious effect on the formation of acicular ferrite, even in the presence of favourable Ti rich inclusions [7]. In weld metals, other alloying elements with marked nitride forming abilities, principally titanium, vanadium and boron, are present but there is a strong case for aluminium to be present in solid solution in ferrite. It is not obvious, as to what level the microstructure is affected, but the work by Thewlis indicates a marked change when the total aluminium content is greater than about 0.03 wt%, implying a reduction in the soluble aluminium content [8]. There have also been reports, that trace additions of aluminium less than 50ppm, can have a marked effect on the microstructure of laser welds [9]. Several other instances of a pronounced effect of aluminium on transformation behaviour have been documented, mainly in steels containing chromium and/or molybdenum additions [10-11]. No universal explanation has been offered for such effects.

One interpretation suggests that partitioning of aluminium to the advancing ferrite/austenite interface reduces the boundary mobility leading to an increase in the hardenability of the steel [8,12]. Other workers have suggested that aluminium, by allowing aluminium nitride formation, causes any boron in the steel to become active in terms of affecting hardenability by releasing "free" boron [13]. Apart from these examples, there is little information on the effects of aluminium on transformation behaviour on conventional steels, apart from a study by Aaronson et al which showed that large additions of aluminium (>1wt%) raised the transformation temperature [14]. This study however, was based on high purity steels held under isothermal conditions, from which no evidence of a marked change in ferrite morphology was reported.

In view of the renewed interest in aluminium free steels for high toughness HAZ applications and the marked microstructural effects apparently associated with the presence of "free" or soluble aluminium in weld metals and Cr/Mo steels, the influence of aluminium on microstructure have been examined, and this paper presents an interim account of the work to date.

2. Experimental Techniques

The chemical compositions of the steels used in the present investigation are given in **Table 1**. These steels were made as 50kg heats under vacuum at British Steel Swinden Technology Centre, Rotherham, England. The range of aluminium additions chosen were deliberately made larger compared to normal conventional steels simply to exaggerate any kinetic or morphological effects of aluminium, although steel B, shown below would be typical of current levels. The nitrogen contents were deliberately maintained at low values to avoid complication from AlN precipitation [3].

One objective of the experimental work was to investigate the detailed changes in ferrite morphology which may result from aluminium addition and this dictated the need to develop a large, in the order of 300 to 500 microns, prior austenite grain size. Thermal cycling, as outlined by Grange, was used to develop a reproducible grain size in each steel, but it appeared that the steel containing 0.072%Al (steel C), was resistant to grain growth [15]. Additional data was therefore obtained to measure the grain growth kinetics, in comparison with steels A, B, and C, over the range 900-1300°C. Each steel was held at temperatures between 900-1300°C for times of 90s, 5mins, 15mins and 30mins respectively, to study the effect of temperature and time on grain growth.

Table 1. Chemical compositions of steels. (wt%)

SET1	C	Si	Mn	P	S	Cr	Mo	Ni	Al	N
VS2824 (A)	0.11	0.30	1.19	<.005	0.002	<0.02	<.005	<.02	<.005	.0009
VS2713 (B)	0.11	0.30	1.21	<.005	0.003	<0.02	<.005		0.033	.0010
VS2714 (C)	0.11	0.31	1.23	<.005	0.002	<0.02	<.005		0.072	.0009
VS2715 (D)	0.08	0.60	1.21	<.005	0.002	<0.02	<.005		0.034	.0011

Each cast was rolled to 15mm plate and cooled in air. A Theta high speed dilatometer was used to establish continuous cooling diagrams for each of the steels. Hollow cylindrical samples (measuring 10mm long by 5mm diameter with a 3mm bore) were heated under vacuum at 100°C/s to 950°C and then cycled between 950°C and 550°C five consecutive times in order to gain a fine grain size [15]. Upon completion of the final cycle each sample was heated from 550°C to 1300°C at the same heating rate and held for 1minute, to produce a coarse final grain size. Each sample was then cooled at a rate of .20/10/5/1 °C/sec, and 20/5 °C/min respectively. For each sample the dilation curve was analysed to obtain the volume fraction of ferrite transformed as a function of temperature and time, using the methods outlined by Eldis; allowing a comparison of the transformation kinetics as a function of aluminium content [16]. The dilatometer test runs at each cooling rate were repeated several times to obtain an indication of the experimental error involved with each test. Typically the transformation start and finish temperatures can be said to be accurate to with in +/- 5°C, in agreement with [16]. Due to the difficulty of obtaining near identical prior austenite grain sizes at all aluminium levels, the transformation kinetics for only steels A and B have been compared here.

The ferrite morphologies of steels A and B were examined in greater detail by interrupted cooling experiments. Initially "solid" dilatometer samples of (dimensions 10mm long by 5mm diameter) were subjected to an identical ramping program to that outlined above, (time to temperature was slightly altered in the program to take into account the solid nature of the samples), in order to develop a comparable grain size. From the peak temperature of 1300°C each sample was cooled to room temperature under vacuum. Each sample was then removed from the dilatometer, instrumented, heated to 1300°C within a furnace, removed and allowed to air-cool (15°C/sec at 650°C). The cooling curve for each sample was then matched to the CCT diagram so that an appropriate range of interrupted temperatures could be determined, (between 700-630°C). Subsequent samples, after removal from the furnace, were aircooled to the desired temperature, and were quenched in water to retain the initial transformation product (between 1-10% transformed).

All dilatometer and interrupted quenched samples were metallographically prepared by grinding and polishing, to a 1micron finish. Prior austenite grain size was estimated using ASTM charts after etching samples in an aqueous picric acid solution. Optical examination of the ferrite morphology was carried out after etching in 2%Nital. TEM examination was carried out using thin foils, made from steels A and B, having been partially transformed at 660 and 650°C as outlined above to give ~5% transformation. Each foil, was prepared to a thickness of 250microns using a Struers Acutom saw. Manual polishing reduced these to 120-100microns. Final thinning was achieved using a Struers Tenupol twin jet thinner, using an electrolyte of 10% perchloric acid and methanol. All TEM analysis was carried out using a Phillips EM430 transmission electron microscope, with a Link ISIS microanalysis attachment.

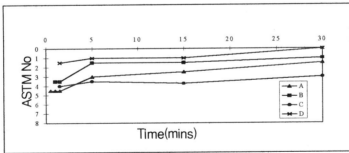

Fig. 1. Grain growth characteristics of experimental steels A-D, treated at 1200°C for between 30-1.5mins, quenched into water.

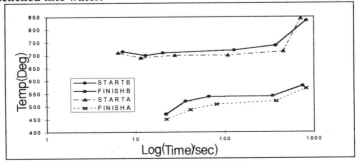

Fig. 2. CCT diagram for steels A and B (actual start/finish temperatures).

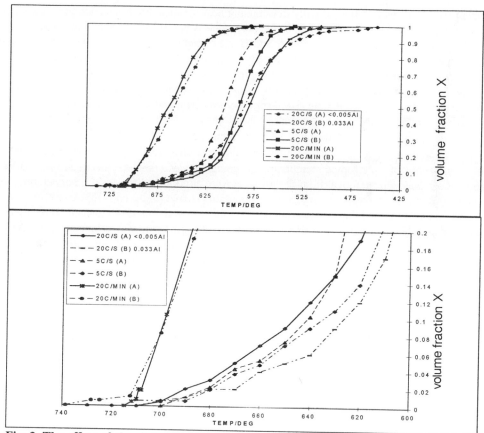

Fig. 3. The effect of aluminium on transformation kinetics for steels A and B

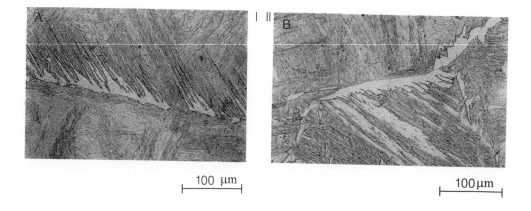

Fig. 4. Optical micrographs. I) steel A<0.005wt%Al, quench at 650°C, II) B,0.033wt%Al, 660°C.

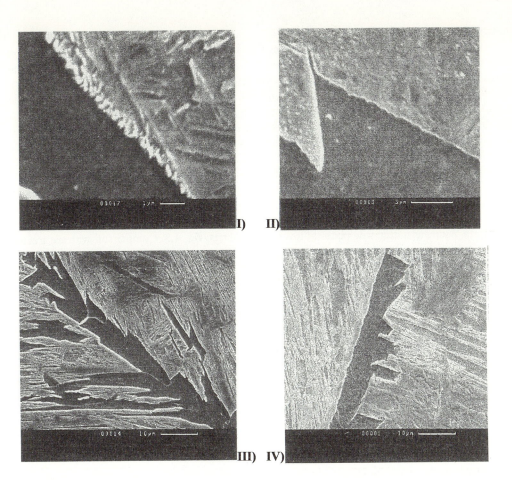

Fig. 5. Sem micrographs illustrating the effect of aluminium on ferrite, austenitised at1300°C for 90 secs, quenched at 660 (0.033Al), 640°C (0.005Al). (I/III 0.033Al, II/IV <0.005Al).

3. Results and Discussion
Grain growth behaviour

From the results obtained, **fig. 1,** provides the most representative example of the trends found. The most striking observation can be seen to be steel C (0.072Al), which maintains a fine grain structure relative to steels A, B and D throughout austenitising treatments of 1200-1300°C for times up to 30mins. According to available solubility data, it would appear that no AlN would be present at temperatures above 1200°C, removing the possibility of stabilising particles causing such an observation. Aluminium may therefore be speculated to influence the mobility of grain boundaries. In addition, comparison of steels B and D suggests an additional effect of silicon[17].

CCT data

Comparison of the CCT diagrams for steels A and B, **fig. 2**, shows that aluminium raises the Ac3 temperature of the steel by approximately 180°C/1wt%Al, over the range of cooling rates studied. Aaronson showed similar effects in high purity alloys after isothermal transformation, 95°C/1wt%Al, and the thermodynamic analysis used by Kirkaldy and Baganis also indicates that a ferrite stabilising element such as aluminium would be expected to raise the Ae3 and, hence the Ac3 temperature[14,18].

Kinetics of ferrite transformation

The data for samples A and B indicate a variable yet significant effect of aluminium on transformation kinetics, even in the small quantities added in this study, **fig. 3**. For clarity, three cooling rates are shown, indicating the trend found. At the most rapid cooling rate (20°C/s), aluminium can been seen to have little effect on the kinetics other than slightly depressing the transformation start temp by 10-15°C. Upon lowering the cooling rate, aluminium can be seen to delay the transformation to lower temperatures and alter the kinetics to completion. This trend remains consistent up to a cooling rate of 20°C/min, at which point the effect of aluminium decreases and the kinetics in both the free and treated steel appear almost identical, **fig. 3**.

Microstructural effects of aluminium on ferrite morphology

To enable direct comparison of the partially transformed microstructures, it was important to take into account the ~15°C difference in transformation start temperature between the aluminium free and treated steel, **fig. 2**. In the aluminium free steel (A), allotriomorphs of ferrite appear to rapidly form Widmannstatten side plates at low undercoolings, **fig. 4,5**. Comparison with the treated steel, **fig. 4,5**, shows this to have a more ordered, crystallographic allotriomorphs, displaying more regular interfaces. Consistently, steel A showed larger protuberances on numerous allotriomorph edges, with high concentrations of irregular steps or ledges. In comparison, the aluminium treated steel, (addition of 300ppm Al) displayed far fewer, **fig. 4,5**. This suggests that aluminium interferes with the growth of the ferrite/austenite interface in some form. These optical observations may be linked to the segregation of aluminium to the advancing interface. This pattern of behaviour appears to be similar to that exhibited by boron [19]. The atomic size of aluminium (2.85Å), shows this to be a large substitutional element. The approach of Townsend and Kirkaldy based upon Widmanstatten growth being a function of carbon diffusion moderated by the effect of anisotropic surface tension, suggests, if aluminium raises the surface tension over the interface, the wavelength of the initial protuberance which leads to Widmanstatten plate formation also increases, which reinforces the experimental observations [20]. Niobium has also been shown to have an influence on Widmanstatten ferrite morphology [21]. It is interesting to note the similarity between the atomic sizes of Al (2.85Å) and Nb (2.8Å), [22].

Based on the work by Hillert, the observations made may be based upon solute/interface interaction and solute drag effecting grain boundary mobility, (in the case of steel C)[12].

Another explanation, similar to that of boron, could be if aluminium was thought to exist at the interface boundary as a mono layer, until such a time as the surface area of the developing interface, through sideplate formation and general allotriomorph growth, is such that a monolayer can no longer exist, since aluminium is spread more thinly. At such a point, the large driving force for transformation, due to the retardation of the interface by Al, creates rapid sideplate growth. Such a theory may be employed to fit the kinetics outlined above. In the same manner aluminium may influence grain size development, by influencing grain boundary mobility [12,19]. Rapid rates of cooling would appear to minimise any effect. In addition, at slow cooling rates aluminium may have time to diffuse significantly within the lattice, such that any effect is also lost. It has been assumed that aluminium acts as a fine film at the boundary, however, aluminium may also exist in small clusters over the interface, located on specific growth ledges, as shown in **fig 6**. Reducing the mobility of certain ledges/steps, would influence the development final ferrite morphology.

TEM Analysis

From the observations made optically it was assumed that aluminium was in some way segregating to the advancing interface and influencing its movement. TEM was employed to investigate the merits of this theory. Samples of steel A and B, transformed to 10% volume fraction of ferrite, (quenched from 660°C, in the Al steel and 640°C in the Al free steel) were studied. **Figure 6** indicates two examples of ferrite/austenite interfaces in the treated aluminium steel. Microanalysis was conducted at points across the interface, either side and on the interface (shown by the dark carbon deposits across the interface, XYZ, **fig. 6B**), in an attempt to determine whether aluminium had segregated across the boundary.

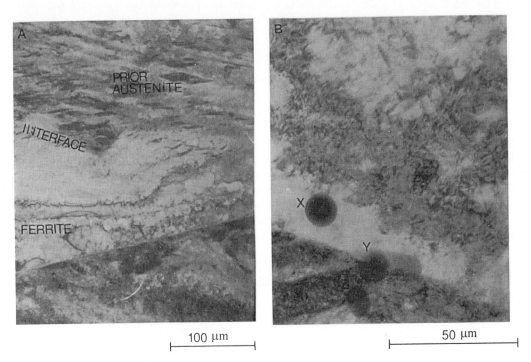

Fig. 6. TEM micrographs. A) Growth steps across interface, B) Location of microanalysis.

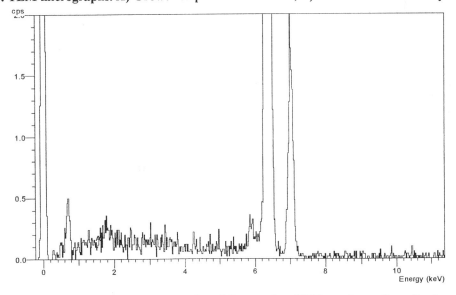

Fig. 7. Characteristic X-Ray spectra produced from points XYZ, as shown from fig. 6.

From the x-ray spectra produced, **fig. 7**, no aluminium segregation across the interface was found. This does not rule out the theory of aluminium segregation at the advancing interface, yet outlines the difficulty in a) finding the growth interface of the ferrite in such a low transformed, large grained sample under TEM, and b) observing segregated areas of aluminium in matrix concentrations of 3 parts per 1000. At this stage of the study TEM has proved inconclusive as to the location of aluminium at the transformation front.

5. Conclusions

• Aluminium has been observed to raise the transformation start and finish temperatures for the steels tested.

• Aluminium changes the transformation kinetics depending on cooling rate in the range <0.005-0.033wt% Al.

• An optical effect of such an addition has been observed optically, apparently reducing sideplate formation, creating a more ordered interface structure.

• One theory suggests that aluminium segregates to the advancing ferrite/austenite interface, yet analysis of the interface using TEM has proved inconclusive at this stage.

•The details of the effect of aluminium on the morphology and transformation kinetics of ferrite remain inconclusive, requiring further experimentation to clarify the mechanisms proposed in this paper.

6. References

[1] T. Gladman, Aluminium Nitride in Steel, British Steel Report PH/R/S1104/4/88/A.

[2] S. L. Case, K. R. Van Horn, (1953), J. Wiley and Sons Inc.

[3] T. Gladman, "The Physical Metallurgy of Microalloyed Steels", Instit Mat, London, UK. (1997).

[4] F. J. Barbarro, PhD thesis, University of New South Wales, Austrailia, (1990).

[5] D. J. Senogles, PhD thesis, University of Leeds, England, (1994).

[6] S. St-Laurent and G. L. Esperance, Mater. Sci. Eng., A149 (1992), p.203.

[7] G. Evans, IIWDoc. II-A-904-93, (1993).

[8] G. Thewlis, Mater. Sci. Technol., 10(1994), p.110.

[9] R.C.Cochrane, D. J. Senogles, L. J. Lloyd, ECSC Research Project.
    ECSCNo:7210.MC/802(FS4/91). Final Report British Steel Ref: FR S337-7941, (Nov 1994).

[10] H. Mabuchi and H. Nakao, Trans. Iron. Steel. Inst. Jpn., 21 (1981), p.495.

[11] H. Mabuchi and H. Nakao, Trans. Iron. Steel Inst. Jpn., 23 (1983), p.514.

[12] M. Hillert, Met. Trans., 6a (1975), p.5.

[13] Y. Jung, H. Ueno, H. Ohtsubo, K. Nakai, Y. Ohmori, ISIJ. Inter., 35 (8) (1995), p.1001.

[14] K. R. Kinsman and H. I. Aaronson, Met. Trans., 4 (1973), p.959.

[15] R. A. Grange, Metal. Trans., 2 (1974), p. 65.

[16] G. T. Eldis, "Symp Hardenability Concepts with Applications to Steel", Chicago Sheraton, Oct 24-26, (1977), Ed D. V. Doane and J. S. Kirkaldy, Met. Soc. AIME, p 126.

[17] G. R. Wang, T. H. North and K. G. Leewis, Weld. Journal. Res. Supp., 69 (1990), p.145.

[18] J. S. Kirkaldy, E. A. Baganis, Metal. Trans., 9a (1978), p 495..

[19] J. E. Morral, T. B. Cameron, "Symp Boron in Steel", Mlwaukee, Wisconsin, Sept18, 1979, Ed S. K. Banerji, J. E Morral, Met Soc AIME, p.19.

[20] R. D. Townsend, J. S. Kirkaldy, Trans. ASM, 61, (1968), p.605.

[21] R. E. Dolby, Weld Journal Res Supp., 6 (1979), p225.

[22] W. Hume-Rothery, G. V. Raynor, "The Structure of Metals and Alloys", Institute of Metals, (1962).

Correspondence Address: msemfe@ecu-01.novell.leeds.ac.uk. FAX: +44 0113 2422531

Materials Science Forum Vols. 284-286 (1998) pp. 225-230
© 1998 Trans Tech Publications, Switzerland

Refinement of Ferrite-Pearlite Structures through Transformation from Heavily Deformed Austenite in a Low Carbon Si-Mn Steel

S. Torizuka, O. Umezawa, K. Tsuzaki and K. Nagai

Frontier Research Center for Structural Materials, National Research Institute for Metals, 1-2-1 Sengen Tsukuba, Ibaraki 305-0047, Japan

Keywords: Grain Refinement, Ferrite-Pearlite Structure, Welding Structural Steel, Si-Mn Steel, Thermomechanical Treatment, Controlled Rolling, Crystallographic Orientation, High-Angle Boundary

Abstract -- In a low-carbon Si-Mn steel without any other alloying elements, an ultra-fine ferrite-pearlite microstructure with a ferrite grain size of 2.3 μm was created through thermomechanical treatment. The feature of the process was one-pass heavy reduction at a low temperature in unrecrystallized austenite region followed by controlled cooling at 10 K/s. Crystallographic orientation of each ferrite grain was randomly distributed and most of the grain boundaries were high-angle ones.

1. INTRODUCTION

We are challenging to develop 800 MPa class high-strength steels with good weldability as conventional 400 MPa class welding structural steels have. The conventional 400 MPa class steels have low-alloy Si-Mn compositions and their microstructure is ferrite-pearlite. Since welding structural steels are mass-product, the cost and recyclability are of great interest. Thus, strengthening by the addition of alloying elements cannot be used from the viewpoint of not only weldability. Low temperature transformation products such as bainite and martensite are effective for strengthening steels, but to obtain them relatively high cooling rates are necessary. The cooling rate in an accelerated cooling process depends on plate thickness and the highest cooling rate in a typical facility is 20K/s for 10 mm thick plates [1]. Since our aim is to develop welding structural materials with thickness beyond 10 mm, the cooling rate has to be lower than 20 K/s. At these low cooling rates bainitic and martensitic transformations cannot be expected in low-alloy Si-Mn steels with low hardenability. On these conditions, we have to double the strength of ferrite-pearlite steels having the compositions of the conventional 400 MPa class steels.

Grain refinement is a promising method to strengthen steels without addition of alloying elements deteriorating weldabilty and recyclability. In order to achieve the doubled strength based on the Hall-Petch law, the ferrite grain size must be refined into 1 μm or finer. There have been many studies on ferrite grain refinement especially by thermomechanical processing [2-7]. Figure 1 summarizes the ferrite grain sizes of ferrite-pearlite structures obtained from work-hardened austenite in a 0.03%Nb steel [3]; the steel plates were air-cooled after deformation at 1123 K, below recrystallization temperature of austenite. The ferrite grain size linearly decreases with an increase in reduction, irrespective of austenite grain size. This figure suggests that heavier deformation than 70% in reduction or finer austenite grain size than 28 μm is required for obtaining finer ferrite grain sizes than 5 μm. However, effect of heavy deformation beyond 70% in reduction on ferrite grain refinement has been scarcely reported.

In the present study, we have thus investigated the possibility of ferrite grain refinement in a low-carbon Si-Mn steel through heavy deformation in unrecrystallized austenite region. Apparent grain refinement is not always successful in increasing strength. Grain boundary character and the distribution of grain orientation have an important influence on mechanical properties. Hence, we have paid particular attention to the crystallographic characteristics of a fine grained structure.

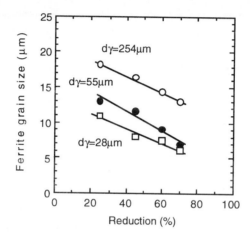

Fig.1   Effects of deformation of austenite at 1123 K (below recrystallization temperature) and initial austenite grain size on ferrite grain size in a 0.03%Nb steel (0.16%C, 0.36%Si, 1.41%Mn) [3]. The steel plates were air-cooled after the rolling at 1123 K.

2. EXPERIMENTAL

The chemical composition of the steel prepared by vacuum induction melting is listed in Table 1. The ingot was homogenized at 1523 K for 3.6 ks followed by hot forging and rolling to a plate with a thickness of 15 mm. The plate was machined into specimens with dimensions of 20 mm in length, 18 mm in width and 12 mm in thickness. Thermomechanical treatment was carried out with a plain compression simulator. A sample was first austenitized at 1173 K for 60 s, resulting in a small austenite grain size of 17 μm. The specimen was then cooled to 1023 K at a rate of 10 K/s and held for 5 s, and subsequently compressed by 73% in nominal reduction at a strain rate of 10/s.

After deformation the sample was immediately cooled to room temperature. The cooling rate was set at 1, 3 or 10 K/s. These cooling rates could be obtained without any cooling mediums such as gas or water. Water quenching (300 K/s) was also used for comparison. The austenite to ferrite transformation start temperature (Ar3) at 10 K/s in undeformed samples was 933 K, indicating that the deformation was applied before the transformation.

Table 1   Chemical composition of the steel (mass %)

C	Si	Mn	P	S	t-Al	t-N	O	Fe
0.17	0.3	1.5	0.02	0.001	0.03	0.001	0.001	bal.

In the compressed sample, the region neighboring a compression anvil was almost free from straining and becomes a dead metal zone due to friction between the sample and the compressing anvil. The actual reduction in the center region of a sample was 90% when the nominal reduction was 73%. The amount of reduction will be hereafter described as the actual value of 90%.

Microstructures were examined with SEM after polishing and etching with a nital. Observation planes were parallel to the compressive direction and the longitudinal direction of a sample. The mean liner intercept method was used to determine the ASTM grain size according to d=1.128L, where L is the mean linear intercept measured on SEM micrographs. Crystallographic orientations of ferrite grains were characterized by electron back scattering diffraction (EBSD) analysis.

3. RESULTS

3.1 Microstructural observation

Figure 2 represents the change of SEM microstructures as a function of final cooling rate after 90% deformation. The longitudinal directions of the samples are parallel to the micron markers. Fine ferrite-pearlite structures were obtained when the cooling rate was not higher than 10 K/s (Fig.1(a), (b) and (c)), but a ferrite-martensite structure was obtained in the water-quenched sample (Fig.1(d)). Ferrite grain size decreased with an increase in cooling rate. Accordingly, the cooling rate of 10 K/s produced the finest ferrite-pearlite structure. The pearlite had evidently a lamellar structure. Equiaxed ferrite grains were obtained for rapidly cooled samples, and elongated ones for slowly cooled samples.

Figure 3 shows the average ferrite grain size as a function of cooling rate after 90% deformation. An ultra-fine ferrite-pearlite structure with the ferrite grain size of 2.3 μm was obtained at the cooling rate of 10K/s. It is noteworthy that the ferrite grain size was 1.5 μm even in the water-quenched sample where the growth of ferrite grains was effectively suppressed by the rapid cooling at 300 K/s. This indicates that the heavy deformation of austenite at 1023 K was still not enough to introduce the nucleation sites necessary for the ferrite grain sizes finer than 1 μm, and other devices should be examined.

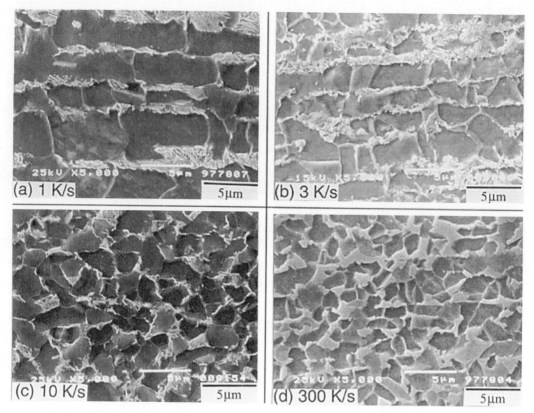

Fig.2   SEM micrographs showing the effect of final cooling rate on microstructures formed from the 90% deformed austenite. The micrographs were taken from the planes parallel to the compressive direction and the longitudinal direction.

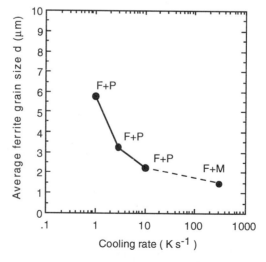

Fig.3   Effect of final cooling rate on the ferrite grain sizes.   The specimens were deformed by 90% in reduction at 1023K and then cooled to room temperature.

3.2 Crystallographic orientations of refined ferrite grains

The most promising sample cooled at 10 K/s after 90% deformation was analyzed by EBSD. Figure 4 is the inverse pole figure for 28 ferrite grains shown in Fig.2(c). The normal direction (ND) to the compressive plane is plotted. Not strong texture but almost random distribution exists in the sample. Misorientation angles between neighboring ferrite grains were also analyzed in a region of 50 μm x 50 μm square. When high-angle boundaries are defined as ones having the misorientation angle not less than 15 degree, 97% of the total ferrite-ferrite boundaries were high-angle ones.

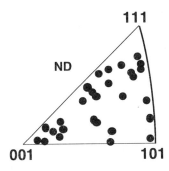

Fig.4   Inverse pole figure for 28 ferrite grains in the sample cooled at 10 K/s after 90% reduction at 1023K. The normal direction to the compressive plane is plotted.

4. DISCUSSION

The present study has shown that the combination of 90% deformation and 10K/s cooling produced an ultra-fine ferrite-pearlite structure with a ferrite grain size of 2.3 μm in a Si-Mn steel without any other alloying elements. It should be emphasized that not only the fine-grained structure was achieved but also most of the ferrite grain boundaries were high angle ones. This result indicates that heavy deformation in unrecrystallized austenite region is very effective for ferrite grain refinement and strengthening of ferrite-pearlite steels. Since the austenite grain size is 17 μm, the thickness and length of austenite grains becomes 1.7 μm and 153 μm, respectively, when the austenite is deformed by 90 % in reduction. This thickness, 1.7 μm, is comparable to the ferrite grain size, 2.3 μm, suggesting that most of ferrite grains nucleated on austenite grain boundaries.

It has been reported that precipitates formed on a grain boundary tend to choose a single variant in the case of undeformed materials [8,9]. The present result suggests that this rule of single variant selection in grain boundary precipitates was broken and many variants formed in the heavily deformed sample. How to brake the single variant selection is important for obtaining a fine-grained structure and the reason for the formation of many ferrite variants is of great interest. This multi-variants formation in the present study is apparently associated with a deformation structure. Detail results and discussion for that will be presented elsewhere [10].

5. CONCLUSIONS

An attempt to create a very fine ferrite-pearlite microstructure in a low-carbon Si-Mn steel by one-pass reduction at a low temperature in the unrecrystallized austenite region has been undertaken. The following conclusions were obtained.

(1) The combination of a heavy deformation of 90% reduction at 1023K and subsequent cooling at 10K/s produces an ultra-fine ferrite-pearlite structure with a ferrite grain size of 2.3 μm.

(2) Crystallographic orientation of each ferrite grains is randomly distributed and most of the grain boundaries are high-angle ones.

REFERENCES

[1] K.Tsukada, K.Matsumoto, K.Hirabe and K.Takeshige, Iron and Steel Maker, (1982), July, p.21.

[2] M.Fukuda, T.Hashimoto and K.Kunishige, Tetsu-to-Hagane, 58 (1972), p.1832.

[3] I. Kozasu, C. Ouchi, T. Sampei and T. Okita, Microalloying 75, ed. by M.Korchinsky, (1975) , p.120.

[4] C. Ouchi, T. Sampei and I. Kozasu, Tetsu-to-Hagane, 67 (1981), p.143.

[5] H.Inagaki, Testu-to-Hagane, 70 (1984), p.412.

[6] C.Ouchi, High Strength Low Alloyed Steels, ed. by D.P.Dunne and T.Chandra, (1984), p.17.

[7] A.J.DeArdo, Microalloying 95, ed. by M.Korchinsky, (1995), p.15.

[8] K.Ameyama, T.Maki and T.Tamura, J. Jpn. Inst. Met., 50 (1986), p.602.

[9] T.Furuhara, S.Takagi, H.Watanabe and T.Maki, Metall. Mater. Trans. A, 27A (1996), p.1635.

[10] S.Torizuka, O.Umezawa, K.Tsuzaki and K.Nagai, to be published in ISIJ Inter.

Materials Science Forum Vols. 284-286 (1998) pp. 231-236

Effect of the Microalloying Elements on Nucleation and Growth Kinetics of Allotriomorphic Ferrite in Medium Carbon-Manganese Steels

C. García de Andrés[1], C. Capdevila[1] and F.G. Caballero[1,2]

[1] Department of Physical Metallurgy, Centro Nacional de Investigaciones Metalúrgicas (CENIM), Consejo Superior de Investigaciones Científicas (CSIC), Avda. Gregorio del Amo 8, E-28040 Madrid, Spain

[2] Department of Materials Science and Metallurgy, University of Cambridge, Pembroke Street, Cambridge CB2 3QZ UK

Keywords: Microalloyed Steels, Allotriomorphic Ferrite, Nucleation, Growth Kinetic

ABSTRACT

The influence of alloying elements on nucleation and growth kinetics of allotriomorphic ferrite has been studied using dilatometric techniques and microstructural analysis in four medium carbon-manganese steels (0.3%C - 1.4% Mn). A careful comparison between a C-Mn steel and microalloyed steels containing V, Ti and Mo subjected to isothermal transformation confirmed that all these elements delay the allotriomorphic ferrite transformation. More significant effect has been found by adding Mo, compared with the effect of the V and Ti.

INTRODUCTION

Titanium and vanadium microalloyed steels are used extensively in forged automotive components [1-2]. Titanium is added with the aim of refining the microstructure through the inhibiting effect to grain coarsening exerted by small TiN precipitates [3-5]. Vanadium is selected due to its precipitation hardening capability, with a view to improve the toughness properties [6-8]. It has been found that toughness and strength can be improved simultaneously by transforming the austenite mainly in a fine acicular microstructure [9-12]. Consequently, there has been considerable effort towards maximizing the amount of acicular ferrite in the final microstructure.

Several authors showed that the presence of a uniform layer of allotriomorphic ferrite along the austenite grain surface induces the transformation of austenite in acicular ferrite instead of bainite [13-15]. In this sense, allotriomorphic ferrite formation plays a particular and important role in influencing the development of acicular ferrite in mixed microstructures.

The purpose of the present study is to clarify experimentally the influence of Ti, V and V-Ti additions on the allotriomorphic ferrite transformation and indirectly, on the development of the intragranular transformation of acicular ferrite in microalloyed forging steels. The effect of molybdenum on the allotriomorphic ferrite transformation in a V-Ti microalloyed steel has also been studied.

MATERIALS AND EXPERIMENTAL PROCEDURE

Four medium carbon manganese steels were used in the present study and their chemical compositions are shown in Table 1. Carbon (≈0.3 %) and manganese (≈1.4%) contents are similar in all the steels. Three of them are microalloyed steels with different vanadium and titanium contents. In addition, V-Ti-Mo steel contains 0.12 % Mo.

Table 1. Chemical compositions (mass %).

Steel	C	Mn	Cu	Cr	S	Si	Al	Ni	V	Ti	Mo
C-Mn	0.31	1.22	-	-	0.011	0.25	-	0.10	0.004	-	0.03
V	0.33	1.49	0.27	0.13	0.002	0.25	0.027	0.11	0.240	0.002	0.04
V-Ti	0.32	1.39	0.13	0.13	0.021	0.33	0.027	0.14	0.130	0.039	0.03
V-Ti-Mo	0.30	1.51	0.20	0.25	0.030	0.32	0.020	0.14	0.110	0.020	0.12

An Adamel Lhomargy DT1000 high resolution dilatometer has been used to determine the nucleation curves corresponding to the isothermal transformation of austenite of these steels. The dimensional variations of the specimen are transmitted via an amorphous silica pushrod. These variations are measured by an LVDT sensor in a gas-tight enclosure enabling testing under vacuum or in an inert atmosphere. The heating and cooling devices of this dilatometer were also used to perform all heat treatments. The DT1000 dilatometer is equipped with a radiation furnace for heating. The power radiated by two tungsten filament lamps is focussed on a specimen of small dimensions (2 mm thickness and 12 mm length) by means of a bi-elliptical reflector. The temperature is measured with a 0.1 mm diameter Chromel-Alumel (Type K) thermocouple welded to the specimen. Cooling is carried out by blowing a jet of helium gas directly onto the specimen surface. The helium flow rate during cooling is controlled by a proportional servovalve. The excellent efficiency of heat transmission and the very low thermal inertia of the system ensure that the heating and cooling rates ranging from 0.003 K/s to approximately 200 K/s remain constant.

Dilatometer specimens were austenitized and subsequently isothermally transformed at temperatures in the range of 873 to 973 K at different times. After isothermal transformation, specimens were gas-quenched. They were polished and etched in the usual way. The allotriomorphic ferrite volume fraction (*vfAF*) was estimated from optical micrographs by an unbiased systematic manual point counting procedure based upon stereological principles [16].

RESULTS AND DISCUSSION

As it is well known, prior austenite grain size (*PAGS*) exerts an important influence on the decomposition of austenite [17-18]. The *PAGS* parameter affects directly the allotriomorphic ferrite growth kinetic obtained by isothermal decomposition of the austenite. Nevertheless, there is no influence on the nucleation time of this phase [19-21].

Austenitization conditions were selected to achieve the same *PAGS* in all the steels, thus avoiding the influence of the austenite grain size on the allotriomorphic ferrite formation. This enables us to study specifically the effect of the microalloying elements on the growth kinetics of this phase. Since the growth rate of allotriomorphs is higher in finer *PAGS*, a coarse *PAGS* of 70 μm approx. was selected in order to facilitate the study of the growth kinetic of allotriomorphic ferrite. Austenitization conditions and the achieved *PAGS* are shown in Table 2.

Table 2. Austenitization conditions.

Steel	Austenitization Temperature, K	Holding Time, s	*PAGS,* μm
C-Mn	1473	120	70
V	1473	300	75
V-Ti	1523	180	72
V-Ti-Mo	1473	120	75

Effect of microalloying elements on the nucleation time is given in Fig. 1. It can be seen that C-Mn steel presents the shortest time for the nucleation of the allotriomorphs. All the microalloying elements delayed the nucleation of this phase. Comparison of the nucleation curves of V steel (0.24% V, 0.002% Ti) and V-Ti steel (0.13% V, 0.039% Ti) suggests that the effect of vanadium on the nucleation time is more important than the one of titanium. Moreover, V-Ti-Mo steel presents the longest nucleation time. Bearing in mind the titanium and molybdenum contents of V-Ti steel (0.039% Ti, 0.03% Mo) and V-Ti-Mo steel (0.020% Ti, 0.12% Mo), it can be concluded that molybdenum exerts a strong influence on the nucleation of the allotriomorphic ferrite. The time needed for the nucleation of ferrite in V-Ti-Mo steel is longer than the one in V-Ti steel, though the titanium content in the first steel is lower.

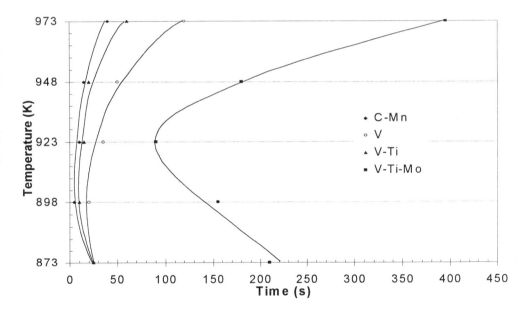

Figure 1. Nucleation of allotriomorphic ferrite.

The temperatures (T_i), at which the incubation time of allotriomorphic ferrite (t_i) is minimum, are reported in Table 3. This temperature is approximately the same in C-Mn, V and V-Ti steels. In this sense, vanadium and titanium do not seem to exert any influence on this temperature T_i. However, the molybdenum content in V-Ti-Mo steel raises T_i temperature at 923 K, which is consistent with the work of Kinsman and Aaronson [22].

Table 3. Temperature and time of the nose of nucleation curves.

Steel	Isothermal Temperature T_i, K	Nucleation Time t_i, S
C-Mn	898	5
V	898	10
V-Ti	898	20
V-Ti-Mo	923	90

With the aim of studying the growth kinetic of allotriomorphic ferrite at the same conditions in all the steels, this work has been carried out at the temperatures T_i reported in Table 3. Experimental kinetic results for the growth of allotriomorphic ferrite at T_i temperature are shown in Fig. 2 for the four steels. In this figure, the incubation time (t_i) of every steel is taken as origin of time. The growth curves of Fig. 2 are plotted exclusively in the range of times for which the allotriomorphic ferrite is the only transformation that takes place. In C-Mn steel, Widmanstätten ferrite appears a few seconds after allotriomorphic ferrite is nucleated. However, in order to evaluate the effect of the microalloying elements on the growth kinetic of allotriomorphic ferrite, the curve of this steel has been plotted for the times below 15 s, at which pearlite starts forming and a volume fraction of Widmanstätten ferrite lower than 2 % is detected. Widmanstätten ferrite has been found in none of the microalloyed steels at their corresponding temperatures T_i. Table 4 reports the time required to nucleate pearlite in all the steels, taking as origin the corresponding incubation time t_i for allotriomorphic ferrite.

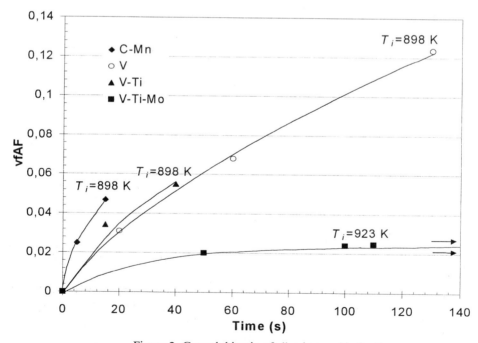

Figure 2. Growth kinetic of allotriomorphic ferrite.

Table 4. Nucleation time of pearlite.

Steel	Nucleation Time, s
C-Mn	15
V	40
V-Ti	130
V-Ti-Mo	>900

The influence of the microalloying elements on the growth rate of allotriomorphic ferrite is clear from the isothermal kinetic results shown in Fig. 2. Vanadium decreases the growth rate of ferrite more severely than the titanium does. As in nucleation, molybdenum exerts an important influence on the kinetic growth of ferrite. The growth rate in V-Ti-Mo steel is significantly slower than in the other three steels. Moreover, molybdenum is responsible for the delay in the formation of pearlite. The nucleation time of this phase is found to be much longer in this steel (Table 4). Vanadium also seems to have a certain influence on the nucleation of pearlite.

CONCLUSIONS

1. Vanadium and titanium delay the nucleation of allotriomorphic ferrite in 0.3 C – 1.4 Mn steels. The effect of vanadium in the nucleation of ferrite is more important than the one of titanium. Experimental results show that the addition of 0.12% Mo in a microalloyed steel containing vanadium and titanium significantly retards the nucleation of allotriomorphic ferrite.
2. Molybdenum raises the temperature at which the incubation time of allotriomorphic ferrite is minimum. Microallying elements such as vanadium and titanium do not seem to exert any influence on this temperature.
3. Formation of Widmanstätten ferrite is inhibited by the microalloying elements considered in this work, whereas an significant volume fraction of this phase has been found in C-Mn steel.
4. Vanadium decreases the growth rate of allotriomorphic ferrite more severely than the titanium does. However, molybdenum combined with vanadium and titanium is the element that exerts the most significant influence on the growth rate of this phase.

ACKNOWLEDGEMENTS

The authors acknowledge financial support from the CICYT- Spain (project PETRI 95-0089-OP).

REFERENCES

[1] F. Peñalba, C. García de Andrés, M. Carsí and F. Zapirain, ISIJ Int. 32 (1992), p. 232.
[2] F.B. Pickering, 'Physical Metallurgy and the Design of Steels', Applied Science Publishers, London (1978), p. 1 .
[3] M.A. Linaza, J.L. Romero, J.M. Rodriguez-Ibabe and J.J. Urcola, Scripta Metallurgica et Materialia 29 (1993), p. 1217.
[4] B. Yunguang, G. Qiang and Z. Lianwei, 'HSLA Steels'95'. Proceedings of the 3rd international conference of HSLA steels held at Beijing, China, organised by The Chinese

Society for Metals on 25-29 October 1995, China Society and Technology Press (1995), p. 259.

[5]   P.A. Manohar, D.P. Dunne, T. Chandra and R. Killmore, ISIJ Int. 36 (1996), p. 194.

[6]   M. Zhang and D.V. Edmonds, 'HSLA Steels'95'. Proceedings of the 3rd international conference of HSLA steels held at Beijing, China, organised by The Chinese Society for Metals on 25-29 October 1995, China Society and Technology Press (1995), p. 133.

[7]   F. Peñalba, C. García de Andrés, M. Carsí and F. Zapirain, Journal of Materials Science 31 (1996), p. 3847.

[8]   J.L. Romero, 'Mejora de las Propiedades Mecanicas en Aceros en Contenido Medio en Carbono Microaleados con Ti y/o V para Forja y Laminacion, mediante Tratamientos Termomecanicos', PhD Thesis, Universidad of Navarra (1996), p. 71.

[9]   R.E. Dolby, 'Factors Controlling Weld Toughness-The Present Position, Part II-Weld Metals', Welding Institute Research Report, No. 14/1976/M,Welding Institute, London (1976), p. 75.

[10] H.K.D.H Bhadeshia, 'Mathematical Modelling of Weld Phenomena', Materials Modelling Series, Institute of Materials, Edited by H. Cerjak, London (1995), p. 71.

[11] M.A. Linaza, J.L. Romero, I. San Martin, J.M. Rodriguez-Ibabe and J.J. Urcola, 'Microalloyed Bar and Forging Steels', Edited by C.J. Van Tyne, G. Krauss and D.K. Matlock, The Minerals, Metals and Materials Society (1996), p. 311.

[12] M.A. Linaza, J.L. Romero, I. San Martin, J.M. Rodriguez-Ibabe and J.J. Urcola, Scripta Metallurgica et Materialia 29 (1993), p. 1217.

[13] H.K.D.H. Bhadeshia, 'Bainite in Steels', The Institute of Metals, London (1992), p. 174.

[14] H.K.D.H. Bhadeshia and S.S. Babu, Materials Transactions JIM 32 (1991), p. 679.

[15] H.K.D.H. Bhadeshia and S.S. Babu, Materials Science and Technology 6 (1990), p.1005.

[16] G.F. Vander Voort, 'Metallography. Principles and Practice', McGraw-Hill Book Company, New York (1984), p. 427.

[17] J. Bardford and W.S. Owen, J. Iron Steel Institute 197 (1961), p. 146.

[18] A.K. Sinha, 'Ferrous Physical Metallurgy', Butterworths, Boston, USA (1989), p. 379.

[19] K.C. Russell, Acta Metall. 16 (1968), p. 761.

[20] K.C. Russell, Acta Metall. 17 (1969), p. 1123.

[21] H.K.D.H Bhadeshia, Metal Sci. 16 (1982), p. 159.

[22] K.R Kinsman and H.I Aaronson, 'Transformations and Hardenability in Steels', Climax Molybdenum Co., Ann Arbor, Michigan (1967), p. 39.

Dr. C. García de Andrés
E-mail address: cgda@cenim.csic.es
Fax number: 07-34-1-5347425

Materials Science Forum Vols. 284-286 (1998) pp. 237-244
© 1998 Trans Tech Publications, Switzerland

The Effect of Si on the Microstructural Evolution of Discontinuously Cooled Strip Steels

E.V. Pereloma[1], I. Timokhina[1] and P.D. Hodgson[2]

[1] Department of Materials Engineering, Monash University, Clayton, Victoria 3168, Australia

[2] School of Engineering and Technology, Deakin University, Geelong, Victoria 3217, Australia

Keywords: Thermomechanical Processing, Microalloyed Steel, Accelerated Cooling, Microstructural Characterisation, Phase Transformation, Austenite, Ferrite, Bainite, Retained Austenite

Abstract

The effects of silicon (0.16 and 1.4 wt%) and thermomechanical processing schedules on the transformation history and final microstructure of a low carbon, Nb microalloyed steels were investigated. Two initial austenite conditions were produced in the steels: recrystallised and non-recrystallised. Complex microstructures consisting of ~50% ferrite, granular bainite and/or acicular ferrite were obtained using discontinuous cooling schedules. The bainite transformation temperatures were ~50-70°C higher in the steel with low Si content, regardless of the thermomechanical processing schedule. Both the Vickers hardness and the microhardness of second phase were higher in the steel with the higher Si content due to a solution strengthening and a finer microstructure. Increased silicon content promotes formation of carbide free phases: polygonal ferrite and acicular ferrite.

Introduction

Recently the development of higher strength hot rolled steels has focused on the control of the final microstructure. Their attractive mechanical properties are result of the multi-phase microstructure consisting of polygonal ferrite, pearlite, bainite, martensite/austenite microconstituent and acicular ferrite. The formation of each phase depends on the steel composition, the austenite condition prior to the transformation produced during finish rolling and the cooling history. The mechanical properties are controlled by the volume fraction and distribution of the phases. Given the complexity of the final structure, and its importance, it is obviously necessary to be able to quantify the transformation behaviour during cooling after rolling.

Most of the studies to date have only considered transformations during continuous cooling or isothermal holding, whereas in practice the cooling path is generally discontinuous. Even the prediction of single phase microstructures under these complex cooling conditions is difficult. This also makes the task of prediction of multi-phase structure nearly unachievable. Further modifications in processing schedules and steel compositions require detailed knowledge on the effect of chemistry and processing on the transformation kinetics. Silicon plays an important role in achieving final microstructure of thermomechanically processed microalloyed steel. Si is a ferrite stabiliser, it suppresses pearlite formation, inhibits carbide formation and encourages the formation of carbide-free bainite [1].

The aim of current work was twofold: (i) improve the understanding of the kinetics of the phase transformations during discontinuous cooling of hot-rolled microalloyed steels and (ii) correlate processing parameters and final microstructure with silicon content of steel.

Experimental

Two experimental microalloyed steels, chemical compositions of which are shown in Table 1, were studied. Both steels were obtained in as-rolled condition from BHP Research-Melbourne Laboratories. Simulations of thermomechanical processing were carried out on the MMC quench-deformation dilatometer at BHP Research. Cylindrical samples used for dilatometer tests were 4 mm diameter and 8 mm long.

Table 1.   Steel compositions (all elements in wt%)

Element	Low Si	High Si
C	0.10	0.105
Mn	1.38	1.38
Si	0.16	1.4
Ni	0.019	0.022
Cr	0.017	0.017
Mo	0.003	0.014
Cu	0.013	0.026
Al	0.024	0.026
Sn	0.002	0.002
Nb	0.005	0.005
Ti	0.003	0.004
V	0.003	0.003
B	0.0003	0.0003
Ca	0.0005	0.0005
P	0.019	0.017
S	0.011	0.011
N	0.0056	0.0038
O	0.0042	0.0041

All samples were solution treated at 1150 °C for 2 min, cooled at 2 K/s to 950 °C and deformed in compression to 30% reduction and held for 20 s. This treatment produced the recrystallised prior austenite microstructure. Non-recrystallised (pancaked) austenite grains were obtained in other experiments by a further 30% deformation at 850 °C (low Si steel) and 800 °C (high Si steel). The continuous cooling transformation (CCT) diagrams were constructed for each austenite condition by cooling after deformation to room temperature at rates from 0.3 to 90 K/s.

Subsequent tests involved slow cooling at 0.3 K/s after the final deformation into the ferrite transformation region to form ~50% of ferrite in the microstructure. This temperature was 773 °C for high Si steel. For low silicon steel with recrystallised austenite condition the cooling rate was changed at 715 °C, while in the non-recrystallised samples this was at 725 °C. For all tests the cooling rate was increased to 3, 10 or 15 and 30 K/s to room temperature at the cooling rate change temperature. Typical dilatometer schedules are shown in Fig. 1.

All samples were characterised using optical metallography, image analysis, the Vickers hardness, microhardness and transmission electron microscopy (TEM) techniques. Thin foil TEM was performed using a Phillips CM20 microscope operated at 200kV in bright field and dark field modes.

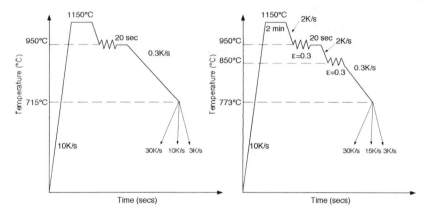

Fig.1 Representative dilatometer schedules: a - recrystallised austenite condition, low Si steel; b - non-recrystallised austenite condition, high Si steel.

Results

The initial austenite microstructure obtained in the recrystallised samples consisted of equaixed grains of 68 ± 6 μm diameter in low Si steel and 58 ± 6 μm diameter in high Si steel. Non-recrystallised austenite microstructure obtained as a result of the two-deformation schedule consisted of elongated (pancaked) grains with a mean width of 30 ± 6 μm and 35 ± 6 μm, respectively, for the high and low Si steels.

CCT diagrams were constructed based on transformation temperatures defined from dilatometer curves and microstructural characterisation. They are given and discussed in detail in [2]. The microstructure in continuously cooled samples from recrystallised austenite changes from polygonal ferrite and pearlite at slow cooling rates to bainite at intermediate cooling rates and martensite at 90 K/s cooling rate. In all samples transformed from non-recrystallised austenite, only polygonal ferrite was present for the cooling rates studied. The ferrite and bainite transformation temperatures were consistently higher in the high Si steel. An increase in the ferrite and bainite transformation start temperatures was observed in the non-recrystallised samples compared with those recrystallised, although the trend in the latter was inconsistent. Vickers hardness of the low Si steel samples was 20-40 VHN lower than in similarly processed samples for the high Si steel.

Microstructures obtained in all samples after discontinuous cooling are presented in Figs. 2 and 3. While the aim was to obtain ~ 50% of ferrite in final microstructure, it could be clearly seen that amount of ferrite is much higher (~70%) in samples with pancaked austenite structure cooled at 3K/s from the cooling rate change temperature. In the low Si steel approximately 60% ferrite was also formed in the pancaked sample cooled at 10 K/s. Only the volume fraction of ferrite in rapidly cooled samples (≥30 K/s) indicates the amount of ferrite formed before the cooling rate change, because at slow cooling rates, nucleation and growth of ferrite continues below this temperature. In both steels the scale of microstructure became finer with increasing cooling rate. The scale of the microstructure is also finer in the non-recrystallised samples. In all samples the prior austenite grain boundaries are decorated with polygonal ferrite. For transformation from pancaked austenite, polygonal ferrite is also present within the austenite grains due to intragranular nucleation on

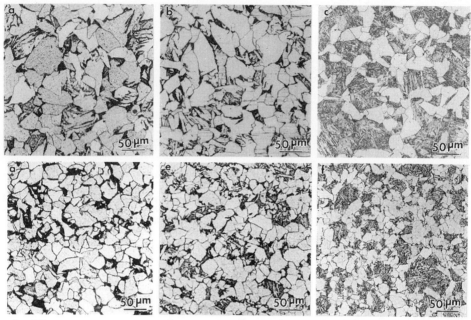

Fig. 2   Microstructures of discontinuously cooled samples of low Si steel that formed from recrystallised (a-c) and non-recrystallised (d-f) austenite. The cooling rate after the cooling change temperature: a, d - 3K/s; b, e - 10K/s and c, f - 30 K/s.

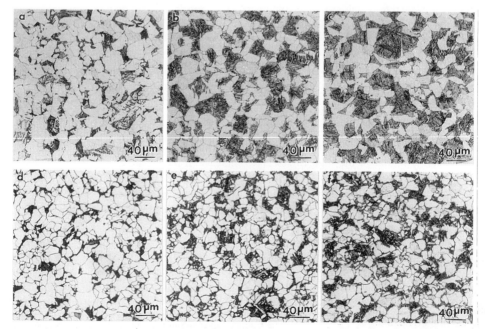

Fig. 3 Microstructures of discontinuously cooled samples of high Si steel transformed from recrystallised (a-c) and non-recrystallised (d-f) austenite. The cooling rate after the cooling change temperature is a, d - 3K/s; b, e - 10K/s and c, f - 30 K/s.

deformation bands, twins, etc. The second phase formed in recrystallised and non-recrystallised samples of low Si steel cooled at 3 K/s and 10 K/s is predominantly granular bainite with small additions of pearlite, while in similar treated samples of high Si steel (10 K/s) acicular ferrite was also formed. In all samples cooled at 30 K/s the predominant second phase formed was acicular ferrite. The bainite transformation start temperatures were 50 -70 $^\circ$C higher in the low Si steel. In both steels the bainite start transformation for the pancaked austenite structure is 10 - 15 $^\circ$C lower than from recrystallised austenite.

TEM studies confirmed the optical metallography observations and allowed more detailed investigations of phases present. In both steels polygonal ferrite with areas of pearlite and degenerated pearlite (Fig. 4, a and Fig. 5, a) are the main microstructural components in samples cooled at 3 K/s. Granular bainite, consisting of ferrite plates with small blocky shaped retained austenite regions, is the second phase present after this treatment (Fig. 4, b). Some small retained austenite grains were observed trapped inside the polygonal ferrite grains (Fig. 4, a).

At intermediate cooling rate (10-15 K/s) the second phase formed in high Si steel is acicular ferrite, consisted of carbide-free ferrite laths with interlayers of retained austenite and/or martensite (Fig. 5, c). More pearlite (Fig. 4, c), in addition to granular bainite (Fig. 4, d), is present in microstructure of low silicon steel than in high Si steel (Fig. 5, b).

In all samples cooled at 30 K/s cooling rate, the formation of acicular ferrite as second phase was observed (Figs. 4, e, f and Fig. 5, d). The laths of the acicular ferrite were coarser in the low Si steel with, correspondingly, thicker interlayer films. In both steels the laths and films were finer in the acicular ferrite transformed from pancaked austenite. The interlayer phase was predominantly retained austenite and martensite. Carbides were also observed between the laths (Fig. 4, f). The amount of interlayer carbides was much higher in the low Si steel. In addition, in the low Si steel there was a significant amount of intralath carbides present. In the low Si steel pearlite forms even at 30 K/s cooling rate. In addition to the above phases, the formation of lower bainite was observed in low Si steel.

The microhardness of the second phase in the high Si steel was always 80-90 μH higher than in similarly processed low Si samples. As anticipated, in both steels the microhardness of the second phase increases with increasing cooling rate due to transition from granular bainite to acicular ferrite (Fig. 6). For the same cooling rate the second phase formed from pancaked austenite was always harder than that formed from recrystallised austenite. In both steels the microhardness of correspondent phases in continuously cooled samples is lower than in those discontinuously cooled.

Discussion

The prior austenite grain size was slightly finer in the high Si steel, which may have contributed to finer final microstructure in this steel. During continuous cooling increasing Si from 0.16 to a 1.4 wt% increased the ferrite start temperature by 35-60 $^\circ$C. This is consistent with the role of silicon as a ferrite stabiliser. A similar effect was also observed in discontinuously cooled samples. However, the effect of silicon on the bainitic reaction in continuously cooled samples is marginal. Transformation start temperatures are slightly lower (by 5-20 $^\circ$C) for the low Si content. In both steels the austenite-to-ferrite transformation start temperatures are higher in pancaked samples than in recrystallised ones. This is related to the additional nucleation sites such as austenite grain boundary serrations, deformation bands and twins. This also leads to the refinement of ferrite grain size, observed in the pancaked samples.

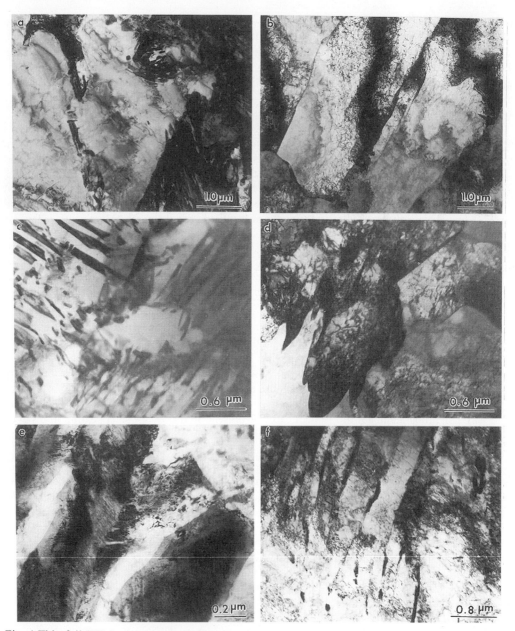

Fig. 4 Thin foil TEM micrographs of the low Si steel samples transferred from the recrystallised (a-e) and non-recrystallised austenite (f) cooled at 3 K/s (a, b), 10 K/s (c,d) and 30K/s (e,f).

The observation of pearlite in microstructures of low Si steel samples cooled at 30 K/s, whereas this was eliminated in high Si steel samples, is in agreement with other work [4, 5] where Si inhibits the formation of Fe_3C and delays pearlite formation. The presence of lower bainite and higher fractionof interlath and intralaths carbides in the discontinuously cooled low Si steel is due to the same effect. Silicon also accelerates the formation of acicular ferrite, which was observed in discontinuously cooled at 10 K/s samples of high Si steel.

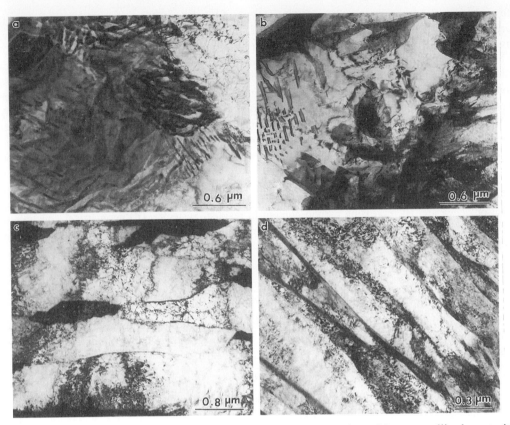

Fig. 5 Thin foil TEM micrographs of the high Si steel samples with recrystallised austenite condition cooled at 3 K/s (a), 10 K/s (b, c) and 30 K/s (d).

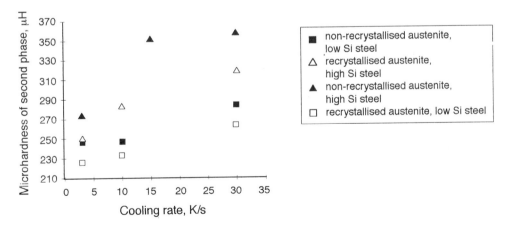

Fig. 6 The effect of cooling rate on the microhardness of the second phase

At lower cooling rates the retained austenite has a blocky shape, while its predominant morphology in faster cooled samples is interlayer films. The change in TMP leads to a change in the kinetics of transformation, which affects the morphology of the phase being formed. The blocky retained austenite is associated with the diffusional formation of the phases at higher temperatures (granular bainite), while retained austenite films appear lower temperatures phases (acicular ferrite, martensite), which have a displacive mechanism of phase transformation [5, 6].

The results have shown that both austenite condition and transformation path are responsible for final microstructure and properties. The microstructure in pancaked samples is always finer and less homogeneous. Formation of ferrite is accelerated in these samples, especially in the high Si steel, as expected. The hardness of the high Si steel samples is always greater than hardness of similarly processed low Si samples, due to the solution strengthening effect by Si and the finer microstructure. Those microhardnesses of identical second phases in discontinuously cooled samples are higher than in continuously cooled. This could be attributed to the increased carbon content in residual austenite due to it rejection from the pre-formed ferrite and due to the decrease in the scale of second phase (acicular ferrite lath width, size of granular bainite plate). There are several reasons why the second phase microstructure becomes finer with increased ferrite volume fraction. They are: (i) increasing carbon content in residual austenite, (ii) geometrical constraints: decrease in available space for growth, and (iii) limitation of nucleation sites due to the decoration of all grain boundaries with ferrite that leads to increased undercooling.

Conclusions

The effect of silicon on the transformation behaviour of two microalloyed steels under continuous and discontinuous cooling conditions was studied. It has been shown that steel composition, initial austenite condition and transformation path control the final microstructure and properties of the steel after thermomechanical processing.

The results have confirmed that in complex thermomechanical processing schedules silicon acts as ferrite stabiliser, solution strengthening element and accelerates formation of acicular ferrite. The work has shown that the prediction of final microstructure and properties of thermomechanically processed steel is a difficult task and additional factors need to be incorporated in existing models.

References

[1]   G. Thomas and J.-Y. Koo, in Proc. 'Structure and Properties of Dual-Phase Steels' (eds. R.A. Kot and J. W. Morris), AIME, (1979), p. 183-201.
[2]   E.V. Pereloma and P.D. Hodgson, Mat. Sci. and Eng., 1998 (submitted for publication).
[3]   A P. Goldren and G.T. Eldis, J. Met. 32 (1980), p. 41.
[4]   G.T. Eldis, A.P. Goldren and F.B. Fletcher, "Alloys for the 80's", Climax Molibdenium Co., Ann Arbor, Mich., (1980), 4.1.
[5]   H.K.D.H. Bhadeshia, Bainite in Steels, The Inst. Mater., London, (1992), 124.
[6]   Martensite: a Tribute to Morris Cohen (eds. G.B. Olsen and W.S. Owen), ASM (1992).
[7]   T. Minote, S. Torizuka, A. Ogawa and M. Niikura, ISIJ Int., 36 (1996), p. 201.

Acknowledgments

The authors are grateful to BHP Research for providing steels and access to the deformation dilatometer. This work was supported by Engineering Faculty New Staff Member Research Fund, Monash University.

For correspondence: elena.pereloma@eng.monash.edu.au

Materials Science Forum Vols. 284-286 (1998) pp. 245-252
© 1998 Trans Tech Publications, Switzerland

Acicular Ferrite Microstructures and Mechanical Properties in a Low Carbon Wrought Steel

M. Díaz-Fuentes, I. Madariaga and I. Gutiérrez

Centro de Estudios e Investigaciones Técnicas de Guipúzcoa (CEIT)
and Escuela Superior de Ingenieros Industriales (ESII)
P° Manuel Lardizábal 15, E-20009 San Sebastián, Basque Country, Spain

Keywords: Acicular Ferrite, Nucleation, Heat Treatment, Inclusions, Mechanical Properties

ABSTRACT

In this work, the formation of acicular ferrite in a low carbon wrought steel has been investigated. After austenitizating at 1250°C for 45 minutes, the steel has been subjected to isothermal heat treatments in the range of 550-350°C for different times. The microstructure obtained is mainly composed of acicular ferrite. Widmanstätten and allotriomorphic ferrite have also been found nucleating at the prior austenite grain boundaries. The volume fraction of these phases decreases remarkably when the isothermal treatment temperature is diminished. It has been observed that acicular ferrite is nucleated at non-conventional metallic inclusions such as multiphased inclusions composed of Al, Si and Mn oxides and MnS inclusions, which are covered, in both cases, by a shell of a hexagonal CuS phase. The mechanical properties of the steel in the as-rolled condition and after the generation of acicular microstructures have been measured showing the improvement obtained with acicular ferrite.

INTRODUCTION

It is well established that a refinement of the microstructure promotes the improvement of the mechanical properties. In the particular case of welding, an improvement can be obtained with the development of acicular ferrite microstructures enhancing both the toughness and the strength. However, in the heat affected zone (HAZ) coarse austenite grain sizes can develop, in the absence of small second phase particles to inhibit grain growth. Also, the aim is to produce an acicular ferrite microstructure in this zone. Acicular ferrite nucleation at inclusions is a process competitive with bainite nucleation at grain boundaries, both transformations taking place by the same displacement mechanism, within approximately the same temperature range [1, 2, 3]. The presence of bainitic zones in an acicular ferrite microstructure can be detrimental to the mechanical properties of the steel. Several factors are known to inhibit, partially or completely, the bainite nucleation and thus favour the nucleation of an acicular microstructure: large austenite grains [2, 4], a thin layer of allotriomorphic ferrite [1, 5] and grain boundary segregation [6, 7]. Finally, the presence in the steel of a distribution of suitable second phase particles is recognised to promote the formation of acicular ferrite, to the detriment of bainite [8, 9]. This is why particulate inoculated steels have been recently developed [10]. In the present work massive acicular ferrite microstructures have been produced in a wrought coarse grained microalloyed steel by nucleation on conventional particles as an alternative to particle inoculation.

EXPERIMENTAL PROCEDURE

The chemical composition of the steel, supplied by Aristrain (Olaberria, Spain) is given in table 1.

Table 1. Chemical composition of microalloyed steel (Weight Percent)

C	Si	Mn	P	S	Cr	Ni	Mo	Cu	Sn	Al	N	V
0.10	0.19	0.96	0.014	0.009	0.12	0.28	0.066	0.38	0.026	0.006	0.104	0.03

Cubic Samples of 10 mm were austenitized at 1250°C for 45 minutes followed by direct quenching in a salt bath at 550, 450 and 350°C and holding for different times between 3 seconds and 20 minutes and finally, quenched in water to room temperature. The samples were cut, polished and etched in nital to be observed by optical microscopy. Transmission electron microscopy was used to analyse the active inclusions. Carbon extraction replicas mounted on gold grids were obtained from the sample isothermally treated at 450°C for 5 seconds. The replicas were examined using a Phillips CM12 transmission electron microscope operated at 100 kv and equipped with an energy disperse spectrometer EDAX 9190. Hardness measurements were carried out using a LECO M-400-G2 hardness tester. Rectangular samples of 10 x 10 x 55 mm were used to perform Charpy tests (ASTM standard [11]) and tensile tests were performed in a INSTRON 4505 machine.

RESULTS AND DISCUSSION

Microstructure

An austenite grain size of 182±10 μm, has been obtained after the austenitization treatment.

The final microstructures after isothermal treatment, are mainly composed of acicular ferrite. However significant volume fractions of other phases are also present, as can be seen in Fig. 1. The

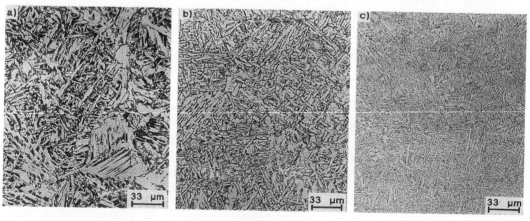

Fig. 1.- Microstructures obtained with isothermal treatments at (a) 550°C, (b) 450°C, (c) 350°C.

optical analysis of the microstructure at 550°C revealed significant amounts of allotriomorphic and Widmänstatten ferrite covering the prior austenite grain boundaries, which were formed before the beginning of the acicular ferrite transformation. At this temperature, as the austenite becomes enriched in carbon and the treatment time increases, pearlite is formed. Pearlite was only found at 550°C. At 450°C, the microstructure exhibited a reduction in the volume fraction of allotriomorphic and Widmänstatten ferrite which were partially replaced by bainite. The microstructure observed at

350°C was composed of acicular ferrite and infrequent bainite sheaves. The presence of allotriomorphic and Widmänstatten ferrite was now significantly lower than in the preceding cases. The measurements of the volume fraction of ferrite nucleating at grain boundaries showed a reduction from 29% to 12% when the isothermal treatment temperature was reduced from 550°C to 450°C and diminished to 8% at 350°C.

An approximate shape of the TTT[1] curves of the present steel is shown in Fig. 2. As can be observed, the beginning of the proeutectoid ferrite transformation takes places after very short incubation times at temperatures in a range of 650-750°C allowing the formation of allotriomorphic ferrite during quenching from 1250°C to the isothermal treatment temperatures. On the other hand, the quenching rate is increased as the isothermal treatment temperature is reduced, thus diminishing the time period in which allotriomorphic ferrite can be formed. In consequence, the volume fraction of allotriomorphic ferrite is reduced notably with decreasing isothermal treatment temperature. Taking into account that allotriomorphic ferrite acts as a substrate for the growth of Widmänstatten ferrite [12, 13] a similar reduction of this latter phase was also observed.

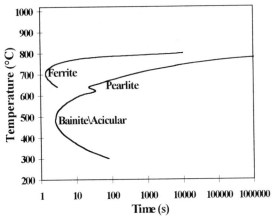

Fig. 2.- TTT curve of the steel analysed

The time evolution of the microstructure formed intragranularly for each isothermal treatment is shown in Fig. 3. The beginning of the intragranular nucleation takes place at inclusions as can be seen in Figs. 3.a, 3.d and 3.g. Some refinement of the microstructure was obtained as the isothermal treatment temperature was reduced. This refinement is directly related, as in the case of the formation of bainite, to carbon partitioning which takes place during the growth of acicular ferrite and enriches the untransformed austenite. As a result, the austenite surrounding the acicular ferrite is more stable at low temperatures, making the growth of the ferrite more difficult with a consequent refinement of the microstructure.

The observations performed by optical microscopy in the present steel showed that both the beginning and the end of the austenite transformation takes place at short times due to the low carbon concentration of the steel in agreement with the diagram in Fig. 2. A measurement of the austenite transformation rate can be obtained with aid of the hardness measurements as shown in Fig. 4. At short isothermal treatment times the austenite, which remained untransformed, becomes martensite during water quenching. As the transformation evolves, the volume fraction of acicular ferrite increases, whereas the amount of martensite diminishes. This fact is reflected in a rapid decrease of

---

1 .-Calculated using the TTT program developed at the Department of Materials Science and Metallurgy, Univ. of Cambridge (U. K) By Dr. H. K. D. H. Bhadeshia and co-workers.

the hardness with time, as can be seen in the graph of Fig. 4. After a treatment time between 10 and 25 s the hardness becomes constant and the transformation can be considered complete.

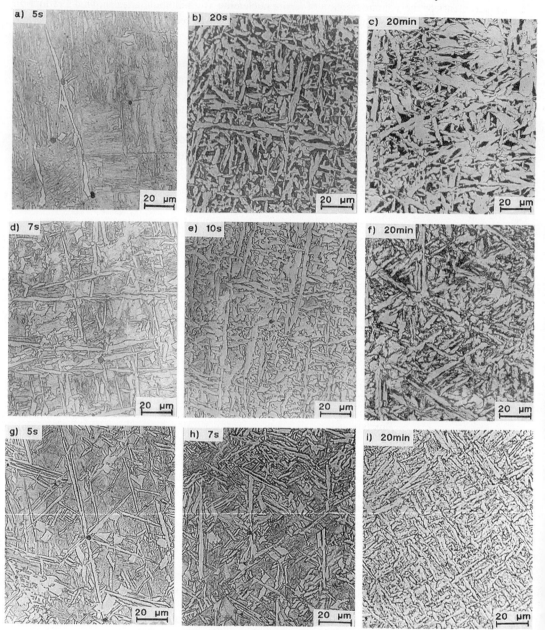

Fig. 3.- Evolution of the acicular ferrite structure with isothermal treatments at 550°C (a-c), 450°C (d-f) and 350°C (g-i)

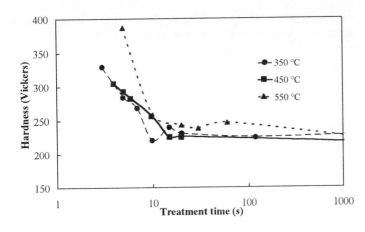

Fig. 4.- Hardness measurements versus treatment time.

The nucleation of acicular ferrite:

The micrographs in Figs. 5 and 6 show the two types of active inclusions that were observed as being effective for acicular ferrite nucleation. The results of the EDS analyses indicated

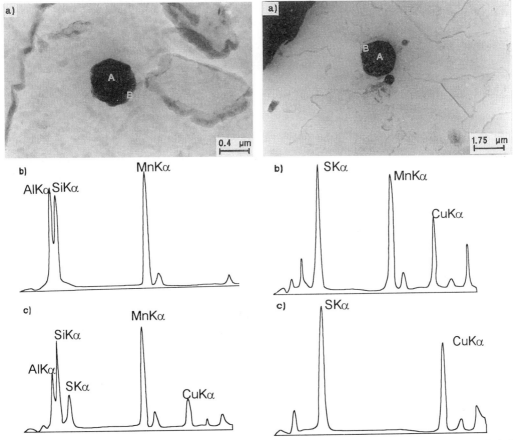

Fig. 5.- Al, Si, Mn, multiphased inclusion covered by a CuS shell. a:TEM image, b:EDS of zone A, c:EDS of zone B.

Fig. 6.- MnS inclusion covered by a CuS shell. a:TEM image, b:EDS of zone A, c:EDS of zone B.

that one type of inclusion has a core rich in Silicon, Manganese and Aluminium covered by a shell of CuS (Fig. 5). It may be assumed that most of the elements in the core are in the form of oxides, such as MnO, SiO_2, and Al_2O_3 [9]. The second type of inclusion, found nucleating acicular ferrite, is composed of a MnS core with a CuS coating (Fig. 6).

Manganese and copper sulphides have also been reported by other authors to act as nucleation sites for acicular ferrite formation [9, 14]. The formation of these inclusions can be understood by comparing their melting points. The melting temperatures are 2015°C for Al_2O_3, 1713°C for SiO_2, 1650°C for MnO, 1620°C for MnS and 1125°C for CuS [9, 15]. Therefore, Al_2O_3, SiO_2, MnO will form much earlier than CuS during steel processing, and CuS will precipitate from austenite at the pre-existing oxide particles. A similar process will occur when the core of the inclusion is MnS.

The CuS layers were separated from the inclusion cores using carbon extraction replica techniques, which permitted analysis by transmission electron microscopy. The diffraction patterns and the chemical analyses obtained enabled the identification of the CuS phase. Both are compatible with the hexagonal CuS lattice, with parameters a=0.379 nm and c=1.633 nm which corresponds to the CuS B 18 type of Structure called covellite [16]. The micrograph in Fig 7 shows a shell of CuS on the shadow of a particle on a carbon replica and its diffraction pattern corresponding to a [0001] zone axis. The same phase has been identified to cover MnS particles nucleating acicular ferrite in a medium carbon steel [14].

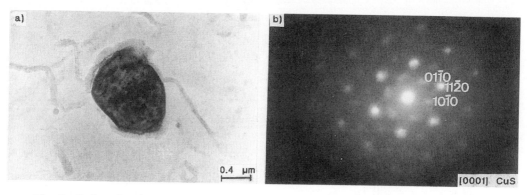

Fig. 7.- CuS surface layer extracted from a second phase particle surface. a TEM image. b Diffraction pattern.

When the transformation of the austenite takes place in the range of temperatures of the present isothermal treatments, there is a competition between intragranular ferrite nucleation at inclusions and grain boundary nucleation. The formation of acicular ferrite or bainite is determined by the global energetic balance. Assuming the variation of the free energy involved in the transformation is the same for both types of structures, the interfacial energies per unit area of interface between inclusion/ferrite $\sigma_{i/\alpha}$, inclusion/austenite $\sigma_{\alpha/\gamma}$, ferrite/austenite $\sigma_{i/\gamma}$ and austenite/austenite $\sigma_{\gamma/\gamma}$ are expected to be determinant in the preferential nucleation of one of them at the expense of the other, see Fig. 8. Ricks et al [17] first studied the global energetic balance of nucleation at inclusions (ΔG_{het}) and grain boundaries (ΔG_{gb}), assuming that $\sigma_{i/\alpha}$ and $\sigma_{i/\gamma}$ are equal, and reaching the conclusion that the latter is always energetically more favourable than the former. However, the nature of the interfaces austenite/inclusion and ferrite/inclusion can be different, making it difficult to maintain the previously mentioned assumption of $\sigma_{i/\alpha}$ and $\sigma_{i/\gamma}$ being equal.

The results obtained from the classical heterogeneous nucleation theory [18, 19] show a marked influence of the global energetic balance on the difference between interfacial energies ($\sigma_{i/\alpha}$ and $\sigma_{i/\gamma}$). If the crystalline structures of the inclusion and ferrite are such that it is possible the formation of low energy ferrite/inclusion interfaces, replacing high energy interfaces between austenite and inclusion, the energetic balance for the nucleation at inclusions is improved. As has been previously reported by several authors [8, 9], a reduction in the interfacial energy $\sigma_{i/\alpha}$ can be achieved with a good lattice matching between inclusion and ferrite. This seems to be the case with the hexagonal CuS phase present in the outside layer of the observed nucleating inclusions. The characteristics of the hexagonal lattice of the CuS, with a high c/a ratio (4.32) causes the precipitation of the CuS on the surface of the inclusions to take place with its basal plane parallel to the surface of the inclusion [14]. This plane presents a good lattice match with the $\{111\}_\alpha$ plane, with the atomic positions of the ferrite coinciding with two corners and the centre of the hexagon in the basal plane of the CuS, allowing a low energy inclusion/ferrite interface. On the contrary, CuS and austenite do not exhibit a low lattice mismatch, even in the most favourable situation when $\{111\}_\gamma$ and $(0001)_{CuS}$ planes are parallel.

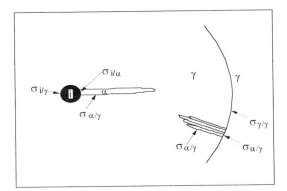

Fig 8.- Competitive process between acicular ferrite and bainite.

Mechanical properties

With the aim of mainly investigating the effect of acicular ferrite on the mechanical properties of the steel, mechanical tests were performed on the samples with the higher acicular ferrite volume fraction, which correspond to the 450°C and 350°C isothermal treatments. The use of these treatment temperatures avoids the influence of other phases present at 550°C. In Table 2 the mechanical properties corresponding to the parent microstructure in the as-rolled condition (ferrite-pearlite microstructure with ferrite grain size 26 μm and pearlite volume fraction 10%) are compared with the results obtained after the 20 minutes at 450°C and 350°C treatments, showing the improvement in the mechanical properties achieved with the presence of acicular ferrite in the wrought steel. It can be observed that a very large increase in the impact energy is obtained without a deterioration of the strength of the steel.

Table 2. Mechanical properties

Treatment	YS (Mpa)	UTS (Mpa)	Hardness (HV)	CVN at 0°C (J)
As-rolled steel	345	519	170	16
20 min. 450°C	479	659	218	125
20 min. 350°C	515	626	225	135

Conclusions

1. Acicular ferrite microstructures have been obtained in a conventional wrought low alloy microalloyed steel by isothermal treatments at 550°C, 450°C and 350°C after austenization at 1250°C.

2. The acicular ferrite nucleation takes place at two kinds of inclusions: multiphased inclusions composed of Al, Si and Mn, and MnS inclusions, in both cases covered by a CuS coating. The good lattice matching between ferrite and the hexagonal CuS phase enables the formation of low energy ferrite/inclusion interfaces aiding the transformation, showing that a residual element like copper can be useful as an alternative to particle inoculation.

3. The generation of acicular ferrite microstructures results in an improvement of the mechanical properties of the steel. This improvement is mostly observed in the toughness, with no deterioration of the strength.

Acknowledgements

This work was carried out in collaboration with Aristrain (Olaberria, Spain). I. M. acknowledges a "Formación de Investigadores" grant from the departamento de Educación y Universidades of the Gobierno Vasco. The authors gratefully acknowledge the leadership and enthusiasm shown by Prof. J. Urcola at the beginning of this project, before his untimely death.

References

1.- S. S. Babu, H. K. D. H. Bhadeshia: Mater. Sci. Engin. A 156, 1992, pp. 1-9.

2.- F. J. Barbaro, P. Kraulis, K. E. Easterling: Mat. Sci. Technol., 1989, vol. 5, pp 1057-1068.

3.- Strangwood M. and Bhadeshia H. K. D. H.: Proc. Int. Conf. Advances in Welding Science and Technology, ASM, Cleveland, OH, 1987, pp. 187-191.

4.- R. A. Farrar, Z. Zhang, S. R. Bannister, G. S. Barritte: J. Mater. Sci., 1993, 28, pp. 1385-1390.

5.- Yang J. R. and Bhadeshia H. K. D. H.: Proc. Int. Conf. Advances in Welding Science and Technology, ASM, Cleveland, OH, 1987, pp. 209-213.

6.- K. Yamamoto, T. Hasegawa, J. Takamura: ISIJ International, 1996, Vol. 36, No. 1, pp. 80-86.

7.- A. O. Kluken, O. Grong and G. Rorvik: Metall. Trans. A, 1990, vol. 21A, pp. 2047-2058.

8.- A. R. Mills, G. Thewlis, J. A. Whiteman: Mat. Sci. Technol., 1987, vol 3, pp. 1051-1061.

9.- Zhang, Z., Farrar R. A.: Mat. Sci. Technol., 1996, vol. 12, pp. 237-260.

10.- J. M. Gregg and H. K. D. H. Bhadeshia: Acta Mater., 1997, vol. 45, No. 2, pp. 739-748

11.-Standard Test Methods For Notched Bar Impact Testing of Metallic Materials, E23-92, Annual Book of ASTM Standars, Vol 03-01, ASTM, Philadelphia 1992, p.205-224.

12.- J.-L. Lee, S.-C. Wang, G.-H. Cheng, 1989, Materials Science and Technology, vol. 5, pp. 674-681.

13.- Y. Ohmori, H. Ohtsubo, Y. Chul Yung, S. Okaguchi and H. Ohtani: Metall. Mat. Trans., vol. 25A, 1994, pp. 1981-1989.

14.- I. Madariaga, I. Gutiérrez: Scripta Materialia, vol. 37, n° 8, 1997, pp. 1185-1192.

15.- Handbook of Chemistry and Physics, 44th edition, 1963, pp. 604-605.

16.- W. B. Pearson: "Handbook of lattice spacing and structures of metals and alloys", 1967, vol 2, pp. 230.

17.- R. A. Ricks, G. S Barritte and P. R Howell: Proceedings of an international Conference in Solid Phase Transformations, 1981, Pittsburgh, Pennsylvania, pp. 463-468.

18.- D. A. Porter, K. E. Easterling: "Phase transformations in metals and alloys", 1981, London, Van Nostrand Reinhold.

19.- K. Easterling: "Introduction to the Physical Metallurgy of Welding", 1983, Norwich, Butterworths & co.

M.Díaz-Fuentes
E-mail: mdfuentes@ceit.es
Fax: (34) 43 21 30 76

Materials Science Forum Vols. 284-286 (1998) pp. 253-260
© *1998 Trans Tech Publications, Switzerland*

Stability of Retained Austenite in a Nb Microalloyed Mn-Si TRIP Steel

D.Q. Bai[1], A. Di Chiro[2] and S. Yue[1]

[1] Dept. of Metallurgical Engineering, McGill University,
3610 University St., Montreal H3A 2B2, Canada

[2] Canadair, Montreal, Canada

Keywords: Retained Austenite, TRIP Steel, Stability

Abstract
The effect of thermomechanical processing on the volume fraction and stability of retained austenite in a Nb microalloyed Mn-Si TRIP steel was investigated. In one set of tests, specimens were isothermally held at 400 °C for a series of different times, then air cooled to room temperature. In another set of tests, specimens were continuously cooled to room temperature at a cooling rate of 2 °C/s or 10 °C/s. The former generated a mixed microstructure of ferrite + bainite + retained austenite, while the latter produced a microstructure of bainite + martensite + retained austenite. It was found that the stability of retained austenite against a strain induced transformation to martensite increases with increasing carbon concentration, decreasing size and hard surrounding phase.

I.   Introduction
Zackay et al.[1] started the pioneering work on TRIP steels about thirty years ago. They found that ductility increased significantly as some metastable austenite transformed to martensite during straining in Fe-Cr-Ni stainless steels. However, such steels were never widely used because large additions of such alloying elements are costly. During the past two decades, the demand on high strength formable steels has stimulated studies on more economically alloyed TRIP grades. A large number of investigations have been conducted by different research groups to develop the low-medium carbon Mn-Si TRIP steels[2, 3]. Most of the work was focused on the intercritical annealing of this type of steel; some concerned the as-hot rolled conditions[4]. It has been reported that a strength level of up to 980 MPa, with an elongation to fracture of over 30% can be obtained by optimizing the amount of retained austenite and its mechanical stability.

It has been recognized that the stability of retained austenite is crucial in terms of enhancing TRIP. The mechanical stability of retained austenite depends not only on its chemical stability, such as carbon and manganese contents, but also on some other stabilizing effects attributed to the differences in particle size, and the surrounding phases. Therefore, it is the purpose of this work to understand the effect of hot deformation processing on the microstructure of a Nb-treated Mn-Si TRIP steel and the stability of retained austenite during compressive deformation.

II.   Experimental Procedure
The chemical composition of the steel investigated in this study is a 0.2C-1.5Mn-1.5Si-0.035Nb TRIP steel. A small amount of Nb was added to enable the material to be processed by conventional controlled rolling, since a moderate deformation in the no-recrystallization region increases the volume fraction of retained austenite. In addition, it was found from the previous investigations that the addition of Nb raises the amount of retained austenite about 25 percent compared to that of a similar chemistry without Nb[5]. Compression specimens with heights of

11.4 mm and diameters of 7.8 mm were machined from the received plate, with their longitudinal axes parallel to the rolling direction.

The thermal cycles and deformation schedules, which were applied to generate different microstructures, are shown in Fig. 1. It can be seen that after austenitizing at 1200 °C for 30 minutes, specimens were cooled to 1050 °C and were subjected to a double-hit deformation with strains of 0.3, strain rate 0.1/s and interpass time of 30 seconds. In the case of Schedule I, the specimens were cooled to 650 °C and held at this temperature for 4 minutes to produce about 35 percent (in volume fraction) polygonal ferrite. This was followed by quenching the specimens into a salt bath in which the temperature was set at 400 °C. After a period of 2, 5, and 10 minutes holding, the specimens were cooled in air to room temperature. In the case of Schedule II, the thermal cycle and deformation parameters were the same as Schedule I, except that the specimens were cooled to 850 °C at a cooling rate of 2 °C/s after deformation, then immediately cooled to room temperature at cooling rate of 2 or 10 °C/s.

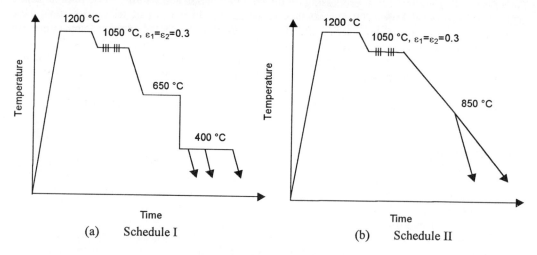

Fig. 1 Testing Schedules.

Specimens for optical microstructure examination were etched either by 2% nital or a solution based on sodium metabisulfate reagents[6]. Specimens for TEM examination were prepared by jet polishing with a solution containing 8% perchloric acid in methanol. The retained austenite measurements were performed by a neutron diffractometer (DUALSPEC power diffractometer) at the NRU reactor of AECL Research, Chalk River Labs. The amount of retained austenite was quantified by analyzing the integrated intensities of (111), (200), (220), and (311) austenite peaks, and those of (110), (002), (112), and (022) planes of ferrite. The carbon concentrations were calculated by using the lattice parameters of the retained austenite as follows:

$$a_\gamma = 3.578 + 0.044\%C$$

III.   Results and Discussion
III.1.   Microstructure and Volume Fraction of Retained Austenite

The microstructures generated from Schedules I and II are shown in Figs. 2 and 3, respectively. The volume fractions and carbon contents of retained austenite in these microstructures are listed in Table 1. It can be seen from Fig. 2 that this microstructure consists of

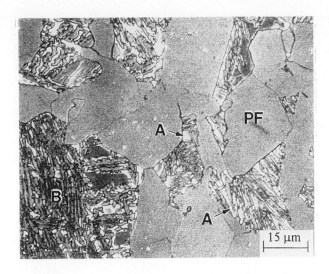

Fig. 2 Microstructure of the specimen held at 400 °C for 2 min.

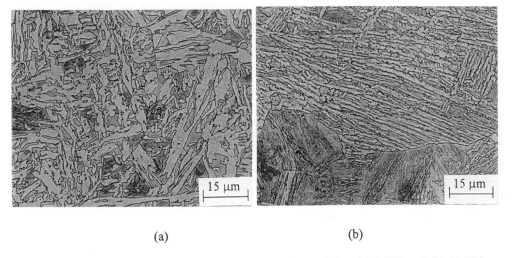

(a)   (b)

Fig. 3 Microstructures of the specimen continuously cooled at (a) 2 °C/s and (b) 10 °C/s.

Table 1 Volume fraction of retained austenite and carbon content

Condition		$f_{\gamma 0}$	C, wt%	M_s, °C
400 °C	2 min	7.7	1.44	-48
	5 min	9.5	1.54	-84
	10 min	6.9	1.28	9
2 °C/s		11.7	0.73	207
10 °C/s		11	0.91	142

38% polygonal ferrite + 52% bainite + 10% martensite / retained austenite (white phase). From X-ray analysis, the amount of retained austenite was estimated to be 7.7%; therefore, the amount of martensite is about 2.3%. Most of the retained austenite is located on ferrite-ferrite and/or ferrite-bainite boundaries, or trapped in between lath bainitic ferrite. In the former case, the retained austenite is mostly blocky or plate shaped, and the sizes range from 0.5 μm to 2 μm. In the latter case, the retained austenite is interlayer lath or film type. The width of the lath type retained austenite ranges from 0.1 μm to 0.5 μm; while that of the film type retained austenite is usually less than 0.1 μm. A small amount of blocky shaped retained austenite was also found inside the polygonal ferrite. The TEM micrographs presented in Figs. 4a) and 4b) show a blocky type retained austenite on the polygonal ferrite grain boundaries, and lath retained austenite trapped in bainitic lath ferrite, respectively.

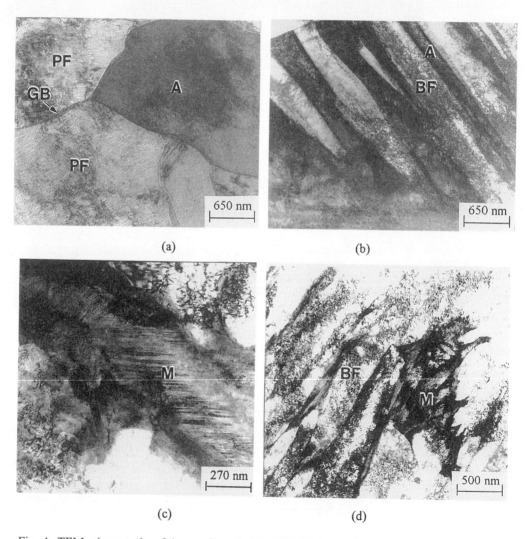

(a)   (b)

(c)   (d)

Fig. 4   TEM micrographs of the specimen held at 400 °C for 2 min. PF-polygonal ferrite, BF-bainitic ferrite, A-retained austenite, M-martensite, GB-grain boundary.

It is interesting to note that both lath (Fig. 4c)) and twinned (Fig. 4d)) martensite were also observed in the specimen isothermally transformed at 400 °C for 2 minutes. The presence of martensite indirectly explains the lower volume fraction and carbon content of the retained austenite observed in this case compared to one isothermally transformed for 5 minutes at 400 °C. It is well known that the further carbon enrichment through bainite transformation can result in a higher volume fraction of retained austenite. This is because after lath bainitic ferrite was formed by a shear mechanism, carbon started to diffuse from supersaturated ferrite to the nearby austenite[7]. It takes a few minutes for the untransformed austenite to get maximum enrichment in carbon. Therefore, a short holding (2 minutes) at the bainite transformation temperature not only led to some of austenite transforming to martensite but also lowered the carbon content in the retained austenite. On the other hand, holding too long (10 minutes) at this temperature also led to the decrease in the carbon content, as listed in Table 1, because of the formation of carbides.

The microstructures shown in Figs. 3a) and 3b) are comprised of mainly bainite. The lath sizes of bainitic ferrite are shorter and wider in the specimen cooled at 2 °C/s than in the specimen cooled at 10 °C/s. The retained austenite particles at lower cooling rate are large and blocky shaped, while those at higher cooling rate are lath and/or film shaped. It is interesting to see that the volume fractions of retained austenite in the continuously cooled specimens are higher than those subjected to the isothermal bainitic transformation. However, the carbon contents are much lower in the former case. This probably resulted from a lack of time for carbon sufficiently to diffuse from bainitic ferrite to austenite. In this case, "trapping" or constraint from surrounding phases plays a role in the retention of austenite since the estimated M_s temperatures are much higher than room temperature.

III.2   Stability of Retained Austenite

In order to examine the stability of the retained austenite, the thermomechanically processed specimens were compressed at a strain rate of $2 \times 10^{-3} s^{-1}$ to different strains. The volume fractions of the retained austenite were then detected by neutron diffractometry. The variations of the retained austenite with strain are shown in Fig. 5. It can be seen that the volume fractions of retained austenite decrease with increasing strain. However, the decreasing rates are different for the specimens isothermally treated for different times. This dependence of retained austenite on strain indicates the stability of retained austenite, which can be generally expressed by the following empirical relationship[3]:

$$\log f_\gamma = \log f_{\gamma 0} - k \cdot \varepsilon$$

Where $f_{\gamma 0}$ is the volume fraction of retained austenite at $\varepsilon=0$, f_γ is the volume fraction of retained austenite after straining, k is constant and a lower k value corresponds to a higher stability of retained austenite against strain-induced γ-to-M transformation. The present k values estimated from data shown in Fig. 5 are 0.45, 0.73, and 0.96 for 2, 5, and 10 min holding, respectively. It means that the retained austenite in 2 min treated specimen is more stable, while that in 10 min treated specimen is less stable. This is understandable since the carbon level in the 10 min treated specimen is lower. However, the higher stability of the retained austenite in the 2 min treated specimen than in the 5 min treated specimen cannot be explained by carbon content, since the carbon level in the latter specimen is slightly higher than in the former specimen. This improved stability in the 2 min treated specimen could be due to the martensite presented in the microstructure.

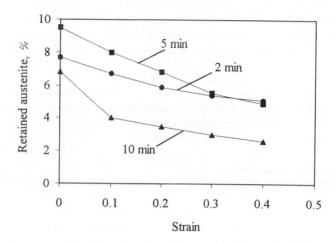

Fig. 5 Variation of retained austenite with strain.

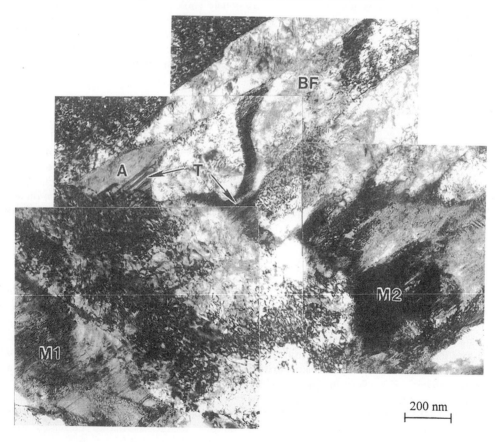

Fig. 6 TEM micrograph showing the deformation induced transformation of retained austenite (A) to martensite (M), $\varepsilon = 0.1$. BF is bainitic ferrite, T indicates deformation twins.

It is known that the stability of retained austenite depends not only on its chemical stability, for example, C and/or Mn enrichment, but also on some other stabilizing effects attributed to the distribution, morphology, and size of retained austenite. The effect of retained austenite particle size on its stability is shown by the TEM micrographs in Fig. 6. It can be seen that two large retained austenite (now martensite, labeled as M1 and M2) have partially transformed to martensite at strain $\varepsilon=0.1$. The twin substructure can be seen in these two particles. The high tangled dislocations, which are accommodated with the martensite transformation, can also be seen in the adjacent ferrite. One relatively smaller retained austenite (labeled as A1) has not transformed to martensite, but deformation twins have been introduced in this particle. Further evidence of deformation twins can be seen in Fig. 7. This suggests that i) smaller retained austenite has higher stability, and transforms to martensite at higher strains; ii) retained austenite (at least those small sized) is plastically deformed before it transforms to marteniste. This might be true especially for those with higher stability (smaller size and higher carbon content), which requires strain accumulation to introduce shear bands or deformation twins as nucleation sites for martensite transformation[8]. For those large retained austenite, stacking faults have been usually observed, and they can also act as nucleation sites for martensite transformation. The critical strain required for initiating martensite transformation will be smaller since there are already some nucleation sites there available.

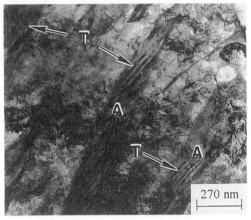

Fig. 7 TEM micrograph showing the deformation induced transformation of retained austenite (A) to martensite, $\varepsilon = 0.1$. T indicates deformation twins.

The work by Jeong et al [8] showed that most of retained austenite with particle size larger than 1 μm has transformed to marteniste at strain about 5%. The size effect on the stability of retained austenite is further illustrated by Fig. 8. The neutron diffraction results for the specimens continuously cooled at 2 °C/s are presented in Figs. 8a) and 8b), and those cooled at 10 °C/s are shown in Figs. 8c) and 8d). It can be seen that the austenite peaks are significantly weakened in the specimens cooled at 2 °C/s as the compressive strain is at $\varepsilon= 0.05$, while those of the 10 °C/s cooled specimens are hardly changed at $\varepsilon = 0.1$. This suggests that the retained austenite in the latter case is more stable than in the former case. It may be argued that the lower stability of the former case is due to the lower carbon content in the specimens. However, it is more likely to be due to the size effect since most of the retained austenite in the specimens cooled at 2 °C/s are larger than 2 μm, while those of the specimens cooled at 10 °C/s are less than 1 μm. Thus it can be concluded that retained austenite with particle sizes larger than 1 μm are relatively unstable, and will probably not contribute significantly to the ductility of the material.

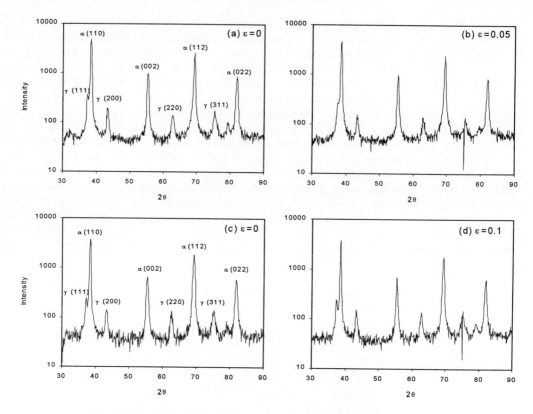

Fig. 8   Variation of the diffraction peaks of the specimens cooled
at (a) and (b) 2 °C/s, and (c) and (d) 10 °C/s.

IV.   Conclusions

Based on the present investigation about the effect of thermomechanical processes on the stability of retained austenite in a Nb treated Mn-Si TRIP steel, it can be concluded that a 400 °C isothermal holding can improve the carbon enrichment in the retained austenite if the holding time is long enough, but does not exceed the point at which the carbides formation starts. Large sized retained austenite is unstable, and transforms to martensite at low strain, while retained austenite with smaller particle sizes (<1 μm) and high carbon enrichment is more stable, and can be retained to higher strains. It is believed that this latter type of retained austenite mostly contributes to the TRIP effect.

References

[1]   V.F. Zackay, E.R. Parker, D. Fahr, and R. Bush, Trans. ASM, vol. 60, (1967), p. 56.
[2]   Y. Sakuma, O. matsumura, and H. Takechi, Metall. Trans. A, vol. 22A, (1992), p. 489.
[3]   K. Sugimoto, M. Kobayashi, and S. Hashimoto, Metall. Trans. A, vol. 23A, (1992), p. 3085.
[4]   I. Tsukatani, S. Hashimoto, and T. Inoue, ISIJ Int., vol. 31, 1991, p. 992.
[5]   A. Di Chiro, J. Root, and S. Yue, 37th MWSP Conf. Proc., (1996), p. 373.
[6]   S. Bandoh, O. Matsumura, and Y. Sakuma, Trans. ISIJ, vol. 28, (1988), p. 569.
[7]   K. Tsuzaki, A. Kodai, and T. Maki, Metaal. Trans. A, vol. 25A, (1994), p. 2009.
[8]   W.C. Jeong, D.K. Matlock, and G. Krauss, Materials Science and Technology, A165, (1993), p. 9.

MECHANICAL PROPERTIES

Materials Science Forum Vols. 284-286 (1998) pp. 263-270
© *1998 Trans Tech Publications, Switzerland*

Influence of Coiling Temperature on Toughness of Hot Rolled Microalloyed Steels

F. Leysen[1], J. Neutjens[1,2], K. Mols[3], S. Vandeputte[3] and Y. Houbaert[1]

[1] Laboratory for Iron and Steelmaking, University of Ghent,
Technologiepark 9, B-9052 Zwijnaarde, Belgium

[2] Now at RDCS, Cockerill Sambre, Blvd Colonster, Liège, Belgium

[3] OCAS-Research Centre of the Sidmar Group, J.F.Kennedylaan 3, B-9060 Zelzate, Belgium

Keywords: Microalloyed Steels, Toughness, Transformation Behaviour, Coiling Temperature

Abstract

The effect of coiling temperature on the toughness of hot rolled microalloyed steels was investigated. Heats with different Nb and/or Ti contents were laboratory hot rolled in austenitic conditions. Coiling temperatures were varied between 400°C and 700°C. A reduction of more than 50% was given below the recrystallization-stop temperature of the steels. Finishing temperature was around 860°C and thickness of the finished strips was about 6 mm. The transformation behaviour of the steels was studied by deformation dilatometry in order to select an appropriate thermomechanical treatment and to correlate the influence of coiling temperature with transformation kinetics.

The influences of grain size, grain boundary cementite, pearlite morphology, coherency of precipitates and solute carbon on toughness of the steels were examined. It was found that the most important parameter in achieving a good toughness in the hot rolled condition is the carbide morphology. By selecting coiling temperatures below the pearlite start temperature of the particular steels, a fine carbide (pearlite) dispersion was obtained with excellent toughness properties. Although a possible harmful influence of the presence of coherent precipitates on the toughness of the microalloyed steels should not be completely excluded, it is thought to have a minor effect at these temperatures.

Introduction

Studies concerning the improvement of toughness of hot rolled microalloyed steels are numerous, especially those directed towards the influence of hot rolling practice. The requirement of the finest possible ferrite grain size to obtain superior toughness properties led to the development of controlled rolling: in steel with rolling finished in the unrecrystallized austenite region, toughness is thought to be governed only by ferrite grain size [1]. Furthermore, this kind of controlled rolling accelerates the formation of incoherent precipitates, thereby suppressing the formation of coherent ones. On this matter, it has already been pointed out that, among others, coherency of precipitates has a detrimental influence on toughness of microalloyed steels [1,2,3]. According to Tanaka, fine incoherent Nb carbonitrides (either formed by strain-induced precipitation after rolling below the austenite T_{nr} or formed in the high temperature ferrite region) cause weak precipitation hardening without impairing toughness. On the other hand, niobium in solution in ferrite would not exert any harmful effect on toughness [1]. Besides, it was shown in literature that an important parameter to increase toughness without loss of strength is the austenite grain size through control of the size of the individual pearlite colonies [4].
The aim of the present study is to screen the influences of metallurgical features connected with coiling temperature on toughness of hot rolled microalloyed steels.

Experimental procedure

The compositions of the steels studied are listed in Table 1. Two Nb bearing steels Nb1 and Nb2 (with different levels of C and Nb) and one Nb-Ti steel were tested. The processing schedule is shown schematically in Fig.1. After reheating for 1 hour at 1250°C, the steels were rolled in austenitic conditions on a reversible laboratory hot-rolling mill from CARL WEZEL (rolls of 320 mm diameter). More than 50% reduction below the recrystallization-stop temperature T_{nr} of austenite was given. Strain rate was about 15 s^{-1}. Thickness of the finished strips was about 6 mm and finishing temperature was around 860°C. Coiling temperatures were varied between 400 and 700°C (increments of 50°C). The cooling rate (water spray cooling) of the strips between the finishing temperature and the coiling temperature was about 16°C/s.
The A_{r3} and A_{r1} temperatures as well as the transformation kinetics were measured using a THETA Dilatronic III S dilatometer equipped with a deformation (compression) device. The compression specimens were solid cylinders $\varnothing 3.5 \times 5$ mm. Temperatures were measured by means of a thermocouple spotwelded to the specimens. From the laboratory hot rolled strips V-notch Charpy specimens were taken in longitudinal direction (notch perpendicular to the strip surface) in order to determine the notch toughness of the steels for temperatures in a range between room temperature and -60°C. Furthermore, samples were taken to perform hardness measurements after isothermal ageing and for optical metallographic examination (microstructure, grain size, cementite distribution). Cementite/pearlite morphology was further examined using a ZEISS DSM 962 scanning electron microscope. Precipitate characterization was carried out on thin foils and extraction replicas using a JEOL JEM 2010 transmission electron microscope. Finally, the solute carbon content of the samples was evaluated by means of APUCOT measurements [5].

Table 1: Composition of the steels (weight%; N, Nb and Ti in ppm)

Steel	C	Mn	Si	P	S	Al_m	N	Nb	Ti
Nb1	0.064	0.502	0	0.009	0.004	0.035	42	189	0
Nb2	0.090	0.700	0.008	0.012	0.005	0.035	47	340	30
Nb-Ti	0.072	0.739	0.011	0.011	0.005	0.025	20	600	350

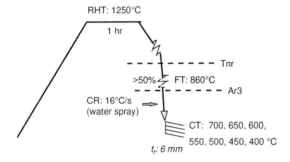

Fig.1: Schematic representation of the laboratory hot rolling cycle.

RHT: reheating temperature
CR: cooling rate
CT: coiling temperature
t_f: final thickness

Experimental results

In order to perform controlled rolling of the steels in the laboratory, the transformation temperatures A_{r3} and A_{r1}, as well as the recrystallisation-stop temperature T_{nr} of austenite had to be established. The results are presented in Table 2. T_{nr} temperatures were evaluated based on the measurement of the hot rolling forces and calculation of the deformation resistance of the steels in successive hot rolling steps.

Table 2: Critical temperatures determined by dilatometry and hot rolling forces

Steel	Nb1	Nb2	Nb-Ti
A_{r3} (°C)	815	770	768
A_{r1} (°C)	532	545	508
T_{nr} (°C)	954	976	970

As the discussion will be focussed on toughness of the steels, the tensile properties are just briefly summarized in Table 3.

Table 3: Tensile properties of the steels

Steel	Nb1	Nb2	Nb-Ti
TS (MPa)	386-425	450-497	537-616
YS (MPa)	330-349	383-424	473-532
El (%) (80 mm)	27.0-33.8	25.0-27.5	20.0-26.2

The lower strength values are obtained for the high coiling temperatures, while the elongation shows a rather limited influence of the coiling temperature considering the large spread on the results. Metallurgical factors affecting toughness in relation to coiling temperature will be discussed below. The results of the impact toughness measurements on V-notch Charpy specimens are presented in Fig.2, where the transition temperatures were correlated with an arbitrary value of absorbed energy of 50 J cm^{-2}. The figure also shows, in the secondary y-axis, the evolution of the transformed fraction of austenite at a cooling rate of 15°C/s as established by deformation dilatometry.

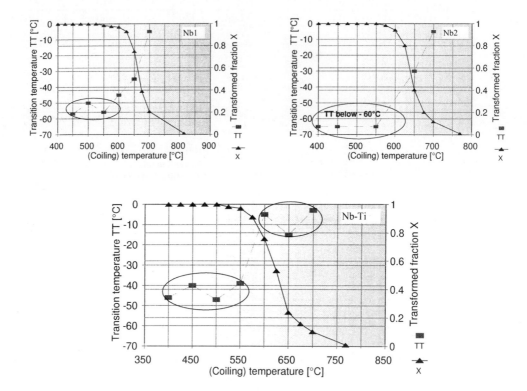

Fig.2: Influence of coiling temperature on toughness (i.e. transition temperature TT) of the Nb1, Nb2 and Nb-Ti steel. The secondary y-axis indicates the evolution of the transformed fraction (X) of austenite as a function of temperature during the cooling stage after hot deformation.

It can be noticed that the coiling temperature exerts a strong influence on the toughness of the steels. As well as for the Nb steels as for the Nb-Ti steel, high coiling temperatures lead to high transition temperatures. For the lower coiling temperatures the Nb-Ti steel shows higher transition temperatures compared to the Nb steels. In relation to the transformation kinetics, the results might suggest that good toughness properties, i.e. low transition temperatures, are reached in case coiling is performed at temperatures close to the termination of the austenite decomposition (transformed fraction of more than 90%) or lower. This aspect will be commented on further in the discussion.

Discussion

Effect of grain size - It is well known that a significant increase in toughness can be obtained through ferrite grain refinement, as for instance by lowering of the finish rolling temperature. In the present research the finish rolling temperature, as well as the cooling rate, was kept the same for all experiments. As a function of coiling temperature, it was observed that the ferrite grain sizes of the steels show little or no variation (0.5 ASTM no. at the most), being about 17 μm for the Nb1 steel, 15 μm for Nb2 and about 12 μm for the Nb-Ti steel. Most probably, the limited influence of coiling temperature can be explained based on the site saturation principle, i.e. it might be supposed that the nucleation process of ferrite takes place in the early beginning of austenite decomposition. On this matter, it is clear from Fig.2 that even at the highest coiling temperature of 700°C more than about 10% ferrite is present at the beginning of coiling, indicating that the applied coiling temperatures do not interact with ferrite nucleation.

Microstructure and carbide morphology - The steels show in all cases a ferrite-carbide(pearlite) microstructure. From the microstructures related to the lower coiling temperatures a volume fraction of 5 to 7% of second phase (pearlite) was estimated, whereas the microstructures at the highest coiling temperature tend to show a smaller pearlite fraction. At the lower coiling temperatures a tendency towards formation of acicular ferrite was observed. However, in Fig.2 this is not reflected in a change of transition temperature. On the other hand, the carbide (pearlite) morphology is strongly affected by coiling temperature. Coarse pearlite particles are formed at high coiling temperatures, while lowering of the coiling temperature leads to a gradual change of carbide distribution over grain boundary cementite to a fine ferrite-carbide aggregate at the lower coiling temperatures. The latter fine ferrite-carbide aggregate we would like to term as a kind of *degenerated* pearlite. These findings are illustrated in Fig.3 for the Nb-Ti steel at two different levels of coiling temperature. In correspondence with the measured toughness properties (see Fig.2), it can be stated that both the occurrence of grain boundary cementite and coarse pearlite lead to poor toughness. The presence of a larger dispersion and volume fraction of pearlite for the lower coiling temperatures is most probably due to a significant undercooling of the austenite-pearlite transformation reaction. In this case the pearlite will be finer and has a lower carbon content. These metallurgical features are favorable for toughness.

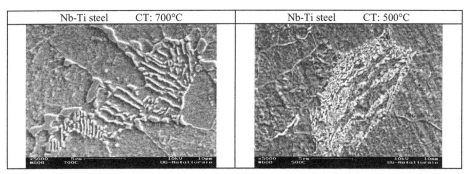

Fig.3: Scanning electron microscope images of carbide(pearlite) morphology occurring in the Nb-Ti steel for two different levels of coiling temperature, i.e. 700 and 500°C.

Precipitation behaviour - Thin foils taken from the Nb-Ti steel and examined with the transmission electron microscope showed that TiN precipitates act as nucleation sites for the Nb-precipitation. However, transmission electron microscopy could not reveal any evidence of the presence of coherent precipitates at the particular coiling temperatures. Yet, as coherent precipitates could exert a detrimental influence on toughness of microalloyed steels [3], it might be important to have some experimental notice about this matter for the coiled strips. Therefore, hardness-ageing curves were established using an ageing temperature of 620°C. In case, a certain amount of Nb and/or Ti would have been kept in solid solution at a particular coiling temperature, the specified annealing temperature should allow coherent precipitation leading to an accompanying increase in hardness during ageing. The hardness-ageing curves for strips coiled at 500, 550, 600 and 650°C are presented in Fig.4. For the Nb1 steel it can be derived from the curves that for coiling temperatures even as high as 600°C a significant hardness increase occurs. Thus a certain amount of Nb will be present in supersaturated solution in ferrite. This would mean that for some part strain induced precipitation, as well as transformation induced precipitation and precipitation in ferrite are not fully completed. As the transformed fraction of austenite at 650°C is over 70% (see Fig.2), it could be accepted that at temperatures well above 600-650°C coherent precipitation is likely to occur.

For the Nb2 steel similar observations can be made, yet the increase in hardness is shifted to lower coiling temperatures, implying that coherent precipitation is to be expected at somewhat lower coiling temperatures in the vicinity of 600°C, where a transformed fraction of more than about 90% was found (Fig.2). The observations, regarding the precipitation behaviour, are in agreement with the expected (based on chemical composition) higher driving force for precipitation in the Nb2 steel.

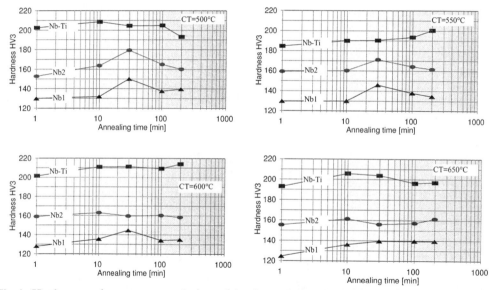

Fig.4: Hardness-ageing curves: evolution of hardness during isothermal annealing of the coiled strips at a temperature of 620°C. Coiling temperatures of 500, 550, 600 and 650°C were examined.

The Nb-Ti steel seems to show no significant change of hardness during the ageing treatment, suggesting that precipitation during processing has been nearly completed. This is most probably due to the high tendency of precipitation in the steel. Firstly, a high contribution of a strain induced precipitation in austenite can be expected. From own experimental results it could be concluded that relatively high finish rolling temperatures are needed in order to avoid complete precipitation in the Nb-Ti steel [7]. Secondly, precipitation of Nb might be enhanced through a possible nucleation effect of TiN. In earlier research however, it was suggested that the tendency of carbonitride precipitation in austenite would be reduced through the combined addition of Nb and Ti [8]. Nevertheless, considering the transformation kinetics (see remark below) and also the higher initial hardness for a coiling temperature of 600°C (the relatively high hardness for a coiling temperature of 500°C may be explained by the appearance of acicular ferrite in the microstructures), coherent precipitation is feasible for coiling temperatures around or somewhat above 600°C.

At coiling temperatures somewhat below 600°C, the (coherent) precipitation will be rather sluggish and on the other hand good toughness properties are observed. In general, if coherency of precipitates would be a metallurgical feature impairing toughness, the temperature range of precipitation is to be situated where decomposition of austenite has progressed to a significant amount (Fig.2). As coherent precipitation within the transformation region might be argued, in earlier research it was found that at high cooling rates and low transformation temperatures, the precipitation behaviour may alter from an incoherent towards a coherent character [6]. Furthermore, it has to be taken into account that transformation kinetics as shown in Fig.2 are applicable for continuous cooling conditions. If for instance the Nb2 steel, which shows a transformed fraction of about 40% at 650°C, is coiled at the temperature of 650°C, it can be realized that the transformed fraction during coiling (thus slow cooling) will be much higher at temperatures close to or somewhat below 650°C. For the tested steels, in general, this would mean that the formation of coherent precipitates at coiling temperatures in the range of about 600°C to 700°C (partly due to the fast completion of the austenite to ferrite transformation during coiling at high temperatures) can not be excluded. Although, at the higher coiling temperatures loss of coherency may occur.

As in Fig.2 for the Nb steel at a coiling temperature of 700°C, the transition temperature is still increasing and since from the ageing experiments it is thought that in this steel coherent precipitation occurs around 600°C, it is however not supposed that coherent precipitation has an important influence on transition temperature.

Interstitial carbon - By means of internal friction (APUCOT) measurements, the interstitial carbon content of the Nb steels for coiling at 550 and 650°C, and of the Nb-Ti steel at 500 and 600°C was evaluated. Only in the Nb steels at low coiling temperature interstitial carbon was found, which may support the results of the ageing experiments.

Summary - From the present research it may be deduced that the coiling temperature applied in the processing of hot rolled microalloyed Nb and Nb-Ti steels exerts a strong influence on toughness properties. By the lowering of coiling temperature a sudden increase in toughness properties is obtained. This behaviour can be explained based on transformation kinetics, i.e. austenite decomposition reaction, of the steels as determined by deformation dilatometry. On cooling to coiling temperatures below specific temperatures where the transformed fraction of austenite exceeds 90%, excellent toughness properties can be established. These specific temperatures can be very well correlated with the pearlite start temperature of the steels; being about 630°C for the Nb1 steel, 600°C for the Nb2 steel and 575°C for the Nb-Ti steel. Coiling below these temperatures leads to a fine carbide (pearlite) dispersion with the observed good toughness. Moreover, the cooling to the aforementioned low coiling temperatures suppresses the decomposition of austenite to within a temperature range where the formation of coherent precipitates is less feasible.

Conclusions

By taking into account the transformation kinetics, i.e. the evolution of austenite decomposition, as determined by means of deformation dilatometry, the influence of coiling temperature on toughness of hot rolled Nb and Nb-Ti microalloyed steels can be clarified. The suppression of transformation during cooling to coiling temperatures below the pearlite start temperature of the steels, leads to a fine carbide (pearlite) dispersion and moreover hinders the formation of coherent precipitates. Both metallurgical features contribute to good toughness properties of the hot rolled steels.

References

1. Tamura, C. Ouchi, Thermomechanical Processing of HSLA Steels, 1988
2. A. De Vito, E. Anelli, HSLA Steels '85, 1985, p. 951-957
3. R. Hamano, Metallurgical Transactions, 1993, p. 127-139
4. R. Lagneborg, Scand. Journal of Metallurgy, 1985, p. 289-298
5. Z.L. Pau, I.G. Ritchie, H.K. Schmidt, Automatic Piezoelectric Ultrasonic Composite Oscillator Technique, User Manual AECL Research White Shell Laboratories, Canada, 1992
6. J.C. Herman, V. Leroy, Thermec '88, Tokyo, 1988, p. 283-290
7. F.Leysen, J. Dilewijns, (private com.) to be published
8. Y. Weixun, Z. Xiaogang, Microalloyed HSLA Steels, Chicago, USA, 1988, p. 521-531

Materials Science Forum Vols. 284-286 (1998) pp. 271-278

Morphology and Mechanical Properties of Bainitic Steels Deformed in Unrecrystallized Austenite Region

K. Fujiwara and S. Okaguchi

Corporate Research Laboratories, Sumitomo Metal Industries, Ltd., Japan

Keywords: Bainite, Morphology, Toughness, Low Carbon Steel, Deformation, Unrecrystallized Austenite

ABSTRACT

Bainite structure in low carbon low alloy steels has played an important role for the practical use of these steels in many constructions[1,2]. By improving the strength and toughness they will be applied to new fields in the future. From the viewpoint of microstructure, it is a packet size that determines the toughness of bainitic steels. Optical micrographs of upper bainite structures exhibit a straight elongated lathlike bainitic ferrite whose growth is determined by austenite grain boundaries. However the application of accelerated cooling after controlled rolling [1,2] makes the microstructures considerably different from those in the conventional quenching and tempering process. Although some of these complicated microstructures have been called " Zw[3] ", the nature of them has never been clarified. It is very important to define an effective way to refine the bainite structure by the deformation in unrecrystallized austenite region in order to improve the toughness of steel plates. This study focuses upon the effects of deformation on the morphology of bainite structure and mechanical properties in low carbon low alloy bainitic steels.

EXPERIMENTAL PROCEDURES

Chemical composition of the steels investigated is shown in Table 1. The ingots were prepared in a 30kg vacuum induction furnace and then rolled into 20mm thick plates. Cylindrical specimens, 8mm in diameter and 12mm in length, were machined from the plates for the examination of TTT diagrams and metallographic analysis. The specimens were austenitized at 1373K for 300s by induction heating in a vacuum chamber, deformed 0 to 75% at various temperatures between 973 and 1173K and were quenched by helium gas to various temperatures where bainite transformation occurs and held isothermally. After various periods, specimens were quenched to room temperature by water spray. The foils for transmission electron microscopy were prepared by a twin jet polishing technique using a 5% perchloric acid and 95% acetic acid electrolyte. The foils were examined by JEM-200CX transmission electron microscope operated at 200kV. For mechanical property tests ingots were rolled into 12mm thick plates and cooled to 800K by using a laboratory mill and water spray.

	C	Si	Mn	P	S	Ni	Nb	Ti	B	sol-Al	N
				Table 1	Chemical compositions					(wt%)	
Steel 1	.18	.20	1.37	.007	.001	.62	.014	.012	.0008	.026	.0021
Steel 2	.10	.20	1.34	.007	.001	.62	.015	.012	.0013	.034	.0004
Steel 3	.08	.20	1.38	.011	.003	.60	.016	.011	.0009	.033	.0011

RESULTS AND DISCUSSION

1. Effects of deformation on the transformation behavior

TTT diagrams clearly consist of two C-curves even in cases of 30 and 50%-deformation, and the upper limit of lower C-curve (i.e. Bs) is clearly found in each case. Metallographic observation revealed bainite and polygonal ferrite formed at temperatures below and above Bs, respectively. It was confirmed that the formation of bainite is much less accelerated by the deformation in un-recrystallized austenite region than that of polygonal ferrite.

2. Effects of deformation on the morphology of bainitic ferrite

SEM micrographs of steel 1 transformed at 848K from 0 and 50%- deformed austenite are shown in Fig.1. They are in the primary stage of transformation where the bainite is partially developed. Fig.1(a) shows typical BI type[4] upper bainite laths formed in parallel with the same crystallographic orientation in the case of non-deformation. In the case of 30%-deformation, the bainitic ferrite laths were shaped like curves of bow. It is noteworthy that deformation more than 50% (b), in contrast, made the structures much finer and complicated their morphology.

The effect of deformation on the length of bainitic ferrite laths of steel 2 is shown in Fig.2. It was confirmed that less than 30%-deformation was less effective on reduction of the length, but deformation more than 50% reduced the length significantly. This remarkable refinement of bainitic ferrite in heavy deformation is considered to be closely connected to the formation of subgrains within deformed austenite grains.

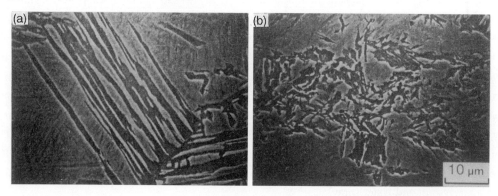

Fig.1 SEM micrographs of steel 1 transformed at 848K ((a)0%, (b)50% deformed at 1173K)

TEM micrograph of partially transformed specimen after 50%-deformation is shown in Fig.3. Coalescence of parallelograms seen in this area are cross sections of the growth axis of the bainitic ferrite laths. As previously reported[5], the bainitic ferrite decomposed from non-deformed austenite grows in <111> α direction with two sets of habit planes between {110} α and {451} α. Trace analysis indicated that the habit planes for the 50%-deformed bainitic ferrite laths also lay in the vicinity of {451} α, and their growth axis was close to the normal of the micrograph. As described above, the electron microscopical studies demonstrate that the bainitic ferrite developed after heavy deformation is of similar crystallographic aspects to those from non-deformed austenite.

3 Effects of deformation on the crystallographic orientation of bainitic ferrite

In the case of non-deformation a large number of bainitic ferrite laths formed in parallel with same crystallographic orientation, as shown in Fig.4(a). In contrast, Fig.4(b) indicates that most adjacent bainitic ferrite has different crystallographic orientations in the case of more than 50%-deformation. (There are at least three different orientations in Fig.4(b), two pairs of laths No.1-2, 3-4 and single lath No.5 show different orientations, respectively.) Therefore, the bainite packet consists of only a few bainitic ferrite laths. It suggests such a heavy deformation enhances the nucleation of bainitic ferrite within unrecrystallized austenite grains. The decrease of both the length and the number of bainitic ferrite laths with the same crystallographic orientation signifi-cantly reduced the size of the bainite packets, which makes the morphology of bainite much more complicated.

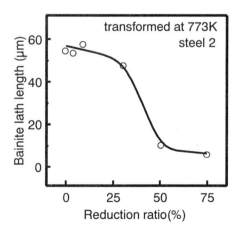

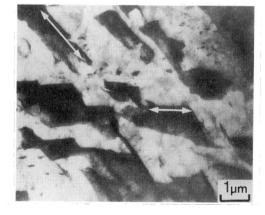

Fig.2 The effect of deformation on the lath length of steel 2

Fig.3 TEM micrograph of steel 2 transformed at 773K after 50% deformation at 1173K

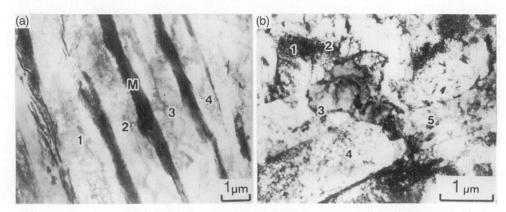

Fig.4 Effect of deformation on the crystallographic orientation of bainitic ferrite transformed at 773K
after (a)0%, (b)50% deformation at 1173K, respectively

4 Refinement of bainitic ferrite laths

In the case of non-deformation, bainitic ferrite laths which nucleated on austenite grain boundaries grow lineally and discontinue their growth when they reach the other austenite grain boundaries. Although they are shaped like curves of bows in the case of slight deformation, bainitic ferrite laths grow and the significant decrease of the length is not found.

On the contrary, heavy deformation reduces their length remarkably and most of adjacent bainitic ferrite laths show different orientations. The decrease of both the length and the number of bainitic ferrite laths formed in the same orientation significantly reduces the size of bainite packets. As generally reported[6], in the case of polygonal ferrite, ferrite grain size decreases linearly with increasing of the amount of deformation. Therefore the refinement manner of bainitic ferrite is fairly dissimilar to that of polygonal ferrite, which is considered to be closely related to the formation of dislocation cell-structures within deformed austenite grains. Therefore, this remarkable change of the length of bainitic ferrite laths with heavy deformation leads to the refinement mechanism which is caused by the formation of cell-structures. Since, in the case of slight deformation, the dislocation density which composes cell-structures is low and the difference of orientation among adjacent cell-structures is small, bainitic ferrite laths which nucleate at grain boundaries are considered to grow beyond the dislocation tangles. Therefore, they are able to grow along the distortions of the austenite lattice. In contrast, in the case of heavy deformation, bainitic ferrite laths are considered to discontinue their growth at the boundaries of cell-structures. Moreover, because these boundaries act as the nucleation sites of bainitic ferrite laths, further amounts of laths which have different orientations develop adjacent to each other.

Roberts et al[7] estimated the subgrain size of low carbon steels (In their report, they treated a cell-structure as a subgrain). This shows that the subgrain size depends on the deforming temperature when strain ratio($\dot{\varepsilon}$) is constant.

$$D \text{ low carbon} = 269 \times \exp (-4770 / T)$$

where,

$$\dot{\varepsilon} = 1/s$$

The relation between the length of bainitic ferrite laths transformed from 75%-deformed austenite and the estimated cell-structure size is shown in Fig.5. This figure clearly shows that the measured length of bainitic ferrite lath relates to the estimated cell-structure size. This means that bainitic ferrite laths nucleate and discontinue their growth on the cell-structure boundaries within the heavily deformed austenite grains. Actually, the high density of dislocation tangles which are considered to be the traces of cell-structure are observed on the interfaces of fine bainitic ferrite laths in the case of heavy deformation (See Fig.6.).

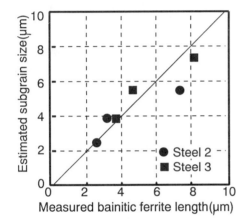

Fig.5 The relation between the length of the lath length and estimated subgrain size.

Fig.6 High density of dislocation tangles of steel 3 transformed at 773K

5 Effects of deformation on the cementite precipitation

Fig.7 shows, bainite structure in the primary stage of transformation at 773K after 50%-deformation. In the case of non-deformation cementite free bainitic ferrite laths formed and carbon enriched austenite region existed between each lath 1-4, as shown in Fig.4(a). It is found that, in the case of 50%-deformation, on the contrary, cementite particles precipitate even in the primary stage of transformation. Therefore, it was also confirmed that carbon enriched austenite regions were not observed between the laths in this case.

Fig.8 shows bainite structure on the final stage of transformation. It was found that coarse inter-lath cementite precipitated in the case of non-deformation, which is considered to lead to reduction of the carbon concentration in the austenite and to the growth of a further amount of bainitic ferrite. In the case of 50%-deformation, as shown in Fig.4(b), inter-lath cementite precipitation also occurred partially. It is noteworthy that there was not observed coarse particles such as in the case of non-deformation.

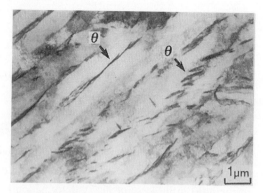

Fig.7 TEM micrograph in the primary stage
of transformation after 50%-deformation

Fig.8 TEM micrograph of the final stage of
transformation after non-deformation

6 Mechanical properties

Figs.9(a) and (b) describe the effect of amount of deformation on the toughness of air
cooled and accelerated cooled plates. The toughness of air cooled plates linearly improves with
increasing of amount of deformation in unrecrystallized region. On the contrary, that of
accelerated plates improves significantly by heavy deformation more than 50%. The toughness of
accelerated plates excels that of air cooled plates at 75% deformation. Microstructures of air
cooled and accelerated plates are mainly polygonal ferrite and bainite, respectively. In the case of
non-deformation, the effective grain size of polygonal ferrite and bainite are of the same level.
However the length of bainitic ferrite lath decreases remarkably by deformation of more than
50%. Figs.10(a) and (b) show the effect of amount of deformation on the strength of air cooled
and accelerated cooled plates. In the case of air cooled plate, tensile strength does not increase by
heavy deformation (Yield strength increases with increasing of the amount of deformation). In
contrast, tensile strength of accelerated plates increases with increasing of amount of deformation.
The tensile strength of bainite structure is increased by the deformation. This means that the
bainite structure inherits the dislocation density from the unrecrystallized austenite grains.

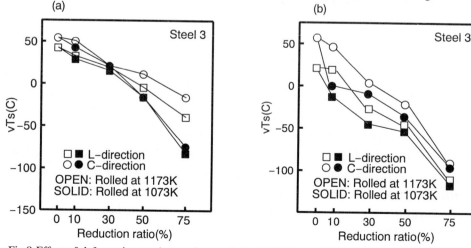

Fig.9 Effect of deformation on the toughness of steel 3 ((a)air cooled, (b)accelerated cooled plate)

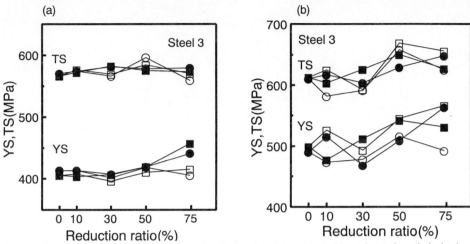

Fig.10 Effect of deformation on the strength of steel 3 ((a)air cooled, (b)accelerated cooled plate)

SUMMARY

(1)Slight deformation is not effective in the reduction of the length of bainitic ferrite laths. In contrast, heavy deformation decreases the length remarkably.

(2)In the case of non-deformation, a large number of bainitic ferrite laths form parallel with the same crystallographic orientation. In the case of heavy deformation, adjacent bainitic ferrite laths have different crystallographic orientation.

(3)The measured length of bainitic ferrite lath is consistent with the estimated cell-structure size. This means that bainitic ferrite laths nucleate and discontinue their growth on the cell-structure boundaries within the heavily deformed austenite grains.

(4)The toughness of polygonal ferrite linearly improves with increasing amounts of deformation in unrecrystallized region. On the contrary, that of bainite improves significantly by heavy deformation of more than 50%. Tensile strength of polygonal ferrite does not increase with heavy deformation. In contrast, that of bainite increase with increasing of amount of deformation. This means that bainite structure inherits the dislocation density from the unrecrystallized austenite grains.

REFEREMCES

[1] D.V.Edmonds and R.C.Cochrane, Metall. Trans.,21A(1990),p1527.

[2] H.Ohtani, S.Okaguchi, Y.Fujishiro and Y.Ohmori, Metall. Trans.,21A(1990),p.877.

[3] T.Araki, M.Enomoto and K.Shibata, Mater.Trans. JIM,32(1991),p.729.

[4] Y.Ohmori, H.Ohtani and T.Kunitake, Trans. ISIJ,11(1971),p.250.

[5] J.D.Watson and P.G.McDougall, Acta Metall., 21(1973),p. 961.

[6]H.Sekine and T.Maruyama, Seitetsu Kenkyu,289(1976),p.43.

[7]W.Roberts and B.Ahlblom, Acta Metall.,26(1978),p.271.

E-mail address: fujiwara-kzk@aw.sumikin.co.jp

Fax number: +(81) 06 489 5757

Materials Science Forum Vols. 284-286 (1998) pp. 279-286

Process History, Microstructure and Weldability Interactions in Microalloyed Steels

D.J. Egner and R.C. Cochrane

Department of Materials, The University of Leeds, Leeds, LS2 9JT, UK

Keywords: Process History, Weld HAZ Simulation, Thermomechanical Controlled Rolling, Microstructure, Impact Properties, Microalloyed Steels

Abstract Weldability testing forms an integral and time consuming part of the fabrication procedures for many high integrity structures, consequently much work has been directed to forming mathematical models to predict the properties of weld HAZ's. Many such models exist but suffer limitations due to the necessity to calibrate the model to each system studied, further they assume there is no significant effect of process history, and hence microstructure, on HAZ formation.

Presented in this paper are the findings of Charpy impact studies on weld HAZ simulated samples of a niobium microalloyed steel in two different rolled conditions. The findings show clearly that HAZ formation and weldability are significantly affected by process history.

Introduction

Prior to and during fabrication of high integrity structures such as oil exploration platforms it is usual to perform extensive weldability testing to indicate the probable range of mechanical properties that might be expected in the welded joint. In the early days of weldability testing of UK offshore platform steels, it was common to assess the HAZ response of a plate with the highest expected CEV or poorest impact toughness in the expectation that the test data may only apply to one particular process route for the steel in question. Whilst, process route had some influence on weldability, the data were often confusing and changes in welding parameters would obscure trends. This led to time consuming and expensive delays to the weldability test programme and failure to achieve the desired level of mechanical properties would result in major difficulties in commissioning of the structure. More valuable and definitive information on trends in HAZ toughness can now be obtained by HAZ simulation and modelling techniques since the principal variables in welding can be combined to predict a characteristic weld cooling time for any given welding procedure. The improved reproducibility with such methods allows the role of steel composition and process route to be identified more clearly than was previously possible.

Several models [1-4] have been developed which attempt to predict microstructure in the HAZ of microalloyed steels by considering the metallurgical changes, such as austenite grain growth and precipitate dissolution and/or re-precipitation, during the welding cycle. More sophisticated versions of such models [5] extend this approach using data on the transformation characteristics to generate microstructural maps as a function of welding variables, primarily weld cooling time. In principle, once the microstructure is predicted the approach of Maynier et al [6] or Adil [7] can then be used to predict the HAZ mechanical properties. All of the present models rely on "calibration" to provide a match to real weld HAZ microstructures and further assume that there is no significant effect of either prior microstructure or process history on HAZ development, thus there should be no changes in HAZ mechanical properties with process history. Further complication results if multiple microalloying additions are present since the initial precipitate distribution or composition is rarely known or considered but has important consequences for particle dissolution rates [8,9].

This paper is concerned with a critical examination of the changes in Charpy toughness following simulated weld thermal cycles and how this relates to existing HAZ microstructural models for a simple microalloyed steel rolled to simulate typical controlled rolling and normal hot rolling schedules.

Experimental.

An experimental steel was cast to the composition given in Table 1. This steel is similar in base composition to that of offshore linepipe, with a microalloy addition of niobium to give single microalloy strengthening, thus enabling a study of a simple microalloy system. The cast was split and rolled under two different rolling conditions. The first condition was similar to that typical of commercial controlled rolling schedules for plate, with the bulk of the deformation occurring in the austenite plus ferrite phase region, thus promoting a fine ferrite structure with coarse, strain induced precipitation. The second rolled condition was that of an as rolled plate in which generally the temperatures or strains are not precisely defined. Such a schedule would be used in the production of sections for example. In such schedules, all of the deformation occurs in the austenite phase region above the niobium carbonitride solvus and the steel is then air cooled, thus promoting fine scale precipitation in ferrite but achieving a coarser more equiaxed grain structure.

Following rolling the plates were characterised using optical metallography and TEM techniques. The rolled plate was also mechanically tested for tensile strength, hardness and Charpy toughness.

Samples were taken from each plate and thermally cycled at three peak temperatures, 950°C, 1050°C (the solvus) and 1350°C, using a Gleeble 1500 weld HAZ simulator. The cooling rate used in the simulation was typical of that experienced during 3kJ/mm submerged metal arc welding of linepipe, at Δt_{8-5} of 70s, however the heating rate was lower than that experienced during welding being limited by the apparatus to 400°C/s.

Each peak temperature simulation was characterised using optical metallography and TEM techniques. Hardness measurements were taken and impact transition curves were produced using Charpy impact testing.

Table 1, Cast analysis of experimental steel.

C	Si	Mn	P	S	Al	N	Ti	Nb	V
0.04	0.3	1.4	0.012	0.002	0.04	0.006	-	0.02	-

The niobium carbonitride solvus temperature was calculated using the equilibrium data presented by Narita [10], and found to be approximately 1050°C.

Results

Initial characterisation by optical metallography and TEM showed marked differences in the microstructures of the steel in the two rolled conditions. The controlled rolled steel of a fine ferrite grain structure, 8.6μm, with coarse microalloy precipitates, Figs. 1&2.

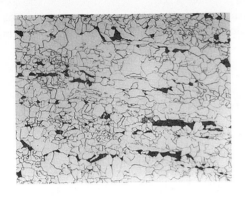

⊢————————⊣ 100μm

Figure 1, Optical micrograph of the
controlled rolled steel.

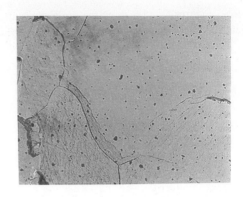

⊢————————⊣ 5μm

Figure 2, TEM image of a carbon replica of the
controlled rolled steel.

In contrast the as rolled steel has a much coarser ferrite grain structure, 25.8μm, and fine rows of
precipitates, Figs. 3&4.

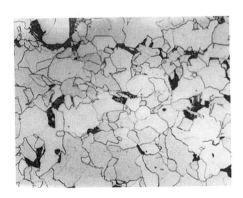

⊢————————⊣ 100μm

Figure 3, Optical micrograph of the
as rolled steel.

⊢————————⊣ 5μm

Figure 4, TEM image of a carbon replica of the
as rolled steel.

Figures 5 and 6 show the same steels after thermal cycling at a peak temperature of 1350°C. Here it can be seen that the bainite colony size of the controlled rolled steel is much larger than in the as rolled steel.

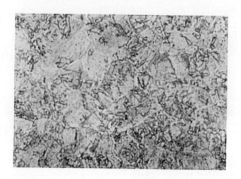

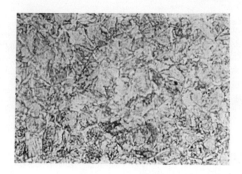

├──────────── 500μm

Figure 5, optical micrograph of the controlled rolled steel after thermal cycling at 1350°C.

├──────────── 500μm

Figure 6, optical micrograph of the as rolled steel after thermal cycling at 150°C.

Tables 2 and 3 show the findings of grain sizing, tensile and hardness tests and a comparison is drawn with a Hall-Petch analysis of strengthening contributions, in which the degree of precipitation strengthening can be determined by $\Delta\sigma$.

Table 2, Base plate strengthening contribution analysis.

Steel	Grain size, d /μm	$K_y d^{-1/2}$ /Mpa	σ_{ss} /Mpa	σ_o /MPa	$\sigma_y$total /MPa	$\sigma_y$test /MPa	$\Delta\sigma$ /MPa	Hardness /H_v
C R	8.6	194	88	70	352	399	47	152
A R	25.8	112	88	70	270	391	121	163

Table 3, HAZ simulated strengthening contribution analysis.

Steel condition	Grain size, d /μm	$K_y d^{-1/2}$ /MPa	σ_{ss} /MPa	σ_o /MPa	$\sigma_y$tot /MPa	Hardness /H_v
C R Base	8.6	194	88	70	352	152
C R 950	6.5	223	88	70	381	164
C R 1050	7.4	208	88	70	366	175
C R 1350	122.0	n/a				183
A R Base	25.8	112	88	70	270	163
A R 950	8.2	199	88	70	357	172
A R 1050	10.5	176	88	70	334	174
A R 1350	70.3	n/a				175

Impact transition curves for the two base plates and corresponding weld simulated specimens are given in Figs. 7&8, in which the transition temperatures are clearly defined, thus demonstrating the effects of weld thermal cycling on plate toughness.

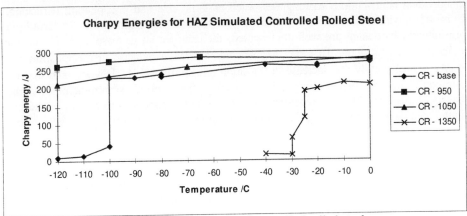

Figure 7, Impact transition curves for HAZ simulated controlled rolled steel.

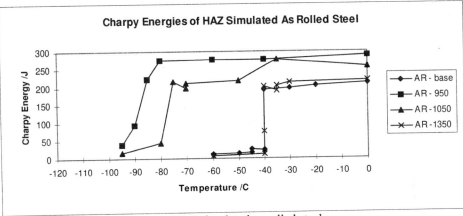

Figure 8, Impact transition curves for HAZ simulated as rolled steel.

The rolling schedules applied to the steel produced the desired microstructures, precipitation distribution and mechanical properties to enable weld simulation studies on two very different process histories of an identical material.

The results presented here show clearly very different behaviour by the microstructure of the base plates as a result of weld HAZ simulations, resulting in various degrees of grain refinement of the ferrite-pearlite structure at peak temperatures of 950 and 1050°C, whereas bainite colonies which differ in size between the two steels are formed after simulation at a peak temperature of 1350°C. Hence the mechanical properties also respond differently. This study has therefore indicated a considerable interaction between the process history and weldability of microalloyed steels.

Discussion

The most striking observation is that in the as-rolled condition there is no shift in toughness from the parent steel to that of the simulated fusion line HAZ, ΔT, the 27J impact transition temperatures remain at -40°C whereas in the case of the controlled rolled condition the 27J ITT shifts from -100°C to -30°C or ΔT is about +70°C. This behaviour is easiest to understand in terms of the individual contributions to toughness such as grain size and precipitation hardening. From studies on steels with a ferrite/pearlite microstructure the embrittlement vectors for ferrite grain-size and precipitation hardening are well documented, for example an increase of 1 $d^{-1/2}$ in grain size embrittles the steel by about 10°C [11,12] whereas a 10MPa increase in precipitation hardening would give approximately 5°C increase in transition temperature [12]. If it assumed that the HAZ microstructure does not depend on the initial state of the steel, as in current models, then it is possible to estimate the shift in transition temperature corresponding to changes in the individual contributions to yield strength using the following equation for impact transition temperature:

$$ITT = A - K\, d^{-1/2} + B\, \sigma_p \qquad\qquad \text{Eq. 1}$$

where A and B are constants, d is the ferrite grain size, K is the Hall-Petch coefficient and σ_p is the precipitation strengthening increment from microalloying additions.

and $\qquad\qquad \Delta T = ITT_{HAZ} - ITT_{pp} \qquad\qquad$ as defined above $\qquad\qquad$ Eq. 2

therefore $\qquad \Delta T = ITT_{HAZ} - A + K\, d^{-1/2} - B\, \sigma_p \qquad\qquad$ Eq. 3

The shift in ITT after weld cycle simulation can therefore be estimated by assuming that ITT_{HAZ} does not change and that there is little or no difference in precipitation in the simulated HAZ condition. The shifts in ITT are +83°C and +2°C for the controlled rolled and as-rolled steels respectively, using the microstructural data in table 2. These figures are consistent with those observed. However, the microstructure of the simulated HAZ of the controlled rolled condition is significantly coarser which is consistent with the somewhat poorer toughness in this condition and assuming that the embrittling effect of bainite colony size is similar to that of ferrite grain-size [13,14] would give rise to a difference of about +10°C based on the values measured.

The effect of prior processing after HAZ simulation at 1050°C (near the microalloying solvus temperature) indicates that despite a similar hardness, and therefore HAZ yield strength for both rolling conditions, the extent of precipitation hardening in the as rolled steel would be expected to be somewhat larger after simulation because of the differences in grain-size. If complete resolution of the precipitates occurred in the HAZ thermal cycle this would result in a 60°C and a 24°C decrease in ITT for the as-rolled and controlled rolled conditions respectively, whereas the improvement in toughness from the additional grain refinement would be -35°C and -8°C respectively. The total expected decrease in ITT for the as-rolled steel (-95°C) is larger than the shift observed after HAZ simulation. This somewhat smaller shift may be a consequence of some additional precipitation hardening during cooling from the weld thermal cycle peak temperature. A similar qualitative study for the controlled rolled steel can be applied if the extrapolated ITT of -135°C is used to judge the shift in transition temperature, -35°C compared to the -32°C predicted. This close agreement between measured and predicted indicates that the niobium precipitates must have either coarsened to such an extent they can no longer contribute a strengthening effect or have dissolved completely.

The effect of HAZ simulation to a peak temperature of 950°C is also to produce a marked reduction in ITT. In the controlled rolled condition, this can be qualitatively explained in terms of a reduction in the extent of precipitation hardening since there is a significant increase in hardness due to grain refinement whereas in the as-rolled condition the improvement in toughness appears to be largely due to the much greater refinement in grain size after HAZ simulation. However the changes in hardness cannot be entirely explained by changes in grain size. In the case of the controlled rolled steel the grain size is refined further by the simulated 950°C weld thermal cycle and the shift in ITT due to this refinement is expected to be of the order of -16°C, and the data in figure 7 suggest the improvement in ITT is at least this great. The increase in hardness for this condition is nevertheless consistent with the measured extent of grain refinement. Unfortunately, it is necessary to extrapolate the Charpy data to obtain the possible shift in ITT, this suggests the ITT after simulation at 950°C is of the order of -155°C, corresponding to a shift of -55°C. This is larger than expected on the basis of grain-size alone and indicated that there has been substantial coarsening of precipitates, indeed that expected from the estimated precipitation contribution to the yield strength of the parent steel before simulation, 47MPa, table 2, would be about -25°C. The overall shift of -40°C (-16 + -25) is therefore in general agreement with extrapolation of the data presented in figure 7, suggesting sufficient coarsening has taken place during simulation to reduce the extent of precipitation hardening appreciably.

The grain refinement in the as-rolled steel after HAZ simulation at 950°C is much greater and would give a decrease of 48°C very close to that observed, ~50°C. The presumption must be that either the fine precipitates remain unchanged due to the Thompson-Freundlich effect or redissolve and reprecipitate to give a similar contribution to strength. Such a conclusion would suggest quite marked differences in behaviour between the two rolling conditions after HAZ simulation, in that the precipitates responsible for strengthening in the controlled rolled condition appear to simply coarsen whereas in the as-rolled steel there is little change. To gain further confirmation of these results tensile testing of the simulated HAZ microstructures and additional electron microscopy are in hand.

An explanation for this differing behaviour is that in the controlled rolled steel diffusion of microalloy atoms will be considerably enhanced by the higher density of lattice defects such as grain boundaries and dislocation substructures, thus providing rapid diffusion paths to enable coarsening and dissolution of the microalloy precipitates.

The trends in impact behaviour can be interpreted in terms of the structural changes observed in the simulated fine grained HAZ, and broadly for the coarse grained HAZ simulation. Nevertheless, there is clearly an effect of initial plate microstructure on the coarse grained simulated HAZ microstructure which is related to differences in the initial particle distributions. The analysis presented also implies that there are differences in precipitation behaviour at or near the particles solvus. One possible interpretation of the data is that in the controlled rolled condition the particles are large enough to coarsen and gradually redissolve during heating to the weld thermal cycle peak temperature and this causes an imbalance between the grain pinning forces from the particles and the driving force for grain growth as temperature increases rather similar to that occurring in abnormal grain growth [4]. The consequence would be that a wide range of grain size is present accounting for the rather larger austenite grain-size and bainite colony size after weld HAZ simulation of the controlled rolled steel. In contrast, the particles in the as-rolled steel appear to redissolve over a rather narrow temperature range and normal grain growth can therefore be maintained over the whole of the heating cycle. Modelling of these effects has proved difficult using existing models (1-4) probably due to the 'calibration' procedure needed to relate HAZ grain-size to a particular particle distribution. Alternatively some of the observations of Pickering on non-equilibrium segregation at austenite grain boundaries [15] may be relevant. Nevertheless, this study has shown that the assumption that the HAZ grain-size and microstructure is solely a function of steel composition for a given set of welding parameters may not be valid. Work is proceeding on more complex systems where it could be anticipated that the discrepancies will be larger.

Conclusions

1) The process history of microalloyed steels, and therefore microstructure, significantly affects the Charpy impact properties after weld HAZ simulation in the temperature range 950-1350°C.

2) Microstructural evolution as a result of weld HAZ simulation is effected by the size and distributions of precipitates present in microalloyed steels, even at peak temperatures close to and above the solvus of those precipitates.

References

[1] M.F.Ashby, K.E.Easterling, Acta. Metall. 30 (1982), p. 1969.

[2] J.C.Ion, M.F.Ashby, K.E.Easterling, Acta. Metall. 32, (1984), p. 1949.

[3] I.Andersen, O.Grong, N.Ryum, Part I & II, Acta. Metall. 43, (7) (1995), p. 2673.

[4] P.J.Alberry, B.Chew, W.K.C.Jones, Met. Tech. 4 (1977), p.317.

[5] D.F.Watt, L.Coon, M.Bibby, J.Goldak, C.Henwood, Part A & B, Acta. Metall. 36 (11) (1988), p.3029.

[6] Ph.Maynier, B.Jungman, J.Dollet, "Hardenability concepts with applications to steel" (edited by P.Doane & J.S.Kirkcaldy) Trans. AIME (1977), p. 518.

[7] G.K.Adil, S.D.Bhole, Cand. Met. 31 (2) (1992), p.151.

[8] H.Adrian, Mat. Sci. Tech. 8 (1992), p.406.

[9] D.C.Houghton, Acta. Metall. Mater. 41 (10) (1993), p.2993.

[10] K.Narita, Trans. Iron and Steel Inst. of Japan, 15 (1975), p.145.

[11] W.B.Morrison, B.Mintz, R.C.Cochrane, British Steel Product Technology Conference. 2 "Controlled Processing of HSLA Steels" York (1976), paper 1.

[12] T.Gladman, "The Physical Metallurgy of Microalloyed Steels" Institute of Materials. London (1997), p.42-46.

[13] M.Nakanishi, Y.Komizo, Y.Fukuda, "The Sumitomo Search" 33 (1986), p. 22.

[14] D.V.Edmonds, R.C.Cochrane, Met. Trans. 21A (1990), p. 1527.

[15] B.Garbarz, F.B.Pickering, Mat. Sci. Tech. 4 (11) (1988), p. 957.

Acknowledgements

The authors would like to acknowledge the help and financial support for a CASE award from British Steel Swinden Technology Centre for this work. Special thanks to Dr P L Harrison for his interest and support in this study.

Correspondence

All correspondence should be addressed to Mr D J Egner, who can be contacted by;
E-mail, msedje@leeds.ac.uk
Fax, +44 113 2422531

Materials Science Forum Vols. 284-286 (1998) pp. 287-294
© 1998 Trans Tech Publications, Switzerland

Low Cycle Fatigue Behaviour of a Continuously Cooled V-Bearing Microalloyed Steel

P.C. Chakraborti[1] and M.K. Mitra[2]

[1] Metallurgical Engineering Department, Jadavpur University, Calcutta - 700032, India

[2] School of Materials Science and Technology, Jadavpur University, Calcutta -70032, India

Keywords: Low Cycle Fatigue, Accelerated Cooling, V-Microalloyed Steel, Bainitic Steel

ABSTRACT

Development of bainitic microstructure through accelerated cooling following hot working of low and medium carbon microalloyed steels for the production of forged automobile components have attracted major interest in recent times. The present work is an attempt to highlight the microstructural variants evolved by spray cooling of a 0.18% carbon V-microalloyed steel from different reheating temperatures (900 to 1300ºC) and the resulting monotonic tensile and cyclic behaviour, especially in the short life regime.

INTRODUCTION

Accelerated cooling of low and medium carbon microalloyed steel after forging at a temperature range of 1100ºC to 1300ºC has opened the possibility of elimination of the separate hardening process for the forged components. It has been well demonstrated that such a processing route becomes most economic for the production of forged automobile components[1] .The resulting microstructure may range from ferrito-pearlitic to bainitic or even martensitic depending on the working temperature and the subsequent cooling rate. In recent years reports[2,3] highlighting the microstructural evolution in low to medium carbon microalloyed steels on continuous cooling have been published. Though conflicting views about bainite being beneficial or detrimental to toughness remain to be resolved, toughness improvement through acicular ferrite morphology in low carbon and V-microalloyed steels has been well documented[4,5]. However, little attention has been paid so far to assess the fatigue characteristics of these microstructural variants. The present work aims to demonstrate the short life strain cycling behaviour of a low carbon V-microalloyed steel continuously cooled from different reheating temperatures resulting variety of microstructure and consequent tensile properties.

MATERIAL AND METHOD

Chemical composition of the steel used in this investigation is given in Table 1.

Table 1. Steel composition in wt%

C	Mn	Si	S	P	V
0.18	1.10	0.24	0.020	0.022	0.13

Cylindrical specimens were reheated for austenitisation (T_R) at 900, 1150, 1250 and 1300°C for 30minutes and subsequently spray cooled using an air-water jet. Tensile properties have been evaluated on 16mm gauge length and 4mm gauge diameter samples in a closed loop servohydraulic universal testing machine of $\pm$ 50 KN load capacity, Instron 8501, using a crosshead velocity of 0.005 mm/s. Fully reversed low cycle fatigue tests were carried out at ambient temperature in total strain control mode using a triangular wave form and a constant strain rate of 10^{-2} s^{-1}. The tests were conducted in Instron 8501 using standard dynamic Instron extensometer with 12.5mm gauge length and $\pm$ 20% range on cylindrical samples of 15mm gauge length and 5mm gauge diameter.

RESULTS AND DISCUSSION

Microstructure evolution : Figs. 1 to 4 represent the variants of microstructure obtained through spray cooling from different austenitising temperatures. Up to 900°C coarsening of austenite grains remain absent and the resulting microstructure consists of very fine ferrite grains of average 10micron size and pearlite unresolved at the optical microscope level. Cabrera et. al.[6] has demonstrated that grain growth in V-bearing steels occurs beyond 950°C and growth rate becomes significant only after 1000°C. Samples austenitised at 1150°C and 1250°C revealed a mixture of degenerated Widmanstatten ferrite and bainite, though bainitic sheaves / plates could be clearly observed only in samples cooled from 1250°C. Microstructure obtained on cooling from 1100°C bears the imprint of mixed austenite grain size whereas at 1250°C austenite grain were predominantly course. The difference in the proportion and morhology of the microconstituents may be atributed to the difference in austenite grain structure at reheating temperature and rate of subsequent cooling. Cooling from 1300°C lead to the development of a highly heterogeneous microstructure containing regions of low carbon martensite in addition to bainitic sheaves and plates. Microhardness of the different phases were evaluated through a Wolpert-Instron V-4911 tester with 50 gm. load and the results are given in Table 2.

Table 2. Knoop Hardness (50gm. load) of the microconstituents

Austenitising Temp. °C	Bainitic Sheaves/Plates	Low carbon intragranular		Degenerated Widmanstatten Ferrite
		Martensite	Bainite	
1150	200	—	260	145
1250	200	—	260	160
1300	200	400	260	—

Tensile Behaviour :Monotonic tensile properties of the samples cooled from different reheating temperatures have been presented in Table 3. It may be noted that sharp yield drop has been noticed for the ferrito-pearlitic microstrucrture (T_R=900ºC) while it is conspicuously absent at higher austenitising temperatures . Bainitic steels are known to exhibit gradual yielding because of mobile dislocations and / or dispersion of relatively hard fine particles (carbides) acting as stress concentration[7]. As expected, Bainitic morphologies (T_R = 1150ºC and 1250ºC) resulted in higher Y.S. & UTS associated with lower % elongation than ferrite-pearlite though %RA remains comparable. However, reheating at 1250ºC produced lower UTS than 1150ºC which may be attributed to predominantly prior austenite grain size and resulting long bainitic plates/sheaves. Similar drop in YS at 1250ºC in V-microalloyed steels have also been observed by Jones et. al.[8]. Korchirsky[9] pointed out that lower reheating temperature of 1150ºC is advantageous especially

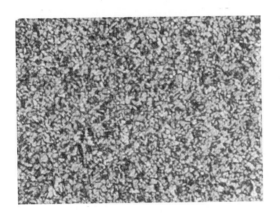

Fig. 1.Optical Micrograph for T_R = 900º C; X 150

Fig. 2.Optical Micrograph for T_R=1150ºC X 150

Fig. 3.Optical Micrograph for T_R = 1250º C; X 150

Fig. 4.Optical Micrograph for T_R=1300ºC X 150

in V-microalloyed steels. Cooling from still higher temperature ($T_R=1300^oC$) produced patches of low carbon martensite resulting sharp increase in strength parameters associated with loss of ductility. Drop in RA was very prominent.

TABLE 3. Monotonic Properties for different reheating / austenitising temperatures (T_R^oC)

Temperature ^oC	0.2 % Proof Stress (MPa)	Ultimate Tensile Strength (MPa)	% Total Elongation	% Reduction in Area
900	580.70 **	609.70	28.30	69
1150	619.80	852.20	13.10	56
1250	609.00	802.60	12.00	65
1350	780.60	1057.00	8.30	15

** Upper Yield Point

Low Cycle Fatigue Behaviour : Figs. 5-8 depict the strain life, Coffin-Manson and Basquin plots for samples cooled from different reheating temperatures. It may be noted that reheating at 1150°C resulted in highest fatigue life (Table-4) at all total strain ranges within the experimental level. Higher load bearing capability associated with high true ductility(ϵ_f) measured in terms of In (100/100-RA) resulting from a uniform fine dispersion of carbides in bainitic/acicular ferrite limits the possibility of local strain accummulation and enhances fatigue life. Ferrite-pearlite and bainitic steels (T_R = 900°, 1150°, 1250°C) followed Coffin-Manson and Basquin relationships but steel cooled from 1300°C exhibited bilinear characteristics typical of heterogenous microstructure[10] and/or very high elastic strain component[11], Basquin relationship, however, holds good. A comparative study of the Coffin-Manson plots reveal that for steels reheated at 900 and 1100°C fatigue life is controlled by plastic strain generated in the sample. Reheating to higher temperature results in increasingly inhomogeneous microstructure promoting localised strain accummulation and restrict fatigue life specially at lower strain levels. In addition, increased flow strain in samples cooled from 1300°C produces a predominant elastic strain component leading to bilinear behaviour. Table-5 includes the cyclic properties determined and some parameters calculated following selected models. Experimental values of fatigue ductility and fatigue strength coefficients compare closely with values calculated following Morrow as long as Coffin-Manson relationship was followed. However, samples reheated at 1250°C resulted significantly higher c and lower b values - indicators for lower fatigue life, especially at low strain ranges, though its high value indicate improved performance at short life regime. Cyclic strain hadening exponent (n') exhibits an inverse relation with reheating temperature with sharp drop beyond 1150°C.

Table 4. Comparative fatigue lives at different reheating/austenitising (T_R°C) temperatures

Austenitising Temperature	Reversals To Failure, $2N_f$					Transition Life in reversals $2N_t$
	Pct. Strain Amplitude,					
T_R° C	0.25	0.375	0.50	0.625	0.75	
900	65,328	20,000	7,690	2,088	—	13,794
1150	1,74,624	23,260	8,740	3,300	2,002	8,255
1250	1,07,876	17,844	6,160	3,430	1.012	3,433
1350	66,798	12,000	5,500	2,300	830	2,390

Table 5. Cyclic Properties for different reheating / austenitising temperatures

Cyclic Properties	Reheating / Austenitising Temperature (T_R° C)			
	900	1150	1250	1300
Strain Hardening Exponent, n'	0.17`	0.14	0.056	0.032
Fatigue Ductility Coefficient, ϵ_f'	0.16	0.20	1.28	0.026
Fatigue Ductility Exponent, c(-)	0.47	0.49	0.75	0.275
c : Morrow	0.54	0.59	0.78	0.86
c : Manjoine & Johnson	0.54	0.57	0.55	0.60
Fatigue Strength Coefficient, σ_f' MPa	800	914	830	1,186
Fatigue Strength Exponent, b(-)	0.083	0.071	0.046	0.085
b : Morrow	0.09	0.082	0.043	0.027
b : Manjoine & Johnson	0.09	0.09	0.09	0.09

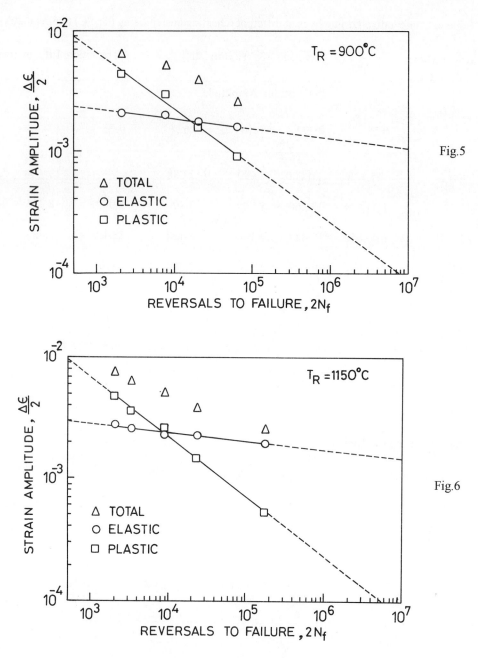

Fig.5

Fig.6

Fig.5   Total life and Coffin - Manson plot for $T_R = 900^\circ$ C

Fig.6   Total life and Coffin - Manson plot for $T_R = 1150^\circ$C

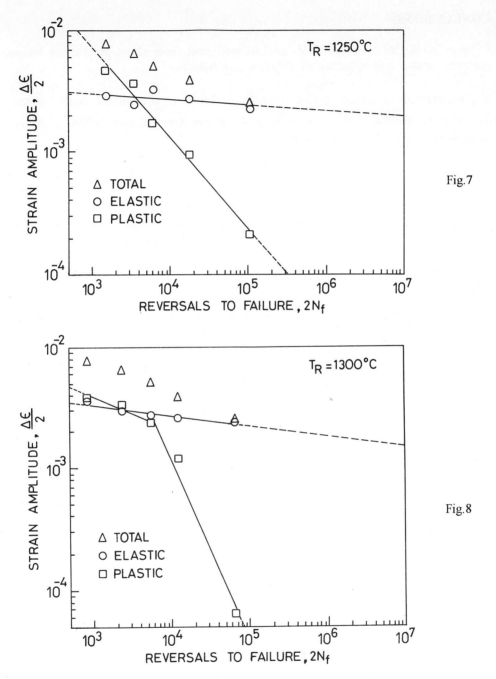

Fig.7

Fig.8

Fig.7   Total life and Coffin - Manson plot for T_R = 1250° C

Fig.8   Total life and Coffin - Manson plot for T_R = 1300°C

CONCLUSIONS

1. Spray cooling of low carbon V - microalloyed steel from elevated reheating temperatures produces bainitic microstructure of different morphologies.

2. Improvement in tensile strength upto T_R = 1250°C occurs without much impairment of true fracture ductility (ϵ_f) .Cooling from higher temperature (1300°C) causes sharp increase UTS with drastic loss in ϵ_f.

3. Fatigue life at all strain levels (up to 1.5 %) is higher for reheating temperatures of 1150°C.

4. Experimental values of LCF parameters b and c compare closely with those calculated from relationships suggested by Morrow as long as Coffin-Manson relationship is followed.

References:

1. S. Engineer and B. Huchtemann, Fundamentals and Applications of Microalloying Bar and Forging Steel, ed : C.J. Van Tyne et. al, TMS (1996), p 61 .
2. G. Krauss and S.W. Thomson, ISIJ International, 35 (1995), p137.
3. D. Drobnjak and A . Koprivica , Fundamentals and Applications of Microalloying Bar and Forging Steels, ed. C.J. Van Tyne et. al.,TMS(1996), p. 93.
4. M. Imagumbai et. al., HSLA Steels : Metallurgy and Applications, ed. J.M. Gray et. al., ASM (1985), p. 557.
5. P. Ishikawa and T. Takahashi, ISIJ International, 35 (1995), p. 1128.
6. J.M. Cabrera, et. al., Fundamentals and Applications of Microalloying Bar and Forging Steels, ed. C.J. Van Tyne et. al., TMS (1996), p. 173.
7. A. T. Joenoes et. al., Fundamentals and Applications of Microalloying Bar and Forging Steels, ed. C.J. Van Tyne et. al., TMS (1996), p. 517.
8. B.L. Jones et. al., HSLA Steels : Metallurgy and Applications, ed. J.M. Gray et. al., ASM (1985), p. 875.
9. M. Korchynsky, HSLA Steels : Metallurgy and Applications, ed. J.M. Gray et. al., ASM (1985), p. 251.
10. S.R. Mediratta et. al., Scripta. Met., 20 (1986), p. 555.
11. V.M. Radhakrishnan, Int. J. Fatigue, 14 (1992), p. 305.

Materials Science Forum Vols. 284-286 (1998) pp. 295-302
© *1998 Trans Tech Publications, Switzerland*

The Role of Carbon in Enhancing Precipitation Strengthening of V-Microalloyed Steels

S. Zajac, T. Siwecki, W.B. Hutchinson and R. Lagneborg

Swedish Institute for Metals Research, Drottning Kristinas väg 48, S-11428 Stockholm, Sweden

Keywords: V-Microalloyed Steels, Carbon, Nitrogen, Yield Strength, Precipitation Strengthening, Carbonitrides

Abstract

The present work has concentrated on the role of carbon in controlling the microstructure and strength of V-N structural steels containing 0.04-0.22%C.

Carbon content has usually been considered not relevant to precipitation strengthening when the precipitation occurs in ferrite because of the very small carbon content in solution in ferrite at equilibrium. We demonstrate that the effective carbon for precipitation in ferrite is much greater than this during the period of phase transformation, which in turn has a great effect on precipitation strengthening. Such behaviour is explained on the basis that the activity of carbon in ferrite is abnormally high in the presence of under-cooled austenite and before cementite nucleation so that profuse nucleation of vanadium carbonitride is encouraged. This new mechanism for precipitation is particularly significant for medium carbon steels typically used for hot rolled bars and sections.

The total carbon content of the steel also contributes to the yield strength by increasing the volume fraction of pearlite. It is shown that the contribution from pearlite may be stronger than generally recognised.

1   INTRODUCTION

The degree of precipitation strengthening of ferrite at a given vanadium content depends on the available quantities of carbon and nitrogen. Since the nitrogen contents of commercial structural steels, up to 200 ppm, are generally below the solubility limit in both ferrite and austenite, all the nitrogen is available for precipitate formation, whether this occurs homogeneously or by the inter-phase mechanism. This means that nitrogen is always present in a form where it can react efficiently with microalloy elements. It was confirmed that nitrogen is a very reliable alloying element, increasing the yield strength of V-microalloyed steels by some 6 MPa for every 0.001% N, essentially independent of processing conditions[1,2,3].

Carbon, on the other hand, is present in larger amounts but has only a very restricted solubility in ferrite. Thus, while the total carbon content might be expected to have some effect on inter-phase precipitation (due to the content dissolved in austenite) it would not be expected to influence homogeneous or random precipitation in ferrite[4]. Indeed, the effect of total carbon content on precipitation strengthening is usually considered negligible in microalloyed steels. In the following sections we shall show that this assumption is not correct, at least in V-microalloyed steels. These new findings are of major significance for the design of steel chemistry intended of hot rolled long products.

2 EXPERIMENTAL PROCEDURE

The investigation was carried out on twenty-one V-microalloyed steels with 0.04-0.22% C, 0.06-0.12% V and 0.0015-0.025% N. The chemical compositions and V/N ratios are given in Table I. They were divided into 6 groups; A - (0.10% C-0.12% V), B - (0.10% C-0.06% V), C - (0.04% C-0.12% V), D - (0.22% C-0.12% V), E - (0.22% C-0.06% V), and G - (0.15% C-0.12% V). All the steels were prepared as 2-kg laboratory ingots (15 x 120 mm in cross section) with a cooling rate during solidification designed to simulate commercial continuous casting.

The criteria for the determination of the heat treatment temperature were that the steels had the same austenite grain size (100 μm) after austenitisation and that the temperature was such that all vanadium (and nitrogen) had gone into solid solution for subsequent precipitation. The actual heat treatment temperatures that were adopted are also given in Table I. Specimens were austenitised for 10 minutes prior to isothermal transformation in salt baths maintained at the appropriate temperature in the range 550-700°C or continuous cooling at 0.5°C/s, 5°C/s or 15°C to room temperature. The holding time at the isothermal temperature was 100 and 500 s followed by cooling at 5°C/s to room temperature.

Table I Chemical composition of V microalloyed steels.

Steel	C	Si	Mn	P	S	Al	V	N	V:N 1/(3.64)	Taust °C
A5	.099	.36	1.38	.006	.008	.005	.129	.0051	6.9	1075
A8	.095	.40	1.38	.008	.010	.003	.120	.0082	4.0	1100
A14	.095	.40	1.38	.008	.010	.003	.120	.0140	2.3	1100
A25	.10	.37	1.36	.007	.009	.004	.130	.0257	1.4	1150
B5	.10	.38	1.39	.008	.010	.004	.056	.0056	2.7	1075
B12	.10	.38	1.39	.008	.010	.004	.056	.0120	1.3	1075
B25	.10	.38	1.40	.007	.010	.005	.062	.0250	0.7	1100
C5	.042	.38	1.39	.008	.010	.003	.122	.0055	6.1	1050
C9	.038	.39	1.39	.008	.010	.004	.119	.0095	3.4	1075
C15	.038	.39	1.39	.008	.010	.004	.119	.0150	2.2	1100
D1	.22	.40	1.37	.008	.014	.006	.120	.0011	109.0	1125
D6	.22	.40	1.41	.008	.015	.006	.120	.0059	20.3	1050
D12	.21	.39	1.39	.006	.014	.006	.122	.0117	10.2	1170
D13	.22	.39	1.38	.008	.016	.006	.119	.0126	9.5	1140
D19	.22	.40	1.40	.008	.015	.006	.120	.0192	6.3	1160
D22	.22	.40	1.40	.008	.013	.006	.123	.0222	5.4	1125
E6	.23	.39	1.38	.006	.013	.004	.056	.0065	8.6	1110
E12	.22	.40	1.39	.008	.016	.006	.058	.0115	5.0	1110
E24	.21	.40	1.38	.008	.014	.006	.089	.0236	3.8	1125
G13	.15	.40	1.43	.009	.014	.005	.120	.0136	8.8	1040
G23	.15	.40	1.39	.008	.016	.005	.128	.0234	5.5	1170

other: Nb<0.002, Ti<0.002 Cr<0.010, Ni<0.018, Mo<0.004, Cu<0.008

The ferrite-pearlite microstructures were examined by optical microscopy after etching in nital. Lower yield strength was evaluated on small compression specimens (φ 5 mm x 5 mm). Precipitate morphology and general structural effects were studied on carbon extraction replicas and thin foils prepared from the isothermally transformed or continuously cooled specimens. These were then examined in a transmission electron microscope (TEM) JEM 2000 EX operating at 200 kV.

3   RESULTS AND DISCUSSION

3.1   *Transformation Characteristic*

Transformation temperatures were determined by dilatometry at three different cooling rates, 0.5°C/sec, 5°C/sec and 15°C/sec and in all cases showed the expected reduction in Ar_3 with increasing cooling rate.   Results which are summarised in Fig. 1 show that a reduction of carbon content from 0.22 % to 0.04 % has a significant influence in raising the transformation temperature as is would be expected from the phase diagram.   There was also some tendency for the Ar_3 temperature to reduce at higher nitrogen levels, especially for faster cooling rates.   This may reflect the inherent austenite stabilising effect of dissolved nitrogen or, alternatively, some effect of V(C,N) precipitation on the kinetics of ferrite growth.

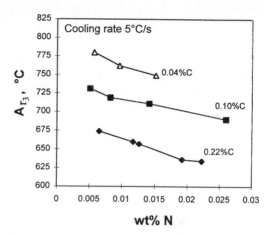

Fig. 1 Dependence of γ-α transformation temperature of 0.12% V steels on carbon and nitrogen contents.

The isothermal austenite to ferrite transformation was more rapid in low C steels than in high C steels, Fig. 2. A small influence of nitrogen content was seen during isothermal phase transformation in the range 550°C to 700°C. Additionally, from Fig. 2(b) it can be seen that the transformation time increases significantly with carbon content. The formation of pearlite was also delayed. The time for the onset of pearlite in the 0.22%C steel was ~3 times greater than for the 0.10%C steel.

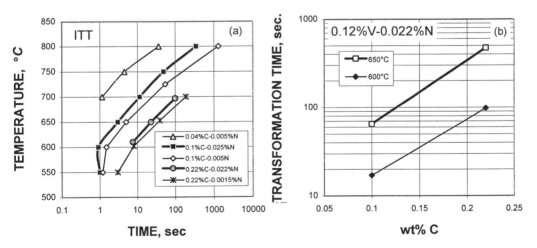

Fig. 2   Effect of carbon on transformation start time (a) and the transformation duration (b) of 0.12% V steels.

3.2   *Microstructures*

Ferrite-pearlite structures with varied pearlite volume fraction depending on the carbon content were characteristic for all the steels continuously cooled at a rate of 0.5°C/s. At 5°C/s the majority of the structure consisted of upper bainite or interlocking Widmanstätten ferrite in finely dispersed pearlite. For the fastest cooling rate of 15°C/s there was a very thin skeleton of ferrite at the prior austenite boundaries but most of the structure consists of lower bainite with some presence of martensite.

The structure of the isothermally transformed steels at 650°C consisted of pro-euctoid ferrite and pearlite, resembling quite closely that of the slowly cooled steel. The amount of primary ferrite decreased on transformation at lower temperatures and adopted a Widmanstätten morphology at 550°C. The remaining constituent was, however, pearlite in all cases.

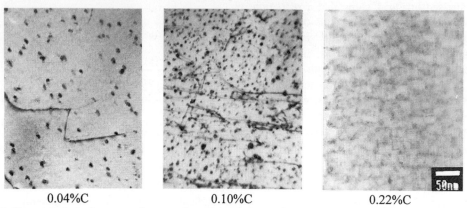

| 0.04%C | 0.10%C | 0.22%C |

Fig. 3 Transmission electron micrographs of precipitates in samples of 0.12%V-0.013%N steels with different carbon contents, isothermally transformed at 650°C.

Vanadium-rich particles were always visible in the ferrite, Fig. 3. This was true for polygonal ferrite, as well as pearlitic and bainitic ferrite. Interphase V(C,N) precipitation, formed at the γ/α interphase boundary has been observed at transformation temperatures 800-700°C. Randomly distributed VN particles predominated in samples transformed at 550-650°C at all carbon contents and no indication of interphase precipitation was found in 0.04-0.22% carbon steels transformed at 650°C and below.

The results of detailed measurements of particle sizes on steels having different carbon and nitrogen contents are summarised in Figure 4 after isothermal transformation at 650°C. As indicated in this figure both nitrogen and carbon caused a decrease in the particle size and at the same time an increase in volume fraction so giving the lowest interparticle spacing. For steel having 0.005% nitrogen and 0.1% carbon the particles were relatively sparse with an average diameter of ~ 11 nm. They became smaller and more dense with increasing nitrogen content, reaching ~ 6 nm at 0.022% nitrogen.

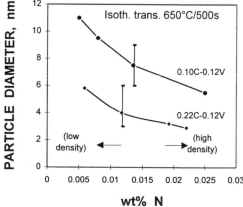

Fig. 4 Average size of V(C,N) at 650°C as a function of nitrogen and carbon contents.

An increase in carbon content from 0.10% to 0.22% caused significant refinement (about a factor of two) of the vanadium-rich particles at all nitrogen levels. These results clearly demonstrate that carbon plays an important role in precipitation of V(C,N).

3.3 *Mechanical properties*

Yield stresses for the different steels and conditions investigated here are summarised in Fig. 5 where results of continuous cooling treatments are shown in (a) and (b) and isothermal treatments in (c) and (d). There is a striking similarity between the figures 5(a), (b) and (c) where strength in plotted as a function of nitrogen content in the steels. In each case of carbon contents the yield strength increases approximately linearly as the nitrogen content is raised, and the slopes of these lines seem to be independent of the carbon content. The dependence on carbon, Fig. 5(d) is much more complicated and to a first approximation shows a parabolic behaviour. This effect of carbon may seem at first sight to be associated with the occurrence of pearlite. At low carbon contents this increase is apparently in agreement with expectations for the effect of pearlite. However, it is clearly shown below that the increase in the yield strength with carbon cannot adequately be explained by the pearlite contribution but rather is due to the influence of carbon on precipitation of V-rich particles. In order to analyse the effect of carbon on precipitation of V(C,N) and precipitation strengthening it is necessary to first evaluate contributions due to ferrite and pearlite so that these can be eliminated.

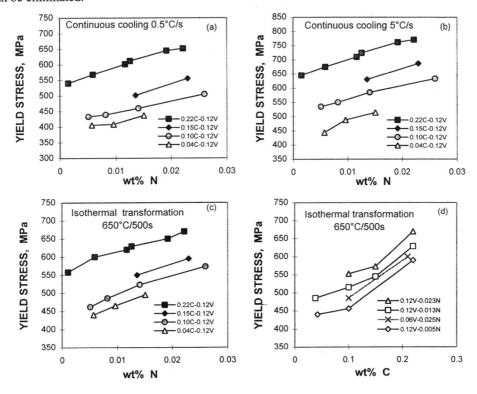

Fig. 5 Yield stress as a function of nitrogen and carbon contents after continuous cooling at 0.5°C/s (a) and 5°C/s (b) and after isothermal transformation at 650°C (c), (d).

3.4   Analysis of strengthening mechanisms

A usual practice when considering the strength of structural steels is to express the yield stress by a series of terms representing (i) the iron matrix, (ii) solid solution effects, (iii) grain size, (iv) precipitation hardening and in some cases also (v) pearlite content[1]. For higher carbon levels Gladman et al's[5] equation is often adopted where the contribution from pearlite is rather weak, i.e. $(1 - f_\alpha^{1/3})\, Re_p$ (where f_α is the volume fraction of ferrite and Re_p is the yield stress of pearlite).

However, an examination of published data for ferrite-pearlite steels[5,6,7] led to the conclusion that the pearlite contribution is much stronger and the most reasonable description of the contributions of pearlite and ferrite to yield stress was as a simple weighted mean, $Re_{C\text{-}Mn} = f_\alpha\, Re_\alpha + (1-f_\alpha)\, Re_p$ (where Re_α is the yield strength of ferrite). A new set of equations relating the yield strength with the chemical composition, ferrite and pearlite, developed for V-microalloyed steels with 0.04-0.25% C is given in Ref. 6.

The precipitation hardening component, ΔRp, of the present steels was obtained by subtracting the yield stress of the base steel, at the measured ferrite grain size and the contribution of pearlite (volume fraction and pearlite interlamellar spacing) from the observed yield stress at each transformation temperature and cooling rate.

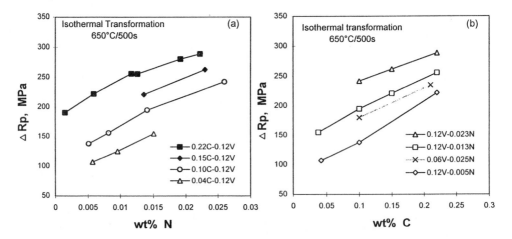

Fig. 6 Deduced values of precipitation strengthening for isothermally transformed V-steels.

Plots of precipitation strengthening against the nitrogen and carbon contents of the steels after isothermal transformation at 650°C are given in Fig. 6(a) and (b) respectively. Each of the six classes of steel chemistry (Table 1) shows a similar dependence of precipitation strengthening on carbon and nitrogen contents. The following remarks can be made concerning these effects;

• the amount of ΔRp increases approximately linearly with N over the ranges investigated ($\sim$ 6 MPa for every 0.001% N). This agrees with the well established effect of nitrogen on ΔRp of V-steels.

• the effect of carbon on ΔRp is unexpectedly strong. For the present steels the strengthening contribution of carbon is approximately 5.5 MPa per 0.01% C.

There seems to be no indication in the literature that the total carbon content of the steel will affect the precipitation strengthening in ferrite, especially when this occurs homogeneously after transformation. Therefore, in order to check that the present findings were not the result of an incorrect evaluation of the effect of the pearlite content, some further measurements were made using a nano-hardness indentor which could be localised purely within the ferrite grains. Between 20 and 50 ferrite grains were measured on samples having four combinations of nitrogen and carbon levels, and the results are given in Table 2. The values in Table 2 show that the strength of the ferrite does depend on the carbon level of the steel as well as the nitrogen level, which therefore confirms that the precipitation strengthening is indeed dependent upon the total carbon content.

Table 2 Average nano-hardness values for ferrite grains after isothermal transformation at 650°C.

Steel	C%	V%	N%	Average nano-hardness, GPa	Standard error ($\frac{\sigma}{\sqrt{N}}$)
D6	0.22	0.12	0.005	3.750	0.083
D22	0.22	0.12	0.022	4.250	0.142
A5	0.10	0.12	0.005	2.785	0.073
A25	0.10	0.13	0.025	3.629	0.054

At first sight there is no evident reason why the total carbon content should affect the precipitation of vanadium carbonitride since the solubility and activity of carbon in ferrite is defined by the iron-carbon equilibrium in the two-phase structure. However, during transformation on cooling the equilibrium phases change from being ferrite + austenite to being ferrite + cementite and there is a substantial change in the carbon solubility in ferrite. Fig. 7 shows the two solubility limits for carbon in ferrite calculated using the Thermo-Calc procedure[8]. At 600°C, for example, the solubility of carbon in metastable equilibrium with austenite is five times greater than the true solubility in equilibrium with cementite. Thus, during the initial stages of transformation the activity of carbon in ferrite is very large and profuse nucleation of V(C,N) will be favoured. The carbon activity will remain unchanged even if V(C,N) is being formed since the diffusion rate is very high and austenite provides an inexhaustible store.

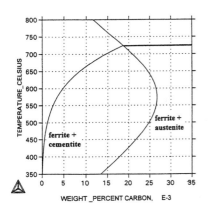

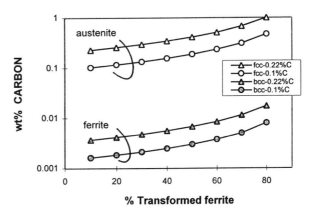

Fig. 7 Solvus lines for carbon in ferrite calculated using Thermo-Calc for equilibrium with cementite as well as austenite.

Fig. 8 Values of carbon contents in ferrite and austenite for two different steels calculated on the assumption that both phases are homogeneous.

In practice the metastable α-γ equilibrium solubilities may not be so large as shown in Figure 7 since diffusion of carbon in austenite occurs away from the interface tending to lower the concentration there. This will be accentuated if the interface migration is hindered by some effect such as solute drag of substitional elements. In the extreme case where boundary mobility is very low and full equilibration of carbon can take place both in ferrite and austenite the new metastable carbon levels can be calculated with the aid of Thermo-Calc (activities of carbon in ferrite and austenite are equal). Figure 8 shows such calculation for two different carbon contents and demonstrates that the level of carbon dissolved in ferrite which governs the nucleation of V(C,N) is indeed a function of the total carbon content. The above situation continues until pearlite is nucleated at which point a new equilibrium with a reduced carbon activity is established.

As the total carbon level of the steel increases, the kinetics of the $\gamma \rightarrow \alpha$ phase transformation are progressively retarded since more time is required for diffusion of carbon in the shrinking austenite. Not only is the carbon activity in ferrite higher but also the period of time during which the ferrite is supersaturated with carbon is prolonged, so allowing more nucleation of V(C,N) particles and accordingly a more dense precipitation. In these ways the positive influence of total carbon content on the precipitation strengthening contribution can be understood.

4 CONCLUSIONS

1. The degree of precipitation strengthening of ferrite at a given vanadium content depends on the available quantities of carbon and nitrogen. The nitrogen content of the ferrite is approximately the same as that of the austenite from which it forms, i.e. the total nitrogen content in steel. It was confirmed that nitrogen is a very reliable alloying element, increasing the yield strength of V-microalloyed steels by some 6 MPa for every 0.001% N, essentially independent of processing conditions.

2. The contribution from carbon was found to depend on the metastable carbon content of ferrite in equilibrium with undercooled austenite as well as on the transformation characteristics. The higher the total carbon content of the steel, the higher is the metastable carbon content of ferrite. Furthermore, increase in the carbon content of steel prolongs the time of this metastable condition. In consequence, for a given vanadium content greater precipitation strengthening is observed in steels with a higher carbon content.

3. The effect of carbon in enhancing precipitation strengthening is particularly significant in steels with medium carbon contents such as in long products where the metastable condition is more extreme and prolonged. Raising the carbon content in the range 0.10% to 0.22% produces a significant increase in the precipitation strengthening contribution in addition to the well known effect of pearlite on the yield strength.

5 ACKNOWLEDGEMENTS

The present endeavour has been supported financially by the U.S. Vanadium Co. /Stratcor/. It is a pleasure to acknowledge the stimulating discussions with Michael Korchynsky of that organisation.

6 REFERENCES

[1] K.J. Irvine, F.B. Pickering and T. Gladman, JISI, Vol. 205 (1967) pp. 161-182.

[2] S. Zajac, T. Siwecki and M. Korchynsky, Proc. Int. Conf. "Low Carbon Steels for the 90's", Eds. R. Asfahani and G. Tither, ASM/TSM, Pittsburgh, USA, (1993) 139-150.

[3] S. Zajac, R. Lagneborg and T.Siwecki, Proc. Int. Conf. "Microalloying'95", Pittsburgh, PA, ISS (1995) pp. 321-340.

[4] W. Roberts, Proc. Int. Conf. on HSLA Steels - Technology and Appl., ed. M. Korchynsky, ASM, Ohio, (1984) pp. 67-84.

[5] T. Gladman, I.D. McIvor and F.B. Pickering, J.Iron Steel Inst., Vol. 210, (1972) pp. 916-930.

[6] S. Zajac, T. Siwecki, W.B. Hutchinson and R. Lagneborg, in press.

[7] H.J. Kouwenhoven, Trans., ASM, Vol. 62, (1969) pp. 437-448.

[8] B. Sundman, B. Jansson and J-O. Andersson, Proc. Conf. User Applications of Alloy Phase Diagrams, ed. L. Kaufman, ASM, Lake Buena Vista, Florida (1986).

Materials Science Forum Vols. 284-286 (1998) pp. 303-310
© *1998 Trans Tech Publications, Switzerland*

Hydrogen Induced Cracking Processes in Structural Microalloyed Steels. Characterization and Modelling

J.A. Alvarez Laso and F. Gutiérrez-Solana

Departamento de Ciencia e Ingeniería del Terreno y de los Materiales
E.T.S. de Ingenieros de Caminos, Canales y Puertos, Universidad de Cantabria,
Avenida de Los Castros s/n, E-39005 Santander, Spain

Keywords: Hydrogen Induced Cracking, Microalloyed Steels, Behaviour, Modelling

ABSTRACT

Microalloyed steels are of great importance in the petroleum industry where they are used in structural and tubular components in pipelines and platform supports. The high toughness they provide and their mechanical resistance to stress corrosion cracking (SCC) processes make these steels extremely useful in the above mentioned fields. Most local environmental conditions present in these types of structures make the high resistance to cracking of these steels unsuitable for characterization using conventional methods based on Linear Elastic Fracture Mechanics.

Once an experimental and analytical methodology based on Elastic-Plastic Fracture Mechanics has been developed [1,2] and shown to be suitable for the characterization of cracking processes in environmental conditions, it has been applied to characterize hydrogen induced cracking (HIC) in two different structural microalloyed steels, offering a quantitative characterization of the cracking behaviour which can be correlated with the fracture micromechanisms that generate the fracture path.

1. INTRODUCTION

The steels used in piping systems and structural components, such as those in petroleum installations or off-shore platforms, have been continually evolving over recent years. This evolution has been aimed at improving the following characteristics: their mechanical behaviour in order to reduce the amount of material used to optimise transport conditions; their toughness to avoid problems caused by brittle fracture in aggressive working environments and temperatures; their resistance to deterioration, including subcritical cracking processes due to fatigue and/or environment; their workability, guaranteeing good weldability for placement; and their cost. Due to the enormous amount of material necessary for fabricating these installations, the increasing demands made on the steels have been accompanied by steadily increasing minimum performance qualities to guarantee safe working conditions [3].

Microalloyed steels with high yield strengths are those which are most commonly used in modern applications with structural responsibility. For example, they are used in 70% of the high strength components in offshore platforms [4]. Their popularity is based on their price/strength ratio, toughness, processing versatility and good weldability which is as good as, or better than, that of carbon steels [5]. This performance has been obtained through the optimization of alloying composition and thermochemical treatments [6,7].

Under working conditions such as in off-shore platforms these steels are subjected to stress corrosion cracking (SCC) problems in marine environments associated with situations involving cathodic protection or sulfide induced cracking due to bacterial activity [8]. In both cases hydrogen plays a fundamental role in the cracking mechanisms and therefore these steels should be shown to be resistant to HIC. The objective of the present work is to thoroughly characterize the behaviour of two families of microalloyed steels in view of HIC phenomena which commonly appear in their off-shore uses.

2. MATERIALS AND CONVENTIONAL CHARACTERIZATION

The materials to be characterized are two microalloyed steels taken from industrial heats. Their chemical composition is presented in Table 1. The first steel, E690, has a high yield strength and is used in the manufacture of self-elevating lift belts for oil platforms. The second, E500, is a steel with a medium yield strength and is used in the manufacture of semi-tubular stabilizers for the above-mentioned lift belts to which they are welded. Both steels are subjected to water quenching, later tempered at 600ºC (E690) or 650ºC (E500), and finally air cooled. These treatments lead to a tempered bainite microstructure in both steels.

The results of the mechanical characterization performed on both steels using hardness and tensile tests are summed up in Table 2. The fractographic study performed on the fracture surface of the tested tensile strength samples shows that in both steels fracture occurs because of microvoid coalescence. The toughness of both materials was characterized by determining the R curve of the J integral following the method proposed by the European Group of Fracture ESIS P1-92 Normalised 25 mm thick compact CT specimens were machined from the available material. Tests were performed on each material at two loading rates under displacement control, v_d, at $4.1 \cdot 10^{-6}$ and $4.1 \cdot 10^{-7}$ m/s. The results show that the materials possess a high toughness which is independent of loading rate. The toughness is much higher in the E500 steel for which $J_{0.2/BL}$ reaches a value of 1017 kJ/m^2 compared with 259 for the E690 steel.

Table 1. Chemical Composition of steels E690 and E500 (% by weight)

Steel	C	Si	Mn	Ni	Cr	Mo	Cu	Sn	Al	V	Ti
E500	0.063	0.23	1.36	0.585	0.115	0.195	0.103	0.003	0.017	0.048	<0.003
E690	0.135	0.241	1.1	1.518	0.496	0.465	0.18	0.009	0.078	<0.003	0.003

Table 2. Characteristic tensile parameters of the different materials

Steel	E500	E690
s_y(MPa)	530	840
s_u(MPa)	640	915
e_{max} (%)	9	6.5

3. BEHAVIOUR IN HYDROGEN INDUCED CRACKING CONDITIONS

HIC processes were characterized by testing 25 mm thick CT specimens in an aggressive environment at a constant loading rate. The aggressive environment was obtained by cathodic polarization at different current densities in a solution of 1N of H_2SO_4. The samples were subjected to a process of continued polarization for 40 hours before and throughout the mechanical test. Crack propagation occurred at this stage and finally led to the specimen's breakage. A later SEM fractographic study revealed the mechanism behind the cracking process.

By applying a suitable analytical methodology [1,2], based on the EPRI procedure [9] for characterizing the behaviour of cracked components in elastic-plastic regime, at any moment it is possible to know the crack length values, a, the crack propagation rate, da/dt, and the J integral. Fig. 1a shows the load-displacement behaviour curve for one of the tests and both Figs. 1a and 1b the characterization obtained from it. The equivalent stress intensity factor, K_J, has been determined from the J integral value. The threshold conditions, K_{th}; the subcritical propagation rate, da/dt_{sc}; the critical fracture condition initiation value, K_c; and the critical propagation rate, da/dt_c, are all defined in Fig. 1b.

Tests were performed under different aggressive conditions by varying both the polarizing current density and the loading rate. Table 3 presents all the results obtained.

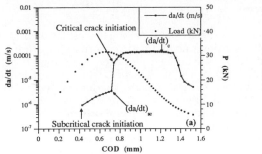 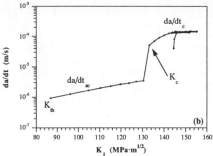

Fig. 1 Characterization of the cracking process through the da/dt-K$_J$ (b) curve obtained from the load displacement curve (a).

Table 3. Characteristic parameters for the HIC processes tested in the two steels

Steel	Current density (mA/cm²)	Displacement rate v$_d$ (m/s)	K$_{th}$ (MPa m$^{1/2}$)	(da/dt)$_{sc}$ (m/s)	K$_c$ (MPa m$^{1/2}$)	(da/dt)$_c$ (m/s)
E500	5	4.1·10^{-7}	105	2·10^{-7}	160	3·10^{-6}
E500	5	8.2·10^{-8}	85	2·10^{-7}	155	6·10^{-7}
E500	5	4.1·10^{-8}	80	8·10^{-8}	142	4·10^{-7}
E500	5	4.1·10^{-9}	68	2·10^{-8}	123	1·10^{-7}
E500	5	8.2·10^{-10}	64	7·10^{-9}	100	1·10^{-7}
E690	1	4.1·10^{-8}	120	5·10^{-8}	260	8·10^{-8}
E690	1	8.2·10^{-10}	70	3·10^{-8}	85	3·10^{-7}
E690	5	4.1·10^{-7}	87	4·10^{-7}	132	5·10^{-5}
E690	5	4.1·10^{-8}	63	2·10^{-7}	120	1·10^{-6}
E690	5	4.1·10^{-9}	55	2·10^{-8}	81	2·10^{-7}
E690	5	8.2·10^{-10}	39	1·10^{-7}	44	3·10^{-6}
E690	10	4.1·10^{-7}	41	2·10^{-6}	90	1·10^{-5}
E690	10	4.1·10^{-8}	35	6·10^{-7}	47	4·10^{-5}
E690	10	4.1·10^{-9}	40	1·10^{-6}	52	1·10^{-4}

4. ANALYSIS: INFLUENCE OF ENVIRONMENTAL CONDITIONS ON THE BEHAVIOUR OF BOTH STEELS AGAINST HIC PROCESSES

The results analysis has been done based on the representation of the parameters which define the behaviour of both subcritical and critical cracking processes in both steels for each environmental condition studied as a function of the loading rate of each test. The areas covered by each type of micromechanism involved in the cracking process are represented in Figs. 2 through 9, in accordance with the SEM observations made of each specimen: intergranular, transgranular by cleavage and transgranular by microvoid formation and coalescence.

4.1 E690 Steel at a **Current Density of 1 mA/cm²**

Figs. 2 and 3 schematically show the variations of the behavioural parameters established in this work based on the K_I and da/dt variables, at the loading rate and hydrogen charging studied in this section, i.e., 1 mA/cm². The behavioural conditions in air have also been established to act as a reference. Subcritical cracking is produced under cleavage mechanisms, which at low loading rates are found together with those of intergranularity. Critical cracking conditions, generated by microvoids, are present for the tests at greater loading rates. The areas corresponding to subcritical and critical cracking, the latter established under hydrogen embrittlement conditions, are thereby separated. Fig. 2 shows the effect of environment on the cracking processes controlled by mechanisms associated with the formation of microvoids. The kinetics of this process are ten times greater in the presence of hydrogen and can be associated with their lower final deformation. By linearly extrapolating the crack propagation rate due to microvoids as a function of loading rate [10], owing to purely mechanical effects, it can be understood that critical phenomena will not be produced during the test under a displacement rate of 10^{-8} m/s because they will take place at lower rates than those of the subcritical cleavage propagation mechanisms typical of a bainitic structure, therefore these latter are always present. Fig. 3 shows the different areas stratified as a function of the local loading state, represented by the stress intensity factor K_I. It can also be seen how a local loading state above that provided by a stress intensity factor of 250 MPa.m$^{1/2}$ is needed in order to reach fracture caused by mechanisms associated with the formation of microvoids. Also, intergranular mechanisms can only appear below a given factor, 80 MPa.m$^{1/2}$, coinciding with the previous modelling of HIC processes in steels [11]. The intermediate local loading state under these environmental conditions provokes cleavage mechanisms to establish the cracking processes, with a greater presence of tearing as local loading, and thereby plasticity, increases.

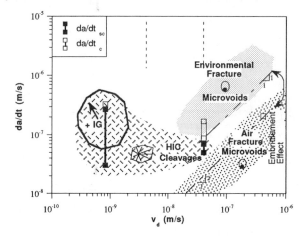

Fig. 2. Map of fracture micromechanisms as a function of crack propagation rate and loading rate in the E690 steel tested at 1 mA/cm²

4.2 E690 Steel at a **Current Density of 5 mA/cm²**

Figs. 4 and 5 show the behavioural graphs at a current density of 5 mA/cm² as a function of the displacement rate used in the tests. The same zones as in the previous case can be seen, although both subcritical and critical propagation is produced at a greater velocity than in the previous case, as well as under local conditions of lower loading, corresponding to the greater embrittling effect of the environment. As can be seen in Fig. 4, the crack propagation rates corresponding to critical processes due to microvoid formation, deduced from the value obtained from the highest loading rate show the embrittling effect of environment in these tests. In the presence of the created environment the propagation rate is more than 100 times higher than that observed in air. It was also observed that the values extrapolated from higher loading rates are close to, although higher than, those observed for the rate of $4.1 \cdot 10^{-8}$ m/s, at which the cracking observed in the area of supposed instability presents cleavage mechanisms together with microvoids. In other words, it is found, as shown by the figure, in the transition zone from subcritical cleavage to critical microvoids. It can also be observed how the subcritical mechanisms of the slowest specimens are quicker than the critical values supposed for the formation of voids, by which said voids never develop under these conditions. Fig. 5, like Fig. 3, shows the cracking micromechanism stratification, with K_J, but, due to the higher embrittlement, lower K_J values are needed at each area for these environmental conditions.

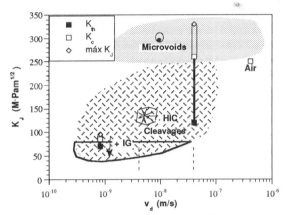

Fig. 3. Map of fracture micromechanisms as a function of local loading and loading rate in the E690 steel tested at 1 mA/cm²

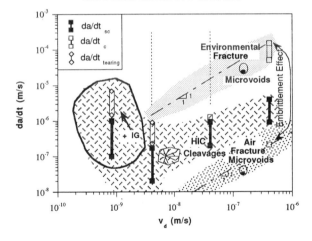

Fig. 4. Map of fracture micromechanisms as a function of crack propagation rate and loading rate in the E690 steel tested at 5 mA/cm²

4.3 E690 Steel at a Current Density of 10 mA/cm²

The overall analysis of the results obtained from the E690 steel tested at 10 mA/cm² is shown in Figs. 6 and 7. Given the high embrittlement imposed by this environment, the material's behaviour as a whole depends little on loading rate. Fig. 6 shows similar subcritical cracking for all three cases, a propagation rate of around 10⁻⁶ m/s typical of cleavages with traces of intergranularity. Instability can be associated to a critical cleavage growth process which is later joined by tearing, for the higher loading rates, or coalescence of voids, for the slowest, typical of a situation reached under a practically elastic regime. Compared to breakage in air, formed by classic microvoids coalescence, the strong embrittling effect of the aggressive environment multiplies the crack propagation kinetics by three or four orders of magnitude. Fig. 7 once again shows the stratification related to the stress intensity factor of the different micromechanisms typical of the cracking process.

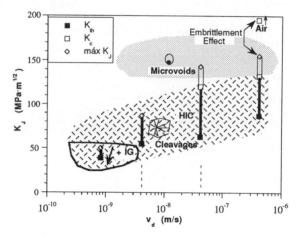

Fig. 5. Map of fracture micromechanisms as a function of local loading and loading rate in the E690 steel tested at 5 mA/cm²

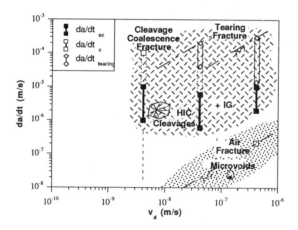

Fig. 6. Map of fracture micromechanisms as a function of crack propagation rate and loading rate for the E690 steel tested at 10 mA/cm²

4.4. E500 Steel at a Current Density of 5 mA/cm²

Figs. 8 and 9 show the evolution of the parameters which characterize the behaviour of this steel for this environmental condition, clearly showing the greater toughness of this steel when compared with the E 690: slower propagation rates and the absence of intergranularity. From both figures it can be deduced that the presence of the environment impedes final breakage taking place by microvoids coalescence: on the one hand the necessary local stress state is not reached, and on the other the subcritical propagation has kinetics greater than expected for the microvoids, given the small embrittling effect this environment has on this steel and the kinetics observed for said voids in crack propagation in air. At the same time one can observe a small effect of loading rate on propagation rate when the fracture mechanisms generate cleavages. The kinetics at which these cleavages are produced impede the development of unstable tearing which would, at first sight, take place at slower rates. The lack of intergranular fracture justifies the trend towards stability in the behaviour as loading rate decreases both at threshold levels from which propagation is produced and in propagation rate.

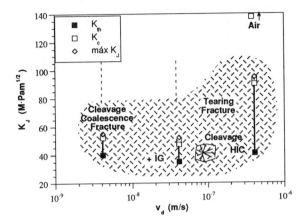

Fig. 7. Map of fracture micromechanisms as a function of local loading and loading rate for the E690 steel tested at 10 mA/cm²

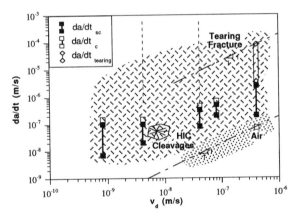

Fig. 8. Map of fracture micromechanisms as a function of propagation rate and loading rate for the E500 steel tested at 5 mA/cm².

CONCLUSIONS

The microalloyed steels with a bainitic structure tested have been shown to be susceptible to presenting HIC processes. This susceptibility follows the rules of a classic dependence:

- On material: apparently those with higher yield strengths are more susceptible when the comparison is made using similar microstructures.

- On environmental aggressiveness: as the concentration of hydrogen present in the material grows its susceptibility to cracking increases and this cracking is produced following more brittle mechanisms.

- On loading rate: which affects the materials susceptibility in given environmental conditions, not through changes in the fracture micromechanisms but through the parameters which define the typical cracking conditions and their kinetics.

The combined action of a given stress state and loading rate determine the time which hydrogen is active in the crack tip process zone, provoking different types of cracking: microvoids coalescence, cleavage, or intergranular. Fracture type depends on hydrogen concentration and on the applied stress intensity factor. These two parameters were varied during the tests producing, as has been demonstrated, changes in the type of crack propagation in the same specimen.

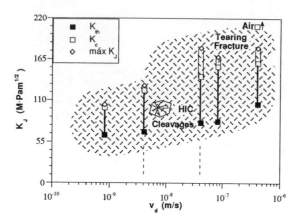

Fig. 9. Map of fracture micromechanisms as a function of local loading and loading rate for the E500 steel tested at 5 mA/cm^2.

ACKNOWLEDGEMENTS

This work is part of the ECSC programme 7210-KB/934 funded by the Spanish Interministerial Commission of Science and Technology by the agreement CICYT MAT 93-0970-CE.

REFERENCES

[1]   J. A. Alvarez, *Aplicación de la Mecánica de la Fractura Elastoplástica a Procesos de Corrosión Bajo Tensión*. Doctoral Thesis, University of Cantabria, (1998)

[2]   F. Gutiérrez-Solana, J. A. Alvarez, in *An. Mec. de la Fract.*, Vol. 14, (1997), pp. 50-68

[3]   H. Chino, M. Abe, K. Katayama, H. Takemiro and H. Akazaki: in *Pipeline Technology Conference*, Oostende, Belgium, (1990), Part A, pp. P.4.1.

[4]   *High-Strength Structural and High-Strength Low-Alloy Steels*, ASM. Metals Handbook, Vol. 1, Tenth Edition, (1990), pp. 388.

[5]   *Metals and their weldability*, Welding Handbook, Vol. 4, 7th Edition. American Welding Society, (1982), pp. 24.

[6]   P.E. Repas: *Metallurgical Fundamentals for HSLA steels in Microalloyed HSLA steels*, ASTM International, (1988), pp. 5.

[7]   E.E. Fletcher: *High-Strength-Low-Alloy Steels: Status, Solution and Physical Metallurgy*, Batelle Press, (1979).

[8]   G. Gabetta and I. Cole: *Fatigue Fract. Engng. Mater. Struct.*, Vol. 16, No. 6, (1993), pp. 603.

[9]   V. Kumar, M.D. German and C.F. Shih: *An Engineering Approach for Elastic-Plastic Fracture Analysis*, General Electric Company, NP-1931, Research Project 1237-1, Topical Report, Schenectady, New York, (1981).

[10]   J. A. Alvarez, G. Méndez, F. Gutiérrez-Solana, I. Gorrochategui and J. Laceur, in *An. Mec. de la Fract.*, Vol. 11, (1994), pp. 413-419..

[11]   F. Gutiérrez-Solana, A. Valiente, J.J. González and J.M. Varona: *Metall. Trans. A*, Vol. 27A, (1996) , pp. 291.

Materials Science Forum Vols. 284-286 (1998) pp. 311-318

Precipitation of NbC and Effect of Mn on the Strength Properties of Hot Strip HSLA Low Carbon Steel

V. Thillou[1,2], M. Hua[1], C.I. Garcia[1], C. Perdrix[3] and A.J. DeArdo[1]

[1] BAsic Metals Processing Research Institute, University of Pittsburgh, PA, USA

[2] Now with EUROPIPE Dunkerque, France

[3] CRDM, SOLLAC Dunkerque, France

Keywords: Niobium, Manganese, Precipitation, Strengthening, HSLA, Thermomechanical Processing, Yield Strength, Modeling, Hot Strip

ABSTRACT:

Two low-carbon microalloyed steels were investigated: 0.07C-0.02Nb-0.33Mn and 0.07C-0.028Nb-1.1Mn. The two steels were hot rolled in a similar way, followed by either air cooling to room temperature (ACRT) or accelerated cooling (AC) to 680°C and held isothermally at 650°C for various times prior to ACRT. The mechanical properties of the steels in their fully-processed condition were evaluated. As expected, the steel with the higher Mn content showed the highest tensile (YS and UTS) properties for all processing conditions. The results of this study also indicated that the tensile properties of both steels remained unchanged, prior to and after the isothermal treatments. The central goal of this study was to conduct a detailed microstructural analysis which could help in the understanding of the observed mechanical properties. The results from the ACRT samples after hot rolling revealed two distinct precipitation sequences: (i) precipitation of fine NbC (5 nm in average size) in austenite, and (ii) interphase and multivariant precipitation of NbC in ferrite. The samples which were AC and then held isothermally also exhibited fine NbC precipitates. All the precipitates observed in these samples had formed in austenite. The absence of the Baker-Nutting (B-N) orientation relationship clearly confirmed that precipitation in ferrite did not take place. Atom Probe analysis of the precipitates revealed that the stoichiometry of the precipitates was close to $NbC_{0.9}$. From the detailed TEM microstructural analysis, all the hardening increments contributing to the yield strength of the steels were computed the results showed a good agreement with the measured tensile properties.

INTRODUCTION

The philosophy of thermomechanical processing (via controlled rolling and cooling) is to properly condition the structure of the austenite to induce the formation of fine ferrite through the γ/α transformation. The presence of microalloying elements in the steel can greatly enhance this microstructural control by their precipitation prior, during or after the γ/α transformation. In this regard, Nb can play simultaneously a multiple role: (i) as initial inhibitor of austenite grain coarsening during the reheating of the slabs, (ii) as austenite recrystallization modifier during rolling, and (iii) as precipitation hardener in conjunction with the γ/α transformation, i.e. by controlling the hardenability of the steel. The roles of Nb are summarized in Fig. 1. Most of the current knowledge on the mechanisms of the kinetics of Nb precipitation and its influence on mechanical properties as depicted in Fig. 1 have been based on studies focused on either the processing of plates and/or steels with high carbon and V and/or Nb additions.

In recent years, several authors have indicated that the precipitation behavior which results from the processing of high strength microalloyed (MA) strip steels might be different than that observed in plate steels. For example, some authors reported that after the hot rolling and coiling of low- carbon, low-alloy steels, lower than expected volume fractions of NbC are formed prior to and after the γ/α transformation [1]. They also have indicated that fine NbC precipitation in austenite may have a contributing effect to the final strength of the steel [1,2]. The present study was directed to assess the precipitation behavior in two high strength MA strip steels with different Mn levels. Of particular interest in this study was to differentiate the precipitation that was formed prior to and after the γ/α transformation and to characterize them quantitatively. In addition to the precipitation, other microstructural features such as the dislocation density and grain size were also measured. The quantitative information from these individual microstructural features was used to calculate the strength properties and to compare these to the measured strength properties of the steels. The results of this analysis will be presented and discussed in this paper.

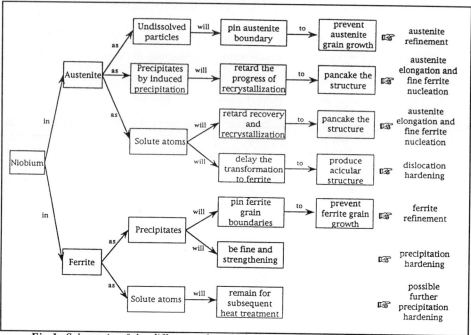

Fig.1: *Schematic of the different roles of Nb during thermomechanical-processing.*

EXPERIMENTAL PROCEDURE

Material. The steels used in this investigation had nominal compositions (wt%) of 0.070C-0.33Mn-0.020Nb-0.004N for steel A and 0.070C-1.1Mn-0.025Nb-0.004N for steel B. These steels were continuously cast and processed into roughed slabs by Sollac.

Processing. The samples used in this study were rolled from 70mm down to 10mm on the experimental hot rolling mill in CRDM, Sollac, France. The hot strip mill processing schedule was as follows: (i) reheated at 1200°C, roughing in three passes from 1140°C to 1030°C, and (ii) finishing in three passes from 928 to 892°C. The specimens were then either air cooled or accelerated cooled to 680°C with a cooling rate of 50°C/s, held at 650°C for 0, 2 or 5 hours and then ACRT. In addition, the transformation temperature, Ar_3, during ACRT was measured to be 835°C for steel A and 790°C for steel B. Specimens from the as-processed steels were sectioned for mechanical testing and microstructural analysis. Sections parallel to the rolling direction from these specimens were machined into standard tensile samples and tested at CRDM.

Optical metallography. The specimen preparation for optical examination involved standard metallographic procedures. These samples were etched with 2% Nital to reveal the microstructure. The stereological parameters of the ferrite grains were investigated by measuring etched and sectioned grain areas with a digitizing tablet linked to a computer controlled image analysis system BioQuant IV for data acquisition and processing.

Electron microscopy. Thin foils were prepared and examined using a JEM 200CX transmission electron microscope operating at 200 Kv. The specimen preparation involved mechanical thinning to a thickness of about 0.2mm and chemical electropolishing (distilled water + H2O2 30%vol + HF // distilled water + HNO3 + HCl + HF). Final foil specimens were prepared from punched discs by twinjet polishing at 50V and 45mA in acetic acid at room temperature. Carbon extraction replica specimens were not used due to the lack of reliability of the extraction efficiency and the impossibility to determine the exact origin of the precipitates [3,4]. In this study, it was essential to use tilting techniques for all the work on the TEM not only to obtain the proper contrast but also to avoid false interpretation of the results and to insure the nature of the precipitation. Specific settings needed for certain imaging conditions could not be carried out routinely but required many attempts and very good foils.

The strain-induced precipitates of NbCN in austenite have a simple cube-cube lattice relationship with austenite: $[100]_{NbCN}$ // $[100]$ & $[010]_{NbCN}$ // $[010]$. The austenite transforms to ferrite with a Kurdjumov-Sachs (K-S) relationship. This allows low energy, immobile interface to develop between fcc and bcc. Ferrite will exhibit that rational orientation relationship to at least one gamma grain, and have a random relationship with the other gamma grains. Consequently, the precipitates formed in austenite would be related to the ferrite by the K-S relationship when observed at room temperature. Because of the multivariance in nature of that K-S relationship, it is very difficult to attribute, without doubt, an orientation relationship to the K-S family.

Niobium precipitates can nucleate in ferrite in two different ways: the interphase precipitation that takes place along with the transformation of γ and the precipitation in the ferrite matrix from a so-called super-saturated ferrite. Ferrite nucleated precipitates can be identified from the orientation relationship between the ferrite and the precipitates reflections in the electron diffraction patterns. NbCN precipitates in the ferrite with the Baker-Nutting (B-N) relationship: $(100)_{NbCN}$ // $(100)_{ferrite}$ & $<100>_{NbCN}$ // $<110>_{ferrite}$ [5]. Although both types of precipitates nucleated in ferrite obey the B-N relationship, interphase precipitates nucleate on only one of the three possible ferrite cube planes as a result of a preferred nucleation effect at the γ/α interphase boundary. The crystallographic orientation adopted will be the one that allows good matching with both phases reducing the interface energy, reducing the free energy of activation for the formation of critically sized nuclei and maximizing diffusion [6]. Thus, because of the trivariant nature of supersaturated precipitation in ferrite, only one third of the total number of precipitates in a given area can be illuminated by a single set of precipitate reflections [7,8]. Superimposed stereographic projections for the different phases studied with their well-defined relationships were used to aid in the interpretation of the diffraction patterns [9].

The quantitative assessment of the dislocation density was done by measuring from a series of TEM micrographs the total projected length of dislocation line per unit area. The dislocation density is taken as $\rho = R / (A.t) = 2N / (L.t)$ with units cm/cm^3, where A is the area of the foil containing the dislocation line length R, t is the foil thickness, N is the number of intersections with dislocations by random lines of length L per unit area A [4,10]. In this study Hilliard's method was used; this consists of concentric circles superimposed randomly on illuminated negatives to avoid any printing contrast errors. Fringes of thickness extinction contours in the two beam diffraction conditions were used to determine the foil thickness [11,12]. The dimensions of the particles were measured from high resolution centered dark field micrographs. Typically around 200 particles in austenite and 700 to 2500 in ferrite were quantitatively measured from each sample condition. Measurements for particle size or interphase row spacings were made by viewing through a magnifying eye piece marked with grids the original micrographic negative plates on a lighted table. Row spacing measurements were made averaging around 150 counts per condition.

Atom probe/Field ion microscopy. APFIM was used to determine both the composition of the Nb precipitates and the distribution of soluble Nb in the structure. This technique is a combination of a field ion microscope and a mass spectrometer of single ion sensitivity. In the FIM, positively charged gas ions generated by the process of field ionization are used to produce images of the atoms on the surface of a solid specimen. Successive atom layers of material may be ionized and removed from the specimen surface by the process of field evaporation. This enables the three dimensional structure of the material to be imaged in atomic detail and also provides the source of ions for analysis by mass spectrometry. Identification of the individual imaged atoms becomes possible. The specimen of the material studied is prepared in a needle-like form with an end radius of typically 50 to 100 nm. Details of the sample preparation and operation of this technique are given elsewhere [13].

RESULTS AND DISCUSSION

Mechanical properties.

The results from the tensile properties are presented in Table 1. As expected, the steel with the higher Mn content exhibited higher tensile properties than the steel with the lower Mn content for a given processing condition. In addition, the results from Table 1 also indicate that the tensile properties did not change with holding time at the coiling temperature, independent of the Mn content. The maximum and minimum holding times during the isothermal treatments represent a time-temperature envelope which could be somewhat representative of the inner and outer wraps of an industrial coil.

Steel	Condition	Yield Strength (MPa)	Tensile Strength (MPa)	Elongation (%)
A: Mn=0.3%	AC+coil (0h)	382	437	31.2
	AC+coil (2h)	385	440	31.2
B: Mn=1.1%	AC+coil (0h)	430	503	31.2
	AC+coil (2h)	428	487	29.7
	AC+coil (5h)	427	490	28.5
A	ACRT	298	365	41.1
B	ACRT	357	455	37.5

Table 1. Measured tensile properties

Microstructural analysis.

The first level of microstructural analysis was conducted using optical metallography. The results from this analysis revealed that the microstructure in steel A after ACRT was equiaxed ferrite with small amounts of pearlite. For steel B, the ferrite was less equiaxed and had more pearlite. Steel B typically exhibited a larger and less uniform ferrite grain sizes than those observed in steel A. The average measured ferrite grain sizes after ACRT and/or after AC are reported in Table 2. The measured grain sizes are in good agreement with those predicted by Structura, a CRDM's computer program [14].

The second level of microstructural analysis was conducted by SEM. The SEM results showed that in steel B, the pearlite is mostly in the form of lamellae up to the 2 hour holding (lamellae spacing = 200 to 350 nm). After the 5 hour holding the lamellae appears to be undergoing spheroidization. In steel A with no holding time, carbides were observed in the form of elongated thick films at ferrite grain boundaries and in the form of a few pearlite colonies. After 2 hours holding, no pearlite is detected. All the carbides are in the form of films at the ferrite grain boundaries. These observations were confirmed by TEM examination.

The third level of microstructural analysis involved the use of TEM. From this analysis the dislocation density of each specimen condition was determined. The results are shown in Table 2.

Steel	Condition	mean intercept predicted by Structura (µm)	mean intercept measured (µm)	Dislocation density : Average (10^9 cm^{-2})	Dislocation density : Standard Dev. (10^9 cm^{-2})
A	AC+coil (0h)	5.2	5.3	5.9	1.2
	AC+coil (2h)	5.3	5.8	8.6	3.4
B	AC+coil (0h)	3.9	3.8	12.3	3.7
	AC+coil (2h)	3.8	3.6	15.7	4.3
	AC+coil (5h)	3.8	4.3	10.0	3.5
A	ACRT	8.1	8.9	4.0	1.4
B	ACRT	6.0	5.6	5.2	1.2

Table 2. Predicted and measured ferrite grain size and dislocation density.

Modes of precipitation observed.

The results from the TEM investigation did not reveal any large undissolved particles or large carbonitride formed during solidification. The large majority of the precipitates observed seem to have been nucleated in austenite and to a much lesser extent in the ferrite matrix. The precipitation in austenite seems to have taken place in two forms; (i) general random precipitation, and (ii) in deformation bands. The general random precipitation in austenite was typical of steel A, while steel B exhibited the two types of precipitation. However, the precipitation of NbC observed on the deformation bands was detected only in the air cooled specimens, Fig. 2. This seems to indicate that this mode of precipitation occurred in the delay time between the finishing rolling temperature (T_{FR}~892°C) and the transformation temperature to ferrite (T_{Ar3}~790°C). This type of precipitation was not observed in the AC specimens because there was no such delay.

Fig. 2 shows a series of dark field TEM micrographs illustrating the lines of precipitates going across several ferrite grains.

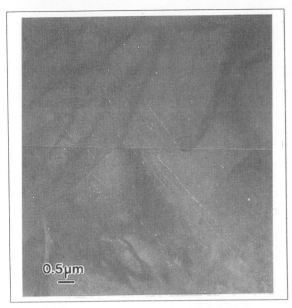

Fig.2: *TEM dark field of precipitates nucleated on γ deformation bands.*

Fig.3: *TEM dark field of interphase precipitation*

No precipitation nucleated in the ferrite was found in the accelerated cooled + coiled specimens. This observation is consistent with previous results from isothermal transformation studies that will be published in another paper [15]. Precipitation formed in ferrite, interphase and general matrix, was found only in the air cooled specimens in both steels. For example, Fig. 3 shows interphase precipitation, while Fig. 4 shows matrix precipitation.The average size and location of the precipitates are given in Table 3. The results from this table indicate that the precipitates are very small, especially the precipitates nucleated on the deformation bands in austenite which are in the order of 3nm. The distribution of the precipitate diameters, where 2400 particles have been counted, is given in Fig. 5. The biggest particles observed regardless of the processing condition were on the order of 45nm.

Steel	Condition	Precipitate type	Average Size (nm)	Standard deviation (nm)
A	AC+coil (0h)	γ grain boundary	7.2	2.8
	AC+coil (2h)	γ grain boundary	5.2	2.7
	AC+coil (0h)	γ grain boundary	5.6	2.7
B	AC+coil (2h)	γ grain boundary	5.1	1.5
	AC+coil (5h)	γ grain boundary	3.9	2.1
A	ACRT	γ grain boundary	NM	NM
		γ matrix	NM	NM
		α interphase	6.6	2.9
		α matrix	6.4	2.9
B	ACRT	γ grain boundary	6.3	2.9
		γ matrix	4.3	1.8
		γ deform. bands	3.0	2.0
		α interphase	5.0	2.1
		α matrix	3.0	1.0

Table 3. *Location and average size of the NbC precipitates observed (NM = not measured).*

The row spacing for the interphase precipitation in ferrite was measured to be approximately 175 +/-39 nm for steel A and 90 +/- 20 nm for steel B.

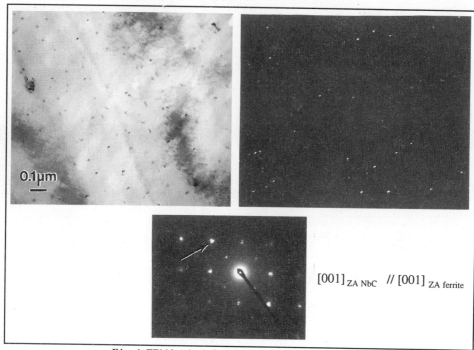

Fig.4: TEM bright & dark fields of matrix precipitation

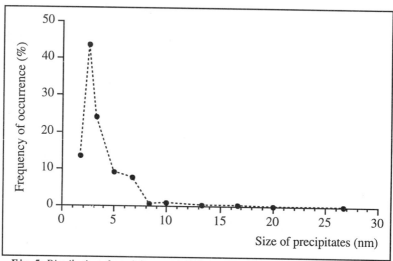

Fig.5: Distribution of precipitates on γ deformation bands (steel B, ACRT condition)

Volume fraction of the precipitates.

Due to the different nature of the precipitates and their heterogeneity in dispersion, it was felt that any attempt in calculating the volume fraction of the precipitates from the TEM measurements would not be representative of a given sample processing condition. Hence, the precipitate extraction residue method was used to get a better idea of the quantity of niobium that was precipitated. The results from this analysis are shown in Table 4. These results are consistent with the TEM observations regarding the density of the precipitates in the different samples investigated.

Steel	Condition	Total Nb in steel composition (ppm)	Total Nb precipitated (ppm)
A	AC+coil	190	90
	ACRT	190	170
B	AC+coil	280	120
	ACRT	280	190

Table 4. Results from the extraction residue method

Type of precipitates.
The fourth level of microstructural analysis was conducted to determine the composition of the precipitates using APFIM. The APFIM analysis revealed the presence of Nb and C in the precipitates which nucleated in austenite. None of the precipitates analyzed contained N atoms. The stoichiometry of the precipitates was close to $NbC_{0.9}$. The analysis also showed that Nb and C atoms were present in solid solution in the matrix and at the grain boundaries. A typical character plot from the APFIM analysis showing individual atoms in solid solution and in clusters is presented in Fig. 6.

	Number of atoms of:		
	Iron (.)	Carbon (c)	Niobium (b)
....	40	0	0
~~			
..ccbb.bbbccc........cccc...bb.cbbccccc	15	16	9
c...	39	1	0
......bbb.........ccc..bbbcccbb.bbcbbbb	18	7	15
bbbbbbbcbbbcccbbbbccccbbcbcc...ccc..bccc	5	17	18
bcccbbbcccbbbbbbbccc......bc..........b	16	10	14
~~			
....	40	0	0
....	40	0	0
.............................cccc..........	36	4	0
....	40	0	0
......................cc...................	38	2	0
....	40	0	0
....	40	0	0
...............................bb........	38	0	2
....	40	0	0
.......cc...................................	38	2	0
....	40	0	0
....	40	0	0
.......bbb.................................	37	0	3
......................b..............	39	0	1
....	40	0	0
.........bb................c...........	37	1	2
....	40	0	0
........cc.................................	38	2	0

Fig.6: Example of character plot of APFIM data for a precipitate and the matrix (steel B, AC+coil 5h)

Mechanical properties assessment.
The different contributions to the strengthening can be calculated from the optical and TEM measurements. In this analysis where the contribution from the pearlite was considered negligible, $\Delta YS_{Peierls-Nabarro} + \Delta YS_{solid\ solution}$ was taken as 48 + 32.5 (%Mn) + 84 (%C) in Mpa [16] . Also, $\Delta YS_{grain\ size}$ was taken as $k_Y * d^{-1/2}$ where k_Y was 18.2 for steel A and 16.5 for steel B [17]. The dislocation hardening can be given by $\Delta YS_{dislocation} = \alpha \cdot G \cdot b \cdot \rho^{0.5}$ [18] where b is the Burgers

vector of a dislocation, b = 0.248×10^{-7} cm, G is the shear modulus, and α is a numerical factor dependent of the crystal structure taken as = 0.38. From these expressions, the calculated YS for the accelerated cooled specimens which showed only precipitation formed in austenite was 375 MPa for steel A and 440 MPa for steel B. These values are in good agreement with the measured YS shown in Table 1. A more detailed approach of the mechanical properties assessment will be published elsewhere [15]. Therefore, from this analysis it appears that the fine precipitation which occurred in the austenite does not have a strong hardening effect. Similar results seem to corroborate this preliminary view regarding the strengthening contribution of the precipitates formed in austenite [19].

CONCLUSIONS

The results from this investigation lead to the following conclusions:
1. Fine NbC precipitation formed in austenite was found in the two steels investigated. Strain induced precipitation formed on prior austenite deformation bands was found only in the ACRT specimens with the higher Mn content.
2. Interphase and general matrix NbC precipitation in ferrite were only observed in the ACRT specimens. NbC precipitation observed in the AC+coil specimens was formed prior to the γ/α transformation. No further precipitation took place during the coiling treatments. This behavior explains the constant tensile properties with coiling treatments.
3. The increase in Mn content lowers the transformation temperature, promotes smaller ferrite grain size, higher dislocation density, and higher pearlite content.
4. Fine precipitation in austenite does not appear to contribute to the strengthening for the processing and steel compositions used in this study.

ACKNOWLEDGEMENTS

The authors wish to thank the R&D division of SOLLAC (USINOR France) and the CRDM -Research center of SOLLAC- in Dunkerque (France) for sponsoring this work and permission to publish this paper. The authors also wish to thank Dr. S.W. Lee from POSCO Pohang (Korea), for carrying out the precipitate extraction experiments. Mrs V. Thillou extends her thanks to Dr. M. Piette and Dr. P. Deshayes from the CRDM, for stimulating discussions.

REFERENCES

1. ITMAN.A, CARDOSO.K.R, KESTENBACH.H-J, Mater. Sc. and Techn., 13 (1997), p.49.
2. KESTENBACH.H.J, Materials Science and Technology, 13 (1997), p.731.
3. HORNBOGEN.E, Strength of metals & alloys, edited by P.HAASEN (1979), p.1337.
4. BAKER.T.N, Yield, flow and fracture of polycrystals (1983), p.235.
5. BAKER.R.G, NUTTING.J, Precipitation processes in steels, ISI special report 64 (1959), p.1.
6. JACK.D.H, JACK.K.H, Materials Science and Engineering A, 11 (1973), p.1.
7. DAVENPORT.A.T, BROSSARD.L.C, MINER.R.E, Journal of Metal, 27 (1975), p.21.
8. HONEYCOMBE.R.W.K, Metallurgical Transactions A, 7A (1976), p.915.
9. CULLITY.B.D, Elements of X-Ray diffraction , Addison-Wesley publishing company(1978).
10. LAPOINTE.A.J, BAKER.T.N, Metal Science, 16 (1982), p.207.
11. EDINGTON.J.W, Practical electron microscopy in materials science, Van Nostrand Co (1976).
12. HIRSCH.P.B. & al, Electron Microscopy of Thin Crystals, Krieger Publishing Co (1965).
13. MILLER.M.K, SMITH.G.D.W, Atom probe microanalysis: principles and applications to materials problems, Materials Research Society (1989).
14. PIETTE.M, POIRIER.F, PERDRIX.C, An integrated model for microstructural evolution in the hot strip mill and mechanical properties prediction of plain and microalloyed C-Mn hot strip, this conference.
15. THILLOU.V, HUA.M, GARCIA.C.I, PERDRIX.C, DEARDO.A.J, results to be published.
16. PICKERING.F.B, GLADMAN.T, ISI special report 81(1963), p.10.
17. PIETTE.M, Private communication on an Internal Report from CRDM-Sollac-Usinor (France).
18. BAKER.T.N, Science Progress, 65 (1978), p.493.
19. SATO.K, SUEHIRO.M, Tetsu-to-hagane, 77 (1991), p.675.

Prof. A.J.DeArdo can be contacted at deardo@engrng.pitt.edu

Materials Science Forum Vols. 284-286 (1998) pp. 319-326
© *1998 Trans Tech Publications, Switzerland*

The Effect of Mo in Si-Mn Nb Bearing TRIP Steels

M. Bouet[1], J. Root[2], E. Es-Sadiqi[3] and S. Yue[1]

[1] McGill University, Dept. of Metallurgical Engineering, M.H. Wong Building, Metallurgical Wing, 3610 University Street, Montreal, Quebec, H3A 2B2, Canada,

[2] Neutron Program for Materials Research, Chalk River Laboratories, Chalk River, Ontario, K0J 1J0, Canada

[3] CANMET, 555 Booth Street, Ottawa, Ontario, K1A 0G1, Canada

Keywords: Transformation Induced Plasticity (TRIP), Strain Induced Transformation (SIT), Thermo Mechanical Processing (TMP), Retained Austenite (RA), Ultimate Tensile Strength (UTS), Total Elongation, Stability, Bainite Transformation

Abstract

Experimental studies were performed to determine the effect of Mo in Si-Mn TRIP steels, as well as its potential for reducing the levels of Si. Three compositions were investigated. Bainite transformation conditions were investigated on the final mechanical properties. Results revealed that Mo has an important retardation effect on the formation of both ferrite and pearlite. The Mo and reduced Si level steel generated excellent mechanical properties (as high as: UTS=1269, T.El.=36%) in the range observed by previous investigators.

1. Introduction

Because in practice, there exists a trade-off between strength and ductility, cold formed highly shaped steel products which require elevated strengths normally must undergo a sequence of energy intensive heat treatment steps following ThermoMechanical Processing (TMP). The cold forming of wire intended for high strength fasteners is a classical example of this problem.

The incessant drive for high quality and cost effective new classes of high strength/high formability steels has sparked a growing interest in the industrial application of TRansformation Induced Plasticity (TRIP) steels. TMProcessed TRIP steels offer the advantage of generating the kind of formability required from the steel as a rolled condition, while obtaining the mechanical properties required directly from the cold forming operation. Hence, stages such as spheroidizing, tempering and quenching stages would no longer be required.

The Strain Induced Transformation (SIT) of metastable Retained Austenite (RA) to the lower energy state martensite phase is accompanied by a local increase in the strain-hardening rate. This locally strengthens the point of plastic instability, postponing it to higher strains. Zackay identified the first steels known to TRIP by the SIT of RA[1].

Investigations done at McGill University have shown that with controlled TMP of Si-Mn TRIP steels microalloyed with Nb, impressively high tensile strengths (above 1200) can be achieved having a wide ductility range (20–55 %T.El.)[2-4]. Furthermore, optimum RA characteristics and mechanical properties were obtained when isothermally processed in the

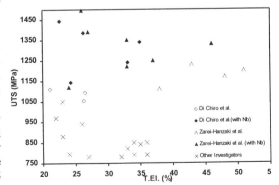

Fig. 1. Range of mechanical properties observed for Si-Mn TRIP steels by various investigators. Effect of Nb addition on the mechanical properties of TRIP steels.

bainite transformation region, as was reported in previous investigations[5-8] (Figure 1). Di Chiro et al. have reported tensile strengths of up to 1350 MPa and total elongations reaching 35%[4].

The mechanical properties of TRIP steels are dictated by the stability of the RA. Intuitively, one would expect optimum TRIP performances with steels containing a wide stability range, suggesting a progressive SIT and a much delayed sample failure. Numerous works have shown that the stability of RA is highly dependent on its particle size[2,8], its neighboring phases[2,9], and its chemical composition[2,4,10]. The carbon enrichment of residual austenite during transformation has been the target of many of these investigations. These have shown a strong correlation of C content on the stabilization of RA to room temperature as well as on its stabilization against mechanical deformation or SIT. Si in high levels (1.5-2.0 %wt.) was reported to stabilize a significant amount of RA[11-13]. Tsukatani et al. have more recently observed similar trends where both R.A. characteristics and mechanical properties were optimized with steels containing 2.0 %wt. Si and 1.5 %wt. Mn[10]. Si is responsible for kinetically inhibiting cementite formation allowing for greater degrees of austenite saturation with C. However Si, in levels above 1.0 %wt., is responsible for an undesirable tenacious oxide layer which is easily rolled into the steel, generating a poor surface quality.

Nb, on the other hand, is generally employed for the purpose of austenite pancaking and grain refinement. It has a strong affinity for C and N and tends to form Nb(CN) precipitates. However, Zarei-Hanzaki et al. and Di Chiro et al.[4], have reported that, when in solid solution, Nb can significantly increase the RA characteristics as well as greatly improve the mechanical properties of the steel[2,3].

In this work, the addition of Mo was considered as an alternative to Si. A number of papers have suggested that Mo has a strong solute drag effect in steel, resulting in important recrystallization and precipitation delays[14,15]. The same is to be expected for the kinetics of ferrite and cementite transformations. Furthermore, Wada et al. reported that Mo increases the solubility of C in austenite, thereby decreasing the driving force of precipitation as well as the diffusivity coefficients of the precipitating or carbide forming species (i.e. Nb, C)[16]. Each of these findings strongly suggests that Mo potentially can be an important addition to TRIP steels. By its presence one could expect: (1) greater austenite saturation levels with C and Nb (i.e. delayed cementite and Nb(CN) formation), thereby enhancing their potential effectiveness; (2) easier process control through delays in transformation, and; (3) lower Si levels required.

The influence of Mo as well as variations in the TMP schedule (i.e. bainite hold time and temperature) on the RA characteristics and resulting tensile properties is reported in this work. Three steels with varying combinations of Si and Mo levels are investigated. RA characteristics (Vol.%, C wt.%) were determined using neutron diffraction. Tensile properties were ascertained using a shear punch testing technique[19].

2. Experimental Procedure

The steels under investigation (Table 1) have similar compositions with differing combinations of Si and Mo levels. This way the effect of each of these constituents can be examined with regard to verifying the feasibility with which Mo can replace Si in TRIP steels. The Si levels in these steels are two to four times greater than typical plain carbon steels. The Mo levels utilized, on the other hand, are typical of industrial microalloyed steel practices. The steels were cast at CANMET and received in the as hot rolled condition.

Mechanical testing was performed using a Material Testing System (MTS) adapted for hot compression[2,17]. Samples were machined into small cylinders, 11.4 mm in height and 7.6 mm in diameter (i.e. 1.5 height/diameter ratio).

Table 1. Steel Chemical Compositions (wt.%)

Steel	C	Si	Mo	Mn	Nb	Al	N (ppm)
A	0.18	1.66	---	1.4	0.025	0.053	40
B	0.20	1.6	0.30	1.4	0.041	0.028	55
C	0.23	1.1	0.33	1.6	0.036	0.036	40

Material characterization was determined by Continuous Cooling Compression (CCC) testing[2,17,20]. The specimens were solutionized at 1200°C for 30 minutes to dissolve all Nb(CN) and AlN precipitates. Subsequently, the samples were cooled to 1050°C, held for a few minutes, at which point they were cooled once again to 550°C and strained at a cooling rate of 0.5°C/s and a strain rate of 0.005/s.

Figure 2 illustrates the TMP schedule utilized for verifying the effect of isothermal bainite hold temperature. The procedure consisted firstly of a solutionizing stage as in the CCC tests. Subsequently, the steel was strained twice by compression at 1050°C, a typical TMP temperature. The samples were then cooled below the austenite-to-ferrite transformation temperature (Ar$_3$) and were isothermally held at 650°C to obtain similar amounts of ferrite in each steel. Following this treatment, the specimens were salt bath quenched and were isothermally held within the bainite transformation region for various temperatures, after which they were air-cooled. Similar experiments were performed for various hold times (i.e. 2, 5, 10 min.) at 400°C. In order to predict the amount of ferrite transformed, isothermal transformation characteristics were determined for each of the steels. This was accomplished by quenching the samples following isothermal holding at 650°C, and by measuring the ferrite content using image analysis, for various holding times.

RA amounts were quantified by analysis of neutron diffraction patterns from the DUALSPEC powder diffractometer at the NRU reactor at Chalk River Laboratories[18]. The neutron beam wavelength was 0.13286 nm. Specimens were rotated continuously. Diffraction patterns were analyzed by the Rietveld profile-refinement method to extract lattice parameters and volume fractions of the austenite and ferrite phases in each specimen. The carbon content of the retained austenite was determined from the austenite lattice parameter, a_{RA}, through the relation:

$$a_{RA} = 3.578nm + 0.044nm \times \%C_{RA}$$

Microstructure was characterized using optical microscopy. Room temperature mechanical properties were determined by a shear punch technique[19]. This technique allows mechanical properties to be determined from small sized specimens. A 3 mm diameter disk is punched out of a thin specimen (~350 μm thick), while the MTS monitors the load/displacement behavior.

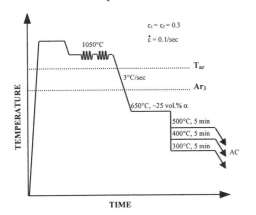

Fig. 2. Schematic illustration of the TMP schedule used for investigating the effect of isothermal bainite hold temperature.

3. Results and Discussion

3.1. Material Characterization

Continuous cooling compression testing was performed for each of the above mentioned steels. Slope inflections in the flow stress versus temperature curves identify where the various transformations occur. Figure 3 illustrates and compares the CCC curves for these three steels. It appears that the addition of Mo increases the flow stress of the steel below 900°C. This can most probably be attributed to the strong solid solution strengthening effect of Mo as well as its effect in retarding dynamic recovery. More importantly, Mo appears to shift the beginning of pearlite formation (i.e. Ar$_1$) to lower temperatures by roughly 35°C (the Ar$_1$ is characterized by the sudden increase in flow stress following the pronounced dip in the curve). This is an agreement with other investigators[14-16], who have shown that Mo has a strong retardation effect on both the kinetics and thermodynamics of carbide formation in steel. Therefore, Mo would appear to be a strong austenite stabilizer. In addition, partial removal of Si from the Mo bearing steel shifts the initiation of cementite formation to even lower temperatures. This can simply be explained by a loss of some of the ferrite stabilizing effect of Si in this steel.

3.2. Ferrite Transformation

Isothermal ferrite transformation curves were obtained for each of the materials under investigation (Fig. 4.). This information was used to obtain similar volume fractions of ferrite prior to quenching into the bainite region during TMP testing. It is shown that the addition of Mo to steel A strongly delays the transformation, as is shown by the negligible amount of ferrite formed for steel B. Consequently, tests involving isothermal ferrite transformation were not performed for this steel. On the other hand, partial removal of Si from steel B accelerated the rate of ferrite transformation, partly negating the effect of Mo, as is shown by steel C. It was necessary to isothermally hold steel A and steel C at 650°C for 4 and 50 minutes, respectively. This way a microstructure containing roughly 25 area% ferrite could be obtained for each.

3.3. Change in Microstructure and Retained Austenite Characteristics with Bainite Hold Temperature

Steels A and B were subjected to the TMP schedule in Figure 2 to generate different conditions for bainite formation. Figure 5 shows the microstructures obtained for these steels following controlled TMP and isothermal bainite transformation for 5 min. at 300, 400, and 500°C. A 3% nital solution was used for etching the steels. The

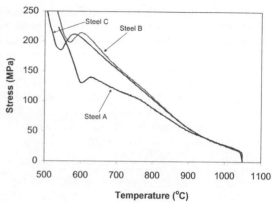

Fig. 3. CCC test for the materials under investigation.

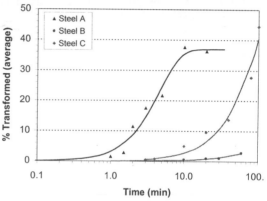

Fig. 4.   The effect of Mo and Si on the   isothermal transformation rate of ferrite at 650°C.

light gray continuous networked phase is ferrite, the plate-like phase is a multiphase structure of bainite, RA and martensite, and the black block type phases are carbides. When comparing these micrographs, it can be quickly ascertained that the microstructure strongly depends on the isothermal bainite transformation temperature. In both steels, it can be seen that the bainite progressively transforms from a coarse to a fine lath structure with decreasing isothermal hold temperature. Though the RA and martensite phases are not readily visible by the etching technique employed (i.e. 3% nital), Hanzaki et al.[21] has shown color etching to be quite effective in distinguishing the RA from the other phases. His investigations have shown that at the higher transformation temperatures (i.e. 500°), the RA is enclosed mostly by thick ferritic platelets, whereas at the lower temperatures (i.e. 300°C), RA is mainly trapped between laths of bainitic ferrite.

The V_{RA} for each combination of steel and isothermal bainite treatment was measured by neutron diffraction. Results show that V_{RA} is optimized at the intermediate bainite temperature of 400°C (Figure 6), where a combination of both types of RA allows for a wide stability range against mechanical deformation and hence a progressive and ongoing TRIP effect. Decreasing the isothermal hold temperature from 500 to 400°C led to a significant increase in V_{RA} for both steels. However, decreasing the temperature even further to 300°C, reversed this trend, generating lower RA levels. This can be attributed to the ease with which carbides form at this temperature. Carbon precipitates as thin films of carbide within the bainitic ferrite platelets, thereby decreasing its availability for stabilization and retention of RA[21]. Steel C has lower measured RA levels almost across the board, with 300°C as the significant exception. Once again, this latter finding specifically demonstrates the important inhibiting ability of Mo on carbide formation. Figure 7 illustrates the variation of the carbon content in the RA with isothermal bainite hold temperature. Both steels exhibit a maximum V_{RA} and C_{RA} at the intermediate hold temperature of 400°C. However, Hanzaki et al.[21] and Di Chiro et al.[4] observed an optimum C_{RA} at 300°C for steel A with a slightly higher Nb level (0.035 %wt.). This discrepancy can be explained by the different hold times utilized (i.e. 5 min. versus 2 min.[4,21]). At 300°C, the RA quickly saturates with C generating unstable levels, which shortly after begin to precipitate out carbides. This in turn initiates a decreasing C_{RA} as well as V_{RA}. Therefore, maximum C_{RA} occurs prior to 5 minutes at 300°C. In addition, steel C has a greater V_{RA} than steel A for similar C contents at 300°C. This can be partly due to the strong carbide inhibiting nature of Mo. However, similar C levels would suggest a similar stabilization effect and, hence, similar RA, which is not the case. It appears that Mo may have a strong solid solution strengthening effect as well, thereby acting as a physical barrier to early on shearing of RA to martensite. The slightly lower Nb levels in steel A may also be contributing to this difference, for this same reason.

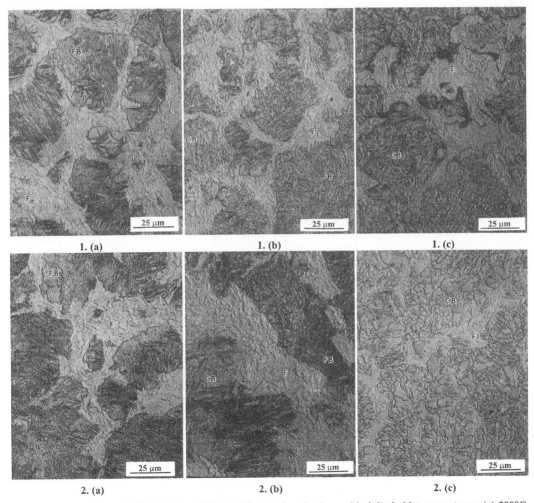

Fig. 5. Microstructure of: (1) steel A and (2) steel C at various isothermal bainite hold temperatures: (a) 300°C, (b) 400°C, (c) 500°C. (F=Ferrite, CB=Coarse Bainite, FB=Fine Bainite)

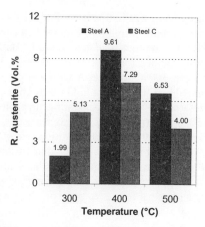 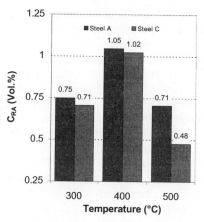

Fig. 6. Effect of bainite hold temperature on the volume fraction of retained austenite. (time =5min.)

Fig. 7. Effect of bainite hold temperature on the C content of retained austenite. (time = 5 min).

3.4. Variation of Mechanical Properties with Isothermal Bainite Hold Temperature

The mechanical properties of the steels, which have undergone isothermal bainite treatment, are displayed in Figure 8. Both steels exhibit an increase in the UTS with decreasing hold temperature, which is largely in part due to the finer bainitic structure[4,21]. The UTS measurements for steel A are significantly lower than those observed by Hanzaki et al.[21] and Di Chiro et al.[4]. The slightly lower Nb level may be responsible for these poorer tensile properties. Steel C, on the other hand, generated values within the range generated by the previous authors.

Steel A exhibits an optimum ductility at 400°C, which coincides with the maximum V_{RA}. The elevated V_{RA}, in conjunction with the wide distribution of stability levels obtainable at this temperature, as explained earlier on, are responsible for the large elongation measurements observed. Conversely, steel C exhibits increasing elongation measurements with decreasing hold temperature. Oddly enough, this coincides with neither the optimum C level (i.e. stability) nor the optimum RA level. The solid solution strengthening effect of Mo, in addition to the effect of C, may be rendering the RA super-stable against mechanical deformation at 400°C. This would allow for optimum ductility to occur only at the lower hold temperature (i.e. 300°C), where C levels are deteriorated by the precipitation of carbides.

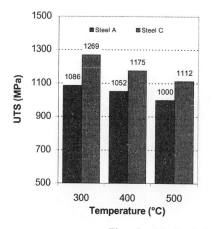 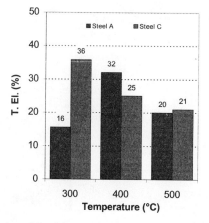

Fig. 8. Mechanical properties following various isothermal bainitic treatments. (hold time = 5 min.)

3.5. Change in Retained Austenite Characteristics with Isothermal Bainite Hold Time

To further understand the contribution of bainite in optimizing TRIP characteristics, different bainite hold times at 400°C were investigated. Figure 9 illustrates the variation of V_{RA} with hold time. Steel A exhibits a maximum at 5 min.. The initial rise of V_{RA} is caused by the stabilization of austenite through its saturation with C rejected from adjacent forming bainite laths. Subsequent supersaturation of the austenite phase leads to the formation of carbides and a drop in the V_{RA}[21]. The V_{RA} for steel C, however, decreases slightly with time. These observations are curious since time appears to have little or no effect on the RA C levels observed (Figure 10), when, in fact, other investigators[4,21] have noticed important differences. Further work must be done to clarify this point.

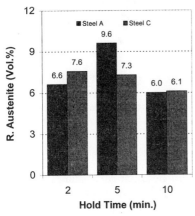 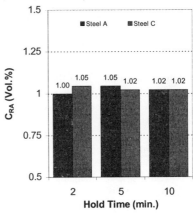

Fig. 9. Effect of bainite hold time on the volume fraction of retained austenite. (temperature = 400°C)   **Fig. 10. Effect of bainite hold time on the C content of retained austenite. (temperature = 400°)**

3.6. Variation of Mechanical Properties with Isothermal Bainite Hold Time

Testing revealed that isothermal hold time has a pronounced effect on the mechanical properties of both steels (Figure 11). In both instances, both the UTS and the total elongation increased with time, the only exception being the total elongation of steel A having dropped slightly for hold a time above 5 minutes. The increase in strength observed, can be attributed partly to the increased amount of bainite and eventually carbide phase being formed[2,21]. Ductility, on the other hand, follows V_{RA}[2,21] while maintaining a certain sensitivity to the stabilization effect of C and other alloying elements (i.e. Nb, Mo, Si). The stabilization effect of these other elements may be contributing to the continuously increasing ductility observed for steel C, which is expected to eventually drop with increasing hold times. This

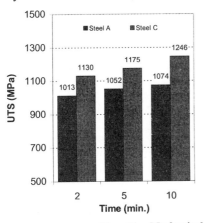 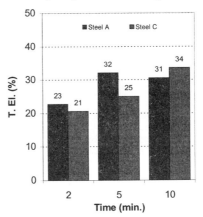

Fig. 11. Mechanical properties following various isothermal bainitic treatments. (temperature = 400°C)

would perhaps explain the apparent insensitivity in ductility for this steel to the declining V_{RA}, past this point.

4. Conclusion

Mo was added to Si-Mn Nb microalloyed TRIP steels to verify the feasibility of it partially replacing Si. This addition showed to generated an excellent combination of strength and ductility comparable to steels without Mo, but with greater Si levels. The mechanical properties of these steels were discussed in relation to their RA characteristics and to their microstructure. The results observed are summarized:

1. CCC testing showed Mo to strongly assist austenite stabilization. The Ar_1 in the presence of Mo was decreased by ~35°C. Furthermore, Mo had an important solid-solution strengthening effect in the flow stress.
2. Isothermal holding at 650°C prior to bainite treatment revealed that Mo has a tremendous retardation effect on both the formation of ferrite and cementite.
3. The mechanical properties of these steels are dependent on the processing conditions in the bainite region (i.e. temperature, time). The UTS is sensitive to microstructure, whereas the total elongation is sensitive to both microstructure and V_{RA}. Steel C, however, contradicts the trend observed by previous investigators[4,21]. Other factors may be involved.

Acknowledgements

The authors would like to thank the Canadian Steel Industry Research Association (CSIRA) and, the Natural Sciences and Engineering Research Council of Canada (NSERC) for their financial support. The authors would also like to thank Rocco Varano for his help and John Root and AECl for their help and the use of their equipment

References

1. V.F. Zackay, E.R. Parker, D. Fahr and R. Bush, Trans. Am, Soc. Mat., 60 (1967), p. 252.
2. A. Zarei-Hanzaki, Ph.D. Thesis, McGill University, (1994).
3. A. Zarei-Hanzaki, P.D. Hodgson and S. Yue, ISIJ International, Vol. 35 (1995), No. 3, p.324.
4. A. Di Chiro, J. Root and S. Yue, 37th MWSP Conf. Proc., ISS, Vol. 33 (1996), p.373.
5. W.C. Jeong, D.K. Matlock and G Krauss, Material Science and Engineering, A165 (1993), p. 1.
6. Y. Sakuma, O. Matsumura, and O. Akisue, ISIJ, Vol. 31 (1991), p.1348.
7. Y. Sakuma, O. Matsumura and H. Takeshi, Metallurgical Transactions A, Vol. 22A (1991), p.489.
8. A. Itami, M. Takahashi and K. Ushioda, High Strength Steels for Automotive Symposium Proceedings (1994), p.245.
9. G.B. Olson and M. Cohen, Metallurgical Transactions, Vol. 13A (1982), p.1907.
10. I. Tsukatani, S Hashimoto and T. Inoue, ISIJ International, Vol. 31, No. 9 (1991), p.992.
11. R. LeHouilier, G. Begin and A. Dobe, Metallurgical Transactions A, Vol. 2A (1971), p.2645.
12. V.M. Pivovarov, I.A. Tanaka, and A.A. Levchenko, Phys. Met. Metallogr., Vol. 33 (1972), p.116.
13. H.K.D.H. Bhadeshia and D.V. Edmonds, Metallurgical Transactions A, Vol. 10a (1979), p. 895.
14. M.G. Akben, B. Bacroix and J.J. Jonas, Acta Metall., Vol. 31 (1983), p. 161.
15. J.J. Jonas, Proceedings of an International Conference sponsored by the Ferrous Metallurgy Committee of the Metallurgical Society of the American Institute of Mining, Metallurgical and Petroleum Engineers (A.I.M.E.) and the Australian Institute of Metals, held at the University of Wollongong, Wollongong Australia (Aug. 20-24, 1984), p.80.
16. H. Wada and R.D. Pehlke, Metall. Trans., Vol. 16B (1985), p.815.
17. A. Zarei-Hanzaki, P.D. Hodgson and S. Yue, 33rd MWSP Conf. Proc., ISS-Aime, Vol. 29 (1992), p.460.
18. I.P. Swainson, N.B. Konyer and J.H. Root, "1998 C2 DUALSPEC Powder Diffractometer User's Guide", (1997).
19. G.E. Lucas, G.R. Odette and J.W. Sheckard, ASTM STP 888, p.112.
20. A. Zarei-Hanzaki, R.Pandi, P.D. Hodgson, and S.Yue, Metallurgical Transactions A, Vol. 24A (1993), p. 2657.
21. A. Zarei-Hanzaki, P.D. Hodgson, and S.Yue, ISIJ International, Vol.35, 1995, 79-85.

[Correspondence: Dr. Steve Yue, e-mail: (steve@minmet.lan. mcgill.ca)]

Materials Science Forum Vols. 284-286 (1998) pp. 327-334
© 1998 Trans Tech Publications, Switzerland

Effect of Niobium on the Mechanical Properties of TRIP Steels

W. Bleck[1], K. Hulka[2] and K. Papamentellos[1]

[1] Institute of Ferrous Metallurgy, RWTH Aachen, Germany

[2] Niobium Products Company, Düsseldorf, Germany

Keywords: TRIP Steel, Niobium Microalloying, Retained Austenite, Bainite Transformation

Abstract

Three TRIP steels have been laboratory melted with 0.17% C, 1.4% Mn, and 1.5% Si as the base composition and additions of 0.02 or 0.04% niobium. After hot and cold rolling the effect of annealing treatment on microstructure development and resulting mechanical properties was evaluated. Niobium retards the isothermal bainite transformation and thus leads to higher retained austenite values. The mechanical properties are dependent on the grain size, the volume fraction of retained austenite, the retained austenite stability, local transformation stresses after bainite formation, and on carbide precipitates. Compared to conventional high strength steels TRIP steels offer an improved strength-ductility relationship and expand the range of cold formable steels to higher strength values.

Introduction

Multiphase steels have been developed for various purposes because of their ability to tailor properties by adjusting the type, the amount and the distribution of different phases. Examples of well-known multiphase steels are carbon steels with a ferritic-pearlitic microstructure or duplex steels with a ferritic-austenitic or ferritic-martensitic microstructure. With regard to light-weight constructions especially in the automotive industry multiphase steels with excellent cold formability gain much interest. During the eighties dual phase steels have been developed, characterized by a microstructure consisting of a dispersion of martensitic islands in a matrix of ferrite. These dual phase steels combine a low yield strength due to the relatively soft ferritic matrix with a high tensile strength as a result of the hard second phase with a volume fraction of about 5 to 15%. Further characteristics are continuous yielding, a high level of strain hardening and the ability to bake hardening.

TRIP steels make use of the transformation-induced plasticity effect, which means that retained austenite is metastable at room temperature and will be transformed to martensite during straining [1]. TRIP steels for cold forming have been developed by using manganese and silicon as the main alloying elements in order to control transformation behaviour in a two step heat treatment [2]. Quenching after intercritical annealing is performed to an intermediate temperature above the martensite-start temperature, which allows the bainite transformation to occur during isothermal holding. The remaining austenite is further enriched with carbon, which shifts the martensite-start temperature below room temperature. A typical phase distribution of TRIP steel in the as-shipped condition is about 50 vol.% ferrite, 40-45 vol.% bainite, and 5-10 vol. % retained austenite. But as deformation develops during press-shop treatment, the retained austenite will transform to martensite which brings about a high resistance to local necking, and thus high uniform elongation values and good formability.

The microalloying element niobium is widely used in steels for several purposes. In the state of solid solution niobium retards recrystallization during hot deformation and the austenite to ferrite transformation. Niobium combines with carbon and nitrogen and forms small precipitates which retard recrystallization grain growth and result in effective strengthening [3].

It is the objective of this investigation to test the effects of small niobium additions on the transformation behaviour, the microstructure and the mechanical properties of TRIP steels.

Experimental procedure

Three melts were produced in the 80 kg vacuum induction furnace of the Institute of Ferrous Metallurgy. The chemical composition of the three heats is given in Table 1. The base steel contains 0.17% C, 1.40% Mn, and 1.50% Si and the two niobium steels had additions of 0.02% and 0.04% Nb respectively. The heats were cast in a stationary mould to ingots with the dimensions 250 x 100 x 100 mm. The laboratory hot rolling was carried out to a final thickness of 2 mm in about 10 rolling steps with a starting temperature of 1200°C, a finishing temperature of about 900°C, followed by air cooling. Cold rolling on a reversing mill led to the final gauge of 0.8 mm, which corresponds to a cold rolling reduction of 60%.

Steel	C	Mn	Si	P	S	N	Nb
TRIP 1	0.17	1.33	1.44	0.006	0.003	0.005	<0.005
TRIP 2	0.17	1.42	1.47	0.006	0.003	0.004	0.023
TRIP 3	0.16	1.44	1.55	0.006	0.003	0.004	0.042

Table 1 Chemical composition of investigated TRIP-steels in mass%

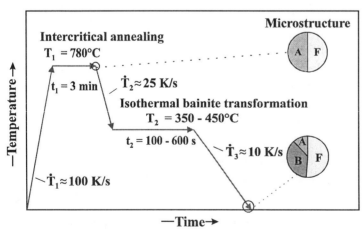

Fig. 1: Laboratory annealing cycle for TRIP steels

Annealing was partly carried out in two different salt baths. Details of this treatment are shown in Fig. 1. The intercritical annealing temperature was 780°C with a holding time of 3 min in order to develop around 50% austenite. After intercritical annealing the cooling rate of 25 K/s was chosen to avoid pearlitic transformation. The second isothermal annealing step was carried out in the temperature range of bainite transformation; temperatures between 350 and 450°C and isothermal holding times between 100 and 600 s were selected to develop different amounts of bainite and to allow different carbon enrichments of the austenite phase. Finally the secondary cooling led to the required microstructure of ferrite, bainite and retained austenite. No temper rolling was applied.

The mechanical properties were measured by means of tensile tests and by forming limit analysis; only the tensile test results are reported in this paper. A complete survey on all test results will be published elsewhere [4].

The transformation behaviour was investigated with a deformation dilatometer "Baehr Dil 805 A/D": The specimens were heated by inductive heating to intercritical temperature and rapidly cooled to the isothermal holding temperature using helium gas. All tests have been performed under inert gas conditions to avoid decarburization.

The microstructure of all samples has been investigated. The volume fraction γ_R of retained austenite has been measured by means of a magnetic method. The magnetic saturation was measured for an austenite free specimen first and then for the specimen after the heat treatment. The difference of both values can be related to the retained austenite volume fraction.

Mechanical properties

The results of the mechanical testing are compared with industrially processed steels of various grades. These are

-   IF-HS; high-strength interstitial free grades, which are characterized by ultra low carbon contents less than 0.005%, a combined microalloying of niobium and titanium, a phosphorus addition for strengthening, and a small boron addition to reduce secondary work embrittlement,
-   DP; commercially available dual phase steels with a carbon content of 0.06 - 0.1%, a manganese content of 0.9 - 1.6%, and a silicon content of 0.9%,
-   HSLA; conventional high-strength low allow steels for cold forming according to EN 10268 with carbon contents less than 0.1%, manganese contents less than 1.4%, and small niobium and/or titanium additions.

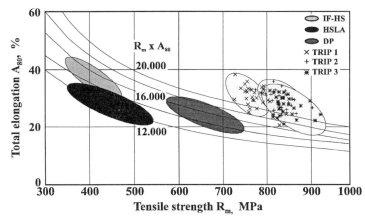

Trip steels offer very attractive combinations of elongation and tensile strength values, Fig. 2. A comparison of different steel families can be made by the product R_m x A_{80} which can be above 20 000 for TRIP steels. The absolute elongation values can be as high as the values of the high-strength IF steels which present the best formable high-strength steel developed so far. The tensile strength range of the investigated TRIP steels span from 700 to 950 MPa and is much higher than today's tensile strength range of cold formable steels.

Fig. 2: Total elongation and tensile strength of TRIP-steels in comparison with conventional high strength steels

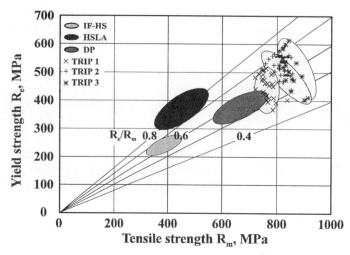

Fig. 3: Yield stength and tensile stength of TRIP-steels in comparison with conventional high strength steels

The yield ratio Re/Rm of the investigated steels TRIP 1, 2 or 3 is lower than in standard HSLA steels and somewhat higher than that of dual phase steels, Fig. 3. The mechanical properties strongly depend on the type and distribution of phases as a result of the different heat treatments. Generally speaking, the microstructures of the investigated steels can result in continuous yielding or pronounced yielding. The first corresponds to a low yield ratio and rather low elongation values while the latter represents the expected TRIP properties of slightly increased yield ratio and higher elongation values.

Microstructure

The beneficial mechanical properties of TRIP steels depend on their amount of mechanically metastable retained austenite γ_R. The optical microstructures of the different steels after hot rolling

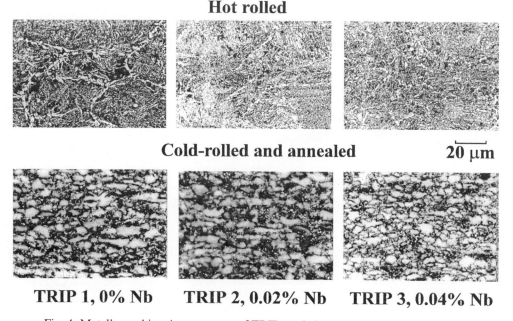

Fig. 4: Metallographic microstructure of TRIP steels in the hot rolled and in the cold-rolled plus annealed condition

as well as after cold rolling plus annealing are shown exemplarily in Fig. 4. It is obvious that niobium causes a grain refinement which derives already from the hot rolled condition. The microstructure after hot rolling consists of bainitic ferrite. After cold rolling and annealing the characteristic TRIP microstructure is developed. Within a matrix of ferrite and bainite small isolated retained austenite islands are visible with an average island diameter of less than 1.5 μm.

The volume fraction of retained austenite is an important parameter to control; the isothermal temperature T_2 and the holding time t_2 are the corresponding process parameters. For the two steels TRIP 1 and 3 the effect of isothermal annealing parameters on the volume fraction γ_R is plotted in Fig. 5 and 6. In the niobium free steel TRIP 1 γ_R values between 1.2 and 6.4 have been measured.

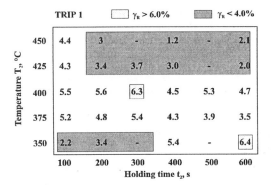

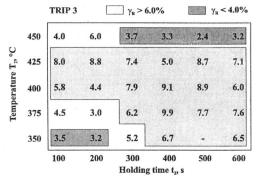

Fig. 5: Relationship between γ_R-% and bainite transformation conditions

Fig. 6: Relationship between γ_R-% and bainite transformation conditions

In the niobium alloyed steel TRIP 3 generally higher amounts of γ_R with a maximum of almost 10 vol% were determined. For all steels temperatures around 400°C lead to the highest γ_R values. The higher γ_R values of steel TRIP 3 is explained by the retarded bainite transformation kinetics, Fig. 7 and 8. The transformation kinetics is monitored by dilatation; the γ_R values, measured by the magnetic method, are attached to the curves.

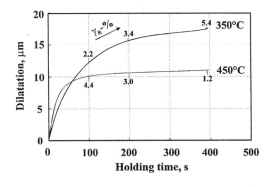

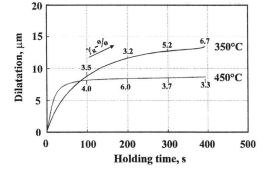

Fig. 7: Bainite transformation kinetics and γ_R-%, TRIP 1

Fig. 8: Bainite transformation kinetics and γ_R-%, TRIP 3

The slope of the dilatation curves gets higher with increasing transformation temperature. That means the bainite transformation is accelerated. It is obvious that the amount of retained austenite increases with progressing bainite transformation. If the bainite transformation is complete, as in the case of 450°C transformation, a reduction of γ_R with increasing holding time is determined. The first phenomenon is due to carbon enrichment in austenite which results in a reduction of the martensite start temperature below ambient temperature. The second effect results possibly from carbide precipitation which reduces the carbon enrichment in the austenite.

Effect of bainite transformation on mechanical properties

The mechanical properties are strongly affected by the bainite transformation conditions, Fig. 9 and 10. For short holding times the tensile strengths of steels TRIP 1 and TRIP 3 depend on the transformation temperature, for longer holding times there is nearly no dependency. The niobium addition of 0.04% in steel TRIP 3 results in a strength increase of about 40 to 60 MPa; steel TRIP 2 with 0.02% Nb shows tensile strength values in between steels TRIP 1 and TRIP 3 with an average tensile strength increase of 30 MPa.

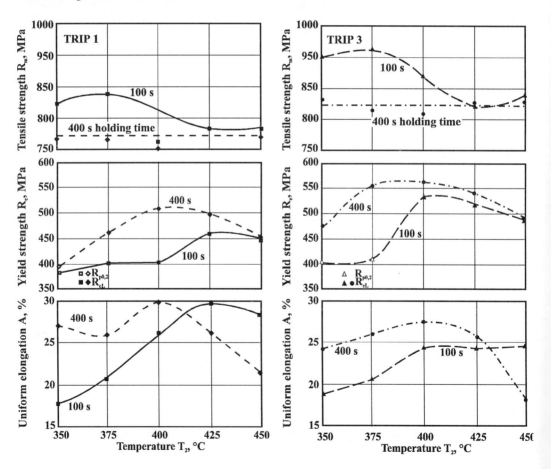

Fig. 9: Mechanical properties over bainite transfomation temperature

Fig. 10: Mechanical properties over bainite transfomation temperature

The yield strength of these steels are also affected by transformation temperature and isothermal holding time. Low temperatures T_2 and short holding times t_2 result in continuous yielding with relatively low yield strength values. For higher temperatures T_2 and holding times t_2 pronounced yielding is dominant which results in higher yield strength values. When it comes to temperatures of 425 or 450°C a slight decrease of yield strength can be noted. For both steels shorter holding times lead to lower yield strength values. Thus the yield ratio $R_{p0.2}$ or R_{eL} over R_m depends on transformation temperature and holding time. Low yield ratios can be attributed to the dual phase effect which is caused by free mobile dislocations due to local transformation stresses. Longer annealing times or higher annealing temperatures during isothermal bainite formation allows these local stresses to reduce. Simultaneously the yield ratio is increased.

Finally the uniform elongation reflects the amount of retained austenite and of carbide precipitation. High uniform elongation values of 25% and more can be obtained for steel TRIP 1 when the amount of retained austenite is higher than approximately 4%, Fig. 11. A further increase up to 10% does not further enhance or decrease the uniform elongation. Not all the experimental data obey this simple rule of thumb, especially results obtained for samples with short holding times. Here local transformation stress fields might cause early localized necking during tensile testing.

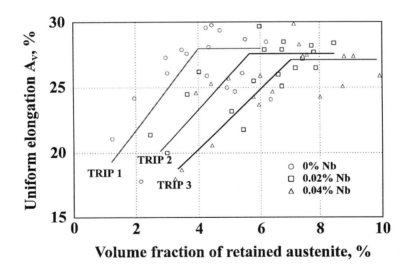

Fig 11: Effect of retained austenite content on uniform elongation

The minimum amount of retained austenite for high elongation values is shifted to higher γ_R fractions if the tensile strength of the material increases. The recommended minimum values are 5.5 and 7.0 % for steels TRIP 2 and TRIP 3 respectively.

The findings of this investigation for the niobium free steel are in good agreement with already published results with regard to the amount of retained austenite and the mechanical properties [5]. Using optimum processing conditions the mechanical properties of niobium free and niobium alloyed TRIP steels can be obtained as presented in Table 2.

Steel	Hot rolling		Cold rolling	Annealing			R_e MPa	R_m MPa	A %
	SRT, °C	FT, °C	Hot/Cold, mm	T_1, °C	T_2, °C	t_2,s			
TRIP 1	1200	900	4.0/0.8	780	400	400	509	750	29.8
TRIP 2	1200	900	4.0/0.8	780	425	200	526	788	28.2
TRIP 3	1200	900	4.0/0.8	780	425	300	542	822	27.4

Table 2: Process conditions and tensile properties of investigated TRIP cold-rolled sheet steels

Conclusions

- TRIP steels offer better strength-ductility relationships than conventional high-strength steels.
- TRIP steels expand the tensile strength range of cold formable steels to values above 750 MPa.
- The amount of retained austenite depends on the bainite transformation, which can be controlled by the niobium content, transformation temperature and isothermal holding time.
- The mechanical properties are affected by the grain size, the amount of retained austenite, local transformation stresses, and possibly by carbide precipitates.
- Niobium retards the bainite transformation kinetics and by this increases the amount of retained austenite during proper annealing. Niobium further increases the yield and tensile strength by approximately 15 MPa per 0.01% addition due to grain refinement.

References

[1] V.F. Zackay et al., Trans. Quart. 60 (1967), p. 252
[2] Y. Sakuma et al., Nippon Steel Techn. Rep. 64 (1995), p. 20-25
[3] H. Stuart, ed.: Niobium, Proceedings of the International Symposium,
    AIME, Warrendale, USA (1984)
[4] W. Bleck, K. Hulka, K. Papamantellos, to be published in steel research
[5] A. Itami et al., ISIJ International, Vol. 35 (1995), No. 9, p. 1121-1127

Materials Science Forum Vols. 284-286 (1998) pp. 335-342
© 1998 Trans Tech Publications, Switzerland

Effect of Microalloying and Thermomechanical Processing on the Structure and Mechanical Properties of Constructional Steel

G. Kodjaspirov

St. Petersburg State Technical University, St. Petersburg, RU-195251, Russia

Keywords: Microalloying, High-Temperature Thermomechanical Processing (HTMP), Mechanical Properties, Microstructure, Rolling, Fracture, Cool Resistance

ABSTRACT

The effect of microalloying with carbide-forming (V,Ti,Nb) and rare-earth elements and of high-temperature thermomechanical processing (HTMP) on the structure, mechanical properties and low temperature behaviour of 38CrSi steel has been investigated. It has been shown that in the case of dissolution of carbides during heating for quench hardening, the tempering resistance of the steel increases. It has also been found that - as a result of HTMP - the susceptibility of the steel irreversible temper brittleness decreases, irrespective of the steel having been microalloyed or not.

The data, originating from tensile testing and impact testing (with the help of laser interferometry) in the temperature range from +20 to -196 degC, has been explained in terms of fractographic analysis. It has been shown that microalloying and HTMP favour the occurrence of ductile microvoids in the fracture. On the other hand, HTMP and rare-earth elements chanches the nature of the fracture at low temperatures, this being in accordance with the data relating to specific energy consumption during fracture.

INTRODUCTION

It is common knowledge, that microalloying in constructional steels has a positive effect on fracture resistance and strength (1). Nb, V and Ti suppress the recrystallization of austenite during rolling and refine the microstructure with simultaneous beneficial effects on strength and fracture resistance (2). On the other hand, there is information regarding the positive effect of rare-earth metals (REM) relating to their absorbtion into grain boundaries, or sites of dislocations congestion, leading to an increased substructure stability, generated during plastic deformation (3). Most of the papers have been devoted to HSLA steels, especially under controlled rolling conditions (2, 4, 5). This effect may be of significance for steels of other carbon and alloy contents, especially under HTMP conditions.

EXPERIMENTAL MATERIALS AND PROCEDURE

The chemical composition of investigated steels is given in Table 1.

Steel	C	Si	Mn	S	P	Cr	V	Nb	Ti
Cr-Si	0,37	1,33	0,62	0,02	0,02	1,58	-	-	-
Cr-Si-V	0,38	1,38	0,62	0,018	0,018	1,60	0,15	-	-
Cr-Si-Nb	0,38	1,39	0,50	0,023	0,020	1,66	-	0,11	-
Cr-Si-Ti	0,37	1,36	0,51	0,023	0,020	1,62	-	-	0,18

Table 1. Chemical compositions of the steels.

Assuming that alloying by carbide forming elements (CFE) would alter the critical points, these were estimated using dilatometry.

As a result, it was established that microalloying has little practical effect on A_{c1} and A_{r1}, but slightly increases A_{c3} and A_{r3}, especially for Ti containing alloy where a 15 degC rise was noted. For the V alloyed steel, the critical point A_{r3} was found to be lower than the unalloyed steel.

Specimens 120 mm long, 28 x 28 mm cross-section, machined from hot rolled and annealed pieces, were heated in an electric oven for 50 min at 920 degC and then rolled in a mill with roll diam 210 mm at 0,3 m/sec. Deformation was performed with one pass, the degree of reduction was 30%. Immediately after hot rolling the rolled specimens were cooled rapidly in industrial oil and tempering at 500 degC with oil-cooling. The specimens without deformation (control treatment) were heat treated according to following the conditions: quenching from 920 degC, cooling in oil and tempering at 500 degC with oil-cooling.

The microstructure investigations were done prepared perpendicular to the rolling axis. The results of the mechanical tests ($\sigma_{0,2}$, δ, ψ, KCU) are given in Table 2. The metallographic structure was studied in the "Neophot-2" optical microscope, fractographic investigations was carried out on the "Philips" scanning electron microscope, dislocation structure - in transmission electron microscope at 200 kV.

RESULTS AND DISCUSSION

Microstructure. After quenching the steel has martensitic or bainite structure, depending on the cooling rate. The typical martensitic structure comprises a set of long, elongated, relatively broad (0,5-1,0 µm) plates and thinner (0,1 µm) weakly misoriented laths, uniformly filled by dislocations of high density ($\rho \approx 10^{11}$ cm^{-2}). As cooling rate falls, the structures of lower or upper bainite arise, characterized by the same high dislocation density and a corresponding distribution of platelike cementite (6). In the thermally improved state the steel has tempered structure arising from the polygonisation of dislocations in the martensitic, or bainitic structure and the precipitation of dispersed carbide particles (M_7C_3 and M_3C) uniformly distributed throughout the matrix. In this case the dislocation density was of the order 10^{10} cm^{-2}. The Ti alloyed steel has the greatest degree of dispersion of the microalloyed steels. The same effect was observed with microalloing in combination with CFE. HTMP further refines the steel structure. A study of the non-metallic inclusions has shown that steel without microalloying, when quenched and tempered contains oxides with a basically globular shape, silicates of a non-regular shape and extended sulphides. Mean sizes of oxides are 20-25 µm, silicates 20-30 µm, sulphides 30-40 µm. Microalloying with CFEs results in a reduction of number of oxides with size reductions down to 20 µm. In Ti and Nb microalloyed steels there are accumulations of $FeO \cdot TiO_2$ and $FeO \cdot NbO_2$ type oxides in the size range 1-2 µm. Also, the size of silicate inclusions decreased to 5-15 µm, and many nitrides and carbonitrides of the 1-2 µm size were evident. At the same time, little change was noted in the sulfide morphology. Microalloying of steel by REM effects the reduction of oxide numbers (from 4 to 1 mark under estimation according maximum mark system). The nonmetallics in the microalloyed CFE with REM steels changed similarly, and in the case of REM and Ti or Nb microalloying the sulfide shape become globular, with the size of 10-15 µm. As a result of HTMP the character and size of nonmetallics practically did not changed. As for dislocation structure, it was changed. High temperature deformation of austenite gives rise to a developed fragmented structure comprising microvolumes which are misoriented by several degrees and have a cross-section of the order 1-2 µm. Torn off dislocation boundaries (partial disclinations) with a misorientation of 1-2 degrees are often observed in the structure, either individually or grouped in a dipole or more complex multipole formations. The austenite which has this deformation-induced structure undergoes a phase transformation during subsequent cooling. In the specific conditions of the present experiment, the transformation involved martensite-bainite reactions. All the structures, typical of the reactions, clearly inherit the distinguishing signs of the fragmented structure of hot-deformed austenite. This is evidenced by refinement of the individual crystallites and the appearance of cross boundaries which intersect larger crystallites, such as lath colonies. This effect has previously been discovered in (6, 7). *Mechanical properties.* The mechanical properties of investigated steels after control heat treatment and HTMP are given in the Table 2.

Steel	$\sigma_{0,2}$ MPa	σ_B MPa	δ %	ψ %	KCU^{+20} MJ/m^2	KC_f^{+20} MJ/m^2	KC_g^{+20} MJ/m^2	KCU^{-60} MJ/m^2	KCU^{-100} MJ/m^2
-	1315	1374	13,0	49,2	0,31	0,15	0,16	0,30	0,13
	1270	1368	13,3	48,6	0,46	0,24	0,22	0,35	0,08
V	1367	1417	9,3	46,3	0,31			0,30	0,17
	1304	1392	12,3	38,3	0,40			0,36	0,12
Ti	1308	1350	11,7	52,4	0,40			0,39	0,16
	1192	1318	12,8	53,0	0,47			0,41	0,15
Nb	1304	1357	10,8	47,1	0,38			0,38	0,21
	1202	1311	13,4	53,7	0,54			0,44	0,19
REM	1270	1373	11,9	53,2	0,41	0,23	0,18	0,38	0,36
	1378	1459	12,2	31,3	0,39	0,19	0,20	0,33	0,32
V-REM	1305	1421	13,2	54,0	0,45	0,25	0,20	0,33	0,37
	1187	1302	11,8	52,0	0,46	0,25	0,21	0,33	0,38
Ti-REM	1242	1356	13,3	53,6	0,45	0,23	0,22	0,42	0,36
	1253	1339	13,2	53,2	0,46	0,22	0,24	0,35	0,32
Nb-REM	1265	1383	13,4	53,6	0,48	0,25	0,23	0,45	0,41
	1247	1349	12,3	45,9	0,45	0,21	0,24	0,38	0,38

Table 2. Mechanical properties of steels.

Notation.
1. Numerator - after quenching and tempering
Denominator - HTMP + tempering
2. Impact strength was determined on the "PSWO-30" impact machine with oscillographed fracture diagram for determination Kc_f and Kc_g ($KCU=KC_f+KC_g$).
KCU - impact strength for specimens with notch R1 mm under +20 degC (KCU^{+20}), -60 degC (KCU^{-60}), and -100 degC (KCU^{-100}).
K_f - fracture formation energy (impact strength)
K_g - fracture growth energy (impact strength)

It can be seen from Table 2 that V-microalloying results in increased strength properties, lower plasticity and the same level of impact strength when compared to the non-alloyed steel. With the addition of REMs to the basic steel, the impact strengths increased by some 30% and the join action of CFE and REM provided an additional increase of 10-15%. The best combination, strength/plasticity/impact strength was measured in the Nb plus REM steel.

The pecularities of CFE microalloyed steels behaviour can be explained in terms of the kinetics of carbide dissolution into an austenite solid solution (during heating prior to quenching), and the subsequent decomposition of this solid solution during tempering. The dissolution start temperature for V-carbides is 950 degC, Ti-carbides 1300 degC and Nb-carbides 1050 degC. Furthermore, it is known that, in the presence of small CFE additions, dissolusion temperatures are lowered.

On this basis, it is possible to assume that some V-carbides are dissolved in the austenite during preheating for quenching. Later tempering at 500 degC results in a decomposition of martensite and a release of the V-containing carbide phase, giving rise to a dispersion hardening effect resulting in a higher temper resistance in this type of alloy. A small reduction in the yield strength, in comparison to the basic steel, observed in the Ti- and Nb-steels may be expained as a result of TiC/NbC re-precipitation during heating prior to hardening, whereby the depletion of C in solid solution results in some softening.

The improved plastic and impact strength properties in alloys with added REM may be explained by their high affinity for oxigen and sulphur, resulting in a change in the nature of non-metallics. Specifically, the pollution of steel by oxides and low melting point sulphides, FeS·MnS, is replaced by high melting point inclusions (Ce₂S₃) which do not easily deform and maintain their globular shape after plastic deformation.

HTMP results in a change in mechanical properties of all the investigated steels. Impact strength of basic and V-, Ti- and Nb-microalloyed steels are increased by 50, 30, 50 and 75% respectively. Simultaneously, the elongation was increased by 30% accompanied by some lowering of yield strength. As a result of HTMP, the strength properties of REM-alloyed steels were increased, but with a decrease of the reduction in area.

For the (V+REM) steels, it was noted that the strength properties were lowered, for the (Nb+REM) steel the impact properties were lower, whilst for the (Ti+REM) steel the mechanical properties were essentially unchanged. The main positive effect of HTMP was observed with the Basic and CFE-microalloyed steels, where impact and elongation properties were improved.

The estimation of ductile brittle transition temperature (TT) had been determined both by the method according to which TT is corresponds to half the impact strength of the room temperature value (0,5KCU^{+20}) (8) and by the original relaxation method using highly sensitive laser interferometry to measure the microplastic deformation rate. In fig.1 are shown the curves for rate of microplastic deformation versus loading temperature from which the $T_{\dot{\varepsilon}}$ was determined.

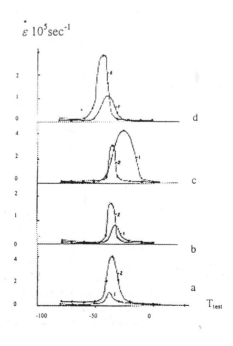

Fig.1. The microplastic deformation - loading temperature relationship of 38CrSi (a), 38CrSiV (b), 38CrSiTi (c), 38CrSiNb (d), treated by thermally improved heat treatment (1) and HTMP (2).

Fig.2. shows the comparison data of the ductile-brittle transition temperature, determined by $0,5KCU^{+20}$ and by the relaxation method.

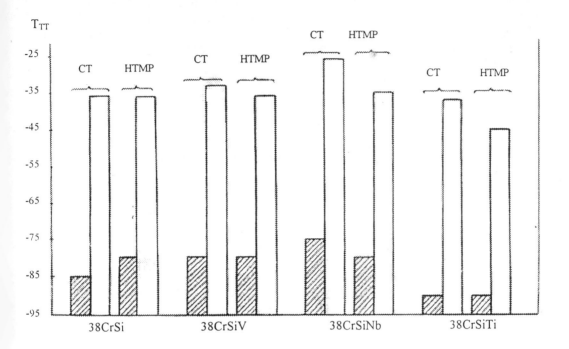

Fig.2. The transition temperature, determined by $0,5KCU^{+20}$ (hatched R1) and by relaxation method.
CT - control treatment (thermally improved).
HTMP - high temperature thermomechanical processing.

These data, as evidenced from the fig.2, show major differences in the value of TT. It is easily clarified: $0,5KCU^{+20}$ - is conventional method, which are used in the case when it is impossible to estimate ratio between ductile and brittle components of the fracture. This method practically does not account for the physical processes which are involved in steels on the microlevel under temperature testing. In contrast the method of determine by $T_{\dot{\varepsilon}}$ is very sensitive method to relaxation of microelastic stresses. As this takes place, comparative estimations, of ductile-brittle transition temperature agree (see Fig.2.). Both as a result of measured T_{TT} by $0,5KCU^{+20}$ and by $T_{\dot{\varepsilon}}$ method, the lowest T_{TT} are corresponding to Nb-content steel treated by HTMP, but the highest T_{TT} is corresponding to Ti-content steel, treated by thermally improved heat treatment.

As a result of the estimation of cold resistance it was established that impact strength of Ti and Nb-content thermally improved steels under the +20 to -60 degC testing temperatures was increased 20-30%, after entering of REM - up to 50%, with the latter level retained up to -100 degC.

As a result of HTMP the impact strength in the +20 to -60 degC temperature range of basic and CFE microalloyed steels were increased in comparison with thermally improved condition, specifically for Ti- and Nb-content steels (15-20%). Further lowering of testing temperatures results in a sharp decrease, more intensive than with thermally improved steels. The latter is attributable to more stressed structure of HTMP treated steel because of high integral density of crystal lattice. As for increasing of impact strength of HTMP treated steels under +20 to -60 degC testing temperatures, it may be explained by forming of a fragmented, with medium angle boundaries, structure in the hot rolled austenite, some features of which were inherited as a result of hardening under rapid cooling by forming martensite-bainite structure. Such type of austenite boundaries create the retardation of crack extension, and this is corresponding to the data of Table 2.

The REM content steel as a result of control treatment alike HTMP have the upper threshold of cold brittleness, at least was 40^0C lower than the other ones.

As a result of fracture analysis, it was established that the Basic steel, thermally improved, tested at less than room temperature has a principally brittle fracture, with regions of intercrystalline and quasi-cleavage and occasional pockets of ductile dimple fractures (up to 10%). V-microalloying increases the fraction of ductile fracture to 50% and Ti - to above 50%. Fractures of the Nb-steel were mixed (50% ductile, 50% quasi-cleavage). Most REM- and (Nb+REM) microalloying steels were ductile (70%) with regions of intercrystalline cleavage (up to 30%).

HTMP increased the ductile fracture in the basic steel from 10% to 40%. With CFE-steels, the fractures became fully ductile. Regions of brittle crystalline fracture, in the REM- and (Nb+REM) steels, disappears, but quasi-cleavage was observed (up to 20%).

With a lowering of the testing temperature to -60 degC, all fractures were essentially brittle.

The thermally improved basic and V-steels had intercrystalline brittle fractures with the small amount of quasi-cleavage, whilst the Nb-steel was approx. 60% of quasi-cleavage. The main fracture mechanism for REM and (Nb+REM) steels was quasi-cleavage, although it was sometimes possible to detect regions of ductile dimples (up to 20%). The fractures of all HTMP treated steels (-60 degC) were a quasi-cleavage mechanism, but the REM steels did exhibit a mixed ductile/quasi-cleavage fracture.

CONCLUSIONS. 1. Microalloying of 38CrSi steel by V, Ti, Nb result in strengthening only in case of dissolving carbides during hardening preheating as this take place plastic properties and impact strength are slightly reduces.
2. HTMP is lowers sensibility of 38CrSi basic and with microalloying to non-reverse tempering brittleness.
3. Microalloying by titanium and niobium allows to increase in 20-30% impact strength for 500 degC tempering temperature of constructional steel 38CrSi type, especially for Nb-content steel.
4. CFE and REM microalloying results in increasing the fraction of energy intensive fracture type.

REFERENCES

1. M.A.Linaza, J.L.Romero, J.M.Rodriguez-Ibabe and J.J.Urcola. Influence of Thermomechanical Treatments on the Microstructure and Toughness of Microalloyed Engineering Steels, Thermomechanical Processing in Theory, Modelling & Practice, Stockholm, Sweden, 1996, pp.351-356.
2. K.Hulka and et. al., Processing, Microstructure and Properties of Microalloyed and other HSLA Steels; ISS, Warrendale (PA), 1992, pp.177-187.
3. V.Gurashov, L.Sokolov. Alloying of steels by rare earth metals. Metal Science and Heat Treatment. March, 1969, № 3, pp.27-64.

4. S.Zajac, T.Siweck, B.Hutchinson and M.Attlegard. Recrystallization Controlled Rolling and Accelerated Cooling for High Strength and Toughness in V-Ti-N Steels. Metallurgical Transactions A, November, 1991, vol.22A, pp.2681-2694.

5. G.Kodjaspirov, E.Gulikhandanov, R.Suliagin. Influence of thermomechanical processing on structure and mechanical properties of HSLA 0,1Mn Ni Nb steel. High Technologies in Advanced Materials Science and Engineering, St.Petersburg, Russia, 1997, pp.17-18.

6. V.Rybin, G.Kodjaspirov and A.Rubtsov. Fine transformation in steel 38KhS during deformation near the tempeature of the pearlite transformations. The Physics of Metals and Metallography, 1986, vol.61, № 1, pp.150-158.

7. V.Rybin, A.Rubtsov, G.Kodjaspirov. Structure transformations in steel during rolling with different degrees of deformation using one and many passes. Physics of Metals and Metallography, 1984, vol.58, № 4, pp.131-134.

8. N.A.Makhutov. Methods of determining critical brittleness temperatures for materials and structural members. Zavodskaya Laboratorya, 1983, № 11, pp.78-80 .

Materials Science Forum Vols. 284-286 (1998) pp. 343-350
© *1998 Trans Tech Publications, Switzerland*

Development Trends in HSLA Steels for Welded Constructions

K. Hulka and F. Heisterkamp

Niobium Products Company GmbH Düsseldorf, Germany

Keywords: High Strength Low Alloy (HSLA) Steel, Thermomechanical Rolling, Heat Affected Zone (HAZ) Toughness, Peritectic Reaction, Fine Acicular Bainite, Accelerated Cooling, Direct Quenching

Abstract

Costs and time savings in the fabrication of constructions by using steel grades with improved weldability and higher strength lead to an increased realisation of steel structures. But the higher strength can only be used, provided the construction is safe, by avoiding any brittle fracture as well as the propagation of existing flaws by ductile fracture.

Besides a high cleanliness of the steel, a low carbon content and the maximisation of grain refinement as strengthening mechanisms are the fundamentals of modern HSLA steel production. As result of the thermomechanical processing of microalloyed steels, high strength levels with a rather low carbon content can be produced. Actually a chemical composition with less than the 0.09% C threshold value is aimed for, in order to avoid the peritectic reaction during solidification, which is responsible for microsegregations and thus a deterioration of the heat affected zone toughness.

A finer effective grain size than typical for thermomechanically rolled steels with a ferritic-pearlitic microstructure is achieved with microstructures of acicular bainite or martensite. The alloy design of such steels, which can guarantee yield strength levels of 690 MPa and more, asks for an overall higher alloy content or a faster cooling rate or combinations of both. In any case, austenite processing before transformation adds to the refinement of these microstructures. The industrial verification relies on interrupted accelerated cooling or direct quenching after thermomechanical rolling. Furthermore, the toughness in the heat affected zone is improved with a microstructure of low carbon acicular bainite.

Background

A reduction in material thickness and weight can be achieved by applying steel with higher strength. This reduction is highest, when only uniaxial stresses occur, such as in pipelines: by doubling the yield strength there, half of the wall thickness is sufficient to carry the load [1]. But also with other stresses deriving from bending or torsion loads the weight reduction is still remarkable [2]. This economic result is demonstrated by the example of beams with different strength levels, Fig. 1 [3], indicating that a higher strength of the steel allows the usage of lighter profiles. Such remarkable weight reduction compensates by far the higher costs for the microalloyed steels. Furthermore, the lower weight reduces fabrication costs such as transportation and handling. Especially the efforts for welding are drastically reduced: related to the reduction in wall thickness, the reduction in the weld metal volume amounts to an exponent of two and the weld metal volume determines the production time needed for the fabrication of the component.

This obvious economic benefit can only be used, if the construction is safe. First of all the steel must not be brittle at working temperature and second it should exhibit sufficient ductility to withstand any crack propagation. Fig. 2 shows results from fracture mechanics tests, indicating that the necessary ductility increases with the square of the yield strength [4]. The European design code for steel constructions [5] describes a safety analysis based on fracture mechanics and a practical approximation by using the more typical Charpy V notch impact test data.

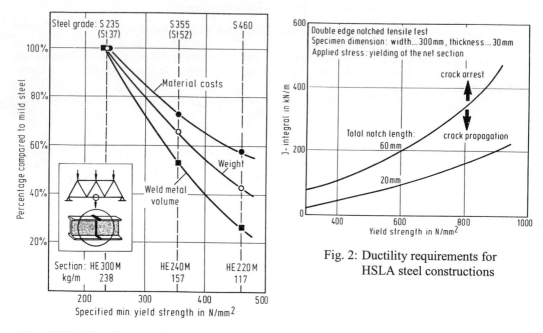

Fig. 2: Ductility requirements for HSLA steel constructions

Fig. 1: Weight and costs reduction by using HSLA steels

An improvement in ductility is obtained with cleaner steel, exhibiting low levels of sulphide and oxide inclusions, and also by reducing the volume fraction of the second phase pearlite, thus asking for a low carbon content, Fig. 3 [6]. In modern high strength low alloy (HSLA) steels advantage is taken of grain refinement as the major strengthening mechanism, since it is the only metallurgical method to improve both, strength and toughness of steel. Fig. 4 [7] indicates the strength and toughness increase obtained with a finer ferrite grain size. All the other strengthening mechanisms, i.e. precipitation, dislocation, or solid solution hardening and especially the traditional means of steel strengthening, using a higher carbon content, have a toughness impairing effect.

Thus the maximisation of grain refinement dominated the steel development in the last decades. It was supported by making use of the outstanding effects obtained by the carbide and nitride forming microalloying elements [8]. Microalloys were used first in normalised ferritic/pearlitic steel. Most of the modern HSLA steels are pearlite reduced or even pearlite free and their processing is dominated by thermomechanical rolling, often applied together with enhanced cooling after rolling. The fundamentals of thermomechanical rolling have been summarised elsewhere [9].

The reduction in non-metallic inclusions is so effective that the possible weldment failure 'lamellar tearing' has become historical. Furthermore, with the lower carbon equivalent of modern HSLA steels the cost effective preheat treatment, necessary to guarantee crack free weldments can be drastically reduced or even avoided [10]. Thermomechanically rolled HSLA steels exhibit also improved weldability and eventually allow the application of more economic welding processes such as high heat input welding.

Safety considerations including toughness requirements concern the whole component including the weldment, weld metal and heat affected zone. Therefore, the steel development towards steels with even higher strength than actually applied has to take into account the poorest toughness observed in a component, which is often the grain coarsened heat affected zone (HAZ), especially in low carbon steels.

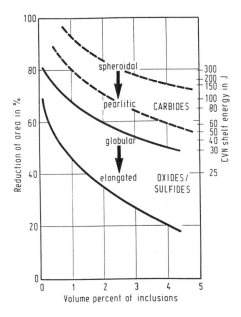

Fig. 3: The effect of second phases on ductility

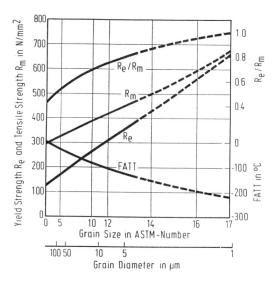

Fig. 4: The effect of ferrite grain size on strength and toughness

The heat affected zone toughness

Besides the non-metallic inclusions two other microstructural features influence the HAZ toughness. As shown in Fig. 5 [11], these are the fracture facet size and the volume fraction of martensite-austenite constituents (M-A). The latter represent local brittle zones and act as initiation sites for brittle fracture, which is more harmful with a coarser effective grain size. In the following, the main influencing factors on both parameters will be discussed.

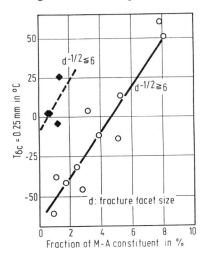

Fig. 5: Influencing factors on the critical CTOD value

Based on test results of a group sponsored research project at The Welding Institute with 50 mm plates of grade 50 steel [12], a correlation has been found between low CTOD data in the HAZ and a carbon content above 0.09%. Metallographic investigations studying the microstructure from the fusion boundary to the base plate indicated, that all steels with more than 0.09% C exhibited pronounced segregations while insignificant segregations were observed in steels with lower carbon content than this threshold value [13].

Fig. 6 explains the reason of such segregations by discussing the solidification behaviour of steel, having in mind that primary segregations are caused by the interdendritic inclusion of the liquid phase, which is naturally enriched in carbon and the other alloying elements. The additional shrinkage, which occurs during the peritectic reaction as a result of the transformation from the cubic body centred delta ferrite into the cubic face centred austenite is responsible for the fact, that 0.09% C became a threshold value. The relatively highest shrinkage occurs with the highest amount of solidified delta ferrite before start of the peritectic reaction, i.e. with steels of about 0.10 - 0.12% C. The same mechanism is known to be responsible for enhanced surface cracks in concast material of steels with such carbon levels.

The most relevant interdendritic enrichment of alloying elements in HSLA steels is of the major alloying element manganese. Such segregation will be only gradually reduced during further solidification and rolling and in the final plate, the segregated zones exhibit manganese levels as high as 2.5%, corresponding to a segregation ratio of about 2. Manganese stabilises the austenite phase and causes an enrichment in carbon in the region of final transformation. Since both elements C and Mn add to hardenability, an increased formation of local brittle phases in the HAZ is observed in the segregated regions. The correlation between segregations in the steel, the amount of M-A constituents and a deterioration of the HAZ toughness has been also observed elsewhere [14].

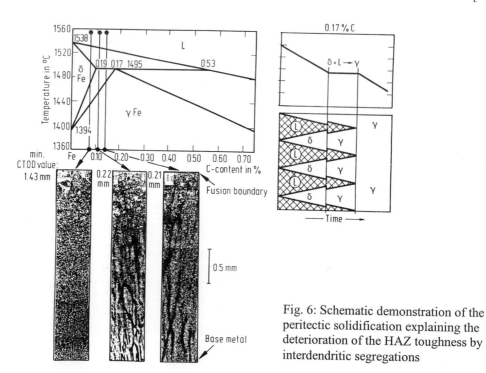

Fig. 6: Schematic demonstration of the peritectic solidification explaining the deterioration of the HAZ toughness by interdendritic segregations

Having acknowledged the outstanding importance of carbon levels below 0.09% for a good HAZ toughness, it is worth mentioning, that a further reduction below this threshold value will reduce also the crystal segregations and improve the homogeneity. It is caused by two reasons:
- First, the temperature range liquidus-solidus becomes smaller with a lower carbon content, thus reducing the formation of crystal segregations already during solidification and
- second, the post solidification cooling takes place in a wider temperature range of the delta ferrite with lower carbon content. The approx. 100 times higher diffusion coefficient in the ferrite compared to the austenite phase at the same temperature adds to post solidification homogenisation.

Besides the outstanding role of a low carbon content, which guarantees that only a small volume fraction of local brittle zones will be formed in the HAZ, there exists an optimum alloy content for each carbon content and welding condition, which allows control of the fracture facet size:

It is known from fundamental investigations [15] that relatively poor toughness is correlated with a coarse grained ferritic-pearlitic microstructure, Widmanstätten ferrite, and especially with a microstructure of granular bainite, where the effective grain size corresponds to the former austenite grain. On the other hand, in martensitic or acicular bainitic microstructures the effective grain size corresponds to the needle width of the microstructural component, which is naturally much finer and typically below 1.5 microns. The latter microstructures guarantee low transition temperatures of ductile to brittle fracture.

In order to improve the toughness in the HAZ, a control of the austenite grain size by particles being stable at the high temperature typical for the region close to the fusion boundary, e.g. 1400° C, such as TiN or 'TiO', is often applied and also intragranular nucleation of the transformation adds to a certain grain refinement. These particles are very effective in such microstructures, where the grain size is directly related to the former austenite grain size, i.e. a ferrite-pearlite microstructure and especially the granular bainite, but by far more effective with regard to grain refinement is the transformation into a microstructure of fine acicular bainite.

By analysing various literature data on HAZ toughness results, Kirkwood has found, that both, the carbon content and the mean transformation temperature determine the HAZ toughness and the best values correspond to an optimum microstructure [16]. By lowering the transformation temperature, the microstructure changes from coarse bainitic and Widmanstätten structures to more acicular microstructures, causing toughness improvement. But with a mean transformation temperature below 350° C the martensitic microstructure observed in the investigated steels brings up a high volume fraction of local brittle phases causing again an impairment in toughness.

The mean transformation temperature depends on the welding conditions (heat input, preheat) and the chemical composition of the steel. Fig. 7 [17] summarises results of welding simulation, indicating that there exists an optimum alloy content for each welding condition and carbon content. The cooling rates of 1° C/s, 10° C/s and 100° C/s correspond to high heat input welding (>10 kJ/mm), submerged arc welding with about 4 kJ/mm heat input and the heat input of manual metal arc welding, respectively. Especially the high heat input welding processes demand a very low carbon content and a rather high alloy content in order to guarantee reasonable toughness in the HAZ. This is a different result as known from the typical carbon equivalent regression formulae, which consider the role of the alloy elements to be additive to the role of carbon.

Since the amount of investigated steels within this study did not allow the development of an own quantification of the influence of the various alloying elements on the hardenability, the factors from the PCM formula [18], developed to describe the cold cracking resistance of low carbon steels, have been applied. This formula does not include the microalloying elements niobium and titanium. There are indications, that both elements in solid solution have a strong transformation retarding effect about twice as effective as vanadium. Therefore they have either a positive or negative influence on the HAZ toughness as all the other elements, since the toughness is a result of the overall steel composition determining the microstructure.

The applied welding simulation process describes the grain coarsened heat affected zone, which is often the area of lowest toughness. It should be mentioned, that also other zones in the HAZ especially in multipass weldments, e.g. the intercritically reheated grain coarsened zone, may exhibit results of reduced toughness. Therefore the above considerations are not valid in all cases and further investigations are necessary to complete the model.

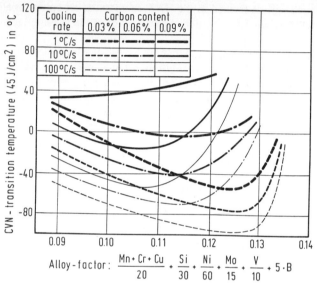

Fig. 7: Impact toughness of simulated grain coarsened HAZ (peak temp.: 1350°C)

Progress in thermomechanical rolling

Actually, the usage of thermomechanically rolled steels is established in steel constructions up to a minimum yield strength of 460 MPa [5]. and their penetration into the market is steadily progressing. The chemical composition of such structural steels is based on < 0.09% C and a manganese content of around 1.5%. Via the accelerated cooling process a microstructure of polygonal ferrite plus some bainite and the 460 MPa yield strength level is obtained even for thick wall. It should be mentioned that all hot strip mills, most of the modern plate mills and even some section mills have an accelerated cooling device at their disposal, which allows a cooling velocity of more that 10° C/s up to about 20° C/s for 20 mm material.

The extension towards higher strength levels, without impairing weldability or using cost effective heat treatments takes into account the above fundamental relationships between microstructure and toughness. A low carbon, fine grained acicular bainite offers the best opportunity to obtain a high strength steel suitable for a safe welded construction. When applying the thermomechanical rolling process before accelerated cooling a high amount of nucleation sites for the γ/α transformation is prepared and the effective grain size of bainite is typically as small as 1 μm [19]. Besides good toughness, the fine grained bainite adds also to the strength increase by both, the fine grain and the high dislocation density.

There are two means of obtaining a ferritic/bainitic microstructure, necessary for yield strength levels of more than 500 MPa, i.e. either increasing the hardenability of the steel by adding more alloying elements or by increasing the cooling velocity. The development of structural steel with 550 MPa yield strength by changing the alloy design in existing facilities has been already described [20]. When the carbon content is low, which is normally the case, the alloy design can make use of higher niobium contents. More niobium dissolved at reheating temperature causes enhanced austenite processing and allows a higher amount of niobium to remain in solid solution at finish rolling temperature, adding to both, increased hardenability and precipitation hardening [17].

Steel with a yield strength level of 690 MPa requires a fully bainitic microstructure and thus higher additions of alloying elements. Table 1 shows the chemical composition and the mechanical properties of such plate material, processed via accelerated cooling after TM processing. A steel composition, which guarantees a fully bainitic microstructure after the accelerated cooling process will exhibit a microstructure of bainite plus martensite with a further increased cooling velocity of 35° C/s or more, Fig. 8 [21]. The process is called direct quenching and results in even more favourable property combinations including also improved weldability, since the carbon content can be further lowered for a given strength level. Table 1 shows such production data, too.

Table 1: Alternative trial production of X 100 and resulting mechanical properties

a.) Accelerated Cooling	
Chemical composition:	
0.07%C, 1.90%Mn, 0.20%Si, 0.30%Mo, 0.20%Cu, 0.20%Ni, 0.05%Nb, 0.015%Ti	CE=0.48
Processing	
FRT=730°C, CST=530°C, CR=20°C/s	
Properties:	
Re=740Mpa, Rm=800Mpa, BDWTT (85%shear) <-20°C	
b.) Direct Quenching	
Chemical Composition:	
0.06%C, 1.90%Mn, 0.20%Si, 0.20%Mo, ----------, -----------, 0.05%Nb, 0.015%Ti	CE=0.43
Processing	
FRT=730°C, CST=350°C, CR=50°C/s	
Properties:	
Re=720Mpa,Rm=880Mpa,BDWTT(85%shear)<-30°C	

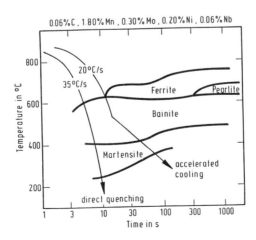

Fig. 8: CCT diagram explaining the microstructure for two cooling regimes

Usually such low carbon steels need no additional tempering process after the direct quenching operation. But it has been shown, that age hardening of directly quenched 0.04% C, copper and nickel bearing steels results in yield strength levels above 1000 MPa together with excellent toughness. The combination of properties is by far superior to the reheat quenched plus tempered condition owing to a finer lath spacing of the martensite as well as smaller particles of the copper and the niobium carbide precipitates [22].

Summary

High strength structural steels are increasingly applied for welded constructions. Their development was based on a steady reduction in the amount of non-metallic inclusions and the reduction in carbon in order to fulfil the safety criteria and the weldability demands. Grain refinement has become the dominating strengthening mechanism and an optimum combination of strength, toughness and weldability is obtained with steels exhibiting a microstructure of low carbon acicular bainite. The economical production of such steel is made by thermomechanical rolling in combination with accelerated cooling or direct quenching. The alloy design has to be balanced with the possibilities in the production facilities and the desired strength level, with microalloys being indispensable.

References

[1] J. Lessels et al., HSLA Steels, IISI, Brussels (Belgium), 1987, chapter 7
[2] D.G. Younger, Metal Progress 107 (1975), No.5, p. 43
[3] TradeARBED S.A., HISTAR, Luxembourg (Lux.), 1989
[4] R. Hubo, VDI-progress report, series 18/book 80, VDI, Düsseldorf (Germany), 1990
[5] Eurocode 3
[6] T. Gladman et al., Effect of second phase particles on mechanical properties of steel, The Iron and Steel Institute, London(UK), 1971, p. 68
[7] K. Lorenz et al., Stahl und Eisen 101 (1981), No. 9, p. 593
[8] Microalloying 75, Union Carbide Corporation, New York (NY), 1977
[9] Niobium Information 7/94, CBMM, São Paulo (Brazil), 1994
[10] R. Hubo and F. Hanus, Stahlbau 63 (1994), p. 84
[11] M. Nikanishi, Y. Komizo and Y. Fukada, The Sumitomo Search 32 (1986), p. 22-34
[12] P.L. Harrison and P.H.M. Hart, The metallurgy, welding and qualification of microalloyed (HSLA) steel weldments, AWS, Miami (Fl), 1990, p. 626
[13] F. Heisterkamp, K. Hulka and A.D. Batte, ibid. Ref. 12, p. 659
[14] J. Tanaka, World Materials Conf., Chicago (Il), 1988, oral presentation
[15] U. Lotter and R. Kawalla, Steel research 61 (1990), No 10, p. 518
[16] A.D. Batte and P.R. Kirkwood, Microalloyed HSLA steels, ASM int., 1988, p. 175
[17] K. Hulka and F. Heisterkamp, Low carbon steels for the 90's, TMS, Warrendale (PA), 1993, p. 211
[18] IIW Doc. IX-576-68
[19] M.K. Gräf, H.G. Hillenbrand and P.A. Peters, Accelerated cooling of steel, TMS, Warrendale (PA), 1986, p. 165
[20] K. Hulka et al., Processing microstructure and properties of microalloyed and other modern high strength low alloy steels, ISS, Warrendale (PA), 1992, p. 177
[21] F. Kawabata et al., Pipeline technology, Elsevere, Amsterdam (NL), 1995, Vol. II, p. 263
[21] J.Y.Yoo et al., ISIJ international, Vol. 35 (1995), p. 1034

For correspondence please contact: Dipl.-Ing. Klaus Hulka
e-mail: kh.niobium@t-online.de   home-page: http://www.cbmm.com.br
fax: +49-211-32 91 76

Materials Science Forum Vols. 284-286 (1998) pp. 351-358
© *1998 Trans Tech Publications, Switzerland*

Cleavage Fracture Micromechanisms in Ti, Ti-V and C-Mn-B Microalloyed Bainitic Steels

A. Echeverría, M.A. Linaza and J.M. Rodriguez-Ibabe

CEIT and ESII, P° Manuel de Lardizabal 15, E-20009-San Sebastián, Basque Country, Spain

Keywords: Bainitic Structures, Cleavage, Ti, Ti-V and C-Mn-B Steels

ABSTRACT

Cleavage fracture micromechanisms have been analysed in four microalloyed steels with bainitic microstructures utilising 4 point bending tests at 77K. In the case of Ti microalloyed steels, the cleavage nucleation has been related to the cracking of a coarse TiN particle. In contrast, with the other two steels (V microalloyed steels) it has not been possible to correlate the cleavage initiation to some microstructural feature. In all the cases a FEM analysis has been considered to study the stress and strain distributions. The results show that in these materials, when there are no TiN brittle particles, some strain is necessary to initiate fracture. In consequence, the critical cleavage fracture stress can be insufficient to characterise the brittle process of the studied microstructures and another parameter including the contribution of strain would be necessary.

INTRODUCTION

In most metals the process of brittle cleavage fracture has been attributed to the nucleation of a microcrack followed by its propagation into the surrounding matrix [1]. It has been found that the cleavage fracture of mild steels obeys the critical tensile stress criterion and that this σ_F critical stress is weakly dependent on temperature [2]. In the case of bainitic steels, the carbide size distribution has been considered to be the microstructural feature that controls cleavage fracture stress value at 77K [3]. Nevertheless, in bainitic microstructures the identification of the controlling microstructural parameter can prove quite difficult (the modification of one parameter promotes the variation of other microstructural features [2]).

In a previous work, the cleavage fracture of Ti, V and Ti-V ferrite-pearlitic and martensitic microstructures has been identified as a process controlled by the propagation into the matrix of a microcrack nucleated at a TiN coarse particle [4, 5]. In this work, four different steels with bainitic microstructures will be studied. Two of the steels are Ti microalloyed containing coarse TiN particles that can act as microcrack nucleation sites. The objective is to compare the cleavage process of bainitic steels, with and without these brittle particles, and to analyse their influence on the critical cleavage fracture stress.

MATERIAL AND EXPERIMENTAL PROCEDURE

The chemical composition of the four steels studied is given in Table 1. As can be observed, there are three types of steels. The first material is a Ti-V microalloyed medium carbon steel (steel A). The second group consists of two V microalloyed steels, one (steel C) with 0.20% Mo. In the third group a C-Mn-B steel with low C content is considered (steel D).

In order to obtain bainitic microstructures, the steels were austenitised for one hour at 1050°C and transformed isothermally for 30 min. at 400°C in the case of steels A and B and at 375°C for steels C and D. As a consequence of the addition of Ti, there exists in steels A and D a fraction of TiN precipitated as coarse particles (> 1 μm). Due to the relevance of these particles in the fracture behaviour, histograms of minimum and maximum dimensions measured by optical microscopy are given in Fig. 1.

Table 1. Chemical composition of the steels in wt %

Steel	C	Mn	Si	S	Al	V	Ti	Mo	B (ppm)	N (ppm)	Ti/N
A	0.21	1.51	0.27	0.002	0.015	0.19	0.017	0.05	-	103	1.65
B	0.24	1.60	0.25	0.002	0.018	0.19	0.001	0.04	-	90	0.11
C	0.25	1.63	0.26	0.002	0.020	0.18	0.002	0.20	-	102	0.19
D	0.09	1.35	0.25	0.004	0.031	-	0.022	-	52	66	3.3

Tensile test specimens were machined from heat treated samples and tests were carried out at 77K. The cleavage fracture strength was measured by means of four point bending tests with V-notched specimens as described by Griffiths and Owen (12.7x12.7x75 mm^3) [6]. In the case of steel B, due to the geometry of the as-received material, the section of the specimen was limited to 11.0x11.0 mm^2. Tests were conducted at 77K at a constant crosshead speed of 2.10^{-3} mm/s. For each condition three different tests were performed.

The stress and strain distributions in the vicinity of the notch in the four point bending specimen were determined by finite element methods considering ABAQUS code [7], for plane strain conditions and taking into account the work hardening behaviour of each analysed microstructure.

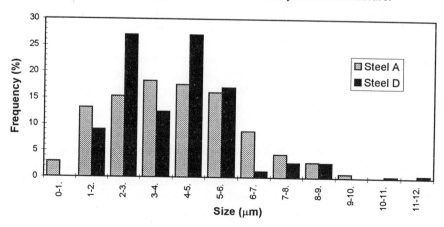

Fig. 1. Histograms of maximum dimension of coarse TiN particles in A and D steels.

RESULTS

Microstructures obtained by isothermal treatment consisted of upper bainite, as can be seen in Fig. 2. The mean austenite grain sizes are : steel A 23.6 μm, steel B 18 μm and steel D 39 μm. As can be observed in Fig. 2, in the case of coarser austenite grain sizes, from each grain several bainite packets are formed. On the other hand, from small austenite grains, the number of packets diminishes.

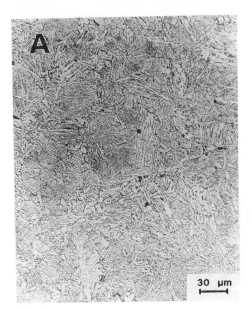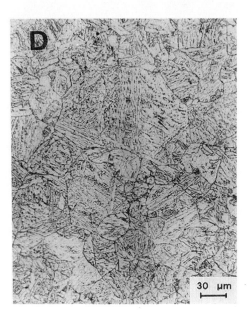

Fig. 2. Bainitic microstructures of steels A and D.

Mean values of tensile test properties measured at liquid nitrogen temperature are summarised in Table 2, together with the constants K and n of the Hollomon equation. The macroscopic cleavage fracture stress, σ_F, calculated from the load at fracture and the finite element analysis, is reported in Table 3. From the fractographic analysis of the specimens, it was possible to identify the cleavage origin of the fracture. In the case of Ti microalloyed A and D steels, the cleavage origin was always associated with a TiN particle. On the contrary, for steels B and C it was not possible to correlate the cleavage origin with some microstructural feature. An example of a cleavage process initiated at a TiN particle and a cleavage origin without an identified microstructural feature are shown in Fig. 3.

Table 2. Tensile test properties at liquid nitrogen temperature

Steel	$\sigma_{0.2\%}$ (MPa)	UTS (MPa)	K	n
A	1038	1176	1579.1	0.081
B	1044	1188	1622.0	0.089
C	1074	1250	1727.0	0.090
D	1011	1094	1404.4	0.066

The distance from the root of the notch to the position of the cleavage initiation was measured (d_{notch}). With this value and considering the stress distribution with distance from the notch root given by the finite elements method, the local cleavage fracture stress, σ_F^*, was determined. The obtained values of σ_F*, d_{notch} and d_{max} (distance at which the normal stress takes the maximum value) are summarised in Table 3. In the table, the size d_p of the plastic zone at the notch plane is also included. As can be observed in the table, in some specimens of steels B and C the cleavage fracture origin was very close to the notch root.

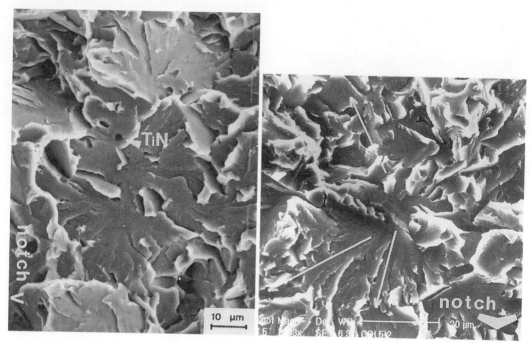

Fig. 3. Cleavage initiation sites in 4PB specimens tested at 77K: a) cracked TiN particle at the origin of the fracture process (steel A) and b) origin without identified broken particle (steel B).

Table 3. Macroscopic and local cleavage fracture stresses

Steel	P_{max} (N)	$\sigma_{nom}/\sigma_{0.2\%}$	σ_F (MPa)	σ_F* (MPa)	d_{notch} (µm)	d_{max} (µm)	d_p (µm)
A	23320	0.94	2070	1943	125	218	283
	25940	1.03	2150	1845	374	243	371
	25280	1.02	2130	2118	207	230	360
B (*)	31980	2.03	2553	1539	0	500	1044
	32860	2.09	2586	2250	225	550	1129
	36710	2.24	2681	1534	0	570	1289
C	45300	1.79	2639	2343	210	447	1168
	40970	1.62	2544	1718	10-15	395	969
	37570	1.48	2444	1687	11	350	789
D	19000	0.77	1823	1650	63.5	165	230
	21860	0.89	1942	1795	119	207	287
	18960	0.76	1813	1793	160	162	225

(*) specimens with 11.0x11.0 mm^2 section

In the case of specimens with TiN particles at cleavage origins, a_{min} and a_{max} minimum and maximum dimensions of the particle and D_{min} and D_{max} dimensions of the first formed cleavage facet were measured. These data and the geometric mean values of both parameters are given in Table 4.

Table 4. TiN particle and first cleavage facet dimensions

Steel	σ_F^* (MPa)	a_{min} (μm)	a_{max} (μm)	a_{mean} (μm)	D_{min} (μm)	D_{max} (μm)	D_{mean} (μm)
	1943	3.0	5.0	3.9	7.0	24.0	13.0
A	1845	1.1	1.4	1.2	12.9	24.3	17.7
	2118	2.0	3.3	2.6	9.4	36.0	18.4
	1650	1.4	2.1	1.7	24.7	32.6	28.4
D	1795	2.3	4.4	3.2	27	48.2	36
	1793	-	-	-	-	-	-

DISCUSSION

Fractographic analysis of Ti microalloyed steels A and D shows that the cleavage origins are related to broken TiN particles. This behaviour is similar to other observations with steels containing TiN coarse particles in ferritic, ferritic/pearlitic and martensitic microstructures [4, 5]. This implies that, contrary to what happens in upper bainitic steels, the nucleation of the microcrack has taken place in a brittle TiN particle and not in a carbide [3]. As a consequence, the TiN coarse particle distribution appears as the dominant microstructural feature in the cleavage initiation process at 77K. Conversely, in steels B and C without Ti addition, it has not been possible to identify the microstructural feature promoting the nucleation of the microcrack. In these circumstances, the possibility arises of considering the nucleation as a process related to the fracture of some small carbide that has not been identified by SEM.

The brittle behaviour of steels A and D show some differences. Although in both cases, cleavage origins are TiN particles, $\sigma_{nom}/\sigma_{0.2\%}$ and σ_F^* values are lower in the case of steel D. In both cases the strain distribution ahead of the notch root is very similar, as seen in Fig. 4, and, from the point of view of TiN coarse particle size distribution and volume fraction, the differences are not so relevant as to have a big influence on cleavage fracture stresses (the size of the particles responsible for cleavage initiation are very similar in both steels as shown in Table 4). Perversely, the first cleavage facet size is coarser in the case of steel D (see Table 4). In Fig. 5 the histograms of cleavage facets distribution for both steels are shown. As can be observed, while in the case of steel A only ~20% of facets have a size coarser than 18 μm, in the case of steel D the fraction is ~35%. This coarser microstructure can contribute to facilitate fracture of TiN particles (the local strain concentration near the particle will be higher). Other authors have also pointed out a possible effect of grain size at a constant brittle particle size (carbides in their case) [8].

In relation to steels B and C without Ti, the values of $\sigma_{nom}/\sigma_{0.2\%}$ and σ_F are higher than in steels A and D. Comparing steels A and B (with similar chemical composition, heat treatment and tensile properties), the addition of Ti promoting the formation of a fraction of coarse TiN particles has contributed to a reduction of 26% in σ_F and a 100% in $\sigma_{nom}/\sigma_{0.2\%}$. These differences cannot be related to the smaller width of specimens of steel B (considering the weakest link behaviour and a Weibull exponent m = 8 [9], differences between fracture loads for B = 11 mm and B = 12.7 mm would be lower than 2%).

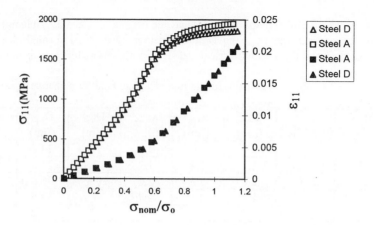

Fig. 4. Stress and strain evolution at 150 μm from the notch root for steels A and D.

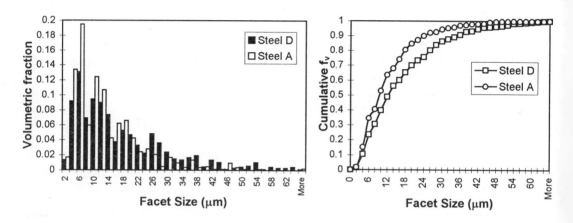

Fig. 5. Facet size distribution of steels A and D: a) histogram of volume fraction and b) its corresponding cumulative frequency.

The local maximum principal stress evolution as a function of $\sigma_{nom}/\sigma_{0.2\%}$, in the zone corresponding to the cleavage fracture initiation for a specimen of steel B (d = 225 μm), shows that the fracture takes place when the stress has reached a saturation value (see Fig. 6). This behaviour is more evident for higher $\sigma_{nom}/\sigma_{0.2\%}$ values. On the other hand, in four of the six tested specimens corresponding to steels B and C, the cleavage origin is very close to the notch tip. In the other two cases, the σ_F^* local cleavage fracture stress value is not so different when it is compared with the mean value measured for steel A (between 14 and 18% higher). As a consequence, even if the critical cleavage fracture stress (σ_F) exhibits higher values for the bainitic steels without TiN coarse particles, taking into account that the fracture has occurred in a stress saturation condition and that there are cleavage origins near to the notch tip (zone in which the strain takes the maximum value), it is necessary to further consider that strain contributes to the cleavage process.

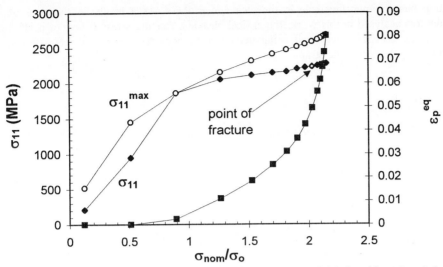

Fig. 6. Stress (σ_{11}) and strain (ε_p^{eq}) evolution at the site of cleavage initiation (d=225 μm) for steel B. (σ_{11}^{max}: maximum principal stress ahead of the notch root).

In the case of pearlitic steels, Lewandowski and Thompson [10,11] showed that when the cleavage initiation was not associated with inclusions, the initiation sites were located closer to the notch, in a region possessing a larger component of strain. This behaviour takes place when, due to the relatively low yield strength of the steel, the maximum stress intensification is not able to promote the brittle fracture of a microstructural feature before other fracture modes intervene. In the case of steels B and C, without TiN coarse particles, it seems that a similar behaviour takes place. In both steels the cleavage fracture has taken place when the stress has reached a saturation value. In consequence, it appears that under these circumstances it is necessary to have some minimum strain able to nucleate a defect in the matrix that will propagate under the acting stress. It follows that the nucleation of the defect is not a stress controlled process.

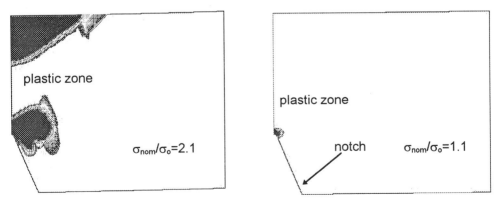

Fig. 7. Strain distribution at notch, showing the plastic zone, for $\sigma_{nom}/\sigma_{0.2\%}$= 2.1 and $\sigma_{nom}/\sigma_{0.2\%}$= 1.1.

In Fig. 7 the plastic zone extension corresponding to $\sigma_{nom}/\sigma_{0.2\%}$= 1.1 and $\sigma_{nom}/\sigma_{0.2\%}$= 2.1 is shown. Both cases correspond to a situation close to the brittle fracture of steels A and B respectively. It is worth emphasising that in the latter case the plastic zone size is notably bigger, showing that plastic deformation has an important role in the cleavage process. As a result, in the type of bainitic microstructures analysed in this work, the critical cleavage fracture stress is not, of itself, sufficient to describe the cleavage fracture, it being necessary to consider other parameters that take into account the contribution of the strain.

CONCLUSIONS

From the work carried out with the four microalloyed steels with bainitic microstructures, the main conclusions are:

- In the case of Ti microalloyed steels with coarse TiN particles, the cleavage fracture process is initiated by the nucleation of microcracks at TiN particles.
- In the alloys without Ti, the critical cleavage fracture stress is higher than in the Ti microalloyed steels. In these steels, cleavage fracture has taken place after the saturation of the maximum principal stress and the acting strain has contributed to the nucleation of defects able to propagate in a brittle way.
- The obtained results show that in these circumstances the cleavage fracture process cannot be correlated with the critical cleavage fracture stress alone and that it is necessary to take into account the strain contribution.

ACKNOWLEDGEMENTS

This work was carried out under a Spanish CICYT research programme. A.E. gratefully acknowledges a research grant from the Department of Education and Research of the Basque Government. The tested material was kindly provided by Tubos Reunidos (Amurrio) and Sidenor.

REFERENCES

[1]   C.J. McMahon and M. Cohen, Acta Metall., **13**, 1965, p. 591.
[2]   J.F. Knott, Reliability and Structural Integrity of Advanced Materials, ECF 9, S. Sedmak, A. Sedmak and D. Ruzic eds., EMAS, **2**, 1992, p. 1375.
[3]   P. Bowen, S.G. Druce and J.F. Knott, Acta Metall., **34**, 1986, p. 1121.
[4]   M.A. Linaza, J.L. Romero, J.M. Rodriguez-Ibabe and J.J. Urcola, Scripta Met. and Mat., **32**, 1995, p. 395.
[5]   M.A. Linaza, J.M. Rodriguez-Ibabe and J.J. Urcola, Fatigue Fract. Engng. Mater. Struct., **20**, 1997, p 619.
[6]   J.R. Griffiths and D.R. Owen, J. Mech. Phys. Solids, **19**, 1971, p. 419.
[7]   A. Echeverría, Ph. D. Thesis, Univ. of Navarra.
[8]   N.J. Petch, Acta Metall., **34**, 1986, p. 1387.
[9]   A. Martín, I. Ocaña, J. Gil Sevillano and M. Fuentes, Acta Metall. Mater., **42**, 1994, p. 2057.
[10] J.J. Lewandowski and A.W. Thompson, Metall. Trans., **17A**, 1986, p. 1769.
[11] J.J. Lewandowski and A.W. Thompson, Acta Metall., **35**, 1987, p. 1453.

For correspondence, please contact Alberto Echeverría, e-mail: aetxebe@ceit.es
or J.M. Rodriguez-Ibabe, e-mail: jmribabe@ceit.es

MODELLING

Materials Science Forum Vols. 284-286 (1998) pp. 361-368
© *1998 Trans Tech Publications, Switzerland*

An Integrated Model for Microstructural Evolution in the Hot Strip Mill and Tensile Properties Prediction of Plain and Microalloyed C-Mn Hot Strip

M. Piette and Ch. Perdrix

Direction Recherche-Développement et Métallurgie, CRDM, SOLLAC Dunkerque, B.P. 2508, F-59381 DUNKERQUE Cedex 1, France

Keywords: Hot Strip Steel, Modelling, Microstructure, Tensile Properties

ABSTRACT

A model has been developed, which computes the microstructural evolution of austenite, phase transformation, final microstructure and tensile properties (YS, TS, El) of plain carbon and microalloyed hot strip steels. This paper describes the principal features of the model and compares the computed and observed tensile properties on a database of hot strip steels extracted from the normal production of the hot strip mill of Sollac Fos.

1. INTRODUCTION

The tensile properties of hot strip steel are directly related to the microstructure of the steel. This microstructure results from both chemical composition and thermomechanical treatment during the processing of the strip.

A great number of works, either quantitative or qualitative, have been published describing the elementary mechanisms taking place during the different phases of the rolling of the strip (soaking, rolling, cooling, coiling).

The modelling of the hot rolling plate and hot strip has been a constant goal of the different research units of the Usinor Group.

In 1992, Sollac, the Flat Products division of the Usinor Group, has started a project for the prediction of the microstructures and the resulting tensile properties of hot strip steel based on the pioneering works have been previously carried out at IRSID [1, 2, 3, 4, 5, 6]. This model contains the quantified metallurgical knowledge we have in the field of hot strip steel. This knowledge is synthesised in a software called STRUCTURA, describing the elementary mechanisms relative to the hot rolling of strips. These models links the process and composition variables to the resulting microstructure, and relates these microstructures to the tensile properties of the steels.

In this first version, the model deals with ferrite+pearlite microstructures in plain carbon steels, and for Nb and Nb+V microalloyed steels.

The objective of this paper is to describe the different submodels contained in STRUCTURA and show its capabilities. The comparison between measured and computed tensile properties will be shown on a database of 270 industrial coils extracted from the normal production of the hot strip mill of SOLLAC Fos.

2. GOALS OF THE MODELLING

The modelling project (computer metallurgy) has different goals, among which are to be able to :
compute the tensile properties (YS, TS, Elongation) of hot strip steel, knowing the precise process parameter of the coil that has been rolled,

optimise the rolling schedule to determine new, cost-effective and robust processing routes. The reduction of the number of steel compositions, i.e. making a wide range of product qualities with a limited number of steel compositions is also a part of this goal.

Take into account fluctuations of one process to modify the process on the next equipment depending on the difference between obtained and target process parameters,

allow the development of new steels, to increase the efficiency of R&D department,

give a synthetic view of our quantitative knowledge of the metallurgy of hot strip steels, for expert system building and metallurgical learning purposes.

3. GENERAL ORGANISATION OF THE MICROSTRUCTURAL MODEL

The model takes into account the following metallurgical phenomena :

Process Step	Metallurgical phenomena	Input variables	Output variables
Soaking	Dissolution of Microalloying elements	Steel composition, Soaking temperature	Nb, V, Ti, Al, N into solution in the austenite
Roughing + finishing	Austenite recrystallization Austenite grain growth	Detailed rolling schedule	Austenite grain size $d\gamma$, cumulative strain ε_γ, recrystallized fraction $X_\gamma(t)$ at each pass.
	Nb strain induced precipitation		$Nb_\gamma(t)$ and before γ/α transformation.
Cooling	Phase transformation	Austenite microstructure at the exit of the finishing	$F_\gamma(t)$, $F_\alpha(t)$, $F_P(t)$ Ferrite grain size $d\alpha$ Final F_α, F_P
Coiling	Precipitation of AlN	Coiling temperature, Al and N content	Nitrogen concentration in the finished hot strip
	Dissolution of pearlite into grain boundary cementite	Amount of pearlite at coiling start	Amount of pearlite in the finished hot strip

4. AUSTENITE EVOLUTION MODEL

4.1 Soaking

During soaking, the re-dissolution of the microalloying additions Nb and V, and the dissolution of Aluminium Nitrides take place. The solubility products for Nb(C,N) [7] and AlN [8] from the literature have been found in good agreement with our laboratory results. Vanadium is considered to be always fully redissolved, which is true in our industrial practice.

The austenite grain size obtained at the end of reheating will not influence the resulting final austenite grain size before γ/α transformation as long as a minimum of 3 to 4 austenite recrystallizations have occurred (see Figure 1). This observation is consistent with the knowledge described in literature [9, 10].We have therefore not developed a model for the prevision of the austenite grain size after reheating.

4.2 Austenite microstructural evolution

The austenite microstructural evolution model is directly a result of the FAST model developed in IRSID [1, 2, 3, 4, 5, 6]. This model has been successfully compared to other models and to experimental results by independent authors [11]. This model computes the recrystallization kinetics after each rolling pass in a non isothermal situation, and computes the resulting recrystallized grain size. Figure 1 shows the different behaviour of a C-Mn steel (a) and a Nb-microalloyed steel (b) for the same rolling schedule. Full recrystallization occurs in the case of C-Mn steel, whereas strain accumulates in the finishing rolling mill in the case of Nb steel. If full recrystallization is obtained between two successive rolling passes, grain growth will occur as long as no TiN or Nb(C,N) precipitates are present. In this case, grain growth is decreased or suppressed, as it is observed during laboratory experiments.

In the case of partial interpass recrystallization, the grain size and the residual strain of the equivalent homogeneous microstructure is computed via suitable laws of mixtures.

a   b

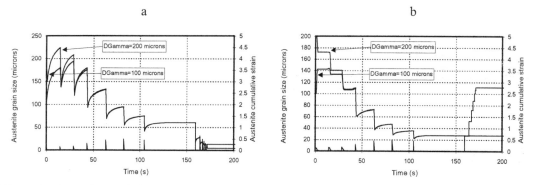

Figure 1 : Austenite evolution computed by STRUCTURA in the case of a plain carbon steel (a) and Nb microalloyed steel (b) for two different initial grain sizes.

4.3 Nb strain induced precipitation

A model describing the Nb strain induced precipitation has been developed. This model is necessary to compute the amount of Nb retained in solution in the austenite, in order to compute the precipitation hardening. The precipitation hardening is assumed to be the result of the Niobium precipitation in ferrite, and the strain induced precipitates in the austenite do not contribute to the hardening as these precipitates are much coarser than precipitates in ferrite.

The strain induced precipitation model is a fitted version of Dutta and Sellars model [12, 13]. The goal of the fitting procedure was to check the numerical values of the coefficient.

The experimental method used to follow the precipitation was an indirect measurement of the amount of Niobium precipitates through their effect on the tensile properties.

5. PHASE TRANSFORMATION MODEL

When the two models describing the austenite microstructural evolution and the strain induced precipitation have run, we obtain the microstructural parameters of austenite at the end of the finishing mill : final austenite grain size $d\gamma$, austenite strain accumulated in the non-recrystallization region ε_γ, Nb in solution in the austenite Nb_γ.

The next step is the phase transformation of the ferrite+pearlite structure from the possibly strained austenite.

The phase transformation model used in STRUCTURA is a result of a study started in IRSID [6] and which has been continued in CRDM.

This model can be split into two sub-models : The Ar3 model, describing the initiation of transformation, and the kinetics model describing the rest of the transformation.

5.1 Ar3 model

The temperature at which the austenite/ferrite transformation starts is a basic starting point of the transformation model. Much effort has been devoted to a precise prediction of that temperature, including the effect of steel composition, cooling rate (even in complex thermal path), austenite grain size dγ and cumulative strain ε_γ. The Ar3 temperature considered in this work is the temperature at which 10% of transformation has occurred.

5.1.1 C-Mn Steels

In the case of C-Mn steels, the only effect considered is the effect of the austenite grain size (dγ) and cooling rate (CR) :

$$Ar3 = A - \beta_{C-Mn} \cdot \sqrt{d\gamma^2 CR} \tag{1}$$

where β_{C-Mn} is a constant depending on the chemical composition. This relationship is in agreement with previously published data [14], where it has been observed that, when varying both cooling rate (CR) and austenite grain size (dγ), the Ar3 temperature depends on the product $CR.d\gamma^2$.

5.1.2 Nb microalloyed steels

In the case of Nb steels, the effects of austenite grain size (dγ), austenite strain (ε) and solute Niobium must be considered. To model these effects, the data from Sun Fu-Yu at IRSID [15] which are in agreement with data published independently by Ouchi [16] have been used. They show that the effect of austenite strain in the non-recrystallization region has been shown to saturate at high strains, whatever the amount of soluble Niobium in austenite. At low strain, the decrease of the Ar3 temperature describes the hardenability properties of Nb in solution in austenite, which vanishes at high strains (Figure 2).

$$Ar3 = A - \beta_{Nb} \cdot \left(1 + f(Nb)\,g(\varepsilon)\,h(d\gamma)\right) \cdot \sqrt{CR} \tag{2}$$

where β_{C-Mn} is a constant depending on the chemical composition. The coefficient A in equation (2) has the same chemical composition dependence as in equation (1).

5.2 Kinetics model

This model is based on the observation of the plot of the CCT transformation kinetics as a function of the instantaneous temperature, showing that the CCT curves obtained are almost parallel to each other (Figure 3). This has already been mentioned by other authors [17]. This means that the knowledge of the starting point of the transformation (Ar3) and a general shape of the transformation curve is enough to obtain the transformation kinetics for any cooling rate.
These observations can be described with the following transformation kinetics :

$$\frac{d\,F_\gamma}{d\,t} = -K.CR.F_\gamma \tag{3}$$

where CR corresponds to the constant cooling rate and F_γ is the austenite volume fraction. When cooling rate changes during cooling, which is very frequent due to the exothermic characteristic of the γ/α transformation, a mean cooling rate is used.
The constant K is related to the transformation rate and depends only on the steel composition. A lighter alloyed steel transforms in a smaller range of temperature than a heavier alloyed one (Figure 4). K therefore increases the less the steel is alloyed. We have found a good correlation between K and the temperature A from equation 1 and 2.

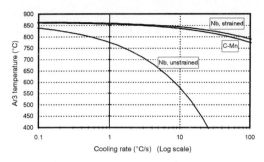

Figure 2 : Computed effect of the cooling rate on the Ar3 temperature, in the case of plain C-Mn steel and Nb steel (both strained and unstrained).

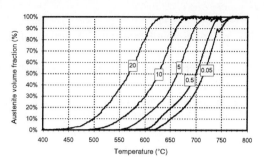

Figure 3 : Measured CCT diagram for a medium Carbon, high Manganese steel cooled at different continuous cooling rates (0.05, 0.5, 5, 10, 20 °C/s) [18].

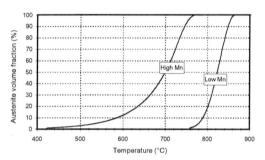

Figure 4 : Computed transformation kinetics of two steels with Mn=300 and 1500.10^{-3}%.

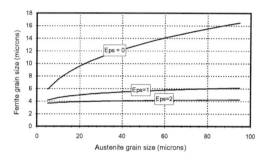

Figure 5 : Computed influence of the austenite grain size on the ferrite grain size, for three different austenite strains (Nb microalloyed steel).

5.3 Ferrite grain size model

The ferrite grain size model is a modified version of IRSID's model [6]. It describes the ferrite grain size as a function of steel composition, austenite grain size (dγ), residual strain (ε) and cooling rate :

$$d\alpha = A(compo).(CR)^{-b}.(d\gamma)^{C(\varepsilon)}$$ (4)

where CR is the mean cooling rate during the first 50% transformed. The coefficient C(ε) depends on the austenite strain accumulated in the non recrystallization region. This coefficient C(ε) decreases as ε increases, which describes the experimental fact that the ferrite grain size does not depend on the austenite grain size when the austenite is highly strained.

Figure 5 shows the effect of austenite grain size and austenite strain on the computed ferrite grain size. We observe that at high strains, the effect of austenite grain size vanishes.

6. TENSILE PROPERTIES MODEL

The tensile properties (YS, TS, El) are computed in a successive way : A first model computes the Yield Strength, a second model relates the Tensile Strength to the Yield Strength, and a third model links the Total Elongation (El) to the Tensile Strength : YS ⇒ TS ⇒ El.

6.1 Yield strength

The Yield Strength model has the form of a law of mixture between the different microconstituents:

$$YS = F_\alpha . YS_\alpha + (1 - F_\alpha) . YS_P \tag{5}$$

F_α and $(1-F_\alpha)$ are the volume fraction of ferrite and pearlite with respective Yield Strengths YS_α and YS_P.

The Yield Strength of pearlite is related to its transformation temperature.

The Yield Strength of ferrite is computed via Hall-Petch equation, combined in a non-linear way to solid solution, Nb, V and Ti hardening models. The non-linear summing rule results from internal studies that show that ferrite strain hardening does not add up linearly with Niobium hardening. This type of equation has already been proposed for the description of the tensile properties of HSLA steels [19] :

$$YS = k_y \left(d\alpha\right)^{-1/2} + \sqrt{\left(\sigma_0 + \sigma_{ss} + \sigma_N\right)^2 + \left(\sigma_{Nb}\right)^2 + \left(\sigma_V\right)^2} \tag{6}$$

It can be noted that this equation simplifies in the case of plain C-Mn steel to a more familiar equation :

$$YS = k_y \left(d\alpha\right)^{-1/2} + \sigma_0 + \sigma_{ss} + \sigma_N \tag{7}$$

This shows that the effect of solid solution and nitrogen strengthening in equation (6) can be modelled through usual coefficients [e.g. 20] that have been determined in non-microalloyed steels. The microalloying precipitation strengthening is computed using a model built after Orowan's equation :

$$\sigma_{Nb} = \alpha_{Nb} \sqrt{Nb_{sol,Ar3}} = \alpha_{Nb} \sqrt{Nb_{reheat} - Nb_{strain-induced-precipitation}} \tag{8}$$

$$\sigma_V = \alpha_V \sqrt{V_{sol}} \tag{9}$$

The coefficients α_{Nb} (= 15...31) and α_V (≈ 20) take into account the various efficiencies of the different precipitates, which depend mainly on the precipitates size. This size is controlled by the temperature at which precipitation occurs.

The observed and computed tensile properties have been compared on a database of coils extracted from the normal production of hot strip coils from Sollac Fos. It should be emphasised that the database used for these graphs is different than the database used to fit the model.

Figure 6 shows the good agreement between computed and observed Yield Strength. The standard deviation of the residuals ($YS_{computed} - YS_{measured}$) equals 18 MPa.

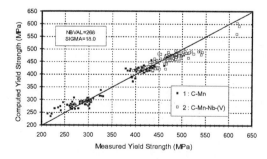

Figure 6 : Comparison between observed and computed Yield Strength.

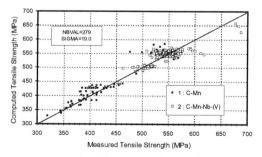

Figure 7 : Comparison between observed and computed Tensile Strength.

6.2 Tensile strength

Two kind of models can be used for the prediction of Tensile Strength. The first has the form a modified version of some Hall-Petch equation, and the second links the Tensile Strength to the Yield Strength using a relation of the form :

$$TS = f(YS, composition) \tag{10}$$

We have preferred the second approach. The performance of this model, when the **measured** Yield Strength is used as an input, allows us to compute the Tensile Strength with a standard deviation of the residuals ($TS_{computed}-TS_{measured}$) equals 14 MPa (Figure 8).

When the **computed** Yield Strength is used as an input (Figure 7), the standard deviation of the residuals becomes 19 MPa, resulting from the error on the Yield Strength.

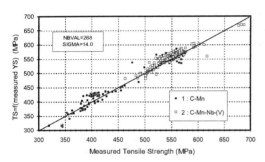

Figure 8 : Tensile strength model compared to measured tensile strength, where $TS_{computed}=f(YS_{measured})$

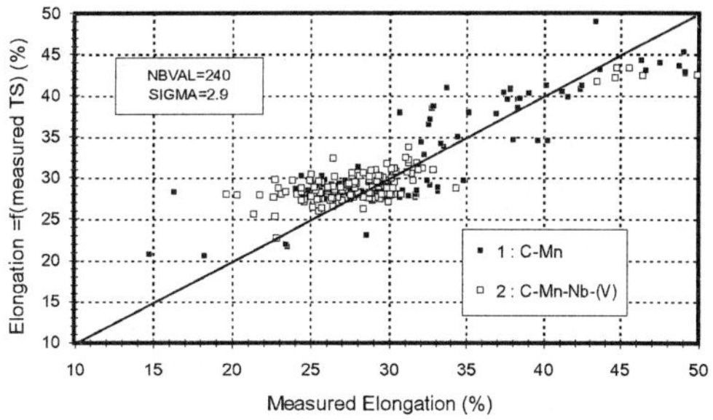

Figure 9 : Elongation model compared to measured elongation, where $El_{computed}=f(TS_{measured})$.

6.3 Total elongation

As for the Tensile Strength model, we have preferred to develop a model for the total elongation based on a previously computed property instead of developing a new fully independent model.

The elongation model is based on the observation that, after correcting the measured elongation from the effect of the tensile specimen geometry [21], the product TS x El depends mainly on the type of microstructure, and slightly on a few process and composition variables. The elongation model is then :

$$El = \frac{k(process, composition)}{TS} . f(geometry) \tag{11}$$

The performance of this model, when the **measured** Tensile Strength is used as an input, allows us to compute the Total Elongation with a standard deviation of the residuals ($El_{computed}-El_{measured}$) equals to 2.9% (Figure 9). This apparently high value must be compared to the variation of elongation measurement within an homogeneous sample, which is about 2%.

7. CONCLUSIONS

A model has been developed, which can predict the microstructure and the tensile properties (YS, TS, El) of plain and microalloyed steels. This is achieved by modelling the microstructural evolution of austenite during rolling, phase transformation and final microstructure.

The model has been compared to a database of 269 industrial coils, combining C-Mn and microalloyed steels. This database is different from the database used to fit the model. The range of tensile properties of the coils is YS=[212, 621] MPa, UTS=[319, 687] MPa, El=[15, 50] % . The precision of the model, measured by the standard deviation of the differences between computed and measured values, gives σ_{YS}=18 MPa and σ_{TS}=19 MPa, σ_{El} = 3.1%.

Its possibilities will continue to be improved as our quantified knowledge of metallurgy improves. The model is used as an help for the design of new steel and for the definition of more robust and cost-effective processing routes.

8. REFERENCES

1   Ch.Perdrix : ECSC Contract 7210-EA/311
2   P.Choquet, A.Le Bon, Ch.Perdrix : Strenght of Metals and Alloys, Proc. ICSMA7 Montreal, ed. H.J.McQueen et al., Canada (1985) Vol.2, 1025
3   P.Choquet, B.de Lamberterie, Ch.Perdrix, H.Biausser : 4th Intern. Conf. on Steel Rolling, Deauville, France (1987), Vol.1, B5.1
4   P.Choquet et al.: Proc. Intern. Conf. THERMEC 88, ISIJ, Tokyo, (1988)
5   Ch.Perdrix, B.Chamont, E.Amoris, H.Biausser : Proc. Intern. Conf. THERMEC 88, ISIJ, Tokyo, (1988), 807
6   P.Choquet, P.Fabrègue, J.Giusti, B.Chamont : Proc.Int. Symph. on Mathematical Modelling of Hot Rolling of Steel, ed. by S.Yue, CIM, Quebec (1990), 34.
7   K.J.Irvine, F.B.Pickering, T.Gladman : JISI Vol.205 (1967), 161
8   W.C.Leslie, R.L.Rickett, C.L.Dotson, C.S.Watson : Trans. Americ. Soc. Metall. Vol.46 (1954), 1470
9   T.Siwecki : ISIJ Internat. Vol.32 (1992), 368
10   T.Senuma, M.Suehiro, H.Yada : ISIJ Internat. Vol. 32 (1992), 423
11   J.J. Cruz-Rivera, J.G. Cabañas-Moreno, M.Umemoto : ISIJ Internat. Vol 36 (1996), 300
12   B.Dutta, C.M.Sellars : Materials Science and Technology Vol.3 (1987), 197
13   B.Dutta, E.Valdes, C.M.Sellars : Acta Metall. Mater. Vol.40 (1992), 653
14   M.Militzer, R.Pandi, E.B.Hawbolt : Metall. and Mater. Trans. Vol.27A (1996), 1547
15   Sun Fu-Yu : PhD Thesis, Centre d'Orsay, Université de Paris-Sud (1981).
16   C.Ouchi, T.Sampei, I.Kosazu : Trans. ISIJ, Vol.22 (1982), 214
17   M.Groslambert, J.Mignon, J.F.Noville, S.Wilmotte : Proc. 4th Internat. Steel Rolling Conf., Deauville (France).
18   F.Roch (IRSID), private communication
19   S.Fu-Yu, X.Wenchong, M.Xiang : Proc. Internat. Conf. on HSLA Steels '85, Beijing (China), ed. by J.M.Gray et al, ASM Internat.
20   F.Pickering : Constitution and Properties of Steels, ed. by R.W.Cahn et al, VCH publishers.
21   ISO 2566/1 (1984)

Materials Science Forum Vols. 284-286 (1998) pp. 369-376
© *1998 Trans Tech Publications, Switzerland*

Modelling the Hot Working of Simple Geometries Employing Physical-Based Constitutive Equations and the Finite Element Method

M. Bini, J.M. Cabrera and J.M. Prado

Departamento de Ciencia de Materiales e Ingeniería Metalúrgica
ETSEIB - Universidad Politécnica de Catalunya, Av. Diagonal 647, E-08028 Barcelona, Spain

Keywords: Medium Carbon, Hot Working, FEM Modelling

ABSTRACT

The prediction of the final microstructure by simulation of the hot plastic deformation usually gives unsatisfactory results. This is partially due to the inadequate constitutive equations employed by the conventional and commercial software available to describe the hot flow behaviour.

In this work the latter limitation is overcome by using physical-based constitutive equations where account of the interaction between microstructure and processing variables is taken.

The hot forging behaviour of a simple piece (a small gear) was simulated in this way employing the software ABAQUS. Additionally it was assumed that strain rate and temperature are not constant during the process. Special attention was paid to the prediction of the final grain size and the appearance of dynamic recrystallization, facts traditionally ignored in numerical modelling.

INTRODUCTION

Over the centuries, hot forming has been employed to provide shape to metals. Nowadays, however, hot working not only produces the desired geometry, but also the mechanical characteristics required.

The development of advanced materials involves in part the improvement of the mechanical properties of traditional materials. In the case of microalloyed steels, the careful control of thermomechanical processing is required for this purpose. In this sense, computer simulations of hot working provide a benchmark to engineers and researchers, and curtail the cost of developing various products and processes.

The simulation of hot working processes by the Finite Element Method (FEM) is a technique which can become in a near future an important tool for designing forging dies, and even to test the adequacy of process parameters such as temperature and punch speed. To accomplish this feature it is necessary to have a correct mechanical model of the hot plastic behaviour of the material, taking into account not only the viscous-plastic behaviour but also aspects such as dynamic recrystallization. With regard to the prediction of the final microstructure, the simulation of the hot plastic deformation usually gives unsatisfactory results. This is, in part, due to the inadequate constitutive equations employed by the conventional and commercial software available to describe the hot flow behaviour. When dynamic recrystallization is included in the simulation, the microstructural changes occurring during hot working can also be modelled. In this way, complete thermomechanical processes can be simulated.

There are few models which couple the typical hot working variables (temperature, strain and strain rate) with microstructural characteristics such as grain size. In this work the latter limitation is overcome by using physical-based constitutive equations where account of the interaction between microstructure and processing variables is taken. The present research group has recently developed these equations for a medium carbon microalloyed steel which are presented in detail elsewhere [1,2]. The hot forging behaviour of a simple piece (a small gear) was simulated employing the software ABAQUS. Additionally, strain rate and temperature were assumed variable during the process. Special attention was given to the prediction of the final grain size and the appearance of dynamic recrystallization, features traditionally ignored in numerical modelling. Diverse forging conditions were simulated, and their effect on the final microstructure was analysed as well as on the stress, strain and temperature distribution.

MATERIAL MODELLING

The chemical composition of the medium carbon microalloyed steel used for this simulation is given in Table I. The mechanical behaviour under hot working conditions of this steel has been previously thoroughly studied [1,2], and only the constitutive equations will be presented here.

%C	%Mn	%Si	%P	%S	%Cr	%Ni	%Mo	%V	%Al	%Ti	N ppm
.34	1.5	.72	.02	.02	.33	.18	.05	.83	.14	.02	114

Table I. Chemical composition of the steel studied.

The rate equations describing the onset of dynamic recrystallization ε_p and the peak stress σ_p are respectively:

$$\varepsilon_p = 7 \cdot 10^{-3} \cdot d_0^{-0.146} \cdot Z^{0.137} \tag{1}$$

$$\sigma_p = \left(\frac{E(T)}{1338}\right) \cdot arcsinh\left[\left(\frac{1}{603}\right) \cdot \left(\frac{\dot{\varepsilon}}{D(T)}\right)^{1/5}\right] + \sigma_{op} \tag{2}$$

where d_o is the initial grain size at the soaking conditions, Z is the Zener-Hollomon parameter, E(T) is the temperature dependent Young's modulus and D(T) is the diffusion coefficient (in both Z and D(T) the self-diffusion energy is employed). The internal stress σ_{op} is only applied when the value of the initial grain size is smaller than approximately 30 μm. This internal stress is also rate dependent according to Eq. (3):

$$\sigma_{op} = \sigma_{ip} + (3.94 + 0.45 \cdot \log(\dot{\varepsilon})) \cdot d_o^{-0.5} \tag{3}$$

Here d_o is in mm. and σ_{ip} is given by:

$$\sigma_{ip} = -15.3 - 1.22 \cdot \log(\dot{\varepsilon}) \tag{4}$$

According to the physical model of Estrin and Mecking [3] and Bergström [4,5], the flow curve σ-ε before dynamic recrystallization is given by:

$$\sigma = \sigma_p \sqrt{\left(1 - e^{-\Omega\varepsilon}\right)} \tag{5}$$

Ω being the term responsible of the softening by dynamic recovery that, on the other hand, depends on the Zener-Hollomon parameter and the initial grain size as shown in Eq. (6):

$$\Omega = 1829 \cdot d_o^{-0.346} \cdot Z^{-0.167} \tag{6}$$

Once dynamic recrystallization is started, the flow curve can be written as follows:

$$\sigma = \sigma_p - (\sigma_p - \sigma_{ss}) \cdot \left[1 - \exp\left(-0.693 \cdot \left(\left[\varepsilon - \varepsilon_p\right] / \left[\dot{\varepsilon}/t_{50\%}\right]\right)^n\right)\right] \tag{7}$$

The second-hand right term represents the volume fraction of material dynamically recrystallized. The Avrami exponent n was evaluated as 1.83, while the steady state stress σ_{ss} is given by:

$$\sigma_{ss} = \left(\frac{E(T)}{1168}\right) \cdot arcsinh\left[\left(\frac{1}{813}\right) \cdot \left(\frac{\dot{\varepsilon}}{D(T)}\right)^{1/5}\right] + \sigma_{oss} \tag{8}$$

where

$$\sigma_{oss} = 3.5 + 0.751 \cdot (d_{rec})^{-0.5} \text{ if } d_{rec} < 30\mu m \text{ and } \sigma_{oss} = 0 \text{ if } d_{rec} > 30\mu m \tag{9}$$

The recrystallization grain size d_{rec} is determined by the expression proposed by Derby [6,7]:

$$\sigma_{ss}/\mu(T) = 2.39 \cdot (d_{rec}/b)^{-\frac{2}{3}} \text{ if } d_o > 30\mu m, \text{ and } \sigma_{ss}/\mu(T) = 0.13 \cdot (d_{rec}/b)^{-0.395} \text{ when } d_o < 30\mu m \quad (10)$$

Here $\mu(T)$ is the temperature dependent shear modulus, and b is the value of the Burgers vector.

The kinetics of the dynamic recrystallization is typically described in terms of $t_{50\%}$, the time for the recrystallization of 50% of the volume. In the current steel, this time was evaluated as:

$$t_{50\%} = 1.8 \cdot 10^{-6} \cdot d_o^{0.24} \cdot \dot{\varepsilon}^{-0.745} \cdot \exp(116125/RT) \quad (11)$$

All these equations have been deduced and validated experimentally by the compression of cylindrical specimens [1,2].

SIMULATION OF THE FORGING PROCESS

The geometry of the piece to be forged is shown in Fig. 1. Due to its axial symmetry only a fourth section will be studied. In order to point out the effect of initial temperature, punch speed and initial grain size, the cases indicated in Table II has been analysed. It must be mentioned that the friction coefficient was set constant and equal to 0.22. The friction effect was therefore not included in the present analysis.

In order to solve the thermal problem, the initial temperature of the upper and lower dies has been set to 300°C. The surroundings of the piece and die were supposed to be at room temperature (25°C). The upper die changes continuously its velocity in such a way that the vertical axial strain rate keeps macroscopically constant during forging.

The plastic model of the considered material has been implemented in the program ABAQUS by using its users subroutine capability. The simulation was done as a coupled thermo-mechanical problem taking into account not only the heat generated during plastic deformation but also dissipated by conduction, convection and radiation. The equations describing the latter phenomena were taken from ref. [8].

The dies were supposed to behave elastically during deformation, but their strain, stress and temperature distributions will not be shown in this work, as this study will be focused on the workpiece. The forging is done in one step, the initial billet being of cylindrical shape. The meshing was fine enough to ensure the convergence of the solution as well as the accuracy of results.

Case	T(°C)	d_o (μm)	$\dot{\varepsilon}$ (s⁻¹)	v
B-1	1100	90	0.1	0.22
B-2	950	13	0.1	0.22
B-3	950	90	0.1	0.22
B-4	950	13	10	0.22

Table II: Initial processing conditions simulated (T: temperature, d_o: initial grain size, $\dot{\varepsilon}$: strain rate and v: friction coefficient).

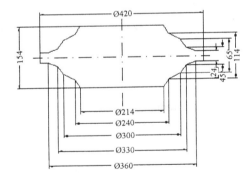

Fig. 1. Sketch of the piece simulated.

RESULTS AND DISCUSSION

The software allows acquisition of the evolution during deformation of the distribution, in the section of the piece, of many process parameters, such as strain, strain rate, stress, temperature, recrystallized grain size and recrystallized area. Although these distributions can be obtained at each step of the process, only those distributions corresponding to the final step will be shown here.

Fig. 2 shows the strain distribution for all the cases. The maximum strain is concentrated in the equatorial part, in the centre of the piece and in the flash region, especially in the area contacting the die. The minimum strain (almost zero) is located at the top of the piece. It must be noted that this non-deformed area is relatively large. No noticeable differences are found between the four cases. This result is logical bearing in mind that the initial billet keeps the same geometry in all the tests, the friction coefficient is also constant, and that a fault criterion for plastic flow has not been included in the modelling. In other words, the strain distribution is mainly dependent on the geometry of the piece.

The strain rate distributions are shown in Fig.3. A high heterogeneity can be observed with a marked peak at the flash area, where the strain rate is over an order of magnitude larger than the nominal one. Although all the distributions show the same aspect, the heterogeneity increases with increasing macroscopic strain rates (compare cases B-2 and B-4 in Fig. 3). No significant effects of the different initial grain sizes (see cases B-2 and B-3 in Fig. 3) or the initial temperature (cases B-1 and B-3) are detected.

The results of the final temperature distribution (see Fig. 4) put in evidence that the thermomechanical process cannot be considered isothermal. In the case B-1 (initial temperature=1100°C), a raise in temperature of 30°C is observed in the equatorial zone. Under the same strain rate and initial grain size conditions, a diminution of the soaking temperature of 150°C (case B-3) promotes a heating of 80°C, especially in the flash area. The effect of a finer grain size on the temperature is negligible as one can see by comparing cases B-2 and B-4 in Fig. 4. However, an increment in the strain rates induces a significant heating. See for example case B-4 ($\dot{\varepsilon}$ =10 s^{-1}), where the peak temperature is 120°C higher than the initial one. In all the cases the top of the piece has slightly been cooled. This is logically due to the heat extraction through the die. This diminution of the temperature is lower at high strain rates (short time for heat extraction) and higher at increasing temperatures (large temperature gradients in the contacting area piece-die favour the heating extraction rate).

In relation to microstructural features, Fig. 5 shows the recrystallized and non-recrystallized areas. The case B-1 presents the major fraction of volume recrystallized (iso-contour >1.5). This observation is in agreement with the expected behaviour (high temperatures and relatively low strain rates as case B-1 promotes dynamic recrystallization). On the other hand, B-4 is the case with the larger non recrystallized area (iso-contour < 1) because of the high strain rate and low temperature. Results are obviously in accordance with the corresponding strain distribution. It must now be remarked that these distributions represent the last step of the forming process. The dynamic recrystallization has taken place in early stages, where the critical strain for the onset of recrystallization is rapidly reached, and the strain rate, grain size and temperature are appropriated to promote this softening mechanism. Similar trends are observed in the final grain size distribution (see Fig. 6). Now those conditions of low temperature and relatively high strain rate give way to a finer grain size distribution. This is particularly clear in the case B-3 where the initial grain size is heavily refined. In the boundary region between recrystallized and non-recrystallized areas, the grain size is coarser because in this region the strain rate is very low.

A final remark must be made on the validity of the model. The constitutive equations of the plastic flow of the current steel were obtained under isothermal and iso-strain rate conditions. An improvement of the model is still required especially to verify what happens when the material undergoes dynamic recrystallization and is subsequently deformed at higher strain rates.

CONCLUSIONS

A general conclusion can be derived from this work. Given a correct model of material behaviour, FEM simulates thermomechanical processes with high accuracy, allowing its better design in a more economical way. The present modelling constitutes an additional step in developing a metallurgical design tool. Further investigations are required to obtain improved constitutive equations, correctly linked to microstructural features. The present approach gives however valuable information on the metallurgical prediction of the given material.

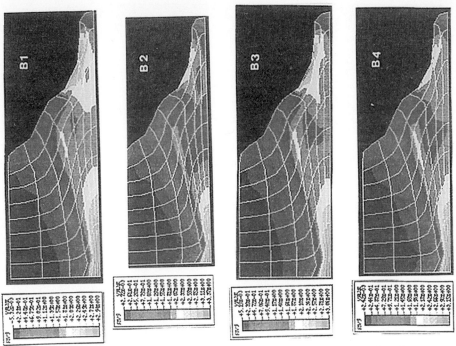

Fig. 2. Contour plot of strain distribution in the four cases analyzed.

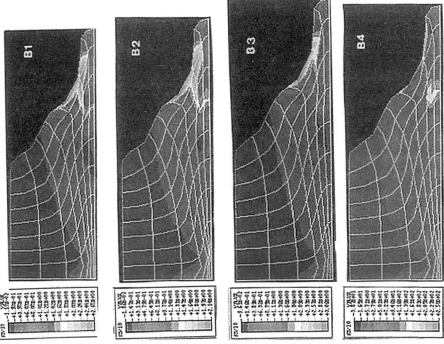

Fig. 3. Contour plot of strain rate distribution in the four cases analyzed.

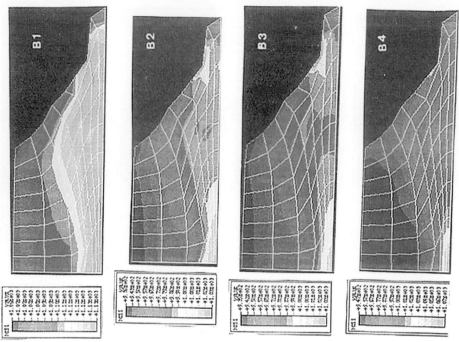

Fig. 4. Contour plot of temperature distribution in the four cases analyzed.

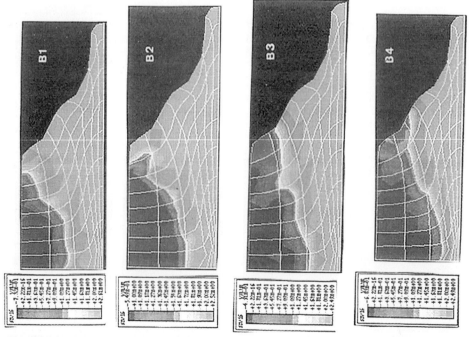

Fig. 5. Contour plot of the recrystallization progress distribution in the four cases analyzed.

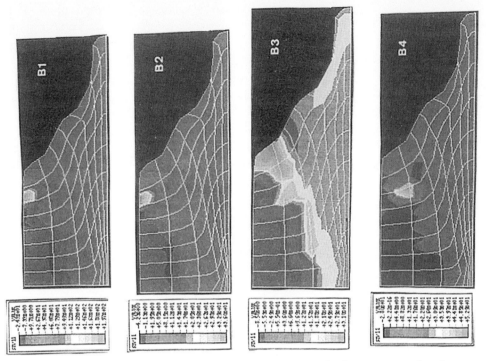

Fig. 6. Contour plot of the recrystallized grain sizes in the four cases analyzed.

ACKNOWLEDGEMENTS

This work was financed by the Comision Interministerial de Ciencia y Tecnología (CICYT), of Spain under contract MAT 97-0827-C02-01.

REFERENCES

[1] J.M. Cabrera, J.J. Jonas and J.M Prado, Mat. Sci. and Technology, 12 (1996), p. 579.

[2] J.M.Cabrera, A. Al Omar, J.J. Jonas and J.M. Prado, Metall. Trans., 28A (1997), p. 2233.

[3] Y. Estrin and H. Mecking, Acta Metall., 32 (1984), p. 57.

[4] Y. Bergström, Mater. Sci. Engineering, 5 (1969-1970), p. 193.

[5] Y. Bergström and B. Aronsson, Metall. Trans., 3 (1972), p. 1951.

[6] B. Derby, Acta Metall., 39 (1991), p. 955.

[7] B. Derby, Scripta Metall., 27 (1992), p. 1581.

[8] J.M.Cabrera, M. Esteller, A. Herrero and J.M.Prado. Proceedings of the V Congreso Nacional de Tratamientos Térmicos, TRATERMAT'92, (1992), Sitges, Barcelona, p. 287.

Correspondance with readers: cabrera@cmem.upc.es

Materials Science Forum Vols. 284-286 (1998) pp. 377-384
© *1998 Trans Tech Publications, Switzerland*

Modeling the Critical Strain for the Initiation of Dynamic Recrystallization during the Hot Strip Rolling of Niobium Microalloyed Steels

F. Siciliano Jr. and J.J. Jonas

Dept of Metallurgical Engineering McGill University,
3610 University Street, Montreal, H3A 2B2, Canada
E-Mail: JohnJ@MinMet.Lan.McGill.Ca

Keywords: Hot Strip Rolling, Niobium Steels, Dynamic Recrystallization, Critical Strain, Mathematical Model, Mill Data

Abstract

The critical strain for the onset of dynamic recrystallization is an important parameter employed in the mathematical modeling of microstructure development and of rolling load. The aim of this study was to derive an equation that expresses the critical strain as a function of a previously developed peak strain equation. Mill logs from four different hot strip mills were used for this purpose, and over 100 logs were analyzed. The Nb compositional range examined in this study was 0.008 to 0.058 Nb, which represents most of the conventional range of Nb levels. The results show that the critical strain/peak strain ratio decreases with an increase in Nb concentration. Decreasing the Mn and increasing the Si levels have similar but smaller effects than that of Nb addition.

Introduction

Knowledge of the critical strain for the initiation of dynamic recrystallization (DRX) is a requirement for the prediction of the operating softening mechanisms in hot working processes. During the hot strip rolling of steels, the rapid softening and intense grain refinement caused by DRX in a particular pass can considerably modify the load behavior in the following pass. Furthermore, once DRX is initiated, softening by means of metadynamic recrystallization (MDRX) will also take place during the interpass interval, as long as sufficient time is available for this mechanism to make a contribution as well.

For the present purpose, it is useful to express the critical strain for DRX (ε_c) as a function of the peak strain (ε_p), as determined from a stress-strain curve. This is because several equations are available to specify the peak strain as a function of the initial grain size, temperature and strain rate for Nb steels [1,2]. Such an approach can then be used to predict the operating recrystallization mechanism, which can be static recrystallization (SRX), or DRX followed by metadynamic recrystallization (MDRX), or some combination of these two static softening mechanisms.

The $\varepsilon_c/\varepsilon_p$ ratio is usually taken as 0.8 for plain C-Mn steels. However, previous workers have reported lower values for Nb steels [1-3]. In this study, the ratio was considered to be a function of the Nb content and was determined by using data from hot strip mill logs. This approach is in contrast to the usual one based on torsion testing, in which the ε_c and ε_p equations are derived directly from experimental data. The expression for ε_c derived in this way can then be used to predict rolling loads and microstructures. The latter aspect of this work is described elsewhere [2-5].

Experimental Procedure

Mill logs from four different hot strip mills were used as input data for this study. They were collected from the following steel plants: Dofasco Steel (Hamilton, Canada), Algoma Steel (Sault Ste. Marie, Canada), BHP (Port Kembla, Australia) and Sumitomo Metal Industries (Kashima, Japan). The chemical compositions of the steels are displayed in Table 1.

Table 1- Chemical compositions of the steels.

Steel	Plant	C	Mn	Si	Nb	Ti	Al
D1	Dofasco	0.06	0.65	0.12	0.020	--	0.035
D2	Dofasco	0.14	0.65	0.23	0.020	--	0.035
D3	Dofasco	0.12	0.85	0.23	0.020	--	0.035
D4	Dofasco	0.12	1.0	0.23	0.020	--	0.035
D5	Dofasco	0.06	0.65	0.12	0.030	--	0.035
D6	Dofasco	0.06	0.65	0.01	0.030	--	0.035
D7	Dofasco	0.06	0.65	0.12	0.045	--	0.035
D8	Dofasco	0.06	0.65	0.01	0.045	--	0.035
D9	Dofasco	0.06	0.45	0.01	0.008	--	0.035
D10	Dofasco	0.06	0.65	0.01	0.008	--	0.035
A1	Algoma	0.05	0.35	0.01	0.035	--	0.043
A2	Algoma	0.05	0.70	0.10	0.053	--	0.045
A3	Algoma	0.06	0.70	0.11	0.058	--	0.045
S1	Sumitomo	0.07	1.12	0.05	0.036	0.016	0.029
S2	Sumitomo	0.09	1.33	0.06	0.023	0.016	0.019
B1	BHP	0.11	1.05	0.01	0.031	--	0.040

The basic pass parameters taken from the mill logs were: temperature, initial and final thickness, work roll radius, work roll rpm, and force. From these data, the mean flow stress (MFS) was calculated according to the Sims method. The spreadsheet used for this calculation is illustrated in Refs. 4 and 5. The MFS also depends on chemical composition, grain size and strain rate [2,3,4]. For this reason, the Misaka equation [6] was modified to take Nb, Mn and Ti additions into account as well as the softening produced by DRX. This model also allows for the occurrence of MDRX when it follows DRX.

The spreadsheet used is shown in Fig. 1. Here, t_{ip}, t_{ps} and $t_{0.5}$ represent the interpass time, precipitation-start time and time for 50% softening, respectively, while ε_a and X are the accumulated strain and fractional softening, respectively. For each pass, there are tests for the initiation (as indicated by ε_c) of dynamic recrystallization. The details of this approach are described elsewhere [2-5]. Note that this model takes into account the occurrence of strain induced precipitation, via t_{ps}, which is assumed here to prevent any further recrystallization [3,7].

INPUTS

pass	d_0	T	$\dot{\varepsilon}$	t_{ip}	ε
	(µm)	(°C)	(s⁻¹)	(s)	
F 1	80	1000	14.2	3.54	0.79
F 2	--	968	25.9	2.23	0.57
F 3	--	952	46.7	1.45	0.54
F 4	--	940	84.3	0.94	0.53
F 5	--	934	141.1	0.61	0.40
F 6	--	921	181.0	0.43	0.27
F 7	--	906	197.1	--	0.16

OUTPUTS

pass	d_0	ε_a	ε_c	DRX ?	t_{ps}	PPTN?	$t_{0.5}$	X	1000/T	Predicted MFS
	(µm)				(s)		(s)		(K⁻¹)	(MPa)
F1	80.0	0.79	0.92	--	47.5	No	7.6	0.28	0.79	129
F2	69.5	0.57	1.12	Y	9.81	No	0.19	1.00	0.81	160
F3	15.4	0.54	0.63	--	11.64	No	0.86	0.69	0.82	153
F4	17.1	0.70	0.78	--	5.47	Y	1.24	0.00	0.82	178
F5	17.1	1.10	0.88	--	2.83	Y	1.36	0.00	0.83	212
F6	17.1	1.37	0.98	--	1.88	Y	1.85	0.00	0.84	236
F7	17.1	1.53	1.08	--	1.20	Y	2.20	0.00	0.85	252

Fig. 1 Spreadsheet for MFS calculation using the present mathematical model.

In this analysis, the critical strain ratio, $\varepsilon_c/\varepsilon_p$, was adjusted to provide the best fit between the MFS's calculated from the mill forces and the ones predicted by the model. The equation used to describe the critical strain is shown below:

$$\varepsilon_c = C\,\varepsilon_p \qquad ...(1)$$

where the peak strain is given by:

$$\varepsilon_p = (1 + 20\,[Nb]) / 1.78 \times 2.8 \times 10^{-4} d_0^{0.5}\left[\dot{\varepsilon} \cdot \exp\left(\frac{375000}{RT}\right)\right]^{0.17} \qquad ...(2)$$

The latter equation was derived from torsion test data [1] and then modified for use with low carbon Nb steels [2]. This approach was applied to more than 100 mill logs for the finish rolling of the compositions listed in Table 1. The method described above can, however, be repeated for other families of steels, using peak strain equations that can readily be found in the literature.

Results and Discussion

Each mill log analysis resulted in an MFS figure, with the "measured" and "predicted" MFS's plotted as functions of the inverse absolute temperature. Each mill log corresponds to a C value, where $C = \varepsilon_c/\varepsilon_p$. The following plots are examples of this type of MFS modeling. These C values were then plotted as a function of the Nb content for the whole range available. It can be seen that the higher the Nb content, the lower the $\varepsilon_c/\varepsilon_p$ ratio. Beyond 0.05% Nb, the C factor approaches a value of around 0.4 according to this analysis.

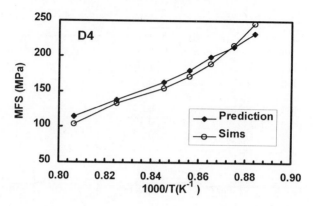

Fig. 2 MFS prediction for the Dofasco D4 grade, showing no DRX.

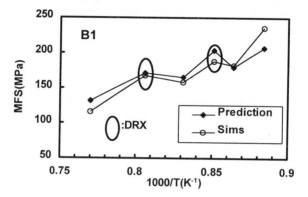

Fig. 3 MFS prediction for the B1 grade, where DRX is considered to occur in passes 2 and 4, followed by MDRX.

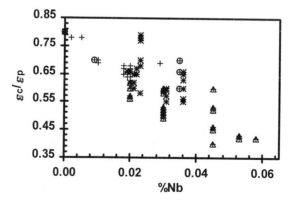

Fig. 4 Initial diagram of the $\varepsilon_c/\varepsilon_p$ ratios as a function of the Nb content.

The effect of Nb on the stress-strain curve is to increase the stress at a particular strain. Concurrently, the peak strain is moved to higher values. Experimental stress-strain curves derived from torsion tests show this effect, see Fig. 5. In this case, the A1 grade (0.035%Nb) was tested together with a C-Mn steel of the same base composition. The peak strain increases with Nb addition more rapidly than the critical strain, as a result of which the $\varepsilon_c/\varepsilon_p$ ratio decreases when the Nb concentration is increased. This suggests that the solute drag due to the presence of Nb has less influence on ε_c than on ε_p (and therefore on nucleation as opposed to growth).

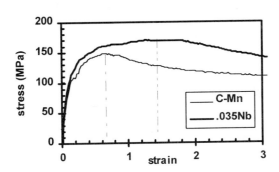

Fig. 5 Effect of Nb addition on the stress-strain curve of a low carbon steel.

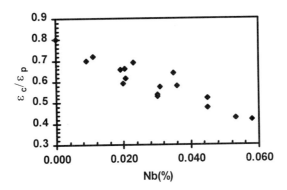

Fig. 6 Mean values of $\varepsilon_c/\varepsilon_p$ ratios (obtained from mill logs) plotted against Nb content.

In order to minimize the spread associated with industrial data, the $\varepsilon_c/\varepsilon_p$ values were averaged for each grade before plotting, see Fig. 6. This diagram illustrates more clearly the dependence of the $\varepsilon_c/\varepsilon_p$ ratio on Nb content. These points, however, still display some scatter. The effects of Mn and Si addition were therefore included in the analysis and the following equation was derived:

$$\varepsilon_c / \varepsilon_p = 0.8 - 13\,[Nb_{eff}] + 112\,[Nb_{eff}]^2 \qquad (3)$$

where:

$$Nb_{eff} = [Nb] - \frac{[Mn]}{120} + \frac{[Si]}{94} \qquad (4)$$

When the data of Fig. 6 are plotted with the aid of this equation, the relationship between $\varepsilon_c/\varepsilon_p$ and the concentrations of Nb, Mn and Si becomes more evident, as shown in Fig. 7. The original uncorrected points are represented by the open squares. Plotting the $\varepsilon_c/\varepsilon_p$ ratio against the *effective* Nb concentration (Eq. 4) results in a clear relationship. Eqs. 3 and 4 are considered to apply to the following ranges: [Nb] 0.01 to 0.06%, [Mn] 0.35 to 1.33% and [Si] 0.01 to 0.23%. They describe the progressive decrease in $\varepsilon_c/\varepsilon_p$ ratio represented by the line in Fig. 7.

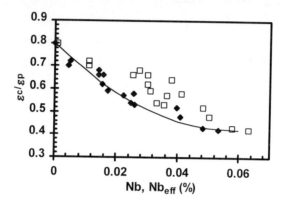

Fig. 7 $\varepsilon_c/\varepsilon_p$ ratios calculated from Equations (3) and (4) as a function of *effective* Nb concentration. The unmodified points are represented by the open squares and plotted solely against [Nb].

At the start of this analysis, the solute effects were expected to be all positive, i.e. all additions were expected to increase ε_p. However, it can be seen from Eq. 4 that the influence of Mn addition is opposite in sign to that of Nb and Si. This indicates that Mn is not making a *direct* contribution to solute drag, but may be exerting, with Si, an indirect effect. A rationalization for the present observations can be based on the report that Mn addition *decreases* the diffusivity of Nb in austenite while the addition of Si *accelerates* Nb diffusivity [8]. Furthermore, it has been shown that Mn in solution has an almost negligible direct solute drag effect on the migration of grain boundaries [9].

The explanation proposed here can be understood more clearly by referring to Fig. 8. Here, the dependence of the solute drag force on grain boundary velocity attributable to Nb diffusion (in Fe) is represented by the full line. When the diffusivity of Nb is increased, as occurs in the presence of Si, the Nb solute drag curve shifts to the right, as indicated by the dotted line. By contrast, when the diffusivity of Nb is decreased, as attributable to the addition of Mn, the Nb solute drag curve shifts to the left, as illustrated by the broken line.

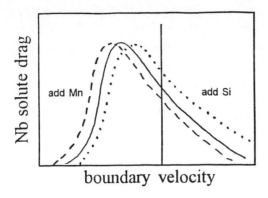

Fig. 8 Effect of boundary velocity on the Nb solute drag exerted on the boundary. The indirect influence of Mn and Si addition is shown schematically.

The influence of adding Si and Mn on the solute drag *attributable to the presence of Nb* can now be deduced with reference to the vertical line that represents a particular grain boundary velocity to the right of the peak, i.e. higher than that associated with the maximum in the drag force. (Most practical examples of solute drag are found to the right of the peak, after normalization of the curve for the diffusivity [9].) It can be seen that the addition of Si *increases* the solute drag with respect to the single addition of Nb, while the addition of Mn has the contrary effect.

According to the above scenario, the decrease in the Nb diffusivity and Nb solute drag associated with Mn addition decreases ε_p (and consequently increases $\varepsilon_c/\varepsilon_p$). On the other hand, when unusually low concentrations of Mn are present, the Nb diffusivity and solute drag are increased, slowing down the rate of softening by DRX and increasing ε_p (i.e. decreasing the $\varepsilon_c/\varepsilon_p$ ratio). This phenomenon can be observed in the work of Akben et al. [10], where a Nb HSLA steel containing 1.90%Mn displayed a lower peak strain than a similar grade that only contained 1.25%Mn. It is important to note here that the influence of Mn addition is only a secondary effect compared to that of Nb and is only observed because it appears to moderate that of Nb.

With regard to the *relative* effects, that of Nb is clearly the greatest, followed by that of Si and then by Mn, in a ratio of 120:1.3:1, respectively. However, due to the relatively high concentrations of Mn and Si generally employed compared to Nb, the *practical* effects on $\varepsilon_c/\varepsilon_p$ are represented by the ratio 13.5:1:2.6, for example, for a steel containing 0.03%Nb, 0.2%Si and 0.7%Mn. The weak effects of Mn and Si on ε_p as well as on austenite strengthening are due to the similarities in their atomic diameters in comparison with Fe.

Conclusions

- The $\varepsilon_c/\varepsilon_p$ ratio decreases with increasing Nb content in the present steels. (Here, ε_c represents the critical strain for the initiation of dynamic recrystallization and ε_p is the peak strain.).

- The elements Mn and Si have small (but not negligible) and opposite effects on the $\varepsilon_c/\varepsilon_p$ ratio.

- The relative effects of Nb, Si and Mn addition on the $\varepsilon_c/\varepsilon_p$ ratio are represented by the ratio 120:1.3:1. This is equivalent to about 13.5:1:2.6 for typical addition levels of these elements, respectively.

Acknowledgments

The authors are indebted to Mr. Brian Nelson of Dofasco Inc. for supplying extensive hot strip mill rolling data. They express their special thanks to Mr. Brian McCrady of Algoma, Mr. Koji Minami of Sumitomo and Dr. Chris Killmore of BHP for the provision of further mill logs. The financial support received from the Canadian Steel Industry Research Association (CSIRA) and the Natural Sciences and Engineering Research Council of Canada (NSERC) is acknowledged with gratitude. FS is grateful to the Conselho Nacional de Desenvolvimento Científico e Tecnológico, CNPq, Brazil, for the award of a Ph.D. scholarship.

References

[1] C. Roucoules, S. Yue and J. J. Jonas: Proc. Int Conf on Modeling of Metal Rolling Processes, The Inst. of Materials, London, UK, (1993) p. 165.

[2] K. Minami, F. Siciliano Jr., T.M. Maccagno and J.J. Jonas: ISIJ Int. **36** (1996) p. 1507.

[3] F. Siciliano, T.M. Maccagno, B.D. Nelson and J.J. Jonas: in T. Chandra (ed.) Thermec 97', Wollongong, Australia, 1997.

[4] F. Siciliano Jr., K. Minami,, T.M. Maccagno and J.J. Jonas: ISIJ Int. **36** (1996) p. 1500.

[5] T.M. Maccagno, J.J. Jonas, S. Yue, B.J. McCrady, R. Slobodian and D. Deeks: ISIJ Int. **34** (1994) p. 917

[6] Y. Misaka and T. Yoshimoto: J. Jpn. Soc. Technol. Plast., **8** (1967-8) p. 414.

[7] P.D. Hodgson and R.K. Gibbs: ISIJ Int. **32** (1992) p. 1329.

[8] S. Kurokawa, J.E. Ruzzante, A.M. Hey and F Dyment: Metal Science, **17** (1983) p. 433.

[9] L.T. Mavropoulos and J.J. Jonas: Canadian Metallurgical Quarterly, **28** (1989) p. 159.

[10] M.G. Akben, I Weiss and J.J. Jonas: Acta Metallurgica, **29** (1981) p. 111.

Materials Science Forum Vols. 284-286 (1998) pp. 385-392

Numerical Simulation of Temperature Distribution and Cooling Rate in Ship Profiles during Water Spray Cooling

V. Olden[1], C. Thaulow[2], H. Hjerpetjønn[3], K. Sørli[4] and V. Osen[5]

[1] SINTEF Materials Technology, N-7034 Trondheim, Norway

[2] The Norwegian University of Science and Technology, N-7034 Trondheim, Norway

[3] FUNDIA PROFILER AS, N-8600 Mo, Norway

[4] SINTEF Applied Mathematics, N-7034 Trondheim, Norway

[5] The Norwegian University of Science and Technology, N-7034 Trondheim, Norway

Keywords: Materials Technology, Numerical Analysis, Steel, Hot Rolling, Ship Profiles, Water Cooling, Heat Transfer Coefficient, FE-Modelling, Cooling Rate, Straightness

Abstract

This report describes a study of transient cooling of bulb flat ship profiles. The study is focused on modelling of temperature profile and displacements during water spray cooling. Laboratory experiments, numerical calculation of the heat transfer coefficient, statistical experimental analysis and 2 and 3D FE-modelling represent the main tools in the analysis. 2D and 3D modelling of temperature profile, stress and displacements, give the opportunity to identify the favourable cooling patterns with respect to desired cooling rate and acceptable resulting straightness.

1. Aims and scope

To be able to meet the strong quality demands for hot rolled steel products, close temperature control is necessary in every production step. Improvements in the production control systems have been necessary. Introduction of the controlled rolling and accelerated cooling (CR+ACC) production process has increased the demand for close production control even further. Control systems based on a combination of empirical data and numerical modelling have been successfully implemented in many mills. The vast majority of hot rolled products are plates and strips. Consequently the main focuses of the production control effort have been put on these products. However, other hot rolled products, such as ship profiles, have to meet the stronger quality demands as well.

In non-symmetrical ship profiles, variations in the temperature field during cooling can result in unwanted and sometimes permanent deformation (hot waves). Selective accelerated water-cooling will give the possibility to achieve a more homogenous temperature field, and thereby minimise the unwanted deformation. There is an increasing demand for high strength ship profiles, with tensile strength in the range of 400 - 500 Mpa. Accelerated water-cooling will give increased strength without additional carbon or expensive micro alloying elements. Thus, the weldability will be maintained.

The work described in this article aims to: 1) identify the main parameters influencing the deformation of CR+ACC non-symmetrical ship profiles (bulb flat) during and after accelerated water spray cooling. A description of relationships between these parameters is a part of this goal. 2) Achieve necessary homogenous temperature distribution and a favourable cooling rate with respect to mechanical properties. Water cooling parameters will be optimised by 2D and 3D finite element modelling (FEM).

2. Testing

To be able to perform accurate modelling of the deformation behaviour, the heat transfer conditions had to be thoroughly investigated. The heat transfer coefficient history during cooling was retrieved by a combination of laboratory experiments and numerical modelling.

3. Equipment

Transient water-cooling laboratory experiments were performed in a specially designed test rig. The test rig includes an electric furnace, a flexible arrangement for water spray nozzles and a sample holder, all placed on a solid stainless steel frame. The furnace and the nozzle holder are movable in x-direction, the nozzles in x, y and z-direction. The nozzle used in the experiments was a flat jet type. Bulb flat ship profiles, 340 mm lengths of 430 x 19 mm, were used as test samples. The steel was a C-Mn steel with chemical composition and mechanical properties as presented in table 1.

Temperature histories were recorded in four thermocouple positions, about 2 mm below the surface. Ø 1.0 mm Type K hot junction thermocouples were used. The location of the thermocouples is indicated in figure 1. For verification of the numerical model a thermocouple was also mounted in the mid thickness of the bulb flat "head." Back and sides of the samples were insulated. To avoid wetting of the insulation a stainless steel shield enclosed the back and sides of the sample during cooling. Temperature histories were recorded by a 60 Hz networked data logger, and stored on a personal computer. Logging frequency was 10 scans per second.

Table 1 Chemical composition and mechanical properties of the steel

C	Si	Mn	Al	Nb	Re	Rm
0.1	0.30	1.30	0.040	0.020	~320 MPa	~450 Mpa

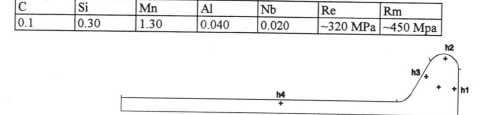

Fig.1 Location of the thermocouples

4 Laboratory experiments

Table 2 Laboratory experiment programme

Exp. no	Water flow l/min	Position of water nozzle x mm	y mm	angle °
1	7.3	130	80	45
3	3.7		80	0
4	10.2		80	0
5	3.7	150		90
6	10.2	150		90

In the laboratory experiment the water flow and the positions of the nozzle were varied systematically. Table 2.

All samples were heated up to 1100°C and held at this temperature for 1 hour in inert atmosphere. The samples then were exposed to the water spray during cooling from 1000 to 50°C in the water-cooled areas. The samples experienced selective water cooling, in the sense that only the bulb flat head was exposed to the spray. See figure 2. The recorded temperature histories then were used as input to a numerical procedure calculating the heat transfer coefficient (HTC). An example of the recorded temperature histories is shown in figure 3.

5. Calculation of the heat transfer coefficient

The crucial boundary condition of the resulting non-linear heat conduction problem is the heat transfer at the part of the bulb surface that is exposed to the water spray. The usual approach is to formulate this type of heat transfer by:

$$-k(T)\frac{\partial T}{\partial n} = h(T)(T - T_\infty)$$

Eq. 1

where k is the heat conductivity, h the heat transfer coefficient, T the surface temperature and T_∞ the surrounding temperature. However, the heat transfer coefficient, h, is unknown. The temperature measurements from the laboratory experiments were used as input to a finite element method with a backward Euler time integrator repeatedly solving the transient non-linear heat conduction problem

$$\rho(T)c_p(T)\frac{\partial T}{\partial t} = \frac{\partial}{\partial x}\left(k(T)\frac{\partial T}{\partial x}\right) + \frac{\partial}{\partial y}\left(k(T)\frac{\partial T}{\partial y}\right)$$

Eq. 2

where ρ and c_p are the material properties of density and heat capacity. All material properties are time dependent and represented by approximated functions.

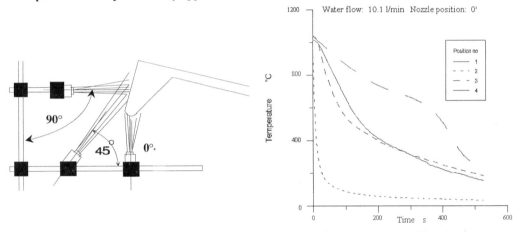

Fig. 2 Positions of the water spray nozzle   Fig.3 Temperature histories

A linearization of Eq. 1 is applied in a two-step procedure. First, a linear extrapolation of T^n and T^{n-1}, denoting the temperature fields at time levels n and $n-1$, is applied to give an estimate of T^{n+1} for the computation of the first estimation of k, c_p and ρ. Then, the resulting linear problem was solved to give a second estimate of T^{n+1}.

Finally, the latter estimate was substituted to give second estimates of k, c_p and ρ, and the resulting linear problem was solved for the final T^{n+1}. The procedure was repeated for each time step. However, remember that the problem also needs to be solved repeatedly for the determination of the unknown h values.

A thorough description of the numerical procedure is given in [1]. Examples of calculated h histories are shown in figure 4 a and b.

6. Numerical simulation of temperature profile and cooling rate

Calculated heat transfer coefficient histories from the laboratory experiments were used as input in 2 and 3D transient numerical simulations of the temperature profile and deformation of the bulb flat.

The FEM-models were generated in the pre processor Patran version, 6.0. For the 2D model linear 4 node quadratic plane elements were used. For the 3D model linear solid 8 node brick elements were used. The 2 and 3D element meshes are given in figure 5. The calculations were performed using ABAQUS Standard. The calculations were transient temperature and stress analysis; temperature coupled both with respect to material properties and of course h.

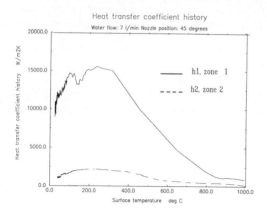

Fig. 4a Calculated Heat Transfer Coefficient Histories; Water flow: 7 l/min, Nozzle position:45°

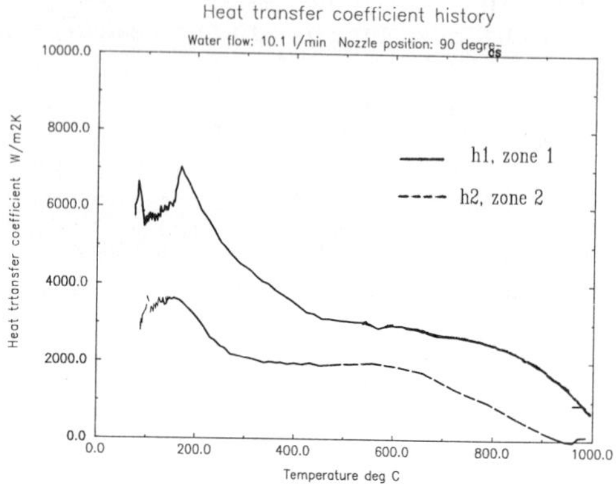

Fig.4b Calculated Heat Transfer Coefficient Histories; Water flow: 10.1 l/min, Nozzle position:90°

Variations in water-cooling start temperature and duration of the water-cooling was investigated numerically. The parts of the HTC-curves representing the chosen start temperature and duration of water-cooling were used as input. The main objective with these investigations was to find the combination of parameters giving a favourable cooling history in the profile, both in the sense of resulting straightness and cooling rate.

Fig. 5   2 and 3D element mesh

The analysis was ruled by the following hypothesis:

- If the temperature histories in the mid-thickness of the profile's long leg and short leg are identical, the temperature-induced stress will be low. This will have a positive effect on the straightness of the profile.

- A cooling rate not exceeding 10 - 15°C/s in the temperature range 800 - 500°C will avoid martensitic transformation

2D analyses were performed first. The "best candidates" from the 2D analyses then where picked for 3D analyses of displacement (straightness).

7. 2D analysis

Initially a statistical analysis [2] of the parameters: water-cooling start temperature, duration of water-cooling, water flow, and nozzle position were performed. The two latter parameters were represented in the simulations by calculated h-curves from the laboratory tests. The measured response was the surface temperature drop during cooling. Water-cooling was applied to the model in the bulb head represented by the surface zones 1 and 2, fig. 6.

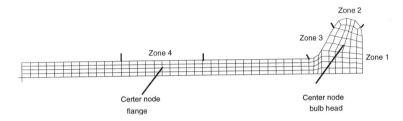

Fig.6 Location of simulated water cooling zones and mid thickness nodes

The statistical simulation programme with the measured surface temperature drop DT is given in table 3. Statistical analyses identified duration of water-cooling and nozzle position as the parameters with the most powerful influence on the surface temperature drop. [3] Cooling start temperature had minimal influence, and was not included in the further analyses.

Table 3　　Statistical simulation programme

Sim. no.	Water flow [l/min]	Nozzle Position [°]	Cooling time [sec]	Cooling start temp. [°C]	Surface temp. drop ΔT [°C]
1	3.7	0	5	850	276
2	10.1	0	5	1000	300
3	3.7	90	5	1000	250
4	10.1	90	5	850	246
5	3.7	0	20	1000	672
6	10.1	0	20	850	750
7	3.7	90	20	850	400
8	10.1	90	20	1000	534
9	6.9	45	12.5	925	794

By multiple regression analysis , the surface temperature drop can be expressed in the following way:

$$\Delta T = 210 + 9.1 W_v + 21.4 T_c - 1.58 P$$

where W_v is the water flow, T_c is the water cooling time and P the nozzle position. However, we have to stress that this formula is valid for this analysis only. If such an expression were to have more general value, the nozzle position or more correct the angle of inclination between the water spray and the steel surface would have to be thoroughly investigated. Figure 7 show a plot of the temperature development for simulation no 9. The temperatures are read in nodes localised in the mid-thickness of the bulb head and in the bulb flange.

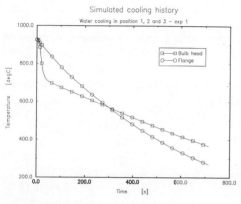

Fig. 7 Simulated cooling history, Water cooling in zone 1, 2 and 3 with 7 l/min and 45° [4]

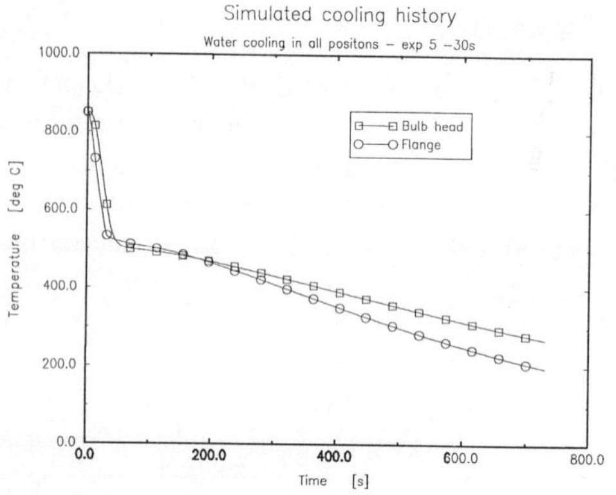

Fig.8 Simulated cooling history, Water cooling in zone 1, 2, 3 and 4 with 10 l/min and 90° [4]

The mid-thickness cooling rate in the temperature range 800 -500 °C is slow, only 1 -2 °C/sec. To reach a cooling rate of 10 - 15 °C/sec it was evident that cooling had to be intensified both in the bulb head and the flange area. A plot from the simulation giving the best result with respect to the temperature development is presented in figure 8. The cooling rate between 800 - 500°C are presented in figure 9.

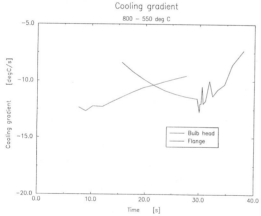

These conditions were achieved by applying water spray cooling from a temperature of 850°C in about 30 seconds with the following heat transfer conditions: Heat transfer coefficient increasing from 3000 - 6000 W/m2K in the bulb head zone 1 and 3, and from 1000 - 4000 W/m2K in zone 2 and the flange zone 4. The heat transfer coefficient curves are presented in figure 4 b.

The resulting cooling rate should be verified both in laboratory and in pilot plant experiments. Nevertheless it is evident that water cooling also must be applied on the flange to reach the aimed cooling rate.

Fig. 9 Cooling rate in best simulated case [4]

8. 3D analysis

The temperature induced stress and resulting displacements in the profile, were examined with a 3D FE-analysis with the same water cooling conditions as for the 2D analysis. The model represents a profile with a length of 2 meters. The water cooling conditions were simplified in the sense that the same conditions (h-curves) were applied in the whole length of the model. Applied temperature dependent yield strength is given in table 4.

Table 4   Temperature dependent yield strength, Re [5]

Temp. [°C]	20	100	200	300	400	500	600	700	800	1000
Re, [Mpa]	315	315	270	240	220	180	110	50	50	40

Stress analysis revealed that the von Mises stress is at its highest level at the end of the water cooling sequence. A 3D plot after about 30 seconds of water-cooling is shown in figure 10. The highest stress level was found in the bulb head and in the transition area between bulb head and the flange reaching about 175 MPa.

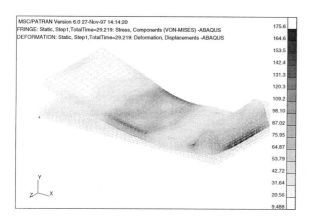

Fig. 10 Plot of stress level after 30 seconds of water cooling

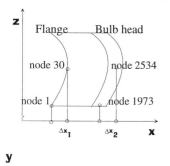

Fig. 11 Relative displacements in the ship profile [4]

The stress level did not exceed the values for tensile strength Re applied in the model. Hence, the displacement is elastic and the profile will in theory regain its initial shape. There is however some uncertainty connected to the values for tensile strength (Re). There is limited knowledge of the tensile strength of structural steel at elevated temperatures, especially above 600°C.

The temperature field is not totally homogenous in the flange and the bulb head area, see fig. 8. The degree of stiffness also differs. As a result, some sideways displacement (x – direction) during cooling will occur. The relative displacement is illustrated in figure 11.

Δx_1 represents the relative displacement between to nodes in the flange and Δx_2 in the bulb head of the model. Where: Δx_1= node 1 – node 30 and $\Delta x2$ = node 1973 – node 2534. The nodes are localised at z=0 and z=100 mm.

Δx_1 and Δx_2 are plotted in figure 12 a and b. The first plot represents the relative displacement from a simulation with water cooling applied only on the bulb head, the other with water cooling also applied on the flange.

The resulting relative displacement have opposite signs, indicating that the profile bends sideways in opposite directions. The displacement is, however, small between -0.07 mm and +0.04 mm. The model represents a profile length of 2 meters. Hence, a ship profile of 20 meters length will have a relative displacement of maximum |0.7| mm.

These results indicate that the straightness of the profiles will be satisfactory within the range of cooling conditions presented in this study. In the next step of the research project, the calculations will be verified in the rolling mill.

9. Further work

The findings of this study must be verified in the rolling mill. This will be an important activity in the near future. The investigations will also include micro-structural analysis, testing of mechanical properties and straightness measurements. Additionally the work on establishing valid relationships between water cooling parameters and cooling rate is planned to be intensified. Different dimensions of the bulb flat profile will also be included in the investigations.

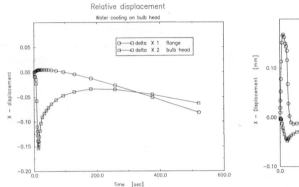
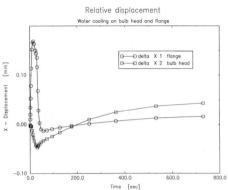

Fig. 12a Relative displacement between nodes: Water cooling of bulb head

Fig. 12b Relative displacement between nodes: Water cooling of bulb head and flange

10. Conclusions

Successful transient water spray experiments on ship profile samples were carried out.

Heat transfer coefficient calculations were performed using a finite element method with a backward Euler time integrator in solving the transient non-linear heat conduction problem.

Statistical analyses identified duration of water cooling and spray nozzle position as the parameters having the most powerful influence on the steel surface temperature drop during water cooling.

Regression analysis established relationships between the water cooling parameters and the cooling rate.

2D and 3D modelling of temperature-profile, stress and displacements identified the most favourable cooling pattern with respect to aimed cooling rate during water cooling and acceptable resulting straightness of the profile.

The findings in this study indicate the aimed cooling rate of 10 –15 °C/sec in the mid thickness of the bulb flat head and flange and acceptable straightness can be obtained by selective water spray cooling on the bulb head and the flange area, represented by heat transfer coefficients in the range of 1000 – 6000 W/ m^2 K.

References:

[1] V. Olden, K. Sørli: "Computation of Heat Transfer Coefficients in Spray Water Cooling of Shipbuilding Profiles", SINTEF Report STF24 A97344, December 22. -97

[2] MINITAB, Release 11, Minitab Inc. 1996

[3] V.Olden: "Utvikling av vannkjølingsteknologi – analyse av forsøksparametre", SINTEF Report STF24 A97355, October 24. –97

[4] V. Olden: "Utvikling av vannkjølingsteknologi – Analyseverktøy med 2 og 3D FE-simulering av avkjølingshastighet og retthet i bulb skipsprofil", SINTEF Report STF24 A97364, December 16. –97

[5] Y. Sakumoto: "Use of FR Steel: Design of Steel Frames to Eliminate Fire Protection", Nippon Steel Corporation, Cat. No. AC June -92

Materials Science Forum Vols. 284-286 (1998) pp. 393-400
© *1998 Trans Tech Publications, Switzerland*

Statistical Modelling of Mechanical Properties of Microalloyed Steels by Application of Artificial Neural Networks

C. Dumortier[1], P. Lehert[2], P. Krupa[3] and A. Charlier[4]

[1] Department of Metallurgy, Polytechnic Faculty of Mons,
56 rue de l'Epargne, B-7000 Mons, Belgium

[2] Data Processing Department, Catholic University of Mons,
151 Chaussée de Binche, B-7000 Mons, Belgium

[3] Cockerill Sambre - Computing Department,
1 rue de l'Usine, B-6090 Couillet, Belgium

[4] Formerly with Cockerill Sambre - R&D Corporate,
C.D.A.M., 10, Quai du Halage, B-4400 Flémalle, Belgium

Keywords: Microalloyed Steel, Mechanical Properties, Statistical Modelling, Artificial Neural Networks, Stepwise Multiple Regression, Principal Components Analysis

ABSTRACT
Neural Networks have been used to determine mechanical properties of microalloyed steels produced at Cockerill Sambre Steel Plant. By comparison with multiple linear regression, this method presents the advantage of dealing with non-linearity effects and to detect easily possible interactions between the predictors. However, statistical multivariate techniques continue to be an interesting tool to improve the quality of the prediction and to explain the results of the modelling.

INTRODUCTION
This work deals with the statistical estimation and prediction of mechanical properties of niobium microalloyed steels produced at Cokerill-Sambre Steel plant on the hot strip mill (Carlam). Extensive literature is devoted to the microalloyed steels which are known to depend on various factors [1] interacting together. It is generally admitted that various mechanisms contribute to explain steel mechanical behavior : contributions relative to solid solution hardening, strengthening by precipitation particles and microstructure, grain size effect and dislocations.
The induced multifactorial data structure is probably the major obstacle to build a theoretical predictive model in general conditions.
The complexity of a model following strict causal considerations asks for a statistical approach [2] , based on observed samples on which some authors suggested to estimate the best precision by linear regression techniques.
However, the various regression formulae available in the literature [3] give a large discrepancy of yield strength (R_e) and ultimate strength (R) estimated values for a given steel composition and a fixed thickness plate. These estimations range from 505 to 656 N/mm^2 for R_e and from 774 to 933 N/mm^2 for R.
To reduce this variability, it is thus justified to establish for a given steel plant its own particular relationships for predicting mechanical properties.
Tne purpose of this paper is to realize this statistical modelling with artificials neural networks and to try to improve the obtained results by use of multivariate analysis.

BACKPROPAGATION NEURAL NETWORKS [4,5,6]
Neural nets are interconnected agglomerates of simple computing
units call neurons or nodes. A symbolic representation of a direct
network is given in figure 1.
The input of any neurone i is given by : $e_i(t) = \Sigma w_{ij}(t)a_j(t) + \Theta_i$
where $a_j(t)$ is the activation state of the j interconnected nodes,
w_{ij} is the input weight from input node i to node j and Θ_i an
offset or a bias term.
A very frequent activation function used is the sigmoid one

$$a_i(t+1) = \frac{1}{1+e^{-ei(t)}}$$

In this work, a feedforward network performing the classical
backpropagation training algorithm with the generalized delta rule
with momentum has been used (momentum rate 0.7, learning rate
0.9).
A three-layers feedforward architecture of this network is
depicted in figure 2.

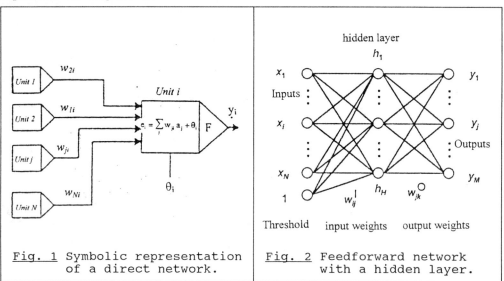

Fig. 1 Symbolic representation   Fig. 2 Feedforward network
       of a direct network.   with a hidden layer.

Neural networks acquire knowledge from learning procedures.
Initially, their associated weights w_{ij} are fixed to random
values.
During the training procedure, the network is said to learn (it
compares the target values t_j with the desired output values y_j)
and the weight values are modified in order to achieve the final
estimation goal.
Initially, the associated weights w_{ij} are fixed to random values.
For a set W of given weight values and for n patterns, the error
between the M target values t_j and the estimated values y_j is
measured by the following expression :

$$E(W) = 1/2 \sum_{i=1}^{n} \sum_{j=1}^{M} (y_{ij} - t_{ij})^2 \quad (1)$$

In the learning procedure, the w_{ij} must be considered as
parameters, whose values are adjusted repeatedly by the
backpropagation algorithm until the root mean square error E(W)

reaches a predetermined value.
The action of the network represented in figure 2 is determined by
its architecture, that is, the number of input, hidden and output
nodes and the values of the weights.
The input and output nodes are fixed by the application while the
number of hidden nodes is a variable which can be adjust by the
user.
Setting the number and size of hidden layers is not a easy task
and many publications have already dealt with this subject.
If too few hidden units are used, the network may fail to learn
properly. Conversely, if too many hidden units are used, the data
may be overfitted and this consequently reduces the network's
ability to generalize for new examples [4].
In fact, for a fixed and sufficient number of iterations, the
error function generally decreases in the learning phase when the
weights number which is related to the number of hidden nodes
increases.
Similarly during the learning step, the error function decreases
when the number of iterations increases.
To deal with the overfitting problem, the ratio of number of
patterns of the learning dataset to the number of hidden units has
been choosen larger than 20 and the number of iterations in
training phase is stopped when the error function calculated on
the validation database is minimized.

STATISTICAL ANALYSIS
Before using estimation techniques such as stepwise multiple
linear regression, it is necessary to have an a priori knowledge
concerning the data reliability as well as the linearity of the
relationships between the criteria and the predictors, the
possible interactions between predictors, and the possible
correlations [2].
When using artificials neural networks, the last considerations
are less critical because the interaction and non-linearity
effects [4] are automatically taken into account by the
multilayers structure.
According to some authors, another advantage of neural networks is
their ability, when encountering new data, of learning, self-
adjusting and eliminating useless data automatically.
However, Yhu-Jen Hwu and Al [7], who have used neural networks to
store and predict the flow stress of HSLA and carbon steels,
observed that the NN capability to extrapolate was not impressive.
In this study, a classical statistical analysis focused on data
reliability has been undertaken.
It consists to detect outliers and extremes by using
monodimensional techniques such as histograms or boxplot [8].
The multivariate approach of Mahanobis distance [9] can also be
used as a complementary outlier detection by defining a possible
outlier as the outside area of a confidence ellipsoïd.
Principal Components Analysis [10] (PCA), whose first purpose is
to give a representation of all variables in a (factorial) space
of limited dimensionality p have also been used to detect the
abnormal cases [11].
In this study, PCA has also been used to represent the cases of
the validation databases in the factorial space constructed from
the training dataset.
The statistical analysis and the neural network results were
carried out with Fortran and C codes running on PC-Pentium and DEC
3000 station.

MATERIAL
The basic material data relate to 12200 carbon and HLSA steel
coils suitable for construction and car production and collected
over 3 years (1995-1997) at Cockerill-Sambre Carlam Steel Plant.
Material data include 30 characteristics relating to :
- mechanical tensile properties (Re, R, El) ;
- mechanical tests (thickness, characteristics and position of the
  tensile specimen) ;
- plate dimensions (thickness, width) ;
- steel chemical analysis (C, Mn, P, S, Si, Al, V, Nb, Cu, Cr,
  Ni, N, Ti, Sn, B contents)
- rolling process (reheating, slabbing, rolling and finishing
  temperatures) ;
Table 1 gives the average, standard deviation, minimum and maximum
value for a 10000 cases database used as training dataset for NN
modelling.

Symbol Variable	Average	St. Dev.	Minimum	Maximum
THI Thickness mm	2.661	1.317	1.15	7.76
C % C x 1000	56.087	29.751	7.00	200.00
Mn % Mn x 1000	38.953	23.171	12.00	153.00
Si % Si x 1000	18.839	36.639	1.00	372.00
Nb % Nb x 1000	4.909	11.113	.00	64.00
V % V x 1000	.808	6.788	.00	88.00
Al % Al x 1000	42.819	10.343	7.00	147.00
N % N x 1000	49.135	19.269	.00	168.00
Ti % Ti x 1000	1.675	7.246	.00	62.00
B % B x 1000	7.649	15.497	.00	71.00
Cu % Cu x 1000	31.288	45.720	5.00	365.00
T°F Reheating T°C	1277.190	33.654	1080.00	1347.00
T°S Slabbing T°	1120.158	40.026	879.00	1183.00
T°R Rolling T° °C	866.865	26.455	761.00	927.00
T°C Coiling T° °C	641.117	36.499	509.00	775.00
Re Yield Strength N/mm^2	302.155	80.721	163.00	563.00
El Elongation %	34.501	6.112	15.00	58.00
R Ultimate Strength N/mm^2	398.485	71.294	249.00	630.00

Table 1. DESCRIPTION OF MAIN VARIABLES.

RESULTS
Among the 30 variables available only 19 have been selected for
the final statistical modelling.
Preliminary regression and neural networks tests have shown that
the remaining ones have no or quasi no influence on the mechanical
properties. Among the deleted variables with slight influence,
belong those asociated with the residual elements (P, S...) which
cannot be useful in any case to enter in a prediction model.
It is important to mention that 2 among those 19 variables are
categorical ones : the position in the coil of the test specimen
(TS_P) with 3 categories (top, bottom, middle) and the type of
normalized specimens (TS_T) with 2 categories (ISO 25/125 and ISO
20/80). To deal with such variables in a regression model, it is
necessary to introduce binary or dummy variables

About 200 cases were a priori exluded to abnormal or missing
values so finally 12197 cases were found eligible for the
analysis.
In a first separation, those 12197 cases were arbitrarily

partionned in two groups of respectively 10697 and 1500 items. This separation respects the fabrication date so the 1500 cases are relative to the last coils produced (june and july 1997). By a randomly procedure, a second separation was operated, the 10697 cases being distributed in two groups, the larger (10000) reserved for the neural network training step while the less important (697) aimed to be used as a first validation dataset.

The NN'architecture used for the mechanical properties modelling is described in figure 3.

```
Input     1     2 3   4   5   6 7   8   9 10  11   12    13    14    15    16    17
          Thi.  C Mn  Si  Nb  V Ti  B  Al  N   Cu  TS_P  TS_T  T°F   T°S   T°R   T°C

Hid. Lay. 1                 o   o   o   o   o   o   o   o   o   o   o   o   o
Hid. Lay. 2                     o   o   o   o   o   o   o

Outputs                                         o
                                      (Re or El or RR)
```

Fig. 3. Symbolic representation of the two hidden layers neural network used for the modelling

ELONGATION

Iter.	r_{CM}	σ %	$E10^{-4}$	r_{CM}	σ %	$E10^{-4}$	r_{CM}	σ %	$E10^{-4}$
	<training Database>			<1st Valid. data >			<2nd Valid. data		
500	.8570	3.154	6391.	.8331	3.094	4864.	.8023	3.418	5894.
1000	.8616	3.108	6391.	.8346	3.080	4792.	.8045	3.402	5837.
1500	.8636	3.086	6467.	.8340	3.086	4786.	.8076	3.377	5748.
2000	.8659	3.063	6419.	.8360	3.069	4731.	.8100	3.358	5690.
2500	.8673	3.048	6451.	.8382	3.050	4667.	.8067	3.384	5792.
3000	.8680	3.039	6293.	.8377	3.055	4690.	.8041	3.405	5879.
3500	.8688	3.032	6110.	.8359	3.070	4761.	.8025	3.417	5948.
4000	.8694	3.025	5967.	.8333	3.092	4853.	.8006	3.432	6028.
4500	.8697	3.022	5914.	.8312	3.109	4921.	.7988	3.445	6088.
5000	.8696	3.022	5933.	.8298	3.121	4958.	.7973	3.457	6129.

ULTIMATE STRENGTH

Iter.	r_{CM}	$\sigma N/mm^2$	$E10^{-4}$	r_{CM}	$\sigma N/mm^2$	$E10^{-4}$	r_{CM}	$\sigma N/mm^2$	$E10^{-4}$
8000	.9844	12.56	850.	.9784	14.45	1048.	.9722	15.33	1233.

YIELD STRENGTH

Iter.	r_{CM}	$\sigma N/mm^2$	$E10^{-4}$	r_{CM}	$\sigma N/mm^2$	$E10^{-4}$	r_{CM}	$\sigma N/mm^2$	$E10^{-4}$
2000	.9750	17.93	1948.	.9687	20.25	2072.	.9542	22.37	2530.

Selected number of iterations

Table 2. Evolution of error, coefficient of correlation r_{CM} and standard error of estimate Std_{est} during training for elongation. Main results for ultimate and yield strength

In this study, the choice of the structure has not really been optimized. In fact, the iterations number must be adapted to the NN architecture and it is possible to obtain good results with other structures. To characterize the NN estimation of the mechanical properties the following parameters have been computed : the error function (1), the coefficient correlation

between calculated and measured values (r_{CM}) for the three
criteria and their standard of estimate ($std_{estimate}$). The results
of the training and validation results are shown in table 2.
Table 2 reveals that standard of estimate are greater for the two
validation datasets but that the results remain good and
significantly better than those obtained by application of a
multiple linear regression model including 3 dummy variables (cf
table 3).
It is also possible to use a neural network with 3 outputs (for
the 3 criteria) but the results are slightly less good (standard
of estimate for Re and R are around 1 to 1.5 N/mm^2 higher and the
same for El) even after using larger numbers of iterations during
the learning phase.

NEURAL NETWORK iterations	2000	2500	8000
Learning dataset n=10000	σ_{Re} 17.9N/mm^2	σ_{El} 3.05%	σ_R 12.6N/mm^2
1st valid. dataset n= 697	σ_{Re} 20.3 /mm^2	σ_{El} 3.05%	σ_R 14.5N/mm^2
2nd valid. dataset n= 1500	σ_{Re} 22.4 /mm^2	σ_{El} 3.38%	σ_R 15.4N/mm^2
MULTIPLE LINEAR REGRESSION			
Learning dataset n=10000	σ_{Re} 22.5N/mm^2	σ_{El} 3.52%	σ_R 16.9N/mm^2
1st valid. dataset n= 697	σ_{Re} 23.2 /mm^2	σ_{El} 3.35%	σ_R 17.1N/mm^2
2nd valid. dataset n= 1500	σ_{Re} 26.0 /mm^2	σ_{El} 3.50%	σ_R 18. N/mm^2

Table 3. Comparison of results obtained by neural networks
and multiple linear regression.

In fact, the σ values are largely influenced by the values of a
limited number of cases wrongly estimated by the NN approach and
this situation is rather independent of the choosen structure.

Figure 5 gives the PCA F1-F2 representation of the 10000 cases of
the training database and the variables whose symbol are given at
table 1.
In the factorial space, the position of two variables V_1 and V_2 is
such that the correlation coefficient r between them is measured
by the cosine of the angle OV_1,OV_2 (O is origin). As a result, two
close variables (or two opposite variables) have an important and
positive (negative) value for r. It is the case for R_e and R which
are strongly correlated (r=0.97) as expected due to their
proximity, and both very close to Nb, all three variables
characterising the first axis.
Factor plot F1-F2 reveals also that elongation has an important
and negative communalty on the first axis, which is due to the
deleterious effect on ductility of an increase on tensile
strengthening by precipitation.
As far as cases representation is concerned, figure 5 reveals that
the training database is not homogeneous since large clusters can
be detected in the multidimensional space.
Moreover, on the same figure the 22 coils characterized by the
largest ultimate strength [abs(σ_R) > 40 N/mm^2] after application
of NN modelling have also been visualized with a different symbol.
The examination of the different factorial plots F1-F2, F1-F3 ...
suggests that the cases characterized by excessive residuals are
situated in p-spaces zones non explored with the training
database used. This situation is illustrated in figure 5 for a set
of four cases having high positive coordinates on axis F1.
Another way to characterize the results is to calculate for each
case of the validation database the number of nearest neighbours

belonging to the learning database and situated in a
hyperspherical region of fixed radius.

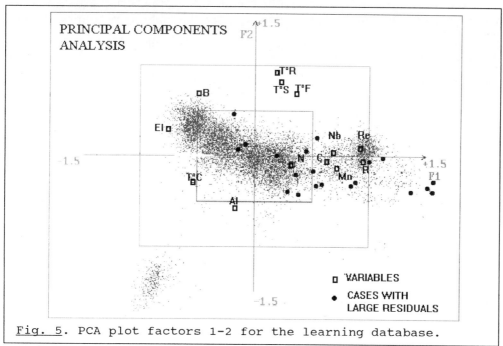

Fig. 5. PCA plot factors 1-2 for the learning database.

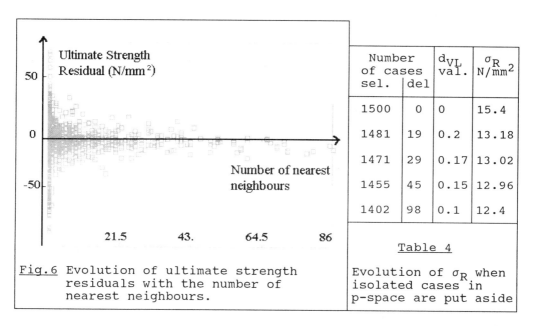

Number of cases		d_{VL}	σ_R
sel.	del	val.	N/mm^2
1500	0	0	15.4
1481	19	0.2	13.18
1471	29	0.17	13.02
1455	45	0.15	12.96
1402	98	0.1	12.4

Table 4

Fig.6 Evolution of ultimate strength
      residuals with the number of
      nearest neighbours.

Evolution of σ_R when
isolated cases in
p-space are put aside

By selecting appropriated radius values for this vicinity, it is
possible to obtain the graph represented in figure 6 which shows
that the mostpart of wrongly modelling cases have no or only very
few neighbours in p-space.
It is also possible to calculate the distance of each case of the

Validation database to the nearest neighbour in the Learning database d_{VL}). Table 4 shows that the standard of estimate of ultimate strength (σ_R) calculated for the validation dataset decreases if the cases corresponding to d_{VL} greater than a fixed value are deleted.

It is obvious that the probability to a case characterized by a large d_{VL} value is higher in p-space zones of low density, so the different concepts previously considered are not independent. In any cases, table 4 shows that it is possible to improve the results by putting aside validation data cases with a high probability of wrong estimation.

For the two validation datasets (2197 patterns), if 68 cases corresponding to $d_{VL} > 0.15$ are deleted, the standards of estimate for R, Re and El calculated on the 2129 remaining cases are respectively 12.5 N/mm^2, 20.2 N/mm^2 and 3.2%, so the quality results is very similar on validation and learning databases.

CONCLUSIONS

Neural networks constitute a powerful tool to deal with large databases and to perform statistical modelling of one or several characteristics.

In this paper, neural networks have been used to predict the mechanical properties of carbon and microalloyed steels produced at Cockerill-Sambre (Carlam) Steel Plant.

The quality of the estimation, which must absolutely be determined on independent datasets, is better than those obtained by linear multiple regression.

The quality of results during the generalization step can be explained by the size of the training dataset and by the use of statistical filtering realized before the analysis.

Multivariate data analysis is also useful to justify and improve the results obtained by neural networks.

REFERENCES

1. C. DUMORTIER, P. LEHERT, P. DONFUT, J.M. DULLIER, J. MAROT, La Revue de Métallurgie - CIT, Novembre 1992, pp 1015-1023.
2. P. LEHERT, C. DUMORTIER, P. DONFUT, Ironmaking and Steelmaking 1996, Vol. 23, N°1, pp 25-30.
3. D. J. NAYLOR, Ironmaking Steelmaking, 1989, 16, pp 246-252.
4. Stephen P. CURRAM and John MINGERS, J. Opl. Res. Soc. Vol 45, N°4, pp 440-450
5. S.T. WELSTEAD, Neural network and fuzzy logic applications in C/C++. John Wiley & Sons, Inc.
6. D.E. RUMELHART, J.E. Mc. CLELLAND, Parallel Distributed Proces.,Vol 1, UHI,MIT Press,Cambridge, 1996.
7. Yhu-Jen Hwu, Yeong-Tsuen Pan and John G. Lenard, Steel research 67 (1996), n°2 pp 59-66
8. P.F. VELLEMAN and C. HOAGLIN, Applications basics, and computing of exploratory data analysis,1981, Boston MA, Duxbury Press
9. N.R. DRAPER and SMITH, Applied regression analysis, 1969 New York, Willey.
10. H. HARTMAN, Modern factor analysis, 2nd Edition 1967, Chicago University Press.
11. J. LEFEBVRE, Introduction aux analyses statistiques multidimensionnelles. Editions Masson 1983.

MEDIUM/HIGH CARBON STEELS AND FORGING

Materials Science Forum Vols. 284-286 (1998) pp. 403-410

Microstructural Evolution during Induction Hardening Heat Treatment of a Vanadium Microalloyed Steel

A.L. Rivas[1,2] and G.M. Michal[1]

[1] Dept. of Materials Science and Engineering, Case Western Reserve University
10900 Euclid Ave, Cleeveland, OH44106 7204

[2] Now at Universidad Simón Bolivar, Venezuela

Keywords: Microalloyed Steel, Induction Hardening, Residual Stresses, Carbide Particles, Austenite Grain Size

Abstract

The microstructural evolution during induction heat treatment applied to a forged 0.1wt. % V microalloyed steel has been investigated. The study examined both as-forged and forged plus normalized starting microstructures. The core of the as-forged steel was comprised of ferrite and pearlite. The prior austenite grain size was about 150 µm and only a faint indication of banding was observed. After normalizing the ferrite/pearlite microstructure was refined. The prior austenite grain size was reduced to 10 µm and banding became more evident. During induction heating, the banded appearance persisted in all but the extreme outer portion of the surface layer that underwent austenitization. Banding was more apparent in the as-forged steel after induction hardening. The austenite average grain sizes developed in both the as-forged and forged and normalized steels after induction hardening were less than 20 µm. High levels of compressive residual stresses were measured in the hardened outer surface layer using an x-ray diffraction technique. Residual stress relaxation due to the application of cycling loading will be discussed

1.   Introduction

Microalloyed steels in the as-forged condition are commonly used in the automotive industry. For heavy-duty applications such as that for diesel engine crankshafts, a surface induction hardening heat treatment is performed in critical regions of the components to enhance their service performance through adding strength and fatigue resistance. The goal of the induction hardening heat treatment is to form a fully martensitic structure in the outer surface of the components to locally increase the hardness and tensile strength. The fatigue strength is raised primarily by the increase in tensile strength. In addition, the fatigue strength is enhanced by the generation of high levels of compressive residual stresses at the outer surface due to the increment in volume that accompanies the austenite to martensite transformation.

Tempering follows the induction hardening heat treatment to increase toughness in the hardened layer. During tempering the induction hardened layer typically contracts and a reduction in the level of compressive residual stresses is observed [1]. Residual stress relaxation can also take place via the cyclic loading that comprises fatigue. Rearrangement and annihilation of dislocation can occur during fatigue leading to a recovery process at room temperature. The recovery mechanism can be retarded by the presence of fine particles, which act as a barrier to dislocation motion.

A recent study evaluated the fatigue resistance of 0.1V-microalloyed steel [2]. This steel exhibited significantly higher fatigue strength at similar hardness levels than a conventionally normalized carbon baseline steel. The increase in fatigue strength was correlated with the higher levels of compressive stresses found in the microalloyed steel. An interesting aspect to determine is the extend of relaxation of compressive residual stresses under cyclic loading conditions for the induction hardening-microalloyed steel. The microalloyed steel contains vanadium carbide particle that may retard the residual stress relaxation during fatigue. The effectiveness of the vanadium carbide particles as barrier to the dislocation motion depends on their size and distribution. It has been observed that shearable particles can be cut by dislocations during fatigue and reverted into solution because of their thermodynamic instability [3]. Vanadium carbide particles are typically strong enough that they are not sheared by dislocation in the ferrite matrix.

The vanadium carbide particle size and distribution can also affect the grain structure characteristics developed during the application of an induction hardening heat treatment. It is well known that the effectiveness of carbide particles as grain growth inhibitors is subjected to the stability of the carbides at high temperatures and to their coarsening characteristics.

The present study was carried out to evaluate the microstructural evolution of a induction hardening vanadium microalloyed steel initially in both the as-forged and forged and normalized starting microstructures as well as the relaxation of residual stresses behavior of those heat treated steels under cyclic loading conditions. Even though microalloyed steels are mainly used in the as-forged condition, eventually they may need a normalizing heat treatment prior induction hardening to restore minimum mechanical property requirements.

2.   Experimental Procedures
2.1   Material and microstructural characterization

The chemical composition of the steel used in this study is given in table 1. The composition corresponds to a medium carbon steel with a vanadium addition of 0.1%. The microalloyed steel is designated as TMS-80, which is a trademark of The Timken Company. After rolling, crankshafts were forged from this material at a temperature of 1315 °C and air-cooled to room temperature. A crankshaft initially in the as-forged condition was normalized at a temperature of 880 to 900 °C for a period of 4 hours. Subsequently, an induction hardening heat treatment was performed on both the as-forged and forged plus normalized crankshafts. During the induction hardening cycle, the surface region of the steel was heated to a maximum temperature of 1040 °C in approximately 22 seconds before being quenched in a polymer solution to room temperature. After quenching the crankshafts were tempered at 246 °C for 3 hours. Portions of the journal regions of the crankshafts were sectioned and metalographically analyzed. Both, as-forged and forged plus normalized general microstructures were characterized as well as the austenite grain size evolution during the induction hardening heat treatment. The general microstructural features were examined using optical microscopy following mechanically preparation with standard metallographic techniques and etching with 2% nital. The prior austenite grain size in the induction hardened region was revealed by etching the samples for about 8 hours in a solution of 100 ml. water, 25 gr. NaOH and 2 gr. picric acid at about 110 °C. A second type of etching was also used. It consisted of 100 gr. of water, 9 gr. of picric acid and 1% of a wetting agent. Volume size distributions were determined from the two dimensional grain size measurements. The method used was that of the Saltykov [4]. The austenite grain size was divided into 15 classes. The measured grain sizes were analyzed with the aid of the computer program "Pro-Plus".

Vanadium carbide particle sizes were measured in the as- forged as well as forged plus normalized starting microstructures. The vanadium carbide particle sizes and distributions were also evaluated along the induction hardened layer of both as-forged and forged plus normalized samples. These measurements were done using carbon extraction replicas. A more detail description of this work can be found elsewhere [5].

Table 1: Chemical Composition of the TMS-80 Steel (Weight Percent)

C	Mn	S	Si	Cr	Ni	Mo	Cu	Al	V	0(ppm)	N(ppm)
0.30	1.38	0.01	0.26	0.11	0.12	0.03	0.16	0.121	0.11	26	88

2.2 Mechanical tests and residual stress measurements

The steel used in this part of the study has a similar composition that that specified in table 1 except that the carbon content was 0.35% and the amount of silicon 0.45%. The steel was forged into round bars that were 76 mm diameter. Change in residual stress relaxation was monitored using c-shaped samples with a T-shaped cross section machined from the forged-round bars. Details of the c-shaped sample geometry and performance can be found elsewhere [6]. The c-shape specimens had their outer rib induction heated, quenched and tempered at 315 °C for 30 minutes. The microstructure at the outer surface was then tempered martensite while that of the inner surface was pearlite ferrite or martensite-ferrite. The induction heat treatment was applied to the microalloyed steel in both its as-forged and forged plus normalized conditions. The specimens were then subjected to compression tests using a 20 kip MTS close-loop servohydraulic machine with a MTS 442 controller. The application of a compressive load put the outer surface in tension while the inner surface was in compression. The samples were tested at increasing load levels until plastic deformation of about 2% strain occurred. The purpose of the compressive tests was to create compressive residual stresses at the outer surface. Fatigue tests were performed and the compressive residual stresses in the circumferential direction measured after a determined number of cycles. The evaluation of residual stresses was performed using x-ray diffraction with Cu Kα radiation. The tests were run at a counting time of 25 second and step angle of 0.03 degree. The samples had their outer surfaces electropolished. The x-ray diffractometer used was a Scintag Model X-1.

3. Results
3.1 Microstructural evolution during induction hardening heat treatment
3.1.1 General Microstructure Characteristics

Figure 1 shows the microstructure of the steels in both the forged and forged plus normalized conditions. The prior austenite grain size in the as-forged steel was estimated to be 150 μm. The ferrite content was measured to be 20 %. The normalized steel exhibited a finer structure than that of the as-forged with a banded ferrite-pearlite distribution. The prior austenite grain size in the as-forged plus normalized steel was measured to be 10 μm and the ferrite content was measured to be 50 %. Figure 2 contains micrographs taken from the transition region near the core of the forged plus normalized sample. The transition zone reflects the starting phase transformation process taking place during induction heating. The phase transformation as expected, starts with the decomposition of pearlite colonies into austenite. The finer initial pearlite ferrite microstructure of the normalized steel leads to a finer austenite grain after induction heat treatment. In the normalized steel the austenite grain size was smaller than 3 μm while in the as-forged steel it was estimated to be about 4.5 μm.

As the temperature increases, a complete transformation of the initial microstructure takes place. The temperature at which complete transformation occurred was estimated as 840 °C (1554 °F) [7]. The main processes occurring above that temperature were the homogenization of the carbon in the austenite and growth of the austenite grains and VC particles. A banded contrast developed within the induction hardened layer after etching, as shown in figure 3. Only the extreme outer portion of the layer that underwent austenitization not show this banding contrast. These etching characteristics reflect Mn segregation.

To follow the austenite grain size evolution during the induction hardening heat treatments, the variations of grain size from the near core region to the outer surface region were evaluated. The grain sizes slightly growth up to a distance of about 1mm from the outer surface in both as-forged and forged and normalized steels. Figure 4 shows the grain structures observed in the outer surface of the as-forged and forged plus normalized steels. The grain size is larger in the as-forged steel

The austenite grain size distribution at the outer surface in both the as-forged and forged-normalized steels was estimated. The results are shown in figure 5. The average grain size was determined as 13.9 ± 7.6 μm for the as-forged and 9.8 ± 5.6 μm for the forged and normalized material.

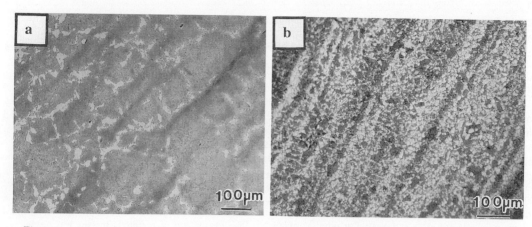

Figure1. a) Microstructure of the as-forged steel. b) Microstructure of the forged plus normalized steel.

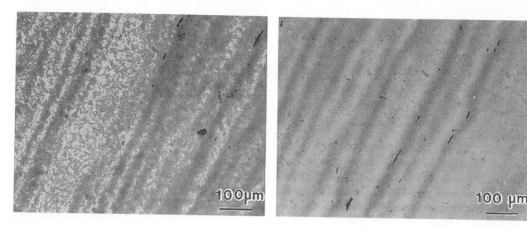

Figure 2. Transition zone near the core of the forged plus normalized steel

Figure 3. Banded type of contrast developed within the induction hardened layer after etching

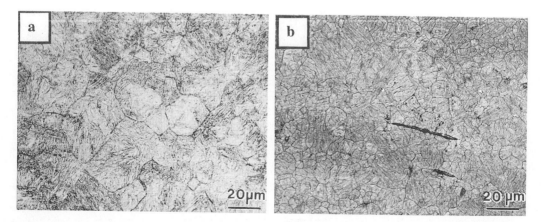

Figure 4. Grain size structure at the outer surface after induction hardening heating.
a) As-forged steel. b) Forged plus normalized steel.

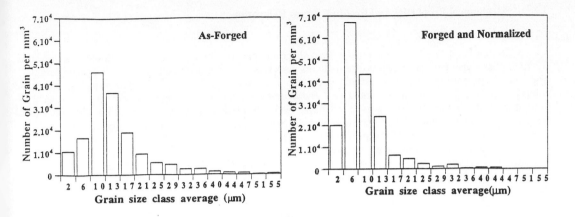

Figure 5. Grain size distribution histogram at the outer hardened surface.

3.1.2. Vanadium carbide particle size

The average vanadium carbide particle size measured in the steel in the as-forged condition was 3.5 nm. The application of the normalizing heat treatment caused an increase in the average particle size of the as-forged steel. The average particle radius estimated in the forged plus normalized specimen was 7 nm. Figure 6 present the variation of VC particle size with distance from the hardened surface for both the as-forged and forged plus normalized starting microstructures. It is observed that the application of an induction hardening heat treatment to the steels in the as-forged and forged plus normalized conditions caused an increase in the average size of vanadium carbide particles throughout the outer layer of the steel that underwent austenitization.

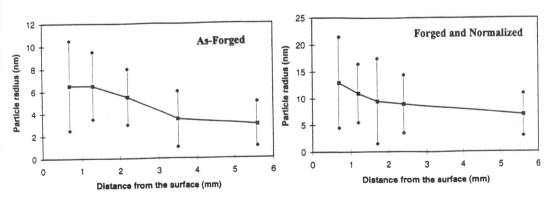

Figure 6. Variation of vanadium carbide particle size with distance from the hardened surface

3.2   Mechanical Properties and Residual Stress Relaxation

Strain-Load curves obtained using the c-shape samples for the two material conditions are shown in figure 7. The upper part of the stress-strain curves corresponds to the outer surface of the c-shaped samples. The lower part corresponds to the stress at the inner surface region of the c-shaped samples, which is comprised by ferrite-pearlite in the forged steel and martensite-ferrite for the forged and normalized steel. The ratio of the outer to inner surface strain in the elastic region of these specimens was about 1.6. This value also represents the outer to inner surface stress ratio since it was measured in the elastic region of the materials. The strain at the outer surface of the steel initially in the forged plus normalized condition was slightly lower than that of the steel initially in the as- forged condition. The yield stress was estimated as 162 ksi for the outer surface of the forged steel and 173 ksi for the outer surface of the forged plus normalized material. At the inner surface, the estimated yield stress of the as-forged steel was 96 ksi and 114.6 ksi for the forged and normalized steel

The variations in residual stresses were monitored on the c-shaped samples subjected to a cyclic load of +/-400 lbs. The strain experienced for the materials at that load in the outer surface was 0.38% and in the inner surface 0.23%. The stresses associated with those strain levels were 112 ksi and 68 ksi, respectively. After loading by 10 cycles the residual stress values in both steels were reduced in about 10 ksi. After 10 cycles of loading, the remaining residual stress levels were -53ksi for the as-forged steel and -55ksi for the forged plus normalized steel. These levels of stresses remained unchangeable up to 10000 applied cycles. At 82000 cycles, the as-forged steel had experienced a lost of residual stress to the level of -45 ksi. The sample fractured at about 93000 cycles. A similar trend was observed in the forged plus normalized steel. The forged plus normalized sample fractured at 90000 cycles.

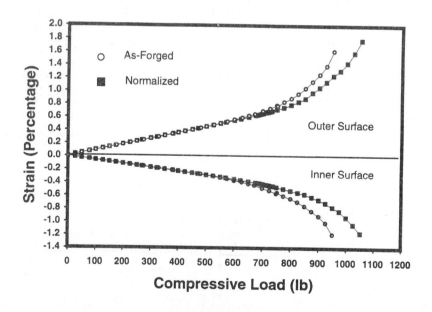

Figure 7. Strain-load curved for both the outer and inner surfaces of the C-shaped samples of both as-forged and forged plus normalized steels.

4. Discussion

The banded ferrite-pearlite microstructures observed in the steels are very common. They are the result of the segregation of Mn in the as-cast microstructure [8]. During solidification, the Mn segregates at the interdendritic space. During rolling the segregated Mn regions and the depleted Mn region are enlarged forming bands with large and small amounts of Mn. Because of the low diffusion coefficient of Mn in austenite, during the normalizing heat treatment the Mn is not homogenized. During cooling, the low Mn areas preferentially nucleate the proeutectoid ferrite and the excess of carbon left by the ferrite formation diffuses toward those areas rich in Mn in where the formation of pearlite occurs. Therefore, in the final microstructure, the low Mn areas are associated with higher levels of proeutectoid ferrite and the high Mn areas with higher levels of pearlite, giving the characteristic banded appearance.

In addition to Mn banding, the possibility of carbon segregation in the steels also exists. An initial evaluation of the rate of diffusion of carbon in austenite indicated that both the as-forged and forged plus normalized steels experienced sufficient combination of time and temperature to homogenize their carbon distributions. However, such an analysis does not take into account the presence of Mn segregation. Because of Mn segregation, there is a change in the driving force for homogenization of the carbon. The activity of carbon in austenite as a function of the Mn content and temperature is given by equation 1, where T is in K and compositions are in weight % [9].

$$log(a_c) = log(\%C) + \frac{2300}{T} - 2.25 + \frac{181}{T}(\%C) - \frac{21.8}{T}(\%Mn) \qquad \text{Eq.1}$$

Equation 1 indicates that regions lower in Mn content will have lower carbon contents when the activity of carbon is equal in the low and high Mn content regions in the austenite. This establishes a paraequilibrium situation. The carbon will remain segregated until homogenization of the Mn occurs.

When realistic numbers for the extent of Mn segregation are inserted into equation 1, the corresponding differences in the carbon contents at paraequilibrium are small. At the temperatures of normalizing the extend of carbon banding amounts to only about 0.01 weight percent. Thus the Mn banding is not able to prevent the carbon from becoming almost completely homogeneous during heat treatment.

The initial small grain size observed in both the as-forged and forged plus normalized steels is caused by the rapid induction heating. In this type of heating process high temperatures are reached very fast and the austenite forms in fractions of second [10]. The initial finer austenite grain size observed in the normalized steel is the result of its finer starting microstructure [11]. It is known that the diffusion of carbon controls the speed of austenite formation. The diffusion distance for carbon in the normalized steel is considerable smaller than that in the as-forged sample. Thus, the austenite formation is accelerated in the forged plus normalized starting microstructure.

Even though the peak temperature for the induction hardening cycle was as high as 1040 °C, a fine austenite grain size was achieved. In addition to the short duration of the heat treatment cycle, the high number density of vanadium carbide particles allowed the fine austenite grain size to be maintained. The normalizing heat treatment did coarsen the vanadium carbide particles relative to those in the as-forged condition. However, the vanadium carbide particle distributions in both the as-forged and forged plus normalized conditions were sufficient to suppress excessive austenite grain growth.

The finer scale of the initial microstructure of the forged plus normalized steel allowed for a faster homogenization of the carbon distribution in the austenite during induction hardening. This led to a more uniform martensite microstructure in the hardened outer layer. This effect was manifested in the flow stress of the tempered outer surface of the forged plus normalized steel being slightly greater than that of the as-forged steel.

The residual stress relaxation behavior of the as-forged and forged plus normalized steels was similar. The increase in fatigue resistance of the steels was substantially increased by the compressive residual stress induced in the steels' surfaces. As a consequence the two steel conditions exhibited similar fatigue behavior.

5.   Conclusions

1. Normalizing yielded a finer austenite grain size both before and after induction hardening.
2. Both the normalizing and induction hardening heat treatment were ineffective in removing the Mn banding.
3. Vanadium carbide particles were larger in the normalized steel compared to the as-forged steel
4. Induction hardening increased the size of vanadium carbide particles
5. The residual stress relaxation behavior in the as-forged and forged plus normalized steels was similar.

Acknowledgement

The authors thank the Cummins Engine Company and the Timken Company for providing the experimental steels and for the financial support to this project. Special thanks to Consejo National de Investigaciones Cientificas y Technologicas, CONICIT, and Universidad Simon Bolivar, Venezuela, for their financial support.

References

[1] A. L. Rivas, G. M. Michal, M. E. Burnett and C. F. Musolff:, 38TH MWSP Conference, ISS, XXXIV, (1997), p.235.
[2] M. E. Burnett, C. F. Musolff, G. M. Michal and A. L. Rivas, Fundamental and Applications of Micoalloying Forging Steels, TMS , (1996), p. 455.
[3] T. Broom, T. A. Mazza and V. N. Whittaker, J. Inst. Met. 86 (1977), p.17
[4] E.E Underwood: In Quantitative Microscopy, R.T. DeHoff and F. N. Rhines, eds., McGraw-Hill, Inc., New York, NY, 1968, p.149.
[5] A. L. Rivas, G. M. Michal, M. E. Burnett and C. F. Musolff. Fundamental and Applications of Micoalloying Forging Steels, TMS , (1996), p. 159.
[6] A. L. Rivas and G. M. Michal, 39TH MWSP Conference, ISS, XXXIV, (1998) to be published.
[7] V. M. Zalkin, Met. Sci. Heat Treat. 1-4, (1986), p. 96
[8] J. W. Martin Stability of Microstructures in Metallic systems, Cambridge University Press, (1980), p. 34
[9] T. Wada, H. Wada, J. F. Elliot and J. Chipman, Metallurgical Transactions, 3, (1972), p. 1657
[10] ASM Metals Handbook, Heat treating, volume 4, (1991), p. 184
[11] J. Wyszkowski, Iron and steel, (1970), p. 77

Materials Science Forum Vols. 284-286 (1998) pp. 411-418
© *1998 Trans Tech Publications, Switzerland*

Microalloying and a New Post Forging Treatment of Medium Carbon Steels

I. González-Baquet[1,2], R. Kaspar[1], J. Richter[3], G. Nußbaum[3] and A. Köthe[3]

[1] Max-Planck-Institut für Eisenforschung GmbH,
Max-Planck-Strasse 1, D-40237 Düsseldorf, Germany
Fax: +49-211-6792333

[2] Now with ACERALIA, Aptdo. 90, E-33400 Avilés, Spain
Fax: +34-85126089

[3] Institut für Festkörper- und Werkstofforschung,
Helmholtzstr. 20, D-01069 Dresden, Germany
Fax: +49-351-4659320

Keywords: Forging, Medium Carbon Microalloyed Steel, Alloyed Steel, Continuous Cooling, Two-Step Cooling, Tensile Properties, Toughness, Environment

Abstract

In the last years, the automotive industry has been demanding a continuous reduction of weight in order to compensate for the introduction of new constructive components according to higher requirements on passenger safety and driving comfort, as well as to accomplish with new strict environmental regulations. The adoption of new materials with improved mechanical properties and/or lower density, together with modern cost effective processing technologies have contributed greatly to reach the above objectives [1]. Particularly microalloyed steels, which may provide a higher strength with small alloying additions and low heat treatment expenses, have evolved into an important category of engineering materials suitable for the manufacturing of hull, frame and motor components.

As an example, the classical quench and tempering (QT) of alloyed steels followed by straightening operations and stress relieving cycles has been progressively substituted by the continuous cooling (CC) of more economical microalloyed steels from the forging temperature. An accurate control of the cooling rate adjusts the properties of the ferritic-pearlitic microstructure obtained. This technology results in lower processing costs and eliminates product rejectances due to quench-cracking. However there are limitations in the mechanical properties achievable by this processing route if compared with the QT-operation [2], and hardly any cooling rate may be encountered which would lead to a balance of strength and ductility comparable to that after QT.

In this context a new post forging treatment (two-step cooling, TSC) has been presented elsewhere [3] as a promising alternative to the CC of medium carbon microalloyed steels. This paper is focused on the particular contributions of different microalloying additions with optimized deformation schedules to improve the mechanical properties achievable through this new technology.

Materials and experimental procedure

Two commercial medium carbon steel grades (VN and TV) were selected as a basis for this study, **table 1**. Both grades belong to the group of steels designed for continuous cooling (CC). They are microalloyed with vanadium which is commonly used for precipitation hardening of ferritie. Steel TV is additionally microalloyed with titanium making use of TiN as an effective austenite grain growth inhibitor at high reheating temperatures. An enhanced content of sulfur provides better machinability which is an important property in view of the considerable machining

Steel	%C	%Si	%Mn	%Cr	%S	%Al	%Ti	%Nb	%V	ppmN
VN	0,24	0,53	1,28	0,15	0,04	0,03	--	--	0,08	152
TV	0,31	0,71	1,54	0,24	0,03	0,03	0,015	--	0,10	98
T2V	0,29	0,60	1,51	0,27	0,03	0,03	0,022	--	0,17	80
T2VN	0,27	0,60	1,51	0,27	0,04	0,03	0,024	--	0,18	154
TVNb	0,31	0,61	1,50	0,27	0,03	0,03	0,020	0,049	0,09	75
41CrS4	0,43	0,25	0,75	1,14	0,03	0,03	--	--	--	--

Table 1. Chemical composition of the steels examined, mass contents in %

commonly required after forging. Moreover, the combined use of TiN and MnS can additionally suppress the extreme coarsening of austenite grains [4].

Three microalloying variants (T2V, T2VN and TVNb) were prepared additionally to the two basic grades. An enhanced content of vanadium in steels T2V and T2VN was designed for the testing of the hardening potential of vanadium through fine-dispersed precipitates formed during the γ-α-transformation. The niobium addition in the TVNb-steel should elevate the recrystallization temperature of austenite and, therefore, offer new possibilities for an optimized thermomechanical treatment. On two steel grades (VN and T2VN) the effect of a higher nitrogen content was tested. The commercial Cr-alloyed steel 41CrS4 was additionally selected, representing the classical quench-tempered forging steels for automotive parts.

The forging process was simulated by carrying out plane strain hot compression tests on the hot deformation simulator (WUMSI) at the Max-Planck-Institut für Eisenforschung at Düsseldorf [5]. An average strain rate of 10 s^{-1} was applied. The specimen temperature was recorded by means of a thermocouple embedded in the specimens during reheating and deformation as well as in the course of cooling. Each WUMSI-specimen provided several secondary samples for mechanical testing by hardness measurements, tensile tests (specimen diameter 5 mm) and Charpy impact tests (3 x 3 mm^2 fracture section) as well as for microstructure investigations.

Two-step-cooling (TSC): concept and control parameters

Figure 1 depicts the principle of the newly developed so called two-step cooling (TSC) strategy after forging as an alternative to the conventional continuous cooling (CC). This new cooling design promotes the ferrite formation during the first slow cooling step with a cooling rate

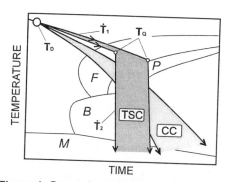

$\dot{T}_1$ (natural cooling on still air). It occurs at rather high temperatures, which implies two consequences: firstly, the ferrite is polygonal (good ductility) with high carbon content (higher strength); secondly, the carbon is able to diffuse from the supersaturated ferrite into the adjacent austenite, the latter being stabilized promoting the formation of bainite or even martensite instead of pearlite. A fast cooling (water quench) in the second step, with a cooling rate $\dot{T}_2$, leads to a bainite/martensite transformation of the remaining austenite, providing a hard second phase. The resulting microstructure with both soft and hard phases is mainly controlled - under the same cooling rates $\dot{T}_1$ and $\dot{T}_2$ - by the forging temperature T_D, which determines the austenitic microstructure,

Figure 1. Processing schedule for the two-step cooling (TSC) with forging temperature T_D and quench temperature T_Q as main control parameters. For comparison: the continuous cooling (CC) with varying cooling rate

and the quench temperature T_Q, which controls the amount of ferrite formed. Lowering the forging temperature T_D leads to a finer or even unrecrystallized austenite structure with enhanced ferrite nucleation, while lowering the quench temperature T_Q (not bellow A_{r1}) prolongs the time available for ferrite formation and so raises its volume fraction.

Effect of microalloying and forging temperature on the austenite microstructure

A fine-grained austenitic microstructure usually results in a fine and homogeneous microstructure after transformation, which contributes greatly to improve the toughness of forged products. The grain growth of austenite during reheating can be effectively controlled by the addition of certain microalloying elements, especially titanium. An earlier work [6] showed that the titanium microalloyed TV-steel exhibits a much finer austenite grain size at high temperatures than the Ti-free VN-steel. Moreover, Ti-microalloyed steels show a slightly finer microstructure even after recrystallization, presumably as a result of a much finer initial grain size. Therefore, the efficient control of the grain coarsening exerted by titanium during reheating has also a positive influence on the recrystallized grain size. According to these results and regarding the fact that forgings usually involve a rather non-uniform distribution of strain, often including some locations without any deformation at all, the presence of titanium is considered to be a recommendable means to obtain a uniform fine grained austenitic microstructure. As reported in [7], no significant grain growth of the recrystallized grains should be expected even during a slow subsequent cooling down to A_{r3}, so that the finer grained austenite in the Ti-microalloyed steels remains preserved until the start of the γ-α-transformation. Besides using Ti-microalloying, lowering the forging temperature T_D also leads to a finer recrystallized austenite structure with an enhanced ferrite nucleation due to a larger specific area of austenitic grain boundaries, as can be seen in **figure 2**.

If deformation occurs bellow the recrystallization temperature, the specific austenite grain boundary area increases even more due to elongation of grains. Moreover, the ferrite nucleation rate grows owing to a raise of the density of defects near the austenite grain boundaries. An additional nucleation can take place on deformation bands within the grains [8]. Nevertheless, this usually requires rather low deformation temperatures, with an increased tool wear and higher forging loads. Microalloying elements (especially niobium) retard the recrystallization both in solid solution [9] and, especially, in form of carbonitrides after deformation-induced precipitation in austenite [10,11]. For this reason, the forging temperatures of Nb-bearing steels need not to be so low in order to

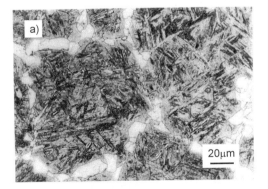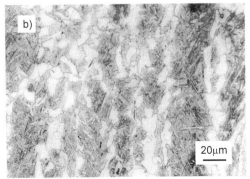

Figure 2. Influence of the forging temperature T_D on the microstructure after TSC (T_Q = 660 °C) of the steel VN: **a)** T_D = 1200 °C, coarse recrystallized austenite (8% ferrite); **b)** T_D = 850 °C, fine recrystallized austenite (33% ferrite)

produce an unrecrystallized austenite [12]. This is demonstrated by a significantly higher recrystallization-stop temperature of TVNb-steel, **figure 3**, in comparison with the Nb-free steel grades. On the other hand, a smaller initial austenite grain size enhances the recrystallization process. This may be a reason for the lower recrystallization-stop temperatures of the fine grained TV and T2V-steels in comparison to that of VN-steel, especially for small strains.

Annealing after two-step cooling

The microstructure of the steels investigated showed rather large ferrite fractions of up to 40% after TSC. Nevertheless, the stress-strain curves monitored during tensile testing did not show any discontinuous yielding, exhibiting rather low values of 0.2%-proof stress [3]. Assuming the existing theories for dual-phase steels, which show a similar behavior [13], the rather low $R_{p0.2}$ values have been attributed to the production of mobile dislocations in ferrite, close to the ferrite-martensite interphases, due to the volume expansion taking place during the austenite to martensite transformation. Such free dislocations are not locked through atmospheres of interstitial solutes, making the onset of plastic deformation easier and avoiding discontinuous yielding. Additionally to this, the steels showed poor values of reduction of area at fracture (Z) after TSC when compared with QT, which can obviously be attributed to the presence of untempered martensite [3]. In order to eliminate both shortcomings (low values of $R_{p0.2}$ and Z) an additional annealing was employed.

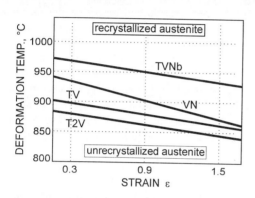

Figure 3. *Temperature ranges for a complete recrystallization of austenite (above the lines) as a function of strain (reheating 1230 °C, 2 min, strain rate 10 s⁻¹)*

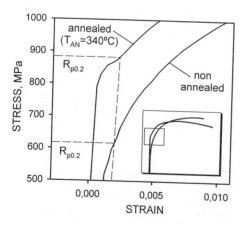

Figure 4. *Influence of an additional annealing treatment (AN) at T_{AN} = 340 °C (1 hour) on the stress-strain curve of steel VN after TSC*

The annealing treatment (AN) brings about significant changes in the initial part of the stress-strain curve towards discontinuous yielding, **figure 4**, with a marked improvement of $R_{p0.2}$. This feature can be related to the pinning effect of the interstitial atoms and, presumably, to the precipitation of vanadium carbonitrides from the supersaturated ferrite nucleated preferentially in the locations with a higher dislocation density [14,15]. An additional contribution to the increase in strength can be associated with the decomposition of retained austenite observed within the temperature range of 300-500 °C in steels with a similar composition (even though with higher Cr content) [16]. Simultaneously, annealing causes an important improvement of Z as a result of the annealing process operating in the hard bainitic-martensitic phases. All in all, it has been ascertained that the annealing raises both strength and ductility of the steels studied, being a very effective tool to adjust the mechanical properties of TSC-treated steels to the desired level. Detailed studies about the influence of the annealing temperature on the evolution of the mechanical properties of the investigated steels [3,17] showed that the best combination of strength and ductility can be attained by annealing at about T_{AN} = 420 °C.

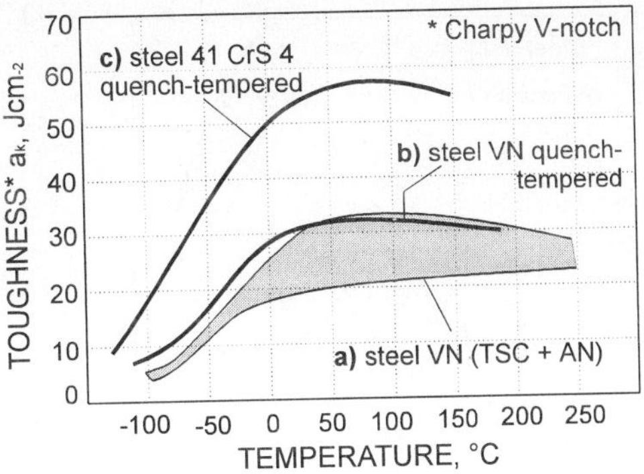

Figure 5. Comparison of Charpy transition curves of the steel VN after a) TSC + AN (T_D = 1200 °C, T_Q = 660 °C, T_{AN} = 420 °C), b) QT (T_D = 1200 °C, T_{AN} = 550-650 °C) and c) of the alloyed steel 41CrS4 after QT (T_D = 1200 °C, T_{AN} = 550 °C)

Homogeneity of properties and toughness behavior

Regarding the industrial use of the TSC + AN technology, it has been shown that even remarkable differences in the ferrite fraction, due to local deviations of T_D and/or T_Q, do not alter the attained strength properties substantially [3,17]. Nearly constant properties in spite of large variations of the processing conditions guarantee homogeneous properties of the product over a wide range of cross-sections and forging parameters.

The toughness properties of the TSC-treated steels have been studied by means of the Charpy test. They have been compared with the properties of the same steels after QT and those of a conventional Cr-alloyed steel (41CrS4) commonly used for automotive applications, **figure 5**. In general, medium carbon microalloyed steels – after both TSC (a) and QT (b) treatments – show lower toughness properties than the QT alloyed steel 41CrS4 (c). Alloying elements like Cr, Mo, and W result in a fine dispersion of stable carbides during annealing of martensite at the temperature range 500-600 °C. The appearance of such particles also influences the precipitation kinetics of iron carbides (ε-carbide and cementite), reducing the embrittlement induced by their preferential formation on former austenite grain boundaries and martensite needles as a continuous net and thus improving toughness [18]. The comparison of TSC + AN with QT of the same microalloyed medium carbon steels tested shows slightly lower upper shelf energies and higher transition temperatures than after QT. Nevertheless, by optimizing TSC-processing conditions comparable toughness values can be achieved, which is of remarkable metallurgical interest, regarding the fact that a ferrite-containing microstructure after TSC not only reaches a similar strength but also comparable toughness properties as a tempered medium-carbon martensite.

Summarized effect of TSC + AN and microalloying on tensile properties

The modified treatment (TSC + AN) after forging as a substitute for the conventional (CC) includes the adaptation of forging temperatures as well as the modification of the subsequent cooling. The arrows in **figure 6a** represent the contribution of annealing to the improvement of the mechanical properties of the VN-steel after TSC, plotted as $R_{p0.2}$ vs Z. In comparison to this, mechanical properties of the same steel after QT are represented in the hatched area. The additional effect of microalloying is given in **figure 6b**. In the case of common high forging temperatures (T_D = 1200 °C) the addition of Ti (finer austenite structure) and of a larger amount of V (precipitation strengthening) leads to an additional increase in $R_{p0.2}$ (small vertical arrow in figure 6b). If lower forging temperatures can be applied (T_D = 920 °C), Nb-microalloying enables the austenite

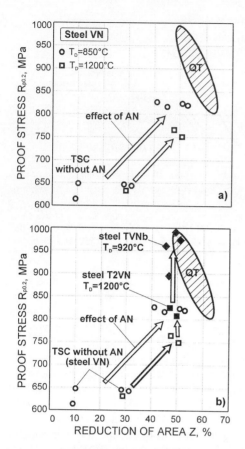

Figure 6. *Improving mechanical properties after TSC **a)** by the subsequent annealing (AN) (T_{AN} = 420 °C) on steel VN and **b)** through the additional microalloying adjusted to the hot deformation temperature T_D*

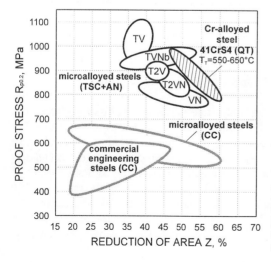

deformation without recrystallization, which leads to a finer and more homogeneous final microstructure and hence to higher $R_{p0.2}$ (the improvement along the longer vertical arrow) combined with a good ductility.

The summarizing view in **figure 7** shows considerably better mechanical properties achieved by using TSC + AN with the various microalloyed variants, compared with the conventional CC of the same microalloyed steels as well as with some commercial engineering steels [19]. Even the properties of the more expensive quenched and tempered Cr-alloyed steel can be obtained by applying this new processing on microalloyed steels deliberately designed. This illustrates the large potential of this modified treatment based on the improvement of ferrite-containing multi-phase microstructures as a substitute for a pure tempered martensite (QT) or a simple ferrite-pearlite (CC).

From the point of view of industrial applicability the sequence of air cooling and water quench is environmentally reasonable and even more simple than controlled cooling by forced air used with CC-treatment. A rather wide technological window for the acceptable quench temperatures T_Q brings about similar or even less demands on process control compared to CC. Some increase in manufacturing costs due to additional annealing (at rather low temperatures) can be justified by a significant improvement of mechanical properties. The design of microalloying can be well adjusted to the forging loads achievable on specific production lines.

Figure 7. *Mechanical properties of the microalloyed steels investigated after TSC + AN compared to those after CC. As a reference, mechanical properties of several commercial engineering steels after CC [19] and of the alloyed steel 41CrS4 after traditional QT are shown*

Conclusions

It has been shown that through TSC + AN noticeably improved mechanical properties, if compared to those of CC-treated steels, are achievable, depending on the microalloying design. A higher vanadium and nitrogen content is favourable even for elevated forging temperatures T_D. If lower T_D can be used, an additional microalloying with niobium facilitates forging bellow recrystallization temperature, leading to a more homogeneous distribution of ferrite and excellent properties. Titanium might be a recommendable means to obtain a uniform fine grained microstructure, even in forgings with a rather heterogeneous distribution of strain.

Acknowledgments

The work has been financially supported by the Bundesministerium für Wirtschaft of the FRG, through the Arbeitsgemeinschaft industrieller Forschungsvereinigungen (AiF) and by the Cominsión Interministerial de Ciencia y Tecnología (CICYT) of Spain which are gratefully acknowledged.

References

[1]   *Adlof, W.W.:* Bruchgetrennte Pleuelstangen aus Stahl, Schmiede-Journal (1996), p. 15-18

[2]   *Kaspar, R.; González-Baquet, I.; Schreiber, N.; Richter, J.; Nußbaum, G.; Köthe, A.:* Application of thermomechanical treatment on medium carbon microalloyed steels continuously cooled from forging temperature, Steel Res. 68 (1997) No. 1, p. 27-31

[3]   *Kaspar, R.; González-Baquet, I.; Richter, J.; Nußbaum, G.; Köthe, A.:* New post forging treatment of medium carbon microalloyed steels, Steel Res. 68 (1997) No. 6, p. 266-271

[4]   *Koyasu, Y.; Takahashi, T.; Ishii, N.; Takada, H.; Takeda, H.:* Nippon Steel Tech. Rep., Overseas 47 (1990), p. 37-45

[5]   *Pawelski, O.; Kaspar, R.:* Materialprüfung 30 (1988), p. 357-60; and Materialprüfung 31 (1989), p. 14-16

[6]   *González-Baquet, I.; Kaspar, R.; Richter, J.:* Conditioning of austenite by hot working of microalloyed forging steels, Steel Res. 68 (1997) No. 2, p. 61-66

[7]   *Kaspar, R.; Mahmoud, N.:* Austenite grain growth during hot forging of medium carbon engineering steels with or without V-Ti microalloying, Mater. Sci. and Tech. 7 (1991), p. 249-54

[8]   *Kaspar, R.; Lotter, U.; Biegus, C.:* The influence of thermomechanical treatment on the transformation behaviour of steels, Steel Res. 65 (1994) No. 6, p. 242-247

[9]   *Zajac, S.; Lagneborg, R.; Siwecki, T.:* The role of nitrogen in microalloyed steels, Microalloying '95 (Pittsburgh, 1995), p. 321-338

[10] *Yamamoto, S.; Ouchi, C.; Osuka, T.:* Proc. of Int. Conf. on Thermomechanical Processing of Microalloyed Austenite, (Pittsburgh, 1981), p. 613

[11] *Weiss, I.; Jonas, J.J.:* Metall. Trans. 10A (1979), p. 831-40

[12] *Hansen, S. S.; Vander Sande, J.B.; Cohen, M.:* Metall. Trans. 11A (1980), p. 387-401

[13] *Öström, P.:* Metall. Trans. 12A (1982), p. 355-57

[14] *Rashid, M. S.; Rao, B.V.N.:* Metall. Trans. 13A (1982), p. 1679-86

[15] *Bhadesia, H. K. D. H.; Waugh, A.R.:* Acta Metall. 30 (1982), p. 775-84

[16] *Takada, H.; Koyasu, Y.:* Fundamentals and Applications of Microalloying Forging Steels, (Warrendale, 1996), p. 143-57

[17] *González-Baquet, I.:* Thermomechanische Behandlung mit neuartigem Abkühlkonzept zur Verbesserung der mechanischen Eigenschaften von mikrolegierten Schmiedestählen, Aachen, 1997 (Dr.-Ing. thesis)

[18] *Honeycombe, R.W.K.; Badeshia, H.K.D.H.:* Steels, microstructure and properties, Chap. 9 (1995)

[19] *Cristinacce, M.; Reynolds, P.E.:* The current status of the development and use of air cooled steels for the automotive industry, Fundamentals and Applications of Microalloyed Forging Steels (Warrendale, 1996), p. 29-43

Materials Science Forum Vols. 284-286 (1998) pp. 419-426

Acicular Ferrite Microstructures and Mechanical Properties in a Medium Carbon Forging Steel

I. Madariaga and I. Gutiérrez

Centro de Estudios e Investigaciones Técnicas de Guipúzcoa (CEIT),
Escuela Superior de Ingenieros Industriales de San Sebastián (ESII), Universidad de Navarra,
P°Manuel Lardizábal 15, E-20009 San Sebastián (País Vasco), Spain

Keywords: Acicular, Transformations, Inclusions, Intragranular

Abstract

A set of different acicular ferrite type microstructures have been developed, with both isothermal treatments between 500°C and 400°C and continuous cooling after austenization at 1250°C, in a medium carbon forging steel. The inclusions responsible for the intragranular nucleation of the ferrite have been studied by means of transmission electron microscopy. The results obtained enabled the identification of a MnS core covered by hexagonal CuS, which seems to make nucleation at inclusions energetically more favourable than nucleation at grain boundaries.

A good combination of strength and toughness has been obtained associated with a refined acicular ferrite microstructure, generated by continuous cooling, and with a mixture of acicular ferrite and pearlite, obtained at 500°C by isothermal treatment.

Introduction

Acicular ferrite as bainite is characterised by a shear diffusional mechanism of growth: the ferrite grows initially without diffusion, supersaturated in carbon, which is rapidly rejected into the non transformed austenite [1-3]. The main difference between both microstructures arises in the site where the nucleation of each phase takes place. While in the bainite, nucleation occurs at austenite grain boundaries, acicular ferrite nucleates intragranularly on inclusions. The difference in nucleation site determines the morphological differences existent between acicular ferrite and bainite. In the case of acicular ferrite, the ferrite units grow individually with a high misorientation between them, and in the case of bainite, sheaves of parallel ferrite plates form emanating from austenite grain boundaries.

The good mechanical properties obtained with acicular ferrite, by virtue of its fine grained microstructure, makes this a desirable phase in both low carbon welding steels [4-6] and medium carbon forging steels [7, 8]. On the other hand, bainite microstructures frequently exhibit poor toughness properties. Taking into account that both acicular ferrite and bainite form in the same temperature range in a competitive way, it is necessary to determine the factors that enhance acicular ferrite formation to the detriment of bainite.

Bainite formation is inhibited by an increase of the austenite grain size [9], the formation of a thin layer of allotriomorphic ferrite [10-12], and the segregation of alloying elements such as B, P,

S [13, 14]. However, the presence of suitable inclusions can improve the global energetic balance of the austenite - ferrite transformation and enhance intragranular nucleation of acicular ferrite [14-16]. In low carbon weldable steels, particle inoculation has been used to promote acicular ferrite formation in the base material [17] .In the present work, it has been demonstrated that acicular ferrite microstructures can be produced in a conventional steel forging leading to an enhancement of the mechanical properties.

Experimental procedure

Table I presents the chemical composition of the steel used in the present work. Cubic samples of 10 mm side were austenitized at 1250°C in a Lindberg furnace using an inert argon atmosphere. After the austenitization, different isothermal treatments and continuous cooling, producing acicular ferrite, were performed. The isothermal treatments consisted of quenching the samples directly into a salt bath at 500, 450 or 400°C. The treatments were interrupted at different times in order to examine the evolution of the microstructure. Continuous cooling was made with aid of jets of air to obtain different rates from 3 °C/s to 6 °C/s, in the range 800°C to 500°C.

Table I. Chemical composition of the microalloyed steel used (% weight)

C	Mn	Si	P	S	Cr	Mo	V	Cu	Al	Ti	N
0.37	1.45	0.56	0.01	0.043	0.04	0.025	0.11	0.14	0.024	0.015	0.0162

The samples were sectioned, so as to analyse the transformation in the bulk, avoiding any surface effects. Mechanical polishing and etching in 2 % nital were used for optical metallography. Thin foil and carbon extraction replicas were prepared for TEM observations following techniques previously reported [15].

Four different microstructures, corresponding to 20 minutes at 500°C, 450°C, 400°C and continuous cooling at 6°C/s, were obtained in square bar ($12x12x60$ mm^3), from which Charpy (ASTM E23 standard [18]) and tensile tests and hardness measurements could be performed.

Results and Discussion

The microstructure obtained after 20 seconds, 30 seconds and 20 minutes of treatment at 500°C, and 10 seconds, 30 seconds and 20 minutes at 450°C and 400°C can be seen in Fig. 1. At 500°C, the micrograph corresponding to 20 seconds of treatment has been presented because at 10 seconds the transformation had not yet begun. As can be seen, the beginning of the transformation is similar at all the temperatures. Isolated ferrite plates nucleate intragranularly at inclusions and no grain boundary nucleation is observed. For example, in the micrograph corresponding to 30 seconds at 500°C, there is a grain boundary that can only be detected by the different main orientations of the ferrite plates in each austenite grain [19]. The transformation at 450°C is faster than at 400°C and 500°C, which indicates that the nose of the region of the TTT curve corresponding to acicular ferrite formation must lie around 450°C.

The subsequent development of the microstructure is different for each temperature. The ferrite plates initially nucleated at inclusions act as nucleation sites from which new ferrite units grow sympathetically. The first microstructural difference arises in the way this sympathetic nucleation proceeds. Whilst in the case of the treatments at 500°C and 450°C it is face to edge sympathetic nucleation (the new ferrite plates grow with a high disorientation angle with respect to the pre-existing one), at 400°C face to face sympathetic nucleation is observed (the new ferrite plates and the first nucleated plates are approximately parallel) [20]. The differences between these two kinds of sympathetic nucleation can be better seen in the TEM micrographs of Fig. 2a and 2b, which correspond to treatments of 10 seconds at 450°C and 20 seconds at 400°C respectively. In Fig. 2a, a new ferrite plate is growing from a pre-existing one, subtending an angle of about 90°; whilst in the case of Fig. 2b a sheaf of parallel ferrite plates, similar to bainite sheaves is observed.

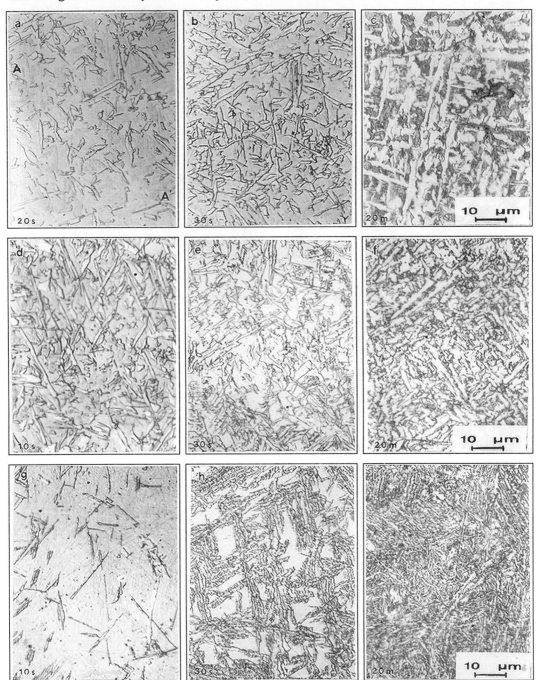

Fig. 1.- Evolution of the microstructure with isothermal treatments. *a.* 20 s at 500°C. *b.* 30 s at 500°C. *c.* 20 m at 500°C. *d.* 10 s at 450°C. *e.* 30 s at 450°C. *f.* 20 m at 450°C. *g.* 10 s at 400°C. *h.* 30 s at 400°C. *i.* 20 m at 400°C

However, the differences with conventional bainite are significant. First of all, the transformation does not initiate from grain boundaries, as above mentioned. Additionally, as this structure develops in a grain previously divided by the initial acicular ferrite plates, the growth of the sheaves is limited and thus a fine microstructure ensues. Another microstructural difference, observed when changing the treatment temperature, has to be mentioned. At 500°C, a mixture of acicular ferrite and pearlite is obtained, see Fig. 1c. The SEM micrograph in Fig. 3 clearly shows the lamellar aspect of the pearlite intragranularly nucleated between ferrite plates. This can be explained as follows, in agreement with the upper part of the bainite/acicular ferrite temperature range of the TTT diagram. Once the initial austenite grains are divided by ferrite plates into small carbon enriched austenite zones, over relatively long times, these zones transform to pearlite, resulting in the final microstructure presented in Fig 1c and Fig. 3.

Fig. 2.- TEM micrographs showing the beginning of the transformation. *a.* 10 seconds at 450°C. *b* 20 seconds at 400°C

At 450°C, the microstructure is fully acicular. Some retained austenite can be found for intermediate treatment times, due to the relatively high Si content. Afterwards, this austenite transforms to ferrite plus carbides [16].

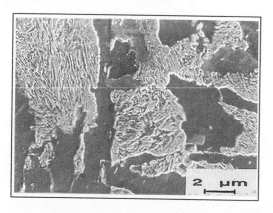

Fig. 3.- SEM micrograph showing pearlite lamellae between plates of ferrite in the microstructure obtained at 500°C.

The microstructure obtained with continuous cooling (6°C/s between 800°C and 500°C) after austenization at 1250°C is shown in Fig. 4. From all the tested cooling rates, 6°C/s was the one giving the highest volume fraction of acicular ferrite, because it is rapid enough to avoid the formation of large volume fractions of pearlite and/or allotriomorphic ferrite, but slow enough to prevent the formation of parallel ferrite plates, as happened with the isothermal treatment at 400°C, or even martensite. In spite of some small equiaxed ferrite grains, the microstructure exhibits the highly misorientated interlocking ferrite plates, characteristic of acicular ferrite, as can be clearly observed in the TEM micrograph of Fig. 4b. When this microstructure is compared with those produced during isothermal treatments at 500 and 450°C it can be seen that it is finer. The long plates, produced by fast growth of the first acicular ferrite units being formed during isothermal treatments, are not present in the samples continuously cooled.

A detailed analysis of the inclusions responsible for the ferrite nucleation was made using carbon extraction replicas as has been reported elsewhere [16]. The analysis, using energy dispersive spectrometry (EDS), of such inclusions revealed the presence of a MnS core covered, or partially covered, by a shell of the CuS hexagonal phase called covellite with lattice parameters a=0.379 nm and c= 1.633 nm [21]. A TEM micrograph of one of these inclusions, together with the EDS analyses of the zones indicated in the micrograph, are presented in Fig 5.

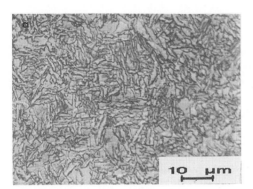 

Fig. 4.- Microstructure obtained with continuous cooling at 6°C/s after austenization at 1250°C. *a.* Optical micrograph. *b.* TEM image

This phase probably forms high energy interfaces with austenite due to the high misfit existing between CuS and austenite, independent of the plane of epitaxy and the chosen orientations within this plane. However, as discussed elsewhere [15, 16], under certain orientations between CuS and the nucleating ferrite, low misfit between both phases is possible. For example, the misfit calculated according to the Bramfitt [22] equation when $\{0001\}_{CuS}\|\{111\}_{\alpha}$ and $<2\bar{1}\bar{1}0>_{CuS}\|<1\bar{1}0>_{\alpha}$ is only 6%.

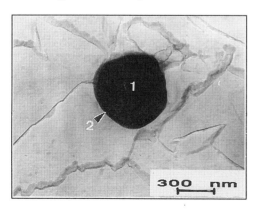 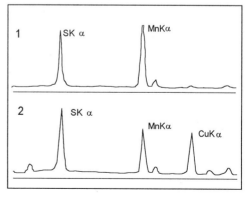

Fig. 5.- TEM micrograph of one of these inclusions together with the EDS analyses of the zones indicated in the micrograph.

The consideration of low misfit, between ferrite and second phase particle lattices, being a factor promoting acicular ferrite formation encounters some difficulties, when the need of ferrite to follow an orientation relationship with parent austenite is taken into account, as discussed by other authors [3]. The presence of the CuS shell changes the situation. The core of the nucleating particle is relevant for its size, which is an important factor for heterogeneous nucleation. However, the CuS shell determines the interactions across the interface with austenite and ferrite. The second phase particles can no longer be "seen" externally as a monocrystal. Accordingly, it is feasible for the ferrite nucleus to form at a site on the CuS shell where it is possible, at the same time, to satisfy one variant of the orientation relationship with austenite and have a low energy interface with the substrate [15]. The following equation (Eq. 1) determines the free energy change in the case of nucleation at inclusions:

$$\Delta G_{het} = -\Delta G_v \, V_\alpha + \Delta G_S \, V_\alpha + \sigma_{\alpha/\gamma} \, S_{\alpha/\gamma} + \sigma_{\alpha/i} \, S_i - \sigma_{\gamma/i} \, S_i \qquad (1)$$

where ΔG_v represents the driving force per unit volume of the $\gamma \rightarrow \alpha$ transformation; ΔG_S is the strain energy per unit volume of ferrite; V_α is the volume of austenite that transforms to ferrite; $S_{\alpha/\gamma}$, S_i, represent the area of the interfaces; and $\sigma_{\gamma/i}$, $\sigma_{\alpha/i}$, and $\sigma_{\alpha/\gamma}$ are the interfacial energies per unit area of interface.

When the difference between the interfacial energies $\sigma_{\gamma/i}$ and $\sigma_{\alpha/i}$ is neglected, the free energy change ΔG_{het} is independent of the nature of the nucleating inclusion and is only affected by the geometrical shape of the inclusion. In this case, nucleation at inclusions is always less favourable than grain boundary nucleation, as calculated by Ricks et al. [23]. However, when this difference is considered, if the free energy change, when nucleation takes place heterogeneously at inclusions (ΔG_{het}), is plotted against the difference in interfacial energies the graph of Fig. 6 is obtained. In the same graph, the free energy change associated with grain boundary nucleation is also presented for comparison. As can be seen, when the quotient $(\sigma_{\gamma/i} - \sigma_{\alpha/i})/\sigma_{\alpha/\gamma}$ is higher than approximately 0.3, nucleation at inclusions is energetically more favourable than grain boundary nucleation. This can explain why, in the present case, no bainite has formed.

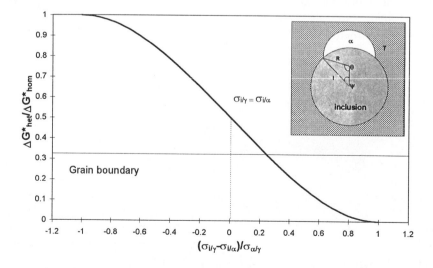

Fig. 6. - Energy barrier for nucleation at inclusions as a function of the difference in interfacial energies.

The mechanical properties obtained with the produced microstructures are presented in Table 2 for the following treatments: isothermal treatment at 500°C, 450°C and 400°C and continuos cooling at 6°C/s following austenization at 1200°C. The surface of the Charpy test fractures has been analysed and the fractions of ductile and brittle fractures have been measured with the results shown in the same table.

Table II. Mechanical properties of the different microstructures with acicular ferrite

Treatment	YS (MPa)	UTS (MPa)	Hardness (HV)	CVN (J)	% Ductile
Aus 1200 20' 500	623.4	910	296	51	56
Aus 1200 20' 450	655.1	912	283	23	15
Aus 1200 20'400	551.7	910.9	321	29	22
Aus 1200 6°C/s	628.5	951.6	290	50	71

An excellent combination of mechanical properties has been obtained with the microstructures developed in the present work. It can be seen that it is possible to achieve good toughness without the deterioration of the strength of the steel. From all the analyzed variants, the best results have been obtained in the case of the sample treated at 500°C and for the continuously cooled material.

In the first case, a combination of acicular ferrite and small pearlite colonies is responsible for the improvement in mechanical properties. As pointed out earlier, acicular ferrite units form intragranularly in the early stages, followed by a pearlitic transformation. The role of the acicular ferrite is to produce a refinement of the size of pearlite colonies in the final microstructure. This refinement can be the result of both, the creation of intragranular nucleation sites for pearlite and the limitations on growth due to the presence of acicular ferrite inside the prior austenite grains, before the onset of pearlitic transformation.

It has been observed that continuous cooling produces a refinement of the acicular ferrite microstructure. The long acicular ferrite plates present in the final microstructure, under isothermal conditions, see Fig. 1, are not observed after continuous cooling, as can be seen in Fig. 4.a. These relatively large size plates can impair the toughness. This can explain, at least in part, the much lower impact energies obtained after the 450°C and 400°C isothermal treatments, as compared with the sample continuously cooled. Another factor in this deterioration could be the possible presence of carbides formed after long isothermal treatment times.

Conclusions

Acicular ferrite microstructures have been obtained with isothermal treatments at 450°C, 400°C and continuous cooling at 6°C/s. Isothermal treatment at 500°C results in a mixture of acicular ferrite plus pearlite.

The onset of the transformations takes place intragranularly at inclusions. The first ferrite plates formed act as nucleation sites for the sympathetic development of new ferrite plates. This sympathetic nucleation can be of two kinds depending on the treatment temperature: face to edge or face to face sympathetic nucleation.

The observed nucleating particles are MnS covered by a CuS shell. The hexagonal structure of this shell promotes the preferential intragranular nucleation of acicular ferrite to the detriment of bainite nucleation.

A good combination of strength and toughness can be obtained with the acicular ferrite microstructures developed by both isothermal and continuous cooling treatments, in conventional medium carbon steels.

Acknowledgements

The authors are grateful for the friendship and leadership of Javier Urcola and acknowledge the influence he had on this work before his untimely death.

The CICYT is thanked for finnancial support within a PETRI project in colaboration with GSB (Legazpia). I. Madariaga acknowledges a 'Formación de Investigadores' grant from the Departamento de Educación y Universidades of the Gobierno Vasco.

References

[1] F. J. Barbaro, P. Kraulis, K. E. Easterling: Mat. Sci. Technol., 5 (1989), pp 1057-1068.

[2] S. S. Babu, H. K. D. H. Bhadeshia: Mater. Sci. Engin. A 156 (1992), pp. 1-9.

[3] H. K. D. H. Bhadeshia: Bainite in steels. Transformations, Microstructure and Properties. Cáp X. (1992), The Institute of Materials.

[4] Y. Ito, M. Nakanishi, and Y. Komizo: Met. Cons, 14 (9) (1982), pp. 472-47.

[5] J. G. Garland, P. R. Kirkwood: Met. Constr., 7 (5) (1975), pp. 275-283.

[6] J. G. Garland, P. R. Kirkwood: Met. Constr., 7 (6) (1975), pp. 320-330.

[7] M. A. Linaza, J. L. Romero J. L., Rodriguez-Ibabe J. M. and Urcola J. J.: Scripta Metall. et Mater., 32 (1995), pp. 395-400.

[8] M. A. Linaza, J. L. Romero, J. M. Rodriguez-Ibabe and J. J. Urcola: Scripta Metall. et Mater., 29 (1993), pp. 1217-1222.

[9] R. A. Farrar, Z. Zhang, S. R. Bannister, G. S. Barritte: J. Mater. Sci., 28 (1993), pp. 1385-1390.

[10] R. A. Ricks, P. R. Howell, G. S. Barritte: Jr. Mater. Sci. 17 (1982), pp. 732-740.

[11] S. Ohkita, H. Homma, S. Tsushima, N. Mori: Conf. Proc. 'Joint Australian Welding and Testing Conference', Perth, (1984), Aust. Weld. Inst./Aust. Inst. NDT, pp. 27.1-27.21.

[12] J. R. Yang and H. K. D. H. Bhadeshia: Proc. Int. Conf. Advances in Welding Science and Technology, ASM, Cleveland, OH, (1987), pp. 209-213.

[13] Y. Ito, M. Nakanishi, and Y. Komizo: Met. Cons, 14 (9) (1982), pp. 472-47.

[14] A. O. Kluken, O. Grong and G. Rorvik: Metall. Trans. A, 21A (1990), pp. 2047-2058.

[15] I. Madariaga, I. Gutiérrez: Scripta Materialia, 37 (8) (1997), pp. 1185-1192.

[16] I. Madariaga, J. L. Romero, I. Gutierrez: Metall. Trans., 29A, (1998), pp. 1003-1015.

[17] J. M. Gregg and H. K. D. H. Bhadeshia: Acta Mater., 45 (2) (1997), pp. 739-748.

[18] Standard Test Methods for Notched Bar Impact Testing of Metallic Materials, E23-92, Annual Book of ASTM Standards, vol. 03.01, ASTM, Philadelphia (1992), pp. 205-224.

[19] S. S. Babu, H. K. D. H. Bhadeshia: Mater. Sci. Engin. A 156 (1992), pp. 1-9.

[20] H. I. Aaronson, C. Wells: Trans AIME, (1956), 1216-1223.

[21] B. Pearson: 'Handbook of lattice spacings and structures of metals and alloys', 2 (1967), Pergamon Press, Oxford, p. 230.

[22] B. L. Bramfitt: Met. Trans, 1 (1970), pp. 1987-1995.

[23] R. A. Ricks, G. S. Barritte and P. R. Howell: Proceedings of an International Conference on Solid - Solid Phase Transformations, Pittsburgh, Pennsylvania, (1981), pp. 463-468.

e-mail: imadariaga@ceit.es

Materials Science Forum Vols. 284-286 (1998) pp. 427-434
© *1998 Trans Tech Publications, Switzerland*

Effect of Vanadium Alloying on the Precipitation Reactions during Pearlite Formation in Medium and High Carbon Steels

G. Fourlaris

University of Leeds, Department of Materials, Leeds LS2 9JT, UK

Keywords: Pearlite Reaction, Vanadium Steel, Precipitation of Vanadium Carbide, Automotive Steels, Interphase Precipitation

Abstract : Vanadium microalloyed medium carbon steels are currently used in the automotive industry due to the improved tensile and impact properties that are achieved with moderate additions of vanadium. So far, from results published in the literature, the improvement in mechanical properties is attributed to the precipitation of VC within the proeutectoid ferrite that constitutes the major microstructural phase of these steels, while the pearlitic component has received relatively less attention. The purpose of the present study is to elaborate on the precipitation of vanadium carbide occurring during the isothermal pearlitic reaction in vanadium alloyed medium and high carbon steels. Since during this isothermal transformation the formation of proeutectoid ferrite precedes that of pearlite, the precipitation of VC within the proeutectoid ferrite was also studied, in order to verify existing literature results and to establish a basis of comparison with the VC precipitation occurring within the pearlitic ferrite during the subsequent pearlite formation. Three vanadium alloyed steels were used during the present research. These included two experimentally produced alloys, and a commercially available steel. The present study confirmed that, vanadium carbide interphase precipitates always exist within the pearlitic ferrite component of the lamellar and within non-lamellar pearlite. In the case of the lamellar pearlite, aligned arrays of interphase VC precipitates are present within the pearlitic ferrite, whereas in the non-lamellar pearlitic ferrite random distributions are most frequently observed. No partitioning of vanadium was observed into the adjacent pearlitic cementite regions even in cases of prolonged ageing at the high temperatures of the isothermal transformation. Morever, the experimental results show clearly that interphase VC precipitation exists not only within pearlitic ferrite but also within the proeutectoid ferrite,that forms in low carbon steels, in a form of well defined linear or curved arrays of regular shaped and uniformly sized particles.

Introduction: It is well known that vanadium has limited solubility in austenite and vanadium carbide (VC) limited solubility in ferrite [1]. Consequently, during the isothermal pearlitic transformation of vanadium alloyed steels, there is a strong tendency for vanadium to partition away from the growing ferrite and to form vanadium carbide. Vanadium microalloyed medium carbon steels are currently used in the automotive industry due to the improved tensile and impact properties that are achieved with moderate additions of vanadium [2,3]. So far, from the research results presented, the improvement in mechanical properties is attributed to the precipitation of VC within the proeutectoid ferrite that constitutes the major microstructural phase of these steels [4-7], while the pearlitic component has received relatively less attention [2,3,7,8,9].

Experimental results, have showed the presence of vanadium carbide precipitation in both proeutectoid ferrite [4-7] and pearlitic ferrite [2,8] during continuous cooling transformation of medium carbon vanadium alloyed steels. These results verified earlier observations [4-7]. Moreover similar results have been claimed to occur during isothermal transformation of similar alloy chemistry steels [7-9]. In isothermally transformed specimens the vanadium carbide intersheet spacing was similar in both proeutectoid ferrite and pearlitic ferrite, while in the case of continuous cooled samples the spacing was greater in proeutectoid ferrite [2,3].

The purpose of the present study was to elaborate on the precipitation of vanadium carbide occurring during the isothermal pearlitic reaction in vanadium alloyed medium and high carbon steels. Since during this isothermal transformation the formation of proeutectoid ferrite precedes that of pearlite, the precipitation of VC within the proeutectoid ferrite was also studied, in order to confirm existing literature results and to compare them with the VC precipitation occurring within the pearlitic ferrite during the subsequent pearlite formation.

Experimental Procedure : Three types of vanadium alloyed steels were used during the present research. Their composition is given in Table 1. The steels cover an extensive range of composition both as far as carbon and vanadium contents are concerned. For the production of the two experimental alloys, high purity materials were used and the melting was performed in a vacuum induction furnace .

<u>Table 1: Composition of steels in w. t. %.</u>

	Fe	C	Si	S	P	Mn	Ti	Al	N	V
1*	97.5	0.4								2
2*	98.1	0.7								1
3	96.97	0.42	0.74	0.070	0.016	1.19	0.016	0.005	0.015	0.15

* Nominal Composition.

Heat treatment included an initial austenitisation step at 1250°C for 15 minutes in an argon atmosphere tube furnace. After austenitisation samples were rapidly transferred into a lead or a lead-tin bath where isothermal heat treatments were performed at temperatures maintained to ± 3°C. Further details for specimen preparation are given in another publication [10].

Results : In the Fe-0.4C-0.15V steel the isothermal transformation at 600°C progresses relatively slowly and this temperature was used to follow the evolution of the reaction products. After 10 minutes of isothermal transformation at 600°C the microstructure consists mainly of proeutectoid ferrite, martensite (formed during the ice brine quenching of the untransformed austenite) and a small amount of pearlite. This pearlite has a variety of morphologies. Classical lamellar pearlite can be seen in Figure 1 together with discontinuous non-lamellar pearlite and pearlite with the carbides of spheroidised morphology. These morphologies, as the TEM study that followed confirmed, are genuine and not due to the consequence of the metallographic plane of polish intersecting the various pearlitic colonies at a variety of angles and thus creating apparently dissimilar microstructures.

During the TEM examination of samples of this type of steel the majority of the thin areas were produced in areas corresponding to proeutectoid ferrite since this phase is the major phase existing within the samples. However, in Figure2a a dark field micrograph is presented of vanadium carbide precipation within both proeutectoid and pearlitic ferrite. The sample was taken from steel Fe-0.4C-0.15V, isothermally transformed at 600°C for 1 hour. It can be seen (Figs. 2a and 2.b) that a very fine dispersion of spheroidal, uniformly sized particles aligned in regular spaced arrays exists within the proeutectoid ferrite. By selected area electron diffraction pattern analysis it was confirmed that these fine particles have an fcc structure and a lattice parameter very close to that of VC. The size of the spheroidal shaped VC particles is approximately 10 nm. Vanadium carbide particles precipitate within ferrite having a common orientation relationship that includes the majority of the precipitated particles within proeutectoid ferrite.

Examination of the pearlitic areas of the same sample revealed (Fig.2c) that a fine dispersion of precipitates exists within the pearlitic ferrite. These precipitates have a small size, approximately 5-10 nm, a spheroidal shape and are distributed in the central area of the pearlitic ferrite lamellae. Their density is relatively low and it appears they possess a regular distribution of uniformly spaced arrays of particles. Selected area diffraction pattern analysis (SADP) proved that these precipitates are VC particles. Their dispersion can be found in the majority of the pearlitic ferrite lamellae of the pearlitic colony.

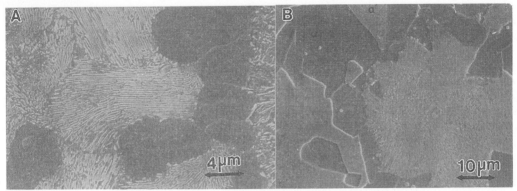

Figure 1: SEM micrographs of pearlite, proeutectoid ferrite and martensite produced in Fe-0.4C-0.15V by isothermal transformation at 600°C for 10 min (A) and 1 hour (B).

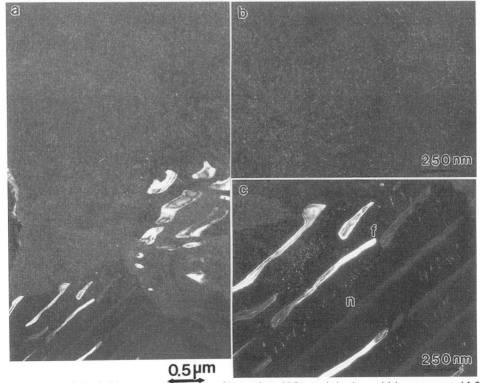

Figure 2 :a) A dark field TEM micrograph of interphaseVC precipitation within proeutectoid ferrite and within pearlitic ferrite, produced in Fe-0.4C-0.15V by isothermal transformation at 600°C for 1h, b) An enlarged micrograph of the proeutectoid ferrite region and c) an enlarged micrograph of the pearlitic ferrite region

In the two experimental steels the vanadium content was significantly higher so that the driving force for precipitation of vanadium carbide was greater. The Fe-0.7C-1V steel has also a higher carbon content. This composition was designed to provide sufficient carbon for the vanadium to form its carbide while the remaining carbon should give the same pearlite volume fraction as in the Fe-0.15V-0.4C steel.

Samples of the Fe-0.4C-2V that were austenitised and isothermally transformed at 600°C for 1h, exhibited primarily a ferritic microstructure, but within certain areas a form of alloyed pearlite was also formed. It appears that cementite has been completely replaced by coarse vanadium carbide particles and this produces the images shown in the SEM micrograph of Figure 3. In Figure 3 the alloyed pearlite exhibits a lamellar appearance with the vanadium carbide particles (their nature verified by EDS-X-Ray microanalysis), forming in some areas lamella of lengths approximately 3 μm. Some spheroidal vanadium carbide particles are also visible, mainly within ferritic areas. These carbides have a size of less than 0.5 μm.

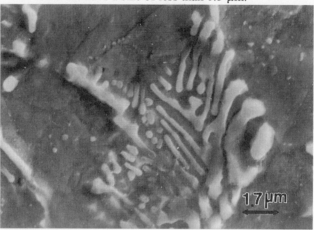

Figure 3: SEM micrograph of the alloyed pearlite formed during in Fe-0.4C-2V by isotherhermal transformation at 600°C for 1h.

Transmission electron microscopical investigation of the same sample shown in Figure 3 revealed the presence of a finer scale precipitation within the ferrite. Using EDS-X-Ray microanalysis and SAD pattern analysis it was confirmed that these precipitates are vanadium carbide precipitates. They appear to be aligned along the major axis of their predominantly lens shaped morphology as can be seen in Figure 4. The lens shaped particles have a size ranging around the typical value of 55 nm x 10 nm. Unfortunately, because the alloyed pearlite areas constituted a very small proportion of the phases present in the examined microstructures and despite extensive specimen preparation efforts, no electron transparent areas could be prepared in regions corresponding to the alloyed pearlite. For the steel Fe-0.4C-2V it was concluded that at this carbon content vanadium absorbs all the available carbon for the creation of vanadium carbide and no carbon is left for the creation of classical pearlitic structures. The only pearlitic structures that were created involve the presence of vanadium carbide that completely replaces cementite.

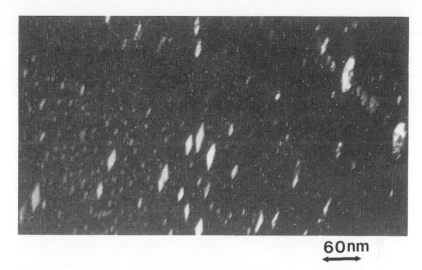

60nm

Figure 4 : A dark field TEM micrograph of interphaseVC precipitation within ferrite produced in Fe-0.4C-2V by isothermal transformation at 600°C for 1h.

The study continued mainly with the examination of the isothermal pearlitic transformation in samples taken from the Fe-0.7C-1V steel. This steel was prepared with this specific composition for two main reasons. Firstly, increasing the carbon content of the steel could reduce the percentage of proeutectoid ferrite at all temperatures of isothermal pearlitic transformation. Secondly, the higher vanadium content guarantees a higher volume fraction of precipitated particles, a factor that would ease the difficulty of detecting vanadium carbide precipitation within pearlitic ferrite.
In Figure 5 an SEM micrograph is shown of specimens of this high carbon steel that was isothermally transformed at 600°C for 40 min followed by ice brine quenching. The pearlite transformation has been completed. It is apparent that the percentage of proeutectoid ferrite is reduced if comparison is made of the structures shown in Figure 5 with the microstructures obtained also at 600°C for the Fe-0.4C-0.15V and shown in Figure 1.

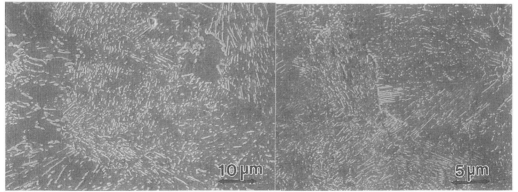

10 µm   5 µm

Figure 5 : SEM micrographs of primarily lamellar pearlite and proeutectoid ferrite produced in Fe-0.7C-1V by isothermal transformation at 600°C for 40 minutes.

TEM examination of thin foils prepared from similarly treated samples revealed the details of the precipitation reaction within proeutectoid ferrite and pearlite. In Figure 6 a dark field TEM micrograph of a sample isothermally transformed at 600°C for 3 hours is provided. It reveals the presence of extensive precipitation within areas of proeutectoid ferrite. These precipitates are vanadium carbide interphase precipitates. The $(200)_{VC}$ reflection was used to create the dark field image that illuminates the majority of the VC particles in the upper proeutectoid ferrite grain. Starting from the right lower part of the dark field micrograph a dense aligned interphase precipitation of VC particles is present. The particles have a uniform size, approximately 2-3 nm and align in parallel arrays that are separated by a spacing of not more than 3-4 nm. Then a discontinuity appears and no VC precipitation occurs within the area this discontinuity spans. At the end of this precipitate free zone, precipitation of VC occurs again but now many elements of uniformity have been lost. The particles are dissimilar in size, their size ranging from 5 nm to 15 nm, the spacing between arrays is not constant and the formation of arrays of particles is not consistent. There are large areas in which VC particles form in a discontinuous manner. Eventually the precipitation terminates at the ferrite/ferrite interface boundary.

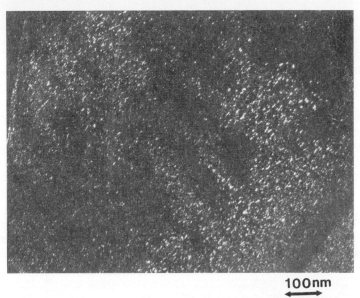

100nm

Figure 6 : Dark field TEM micrograph of VC interphase precipitation within proeutectoid ferrite, produced in Fe-0.7C-1V by isothermal transformation at 600°C for 3 hours. Note the discontinuous nature of the VC precipitates in the ferritic grain.

Vanadium carbide interphase precipitation also occurs within the pearlitic ferrite of the classical lamellar pearlite of a sample isothermally transformed at 600°C for 6 hours.
In Figure 7 it can be seen that a very fine dispersion of interphase VC particles exist within pearlitic ferrite. These spheroidal precipitates have a uniform size, approximately 2-4 nm, and are distributed within the whole area of the pearlitic ferrite lamella of this colony. They form part of concave curved arrays of precipitates that terminate at the ferrite/cementite interface. The $(020)_{VC}$ reflection has been used for the formation of the dark field image. Within the pearlitic ferrite a dislocation network exists that has not interfered with the development of the VC precipitates. EDS-X-Ray microanalysis performed on the cementite lamella failed to show any significant evidence of partitioning of vanadium to the cementite[10].

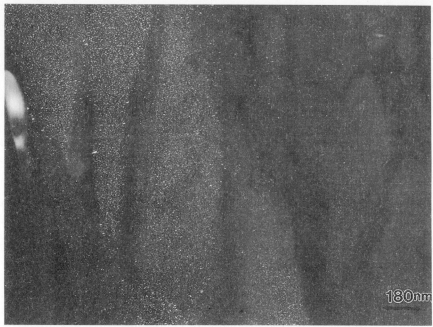

Figure 7 : Dark field TEM micrograph of aligned VC interphase precipitation within pearlitic ferrite, produced in Fe-0.7C-1V by isothermal transformation at 600°C for 6 hours.

Similar precipitation sequences and characteristics were also observed in samples of the Fe-0.7C-1V steel that were isothermally transformed to pearlite at different temperatures.

Discussion: The isothermal pearlite formation in vanadium alloyed steels was preceded by the formation of proeutectoid ferrite and within the proeutectoid ferrite interphase precipitation of vanadium carbide always occurred. The morphology of these precipitates was similar to those found in the studies of low carbon vanadium microalloyed steels, i.e. aligned linear arrays of uniformly sized and shaped vanadium carbide particles. In a few cases, random distributions of vanadium carbide particles within proeutectoid ferrite grains were also observed. The precipitated vanadium particles have a common orientation relationship with ferrite that includes the majority of the precipitated particles within proeutectoid ferrite. However the distribution of vanadium carbide within a proeutectoid ferrite grain is not uniform in the sense that the precipitates exhibit uniformity in some areas of grain but there are other areas within the grain where no vanadium carbide precipitation can be found. In this way a "banded" precipitation of vanadium carbide particles can exist within the proeutectoid ferrite grains where the term "band" denotes a discontinuity or lack of uniformity of precipitation.

When pearlitic transformation occurs two distinctive pearlitic morphologies are created, lamellar pearlite and non-lamellar spheroidal pearlite. Vanadium carbide interphase precipitates always exist within the pearlite ferrite component of the lamellar and non-lamellar pearlite. In the case of the lamellar pearlite aligned arrays of interphase precipitates of vanadium carbide are present. When the amount of vanadium solute is low the precipitation within pearlite ferrite, although it exists, is extremely difficult to detect. When these vanadium carbide interphase precipitates are imaged, they appear to be located in the central area of the pearlitic ferrite lamellae and produce apparent images of precipitate free zones on either side of the pearlitic ferrite/cementite interfaces. However no partitioning of vanadium occurs between the ferrite and the adjacent pearlitic cementite. Moreover, when the concentration of vanadium solute is higher, as in the case of the Fe-1V-0.7C alloy, it is clear that continuous vanadium carbide interphase precipitation exists within pearlite ferrite up to the pearlite ferrite/pearlite cementite interface. In fact the presence of the pearlitic cementite seems to act only as a discontinuity in the development of the vanadium carbide arrays since these arrays appear to continue uninterrupted in the adjacent ferritic lath. Again no significant partitioning of vanadium was observed into the adjacent pearlitic cementite even after prolonged ageing at a high temperature of the isothermal transformation.

The vanadium carbide precipitates within the pearlitic ferrite of the lamellar pearlite are aligned in regular spaced, uniformly sized and shaped concave arrays of precipitating particles. Within each pearlitic colony the vanadium carbide precipitates possess a common orientation relationship that is responsible for the simultaneous dark field illumination of almost every precipitated VC particle within pearlitic ferrite.

Conclusions: The experimental data shows clearly that vanadium carbide exhibits the interphase type precipitation characteristics in the majority of the precipitates formed within the proeutectoid ferrite and pearlitic ferrite of the lamellar pearlite. Both within proeutectoid ferrite and pearlitic ferrite well defined linear or curved arrays of uniformly sized and shaped particles are produced. In the case of the lamellar pearlitic colonies, vanadium carbide precipitation was always observed, as it was mentioned earlier, within pearlitic ferrite. However vanadium carbide does not precipitate within the pearlitic cementite and moreover the pearlitic cementite is not supersaturated in vanadium. During the isothermal pearlitic transformation of vanadium alloyed steels VC interphase precipitation occurs within the proeutectoid ferrite or the pearlitic ferrite of the vanadium steels, within the whole temperature and composition range that pearlite and proeutectoid ferrite form in this system.

The observed precipitation of vanadium carbide within the pearlitic ferrite of the medium to high carbon steels studied in this work is potentially useful in industrial applications where a lack of information regarding the precipitation phenomena and sequences in medium to high carbon vanadium alloyed steels currently exists. Although the Fe-0.7C-1V alloy had a nominal carbon composition of 0.7 wt% it should be regarded as a medium carbon steel such as the commercially available Vanard steel (0.4%C-0.15%V), since the high concentration of vanadium requires a significant proportion of the available carbon for the formation of vanadium carbide. The remaining carbon available for the formation of pearlite in both steels is therefore similar.

References
[1] B. Uhrenius: "Hardenability Concepts with Applications to Steels", ed. D.V.Doane and J.S.Kirkaldy, Warrendale Pennsylvania, ASM, (1976), p. 28.
[2] S.A. Parsons: Ph.D. Thesis, University of Cambridge, UK, 1981.
[3] R.W.K. Honeycombe: "Phase Transformations in Ferrous Alloys", ed. A.R. Marder and J.I. Goldstein, Philadelphia, Pennsylvania, The Metallurgical Society of AIME, (1983),p. 260.
[4] M.Mannerkoski : Acta Polytech.Scand., (1964), ch.26.
[5] A.T.Davenport et al: Met. Sci. Jnl., 2, (1968), p.104.
[6] A.D.Batte and R.W.K.Honeycombe : J.I.S.I., 211, (1973), p.284.
[7] G.L. Dunlop, C.J.Carlsson and G.Frimodig: Metall. Trans., 9A (1978),p. 261.
[8] D.S.Zhou and G.J. Shiflet : Metall. Trans., 22A (1991),p. 1349.
[9] S.A.Hackney and G.J. Shiflet: Acta Met., 35 (1987), 1007.
[10] G.Fourlaris, A.J.Baker, and G.D.Papadimitriou : Acta Met., 43 (1995), p. 3733.

Correspondence: Address all enquiries to Dr G.Fourlaris, Department of Materials, Leeds, LS2 9JT, UK, e-mail: G.Fourlaris@leeds.ac.uk

Materials Science Forum Vols. 284-286 (1998) pp. 435-442
© *1998 Trans Tech Publications, Switzerland*

Warm Forging of a Vanadium Microalloyed Steel

C. García, J.L. Romero and J.M. Rodríguez-Ibabe

C.E.I.T and E.S.I.I of San Sebastian,
Pº Manuel Lardizabal 15, E-20009 San Sebastian, Basque Country, Spain

Keywords: Warm Forging, Vanadium Microalloyed Steel, Precipitation, Ferrite-Pearlite-Bainite Microstructures, Mechanical Properties

ABSTRACT

An experimental study of the warm forging applied to a vanadium microalloyed steel has been performed by plane strain compression tests in the 720-870°C range, followed by air cooling and accelerated cooling (4°C/s). The microstructures obtained are ferrite-pearlite-bainite with very fine ferrite grain sizes. The distribution of vanadium precipitate sizes has been measured and related to the deformation temperature. The microstructures have good strength levels with improved UTS values in the accelerated cooling samples. For deformation temperatures lower that 830°C the material exhibits a complete ductile behaviour at room temperature and at 0°C. On the contrary, samples deformed at 870°C are in the brittle-ductile transition regime. This different behaviour has been related to the microstructural parameters.

INTRODUCTION

Forging grades have been traditionally plain carbon and alloyed steels. Their final properties are achieved after quenching and tempering and their final shape is produced by machining. Distribution of the costs in producing conventional [1] forgings shows that more than 50% of the costs occurs during machining, besides the costs for the base material, the forging process and heat treatment. The last two decades have seen the introduction of microalloyed steels with the aim of reducing associated production costs through the elimination of heat treatments.

Warm forging is the plastic deformation of material at temperatures between $0.5\text{-}0.65T_f$, and in the case of the steels in the range 600-900 °C. It offers some important advantages over traditional forging procedures, with a better dimensional accuracy as well as surface quality in comparison with hot forging and, in the other hand, lower flow stress and therefore lower press loads than in cold forging [2-5].

The objective of this work is to study the application of warm forging to a V microalloyed steel in order to analyse the possibilities of obtaining good strength-toughness combinations through vanadium precipitates.

MATERIAL AND EXPERIMENTAL PROCEDURE

The material analysed is a medium carbon steel microalloyed with vanadium and its chemical composition is listed in Table 1. The steel was supplied in the form of 80 mm diameter round bar obtained by hot rolling of continuous cast 130x130 mm billets.

Warm working simulations were carried out by testing plane strain compression specimens (25x50x10 mm) machined from the bars. After reheating for 10 minutes at test temperature, the specimens were deformed to a strain of 0.3 at constant strain rate of $10s^{-1}$. After the deformation, some specimens were air cooled (~1°C/s) and others accelerated cooled at 4°C/s. Ar_1 and Ar_3 temperature values of the steel are 725 and 810°C respectively (determined by dilatometry tests performed at CENIM), and were taken into account to define the different deformation temperatures: in the austenitic region (870 and 835°C), between Ar_1 and Ar_3 temperatures (800°C) and close to the Ar_1 temperature (at 720°C).

Metallographic samples were cut longitudinally from the plane strain compression specimens and optical and electronic microscopy were carried out. Mechanical properties of the deformed specimens were determined from tensile specimens and V-notch Charpy impact specimens (subsize of 5x10 mm) machined as described in Fig. 1. The absorbed energies were converted to standard specimen dimensions. In order to analyse the size of the precipitates, carbon extraction replicas were prepared from the deformed samples, by conventional methods, and observed in a PHILIPS CM12 scanning transmission electron microscope (STEM) fitted with an EDAX 9100 energy dispersive X-ray system.

Table 1. Chemical composition (wt%).

C	Mn	Si	P	S	Ni	Cr	Mo	V	Cu	Al	Ti	N
0.24	1.56	0.28	0.022	0.003	0.09	0.1	0.04	0.18	0.24	0.019	0.001	0.0099

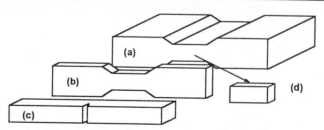

Fig. 1. Scheme of the different geometries machined from the plane strain compression specimens. A: plane strain compression specimen, B: tensile specimen, C: Charpy subsize V-notch specimen and D: metallographic sample.

RESULTS

Average austenite grain size after reheating to 870 °C for 10 min. was 5.0 μm. On the other hand, a small fraction of coarser austenite grains (close to 15 μm) was observed too. Air cooling after forging resulted in the formation of ferrite-pearlite-bainite microstructures. Fig 2a-2f correspond to different examples of the microstructures obtained after deforming at 870, 800 and 720°C and cooling at 1°C/s and 4°C/s respectively. In the range from 870 to 800°C ferrite and bainite volume fraction remain constant and there is a refinement of the ferrite grain size from 4.5 to 1.3 μm. In the case of the 800°C test, there was a 10.6% ferrite volume fraction in the undeformed condition. On the other hand, the application of deformation at 720 °C (near Ar_1 temperature) leads to elongated ferrite grains that coexist with a small fraction of other equiaxed grains (see Fig. 2c and 2f). In this condition the ferrite grain size was determined as the geometric mean value of maximum and minimum dimensions.

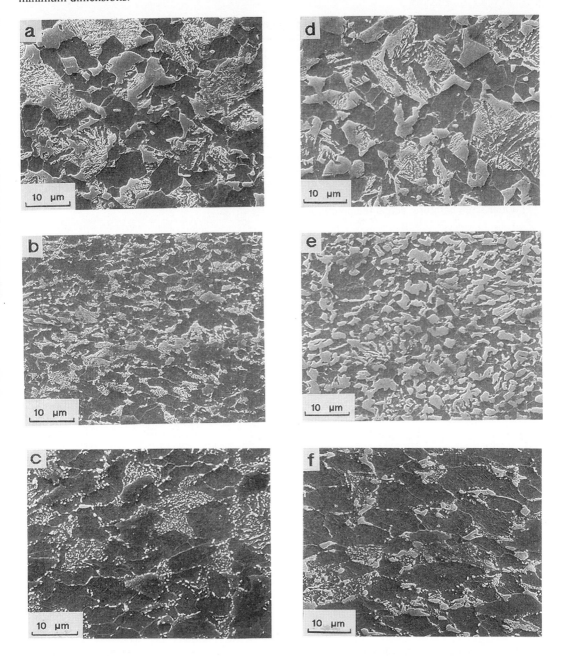

Fig. 2. Microstructures of warm forged simulated samples deformed at a) 870°C, b) 800°C, c) 720°C, d) 870°C + AC, e) 800°C + AC, f) 720°C + AC. (AC: accelerated cooling).

Application of accelerated air cooling after compression modifies the products of the austenite transformation, with an increase in bainite volume fraction (specimen deformed at 800°C) and some refinement in ferrite grain size (specimen deformed at 870°C). In Fig. 2d-f examples of the different microstructures obtained by accelerated cooling are shown. In the case of tests performed at 720°C the pearlite becomes globular, as it can be observed in Fig. 2.c and f.

Table 2. Microstructural parameters and tensile test results.

Temperature (°C)	V_α (%)	D_α (μm)	Bainite (%)	$\sigma_{0.2\%}$ (MPa)	UTS (MPa)	R.A (%)
870	45±3	4.5±0.2	22±3	478.5	804.2	54
835	45±3	3.2±0.2	17±3	529.5	789.0	59
800	41±2	1.3±0.1	17±3	520.5	765.2	67
720	54±2	3.6±0.2	< 7	637.5	765.3	67
870*	39±2	2.7±0.1	21±2	541.0	975.4	42
800*	51±3	2.3±0.1	47±3	457.2	960.9	48
720*(+)	49±3	4.1±0.2	18±3	552.1	983.8	46

* Accelerated cooling
(+) nominal temperature; actual temperature was 741°C

In relation to precipitate sizes, Fig. 3 shows the particle size distribution corresponding to as-received, 870°C deformed and 800°C deformed and air cooled microstructures. In the as-received condition, the average particle size is 8 nm. After the steel has been heated to 800°C and deformed, the mean particle size changes to 13 nm and in the case of 870°C its value is 18 nm. As can be observed, there is a change in particle size distribution, with a displacement of the distribution to coarser particle sizes for higher deformation temperatures.

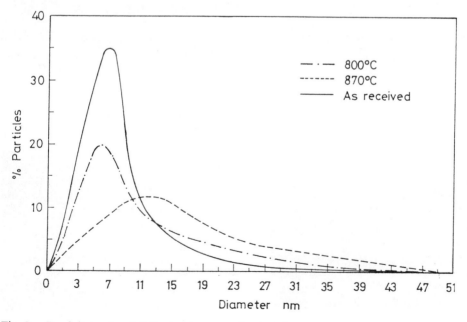

Fig. 3.   Precipitate size distribution in as-received, deformed at 870°C and deformed at 800°C (air cooled) microstructures.

Tensile test results are summarised in Table 2. In the case of air cooled specimens, yield strength increases from 478 to 637 MPa for decreasing deformation temperatures. In relation to the UTS values, the variation is smaller (between 804 and 765 Mpa), the highest value corresponding to the 870°C condition (highest bainite volume fraction). Stress-strain curves corresponding to samples coming from 870, 800 and 720°C deformation conditions and air cooled are given in Fig. 4 and, as can be observed, there are significant differences in the shape of the curves. In the case of the material deformed at 870°C a continuous strain hardening is observed. In contrast, at 800 and 720°C deformation conditions, the curves exhibit well defined yield points.

Accelerated cooled microstructures exhibit, in all the cases, higher UTS values (higher than 960 MPa) than those obtained with air cooled conditions (see Table 2), but with a reduction in ductility (measured as the reduction of area, it is less than 48%). At the three analysed temperatures the curves do not have yield point and, depending on the deformation temperature, the yield strength value (measured as $\sigma_{0.2\%}$) can be higher (870°C) or lower (800 and 720°C) than in the air cooled microstructures.

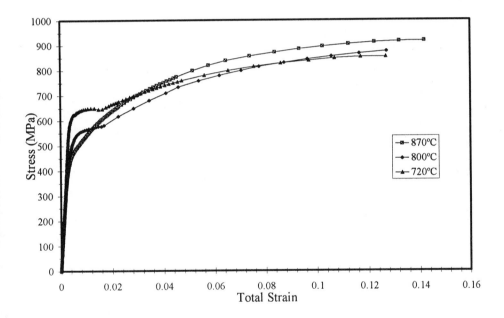

Fig. 4.   Strain-stress curves of tensile test microstructures obtained by deformation at 870, 800 and 720°C and air cooled.

Charpy test results performed at room temperature (20°C) and at 0°C are listed in Table 3. First of all, the results corresponding to the material deformed at 870°C must be considered. The absorbed energy is lower than the data obtained for the rest of the microstructures, and the fractographic analysis shows that the behaviour of the steel corresponds to the lower part of the brittle-ductile transition regime (small ductile propagation before brittle fracture). On the contrary, in the other three conditions, the behaviour of the material is macroscopically ductile with some isolated cleavage islands, completely surrounded by ductile features. Examples of both types of fractures are illustrated in Fig. 5.

Table 3. Charpy test results at 0 and 20°C.

Temperature (°C)	CVN, 20°C (J)	CVN, 0°C (J)
870	46	35.3
835	92.1	95.5
800	114.2	112.2
720	107.3	94.6

DISCUSSION

At 870°C austenitisation temperature the grain size is extremely fine, probably as a consequence of the presence of aluminium nitride and vanadium precipitates. The small value measured is in agreement with the data obtained by Mazzare et al. [6] who observed that, in microalloyed steels, at 900°C aluminium and vanadium retard austenite grain growth.

From the point of view of precipitate size distribution, while in the as-received condition there are 71% of precipitates smaller than 9 nm, after reheating at 800°C, deforming and air cooling, the fraction is ≈40% and in the case of 870°C ≈16%, as can be observed in Fig. 3. The shift of the precipitate size distribution, which is more pronounced as the temperature increases, clearly substantiates that particle coarsening is taking place on reheating. It is known that in the case of vanadium carbides, small size particles can dissolve in relatively short times [7], which explains the disappearance of the lower sized precipitates at 800°C and mainly at 870°C, see Fig. 3. In the studied steel, taking into account that the precipitates are primarily vanadium nitrides and carbonitrides and that they are more stable than carbides [8], it can be assumed that very small amounts of V have been left in solution, and consequently, vanadium will not be available for precipitation during transformation.

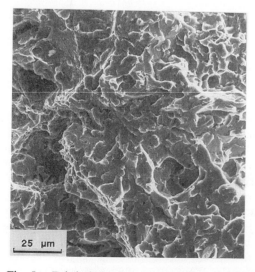 

Fig. 5.　Brittle behaviour of microstructure deformed at 870° C (Charpy test at 20°C) and ductile behaviour of steel tested at 800°C (air cooled samples after deformation).

These vanadium precipitates can have three different effects. As was pointed out before, they contribute to the prevention of austenite grain coarsening during reheating. Secondly, the fine particle volume fraction will maintain the precipitation strengthening effect [9-11]. Finally, as it was proposed by Revidriego et al. [12] with a similar V microalloyed steel, the fine dispersion of undissolved precipitates could retard the austenite recrystallisation.

The microstructures obtained by the different warm forging simulations are in all the cases very fine, as a consequence of the low size of the austenite grains and associated high specific grain boundary area before transformation, in agreement with the results reported by other authors [13]. As shown in Table 2, by this process it has been possible to develop ferrite grains as fine as 2-4 μm in a volume fraction proportion ≥40% in all the cases. The accelerated cooling has promoted some additional refinement in ferrite grain size. This is accompanied in some cases by an increase in the bainite volume fraction.

Finally, the globular pearlite observed in the 720°C samples can be considered to be a consequence of the deformation applied to the pearlite that has enhanced globularisation [14].

The different microstructures developed have a good combination of strength and toughness. The strength values are very similar to the data obtained by Reynolds and Naylor [5] with a V microalloyed steel after soaking at 900°C and finish working at 830°C. Due to the complexity of the microstructures, defined by the presence of different microconstituents (ferrite, pearlite, bainite, particle hardening), it is quite difficult to quantify the contribution of each of them, in comparing with equivalent steels soaked to higher temperatures. Nevertheless, it can be concluded that the precipitation strengthening contribution is probably lower, as a consequence of the small dissolution of V and the observed coarsening of precipitates. On the other hand, the accelerated cooling has contributed to an increase in bainite volume fraction that has conducted to lower $\sigma_{0.2\%}$ values than in the air cooled samples, but with an significant improvement in UTS. Additionally, deforming in the α region, at 720 °C, results in an elongated microstructure with a high density of dislocations, higher in the case of accelerated cooling where it can be considered that the recovery was smaller [13]. In consequence, the microstructure exhibits a deformation strengthening and a precipitation strengthening, assuming that at this temperature the particle distribution is very similar to the as-received material, that is, a very fine particle distribution.

In relation to the toughness results, the material deformed at 870°C has a behaviour corresponding to the lower part of the brittle-ductile transition, both at room temperature and at 0°C. The analysis of the fracture surface of the Charpy specimens shows that there is an significant fraction of cleavage facets greater than 15 μm. This means that unless the mean ferrite grain size is very fine (smaller than 5 μm), there is an important fraction of coarse (> 15 μm) "ferrite units" (microstructural features with the same ferrite crystallographic orientation) that impairs toughness behaviour [15]. Additionally, this characteristic can be exacerbated in the presence of some coarse carbides that promote cleavage initiation. In the other cases, the microstructural refinement achieved has been sufficient to avoid cleavage fracture in the studied temperatures range. This fine microstructure has been able to prevent cleavage microcracks nucleating at some coarser microstructural features (witness the brittle island surrounded by ductile fracture), promoting the macroscopic ductile behaviour of the steel.

CONCLUSIONS

The main conclusions are:

- By warm working simulations in the 720-870°C range it is possible to achieve ferrite-pearlite-bainite microstructures with good strength-toughness combinations.
- These good properties are a consequence of the microstructural refinement obtained by the combination of low working temperatures and vanadium microalloying.
- The size of the vanadium precipitates in the as-received condition is of great importance and they can play different roles depending on processing temperature. The particles can have an influence in austenite recrystallisation/grain growth and in precipitation strengthening.

ACKNOWLEDGEMENTS

The financial support by the Basque Government (UET Programme) is greatly acknowledged. C.Garcia wishes to thank Asociacion de Amigos de la Universidad de Navarra for support through a Research Grant.

REFERENCES

[1]   H.S. Tonshoff, VDI-Z, 124 (1982), p.481.
[2]   R.W. Pollard and L.L. Stainnaker, Tool. Prod., 47 (1981), p. 88.
[3]   T. Suzuki., Int. J. Match. Tools Manuf., 29 (1989), p. 5.
[4]   N.E. Aloi, G. Krauss, D.K. Matlock, C.J. Van Tyne and Y. W Cheng, 36th MWSP Conf. Proc., ISS-AIME, XXXII (1995), p. 201.
[5]   J.H. Reynolds and D.J. Naylor, Mater. Sci. Technol., 4 (1988), p. 586.
[6]   P.T. Mazzare, S.W. Thompson and G. Krauss, Int. Conf. on Processing, Microstructure and Properties of Microalloyed and Other Modern HSLA Steels, A.J. DeArdo ed., ISS (1992), p. 497.
[7]   A.L.Rivas, D.M. Michal, M.E. Burnett and C.F. Musolff, Fundamentals and Applications of Microalloying Forging Steels, Publication of the Minerals, TMS, (1996), p. 159.
[8]   N.K. Ballinger and R.W.K. Honeycombe, Met. Sci., 14 (1980), p. 367.
[9]   T. Gladman, The Physical Metallurgy of Microalloyed Steels, The Institute of Materials, 1997.
[10] W.B. Morrison, Ironmaking and Steelmaking, 22 (1995), p. 453.
[11] D.J. Naylor, Ironmaking and Steelmaking, 16 (1989), p. 246.
[12] F.J. Revidriego, R. Abad, B. López, I. Gutierrez and J.J. Urcola, Scripta Mater. 34 (1996), p. 1589.
[13] Z. Husain, ECSC Information day: Engineering Steels, British Steel (1995), paper 2.6.
[14] K. Hulka, F. Heisterkamp, R. Kaspar and L. Meyer., Processing, Microstructure and Properties of HSLA Steels, A.J. DeArdo ed., TMS (1988), p. 117.
[15] M.A. Linaza, J.L. Romero, I. San Martin, J.M. Rodriguez-Ibabe and J.J. Urcola, Fundamentals and Applications of Microalloying Forging Steels, TMS, (1996), p.311.

Materials Science Forum Vols. 284-286 (1998) pp. 443-450
© *1998 Trans Tech Publications, Switzerland*

Microstructural Characterization and Structure/Property Relations of Microalloyed Medium Carbon Steels with Ferrite-Bainite/Martensite Microstructure

G. Nussbaum[1], J. Richter[1], A. Gueth[1], A. Koethe[1],
R. Kaspar[2] and I. Gonzalez-Baquet[2,3]

[1] Institut für Festkoerper- und Werkstofforschung, Postfach 270016, D-01171 Dresden, Germany

[2] Max-Planck-Institut für Eisenforschung, Postfach 140444, D-40074 Duesseldorf, Germany

[3] Now with Aceralia, Asturias, Spain

Keywords: Forging Steels, Medium Carbon Content, Microalloying, Thermomechanical Treatment, Two-Step Cooling, Low Temperature Annealing, Complex Ferrite-Bainite/Martensite Microstructure, Qualitative and Quantitative Metallography, Macro- and Microhardness, Mechanical Properties

Abstract

Microalloyed forging steels with medium carbon content were hot deformed followed by direct cooling in two steps („slow-fast-cooling") and additional annealing at relatively low temperatures. The development of the complex ferrite-bainite/martensite microstructure during the cooling and annealing treatment was systematically investigated and correlated with mechanical properties. By applying this processing properties comparable with those of quenched and tempered Cr-alloyed steels can be obtained.

1. Introduction

In the production of drop forged parts for the transport industry the classical quenched and tempered chromium alloyed steels have been continuously substituted by vanadium microalloyed medium carbon steels controlled cooled from the forging temperature. However, there are some limitations in 0.2%-proof stress and ductility. Therefore, a new variant of thermomechanical treatment has been developed in order to improve the mechanical properties of microalloyed medium carbon to the level of Cr-alloyed quenched and tempered steels. This modified thermomechanical treatment is described in the contribution of I. Gonzalez-Baquet et al. on this conference [1]. Essentially, it consists in forging at relatively low temperature followed by a two-step cooling. In the first slow cooling step (still air) from forging temperature into the $(\gamma+\alpha)$ two phase region a desired fraction of soft proeutectoid ferrite is formed within a carbon enriched austenite. Accelerated cooling (water quench) in the second step yields hard bainite/martensite phases instead of pearlite and small amounts of retained austenite. An additional annealing treatment at relatively low temperatures of about 400 °C results in a simultaneous improvment of 0.2%-proof stess and ductility due to hardening of the proeutectoid ferrite and tempering of martensite/bainite [2, 3].
The present paper describes the development of microstructure during two-step cooling and annealing and its dependence on the variation of process parameters and microalloying as well as the relations between final microstructure and resulting mechanical properties.

2. Materials and experimental methods

Two commercial medium carbon forging steels of the well known type 27MnSiVS6 were selected as a basis for these investigations (steels VN and TV in table 1). Both steels are microalloyed with vanadium which is predominantly used for precipitation hardening of ferrite. Steel TV is additionally microalloyed with titanium making use of small TiN-particles for effective inhibition of austenite grain growth during reheating and hot forging. Furthermore, three microalloying variants T2V, T2VN, and TVNb were prepared (see table 1). The steels T2V and T2VN have enhanced contents of vanadium (T2VN additionally has an enhanced content of nitrogen), and steel TVNb is microalloyed with niobium to elevate the recrystallization temperature of austenite.

Table 1: Composition of the steels investigated (mass-%)

steel	C	Si	Mn	Cr	S	Al	V	Ti	Nb	N
VN	0.24	0.53	1.28	0.15	0.04	0.03	0.08	-	-	0.0152
TV	0.31	0.71	1.54	0.24	0.03	0.03	0.10	0.015	-	0.0098
T2V	0.29	0.60	1.51	0.27	0.03	0.03	**0.17**	0.022	-	0.0080
T2VN	0.27	0.60	1.51	0.27	0.04	0.03	**0.18**	0.024	-	**0.0154**
TVNb	0.31	0.61	1.50	0.27	0.03	0.03	0.09	0.020	**0.05**	0.0075

The forging process was simulated by plane strain hot compression tests on the hot deformation simulator (WUMSI) at the Max-Planck-Institut fuer Eisenforschung at Duesseldorf [1-3]. Each WUMSI-specimen provided secondary specimens for tensile tests (specimen diameter 5 mm) as well as for the investigation of microstructure.

The evaluation of microstructure during the two-step cooling (TSC) process and the following annealing (AN) was analysed by methods of qualitative and quantitative metallography (measurement of the chord lengths of ferrite grains and bainite/martensite regions, determination of volume fraction of ferrite in the complex microstructure). Scanning and transmission electron microscopy (SEM, TEM) were used to investigate details of the microstructure (morphology of bainite and martensite, dislocation substructure, precipitates). Size and chemical composition of the precipitates of the microalloying elements were investigated by means of carbon extraction replicas in an scanning transmission electron microscope (STEM) with energy dispersive X-ray analysis (EDX). Furthermore, macro- and microhardness measurements were carried out in order to characterize the global hardness of the complex ferrite-bainite/martensite microstructure as well as the hardness of its constituents [4].

3. Results

3.1. Microstructure of hot deformed and two-step cooled (TSC) materials

Fig. 1 depicts the microstructure of two-step cooled WUMSI-specimens of the steels T2VN (a, b) and TVNb (c, d). The specimens were hot deformed at 920 °C (strain $\varepsilon = 0.9$). Accelerated cooling (water quench) started at 680 °C (a, c) and 640 °C (b, d), respectively. As expected, the formation of proeutectoid ferrite begins predominantly at the austenite grain boundaries, and the lower the quench temperature the higher the volume fraction of ferrite. Obviously, in steel T2VN the ferrite formation is enhanced due to the higher vanadium and nitrogen contents and a little lower carbon content. In steel TVNb niobium retards the recrystallization of austenite leading to elongated austenite grains (cf. Figs. c and d). Besides, additional ferrite formation has been observed at deformation bands within the austenite grains of steel TVNb.

Fig. 2 demonstrates the influence of the quench temperature T_Q on the volume fraction of ferrite for deformation temperatures $T_D = 920$ °C (typically for thermomechanical treatment) and $T_D = 1200$ °C (conventional forging). The volume fractions of ferrite are significantly higher if deformation was applied at 920 °C because of the higher ferrite nucleation density due to smaller austenite grain size

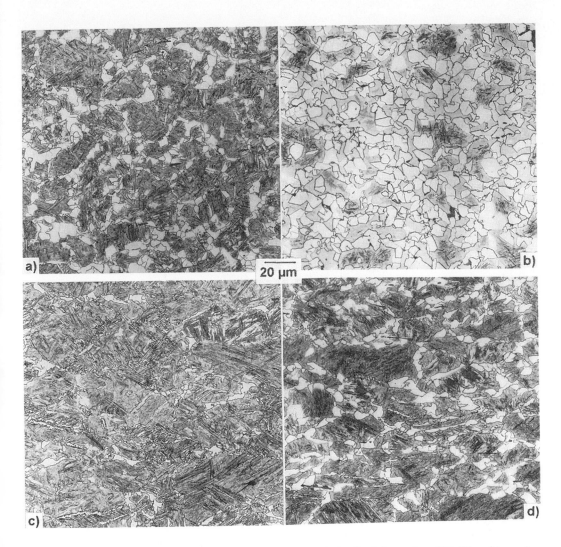

Fig. 1: Microstructure of the steels T2VN (a, b) and TVNb (c, d), deformed at 920 °C and two-step cooled with quench temperature 680 °C (a, c) and 640 °C (b, d), respectively.

and non-recrystallized elongated austenite grains with deformation bands within the grains in the Nb-bearing steel, respectively. As shown in Fig. 2a, deformation at 920 °C and variations of quench temperature and microalloying elements enable varying ferrite fractions from 10 to 50 %. Relatively high and homogeneously distributed volume fractions of soft proeutectoid ferrite are an essential prerequisite to good ductility and toughness.

Fig. 3a illustrates, that the mean chord length of the ferrite grains, d_F, slightly increases with decreasing quench temperature T_Q. This result indicates the growth of the firstly created ferrite grains and the formation of new small grains during holding time in the $(\gamma+\alpha)$ two phase region. As expected, the mean chord length of the bainite/martensite regions, $d_{B/M}$, decreases with increasing volume fraction of ferrite (Fig. 3b). The size of the bainite/martensite regions is significantly smaller in the materials deformed at 920 °C due to the finer austenite grains in comparison with conventional forging at 1200 °C.

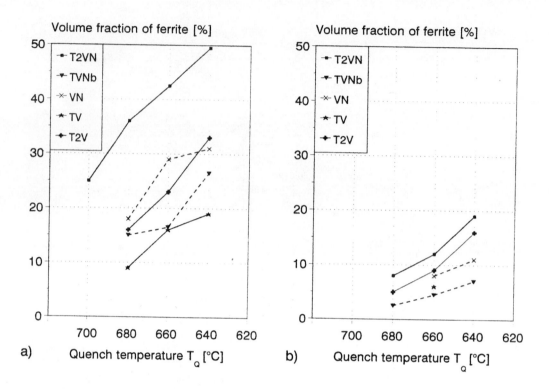

Fig. 2: Influence of quench temperature T_Q on the volume fraction of ferrite for the investigated steels deformed at 920 °C (a) and 1200 °C (b), respectively.

Results of macro- and microhardness measurements of steels TVNb and T2VN deformed at 920 °C followed by two-step cooling with various quench temperatures T_Q are shown in Fig. 4. The macrohardness HV10 represents the global hardness of the complex microstructure consisting of hard and soft phases. HV10 slightly decreases with decreasing T_Q indicating the increase of ferrite fraction (soft phase). The microhardness HV0.05 of the bainite/martensite regions (HVB/M in Fig. 4) slightly increases with decreasing temperature T_Q because of modifications of morphology of hard phases due to an increasing carbon content. The B/M-regions of steel TVNb are significantly harder than those of steel T2VN in spite of its lower carbon content corresponding to the lower ferrite fraction (cf. Fig. 2). Probably, deformation substructure in the non-recrystallized austenite grains of steel TVNb enhances the hardness. The microhardness of ferrite has been tested with forces 9.81 mN (HV0.001) and 0.981 mN (HV0.0001), respectively. The 9.81mN-indentations are about 2 to 3 μm in size - that is the size of the ferrite grains (cf. Fig. 3a). Therefore, the microhardness HVFGB in Fig. 4 represents the hardness of the ferrite with an additional effect of its grain boundaries (pile up of dislocations at grain boundaries, Hall-Petch relation). The decrease of HVFGB with decreasing T_Q corresponds to the increasing ferrite grain size (cf. Fig. 3a). However, the microhardness HV0.0001 (HVF in Fig. 4) reflects the „true" ferrite hardness without influence of grain boundaries. The slight increase of HVF with increasing time between the start of ferrite formation and quenching (decreasing T_Q) is probably caused by enhanced precipitation of V(C,N) in the ferrite grains.

The complex ferrite-bainite/martensite microstructure formed by two-step cooling doesn't yield the desired mechanical properties [2, 3]. Martensite is predominantly formed at the boundaries of

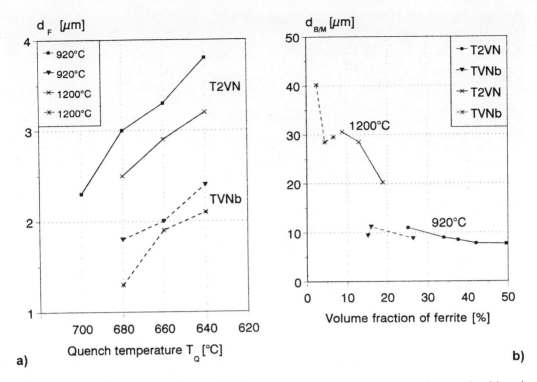

Fig. 3: Mean chord length of ferrite grains, d_F, vs quench temperature T_Q (a), and mean chord length of bainite/martensite regions, $d_{B/M}$, vs volume fraction of ferrite (b). Steels T2VN and TVNb, deformation temperatures 920 and 1200 °C.

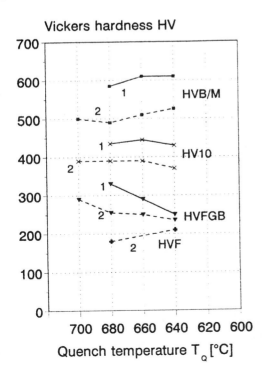

Fig. 4: Influence of quench temperature T_Q on macrohardness HV10 of the complex microstructure and on microhardnesses of bainite/martensite regions (HVB/M) and polygonal ferrite (HVFGB, HVF; for further information see text). Steels TVNb (1) and T2VN (2), deformation temperature 920 °C.

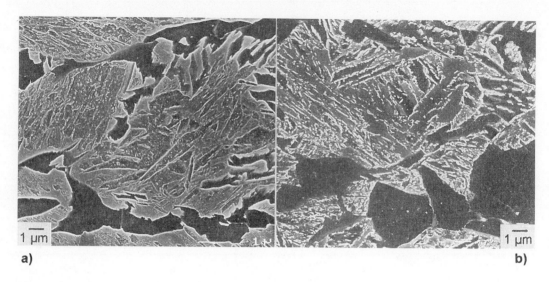

Fig. 5: Microstructure of steel TVNb (a) after two-step cooling with T_Q = 660 °C (deformation temperature T_D = 920 °C) and (b) additionally annealed at 420 °C, 1 h.

proeutectoid ferrite (cf. Fig. 5a). The volume expansion during the austenite to martensite transformation generates mobile dislocations in the adjacent ferrite grains leading to low 0.2%-proof stress. On the other hand, the untempered bainite/martensite phases formed by water quenching are very hard and brittle resulting in low ductility and toughness. The carbon content in the hard phases is highest in steel T2VN because of their higher carbon enrichment due to the high ferrite content. In this steel, after two-step cooling with T_Q = 660 °C a microhardness HV0.01 of about 700 in martensite and 450 in bainite was measured (cf. points M and B in Fig. 6b). Furthermore, about 7 vol.-% retained austenite were analysed by X-ray diffraction in the microstructure of T2VN. In steel TVNb, the corresponding values are: HV0.01 about 600 in martensite and 450 in bainite, respectively, (cf. Fig. 6b), and volume fraction of retained austenite 3 to 5 %.

3.2. Modification of microstructure and mechanical properties by low temperature annealing after two-step cooling (TSC + AN)

Fig. 6 illustrates the effect of annealing temperature T_{AN} (annealing time 1 h) on the macrohardness HV10 of the complex microstructure (Fig. 6a) and on the microhardnesses of its constituents (cf. Fig. 6b). HV10 slightly decreases with increasing annealing temperature up to about 250 °C and considerably decreases above 300 °C (Fig. 6a). The microhardness HV0.05 of the bainite/martensite regions (HVB/M in Fig. 6b) exhibits an analogous dependence on T_{AN}, however, the drop of hardness above 300 °C is more pronounced. The microhardness HV0.01 of martensite strongly decreases with increasing annealing temperature (points M in Fig. 6b), whereas the hardness of bainite only slightly decreases (points B in Fig. 6b). The microhardness HV0.0001 of the ferrite (HVF in Fig. 6b) continuously increases with increasing annealing temperature, indicating the pinning of dislocations and the formation of finely dispersed precipitates of V(C,N) in ferrite. Fig. 7 exhibits small V(C,N) precipitates about 3 to 5 µm in size generated during annealing at 420 °C (annealing time 1 h). These changes of microstructure in ferrite lead to a significant increase of 0.2%-proof stress. The decrease of hardness of the bainite/martensite regions by annealing is essentially due to the formation of

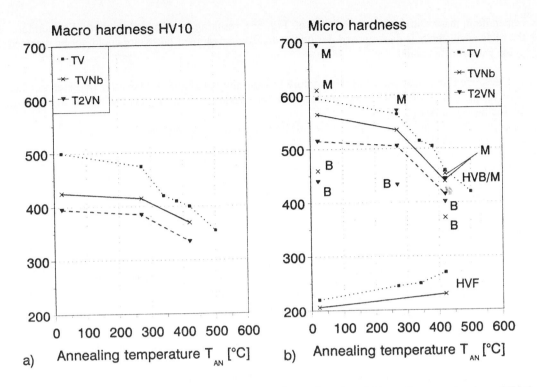

Fig. 6: Influence of the annealing temperature T_{AN} (annealing time 1 h) on the macrohardness HV10 of the complex microstructure (a) and on the microhardnesses of its constituents (b). (M = martensite, B = bainite, HVF = polygonal ferrite; for further information see text). Steels TV, TVNb, and T2VN. Deformation temperature 920 °C, quench temperature 660 °C.

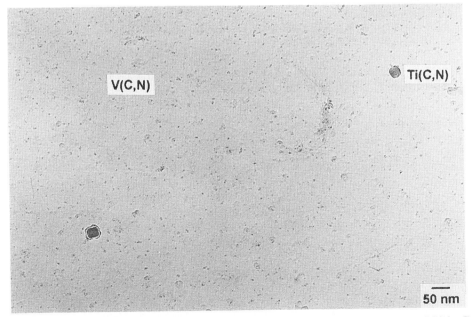

Fig. 7: Finely dispersed V(C,N)-precipitates in ferrite, generated during annealing 420°C/1h. Carbon extraction replica. Steel TV.

cementite in martensite and lower bainite (cf. Fig. 5b), resulting in enhanced ductility. Furthermore, the retained austenite is decomposed during the annealing treatment (austenite could not be detected in the annealed materials).

The described modifications of microstructure by annealing at about 400 °C after two-step cooling have resulted in the following improvements of mechanical properties, depending on the applied microalloying variant [2, 3]:
- increase of 0.2%-proof stress from 650....800 MPa to 820....1000 MPa
- increase of reduction of area from 20....25 % to 40....50 %.
These combinations of high 0.2%-proof stress with good ductility attained by the (TSC+AN) processing are comparable with the property level of Cr-alloyed quenched and tempered steels.

Acknowledgments

The work has been financially supported by the Bundesministerium für Wirtschaft of the FRG (contract no. AiF 9483B). The support given by the „Arbeitsgemeinschaft industrieller Forschungs-vereinigungen" (AiF) and the „Verein Deutscher Eisenhüttenleute" (VDEh) is gratefully acknow-ledged.

References

[1] I. Gonzalez-Baquet, R. Kaspar, J. Richter, G. Nussbaum and A. Koethe, this conference

[2] R. Kaspar, I. Gonzalez-Baquet, J. Richter, G. Nussbaum and A. Koethe, steel research 68, (1997), 6, 266-271

[3] I. Gonzalez-Baquet, R. Kaspar, J. Richter, G. Nussbaum and A. Koethe, steel research 68, (1997), 12, 534-540

[4] J. Richter, G. Nussbaum, A. Gueth, A. Koethe, R. Kaspar and I. Gonzalez-Baquet, Prakt. Met. Sonderband 29 (1998), in press

Materials Science Forum Vols. 284-286 (1998) pp. 451-458
© 1998 Trans Tech Publications, Switzerland

Induction Heating of Microalloyed Forging Steels

W.F. Peschel

Consulting Corporate Technical Director, American Induction Heating Corporation,
33842 James J. Pompo Drive, Fraser, MI 48084, USA

Keywords: Induction, Heating, Forging, Microalloyed, Steels, Billet, Temperature, Process, Control

Abstract: Microalloyed steels are being used more extensively by the forging industry in a production process that requires a more precise control of billet temperature when delivered to the forging press or forging hammer. Although induction has become the prime *billet heater of choice* for most forgers, some of the older induction installations are not able to provide the technology required to successfully control the final product specifications of a microalloyed steel forging.

This paper will present fundamental theory concepts of induction billet heating for forging as well as the conceptual design of an induction billet heater presently being used for various grades of microalloyed steels.

Induction Theory: The basic induction billet heating process utilizes solenoid wound or tunnel type electrical coils through which round or square steel billets are progressively passed at a controlled feed rate. Concurrently, a relatively large value of high frequency alternating electric current is applied to the coil winding that produces an electrical flux field in the coil that couples into the metal billets. The alternating flux field develops an induced voltage in an outer perimeter shell under the billet surface which produces a high value of circumferential current flow that combines with the resistance of the billet material to produce I^2R heating.

The rate of electrical energy input to the billet cross section with induction heating can be substantially higher than with fuel-fired radiant or convection-type furnaces since the heat energy is developed within the billet, rather than being transmitted through the outer billet surface. Depending on the applied power density (watts per square inch) and the frequency (hertz) of the induction coil current, an outer shell section can be heated to forging temperatures within a matter of seconds.

The distribution of temperature over the cross section of a billet being heated by induction is not uniform as 86% of the induced power is in an outer heating shell of the billet for which the heated depth is determined by the induction power frequency (hertz) and resistivity of the billet material (ohm-cm). The central cross section of the billet is heated by thermal conduction within the material and requires a discrete time factor of seconds to minimize the differential temperature between the outer shell and the central core of the heated billet. The two fundamental laws of induction engineering related to these heating characteristics will be discussed in the following sections titled "Current Reference Depth" and "Billet Temperature Control".

Current Reference Depth: The generally accepted formula for the reference depth of the induction current outer shell in a billet being heated, where 86% of the power is generated, is essentially as follows:

$$\text{Current Reference Depth in millimeters} = 50,400x\sqrt{\frac{\text{Resistivity}}{\text{Permeability x Frequency}}} \qquad (\text{Eq.1})$$

The resistivity value units are ohm-centimeters and the frequency is hertz. Permeability is a dimensionless property of a magnetizable substance to the degree in which it modifies the magnetic flux in the region occupied by it in a magnetic field[1]. Since all forging steels are nonmagnetic when heated above 750 °C the permeability value will be equal to 1 and will result in maximizing the current penetration depth when it is most important at the end of the induction heating cycle.

Electrical resistivity of the steel billet will vary with temperature as illustrated in Figure 1, consequently, the current penetration depth will increase as the billet passes through an induction coil and the temperature increases. This factor will also contribute to the fact that maximum current penetration depth will be realized at the final stages of heating when the resistivity is the highest.

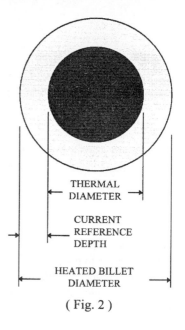

THERMAL DIAMETER

CURRENT REFERENCE DEPTH

HEATED BILLET DIAMETER

(Fig. 2)

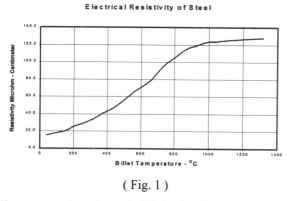

(Fig. 1)

The cross section of a typical induction heated billet diameter is illustrated in Figure 2 with the light shaded area representing the induced current penetration where 86% of the heating energy is developed. The darker center area represents the section where thermal conduction heats the balance of the cross section. The center or "thermal diameter" requires a discrete amount of time to soak the heat to the center and reduce the induction heating surface-to-center temperature differential.

The induction current reference depth of Figure 2 also represents the depth of flux penetration which affects the electrical heating efficiency of the induction heating process. Without trying to be too technical on the subject, a general rule is that the current reference depth should not exceed 25% of the heated billet diameter. Another way to express the rule is that the billet diameter should be at least four times greater than the induced current reference depth of Equation 1.

---

[1] Webster's Ninth New Collegiate Dictionary page 876

Since the electrical resistivity of most microalloyed forging steels follows the characteristics of Figure 1 and the fact that the magnetic permeability factor is unity at the final forging temperatures, the only other process variable of Equation 1 is the frequency of the applied electrical coil current.

Fundamentally it is only the power current frequency factor that can be controlled to realize the optimum induction heating penetration for maximizing electrical coil efficiency and minimizing the thermal diameter and soak heating time. The curves in Figure 3 illustrate the depth of current penetration for nominal power frequency ratings at various temperatures over 800 °C. The nominal power frequency selection for induction billet heater installation would depend on the minimum billet diameter to be heated and a current reference depth representing a value equal or less than 25% of the minimum diameter.

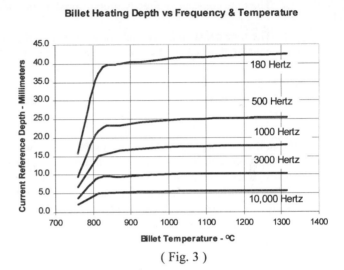

(Fig. 3)

Billet Temperature Control: When induction heating microalloyed forging steels, one of the most important objectives is to have consistent billet temperatures at the forge press or forge hammer. This criteria must be realized without exceeding a temperature during the heating process that will produce detrimental metallurgical grain size or cause incipient melting at grain boundaries. Also some grades of steel high in vanadium may require a minimum time at temperature to allow full dissolution throughout the matrix structure.

Although consistent billet temperature control has always been a prime forging advantage for induction heating, the control limits for microalloyed steels are more stringent and the window for temperature tolerances significantly reduced. Because of the constricted operating temperature criteria it is important to understand the fundamentals involved with induction heating and the process variables that can be logically applied.

When a steel billet is being heated the physical properties of density, specific heat and thermal conductivity are constantly changing with the rise in temperature, in addition to the changing thermal diameter resulting from the increasing current reference depth of Figure 3. The temperature differential from the billet surface through the depth of current penetration is relatively uniform, however, because of the billet center being heated by conduction, there is an inherent temperature differential from the billet surface to the core center. The amount of differential is dependent on the thermal diameter size, the rate of energy input and the physical properties of the billet material.

The billet section diameter being subjected to conduction heating has a specific time constant during which a time-temperature plot of cross section surface-to-center temperatures is of a hyperbolic nature and after such time value the surface-to-center temperature would be constant if all heating parameters were to remain unchanged. Since material properties and heating factors are always in a state of change, the time constant value will vary during the heating cycle with the following general relationships:

1. The time constant is directly proportional to the square of the thermal diameter.

2. The time constant is inversely proportional to the material diffusivity, which is a value equal to thermal conductivity divided by density and divided by specific heat.

In pure theory, if all heating and material properties remain constant the surface-to-center differential temperature for a cylindrical shaped billet section will be equal to the change in mean temperature during the time constant interval.

For instance if the mean, or average, temperature in the thermal diameter section of a billet increases 100 °C during the interval of one time constant, the differential temperature from the outer surface of the billet section to the center will be approximately 100 °C. This hypothesis assumes that the surface temperature is equal to the temperature at the outer thermal diameter which is not totally true because of radiation energy loss and current distribution. Nevertheless the basic concept can be used for understanding the two primary control elements applicable to the induction billet heating process and the following parameters:

1. Lower frequency power heating will result in minimizing the thermal diameter, thus reducing the time constant and differential temperature for a specific heating time.

2. The differential temperature is directly proportional in the rate of energy input to the billet which relates to the applied induction power density (kW per square inch of billet surface) less the surface radiation energy (watts per square inch).

Billet Heater Power Distribution: Because of the magnetic properties of carbon alloy steels being induction heated, the induced power density during the heating cycle for each individual billet will vary considerably and will produce a significantly higher rate of temperature increase at the beginning of the heating cycle than will be realized at the end of the cycle. The relationship of power density input and the resultant temperature rise in an induction heated billet is illustrated in Figure4.

With reference to the preceding criteria for temperature differential control, the higher rate of temperature increase at the first 20% of the heating cycle will result in a substantially large temperature differential from the billet surface to the center core when compared to the heating characteristics at the end of the heating cycle.

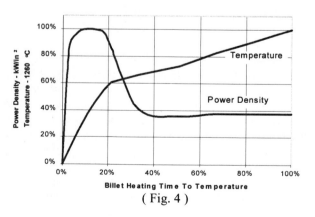

Induction Billet Heater Power Distribution

(Fig. 4)

Heated Billet Temperature: The time-temperature relationship for a typical induction billet heater operating at a production rate of 3000 kg/hour is illustrated in Figure 5 and has a 90 °C surface-to-center temperature differential. As the billet exits the induction coil line it will have a mean temperature of 1260 °C and a surface temperature of 1280 °C. When forging plain carbon steels these temperatures are usually acceptable but for microalloyed steels a tolerance of 25 °C is more desirable

To illustrate the basic heating principle required to meet the more restrictive temperature specifications, Figure 6 represents the same induction heater operating at a 50% production rate of 1500 kg/hour. The resultant surface-to-center temperature differential is 20 °C with an identical mean temperature of 1260 °C but only a surface temperature of 1256 °C that is primarily due to surface radiation effect on the heating process.

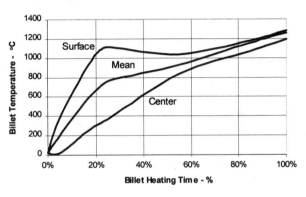

(Fig. 5)

When the close temperature specifications for heating microalloyed forging steels were introduced it was obvious that operating at such reduced production rates was not an acceptable method, consequently, there was a definite need for new technology in the design of induction billet heaters which has been under development for the last several years.

Induction Heater Design: In 1994 a large forging operation in the USA installed a 12,700 metric ton press for forging microalloyed and standard alloy steels. The new installation claimed to have the largest mechanical press in North America for forging billet sizes ranging from 80 mm to 200 mm square and 300 mm to 1680 mm long.

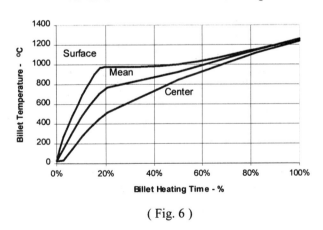

(Fig. 6)

American Induction Heating Corporation in the USA was contracted to supply the induction billet heating equipment with some very specific obligations related to the induction coil designs. The heating capacity requirement was 22,500 kg/hour with 8,000 kW of induction heating power.

The induction heater has a dual line configuration as shown in Figure 7, each with six (6) in-line solenoid coils. In accordance with the equipment contract the design of all coils, for a specific billet size range, had to be identical and interchangeable with other induction billet heaters at the forging complex. Such a requirement precluded the use of conventional coil design techniques for controlling the temperature distribution within the billet heating line. Another consideration was to extend the line length and heating time so as to reduce the surface-to-center temperature differentials discussed earlier, however, this was not practical from either a mechanical design viewpoint or the available facilities floor space.

After extensive design evaluations with computer modeling of billet heating temperature profiles, it was determined that discrete control of the power and heating rates in each of the last three coils would provide any temperature profile required during the last half of the billet heating cycle.

(Fig. 7)

Heating Power Control: The final design for the electrical power configuration at each of the 4000 kW billet heater lines is shown in Figure 8 with individual saturable iron-core reactors installed in series with each of the last three induction coils.

Electrical Power Schematic

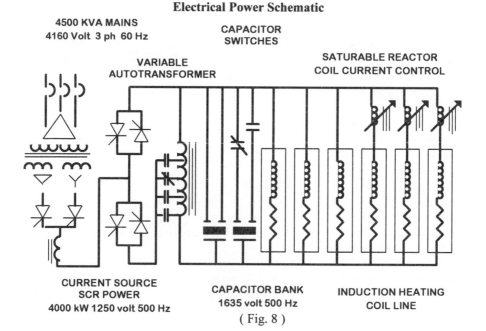

4500 KVA MAINS
4160 Volt 3 ph 60 Hz

CAPACITOR
SWITCHES

VARIABLE
AUTOTRANSFORMER

SATURABLE REACTOR
COIL CURRENT CONTROL

CURRENT SOURCE
SCR POWER
4000 kW 1250 volt 500 Hz

CAPACITOR BANK
1635 volt 500 Hz

INDUCTION HEATING
COIL LINE

(Fig. 8)

There are individual process controllers for each of the three reactors that varies the amount of saturation and consequently a variable value of electrical reactance for each device. The reactors which are connected in series with the induction heating coils provides a variable voltage for each individual coil and the net effect is a controlled power input to the billets being heated in the respective coil. As discussed earlier the input rate of power to the billet being heated directly affects the surface-to-center differential temperature in the billet cross section.

To provide finite process power input control to each final stage heating coil, voltage sensors are connected across each of the last three coils and the capacitor bus that represents the voltage of the first three coils.

Depending on the grade of microalloy steel being heated, menu values of individual "process set-points" for each final stage heating coil is entered into the billet heater's PLC control

(Fig. 9)

logic through the operator terminal shown in Figure 9. The menu value for each process control is a percentage value for the controlled induction coil as related to the first three heating coils. Power to each controlled coil varies as the square of the coil voltage, consequently, a 90% relative setting will represent an 81% rate of heating power input to each billet within the heating coil and control of temperature differential is realized.

The temperature profile for a 180 mm square billet being heated at production rate of 6,500 kg/hour is illustrated in Figure 10 with a final mean temperature of 1260 °C. Uniformity of the cross section temperature is shown in Figure 11 and temperature uniformity from end to end is evident in Figure 12.

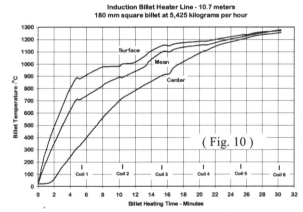

(Fig. 10)

(Fig. 11)

(Fig. 12)

Conclusions: Induction billet heating can be a very precise production process, providing the equipment is capable of operating with accurate control of the following parameters:

1. Material mass or weight being passed through the induction coil line must be constant within a plus or minus 1% tolerance. This means that continuous drive feeds must have electrical feedback speed controls or if the cycle is based on sequential time intervals the control must have sufficient accuracy. In addition, with cycle timer controls, the billet lengths must be accurately controlled. At 1260 °C a 1% feed rate or material weight variance will translate to an 8 °C change in the final heated billet temperature.

2. As previously discussed, the applied voltage to each individual induction coil will directly affect the energy input to the billet and the final heating temperature. In as much as the power is proportional to the coil voltage squared, the voltage control should be within plus or minus 0.5% to limit the heated billet temperature variance to an additional 8 °C tolerance.

3. When operating an induction heating system having individual coil voltage control for modifying the billet temperature or heating rate profile, there are developed techniques that will aid both the heating process and the operating personnel. Since most induction power supplies regulate the output voltage to the main capacitor bus and preheating coils, any final stage heating with variable voltage coils should maintain a proportional voltage value as related to the main bus. Control of the voltage ratio parameters require feedback of coil and bus voltage values into a process controller having provisions for the operator to enter a "process set-point" value.

An induction heating system, with proper design and adequate controls, can be a very efficient and reliable method for heating all grades of microalloyed forging steels.

Correspondence with author: address to American Induction Heating Corporation, P.O. Box 248, Fraser, Michigan, USA. Telephone 810-294-1700. Fax 810-294-2293. E-mail to author at aih@rust.net.

THIN SLAB AND HOT DUCTILITY

Materials Science Forum Vols. 284-286 (1998) pp. 461-468
© 1998 Trans Tech Publications, Switzerland

The Hot Ductility of V, Nb/V and Nb Containing Steels

B. Mintz and R. Abushosha

Dept. Mech. Engineering and Aeronautics, City University, London, England

Keywords: Hot Ductility, V, Nb/V and Nb Containing Steels, Transverse Cracking, Continuous Casting

ABSTRACT

The hot ductility behaviour of a series of V, Nb/V and Nb containing steels having a base composition of 0.1 %C, 1.4%Mn, 0.3 %Si have been examined. Two levels of V, (0.05 and 0.1%), N (0.005 and 0.01%), and Nb, (0.015 and 0.03%Nb) were selected.

Tensile samples were heated to 1330°C and cooled at 50°C/min to test temperatures in the range 700 to 1000°C and tested to failure using a strain rate of $3 \times 10^{-3} s^{-1}$. Heat treatment was chosen to closely follow the commercial continuous casting conditions and the strain rate used was that applied during the straightening operation.

In addition, to make the tests more relevant to continuous casting, tensiles from selected steels were melted, re-solidified and cooled to the test temperature as for the samples solution treated at 1330°C. In all cases a ductility trough was obtained in the temperature range 750 to 950°C. The 0.030%Nb containing steel gave the worst hot ductility of all the steels examined. For samples heated to 1330°C, increasing the V and N levels in the V containing steels caused the ductility to deteriorate but the product of the total V and N concentrations had to be as high as 1.2×10^{-3}, i.e. 0.1%V and 0.012%N, before approaching the low values of reduction of area exhibited by the Nb containing steel. The hot ductility curves for all the Nb/V containing steels at the 0.005% N level were very similar. Adding V at this low N level to the 0.03%Nb containing steel was found to improve ductility in the temperature range 800 to 900°C.

For the as-cast samples, again the V containing steels were found to give better ductility than the 0.03%Nb containing steel.

The hot ductility behaviour was found to be mainly related to the size of the precipitates, the coarser the precipitation the better being the ductility.

INTRODUCTION

Transverse and edge cracking of the strand on straightening during continuous casting of HSLA steels can still be a serious problem and niobium as a microalloying addition is well known for encouraging this form of cracking [1]. Previous work [2] has shown that vanadium containing steels exhibit better hot ductility than niobium over the temperature range over which the straightening operation is carried out during continuous casting. In this previous work [2] tensile samples were solution treated at 1330°C prior to cooling to the test temperature. The depth and width of the hot ductility trough were shown to be related to the product of the V and N concentrations, and V and N levels had to be high (0.1%V and 0.01%N) for the hot ductility to approach that given by a steel containing 0.03%Nb. The simple hot ductility test has been shown to give a good indication of the likelihood of a steel to give transverse cracking [3], and hence it is reasonable to conclude that V containing steels having a product, [V] x [N], less than 10^{-3} will show less tendency to cracking than steels with 0.03%Nb.

The present work is an extension of this work on solution treated steels and examines the influence of V and Nb in combination at low and normal Nb levels on hot ductility. In addition, to make the information from the tensile tests more relevant to the continuous casting operation, tensile samples from selected steels have been cast and cooled directly to the test temperature.

EXPERIMENTAL

The compositions of the solution treated V/Nb and Nb containing steels are given in Table 1, steels 1-6, together with the as-cast steels, steels 7-10.

Table 1. Compositions wt. per cent.

a) Solution treated steels

Steel	C	Mn	Si	S	P	Al	N	Nb	V
1	0.120	1.41	.32	.004	.012	.029	.0042	.015	-
2	0.098	1.40	.30	.004	.017	.027	.0062	.015	.048
3	0.098	1.40	.30	.004	.017	.025	.0057	.015	.100
4	0.090	1.41	.30	.003	.015	.032	.0052	.028	-
5	0.110	1.43	.31	.005	.012	.030	.0046	.030	.053
6	0.110	1.42	.32	.005	.012	.031	.0044	.030	.110

b) As-cast steels

Steel	C	Mn	Si	S	P	Al	N	Nb	V
7	0.110	1.42	.32	.002	.011	.038	.0038	-	-
8	0.100	1.39	.33	.003	.014	.031	.0051	-	.050
9	0.098	1.35	.33	.002	.018	.028	.0100	-	.110
10	0.090	1.41	.30	.003	.015	.032	.0052	.028	-

Compositions of the V/N steels can be found in Ref.2. The steels were supplied by British Steel as hot rolled 13 mm thick plate. The casts all have the same base composition 0.1%C,1.4%Mn,0.3%Si with 0.03%Al. The influence of two levels of V, nominally 0.05 % and 0.1 % and two levels of Nb, 0.015 and 0.03% all at the ~0.005%N level were examined, steels 1 to 6. Lower Nb containing steels, steels 1 to 3 were included for examination because of the trend commercially to lower Nb levels.

Hot tensile samples of 5mm. dia. and 25 mm. gauge length were machined from the plates in the rolling direction. The samples were nickel plated and heated to 1330°C ±5°C in argon gas atmosphere, held for 5 minutes and cooled at 50°C/min to test temperatures in the range 700 to 1050°C ±5°C. The samples were held 5 min at the test temperature and strained to failure at a strain rate of $3 \times 10^{-3} s^{-1}$ on an Instron tensile machine. The samples were ice/brine quenched immediately after fracture to enable the microstructure to be examined. The cooling rate and strain rate were chosen to simulate the cooling rate during the continuous casting operation and the strain rate on straightening respectively. Reduction of area (R of A) values were taken as a measure of hot ductility.

A C-Mn-Al, two V/N steels and the 0.03% Nb steel were also tested in the as cast state, steels 7-10 respectively. The tensile samples were melted by an induction coil and cooled directly to the test temperature using similar cooling and test conditions to those given to the solution treated samples.

The fracture surfaces of all the samples were examined using a Jeol T200, Scanning electron microscope (SEM). Carbon extraction replicas were also taken from selected samples on sections close to the point of fracture and examined with the transmission electron microscope (TEM).

RESULTS

Hot Ductility Curves

1) V and N steels
The hot ductility curves for the previously examined solution treated V/N containing steels are shown in Fig. 1. These curves have been discussed in detail in Ref. 2 and only the salient points are referred to here. It can be seen that increasing the V or N levels leads to worse ductility. However, the V and N levels would have to be in excess of 0.1 and 0.01% respectively to give the poor ductility exhibited by the 0.028%Nb containing steel.

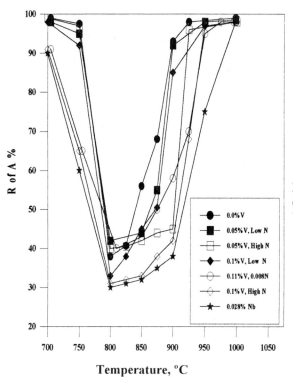

Fig. 1 Hot ductility curves for V/N steels (ref. 2). Also included is the curve for the V free, 0.028%Nb containing steel.

2) As-cast hot ductility curves
The hot ductility curves for the as-cast steels examined are given in Fig.2, and again show that the Nb containing steel gives the worst ductility. Trends were similar to those observed in Fig. 1 for the solution treated state but as has been noted before with as-cast data, the differences in R of A values between the various steels was smaller than for solution treated material, making comparisons more difficult [4].

3) Nb and Nb/V steels
The hot ductility curves for the Nb and Nb/V solution treated steels are given in Fig.3. The low Nb, V free steel gives the best hot ductility behaviour while the steel with the more conventional Nb addition gives the worst. It can be seen that raising the Nb level from 0.015% to 0.028% results in both a significant increase in the depth as well as the width of the trough, Fig.3. However, at this higher Nb level adding V improved the ductility in the trough, i.e. for the temperature range 800-900°C but not in the higher temperature range 950-1000°C where a small deterioration was noted. For the lower, 0.015% Nb containing steel no improvement in ductility occurred on adding V

throughout the whole temperature range from 800-950°C and ductility was seriously impaired in the range 900-950°C. However, the ductility was nevertheless better than that shown by the higher Nb containing steel with otherwise similar composition.

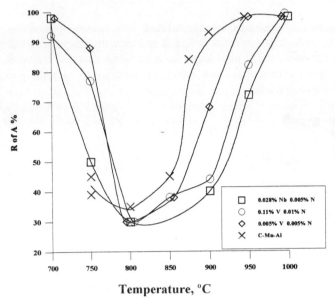

Fig. 2 Hot ductility curves for as-cast C-Mn-Al, C-Mn-V-Al and C-Mn-Nb-Al steels.

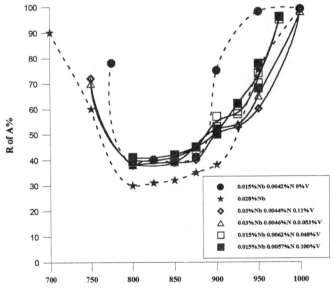

Fig. 3 Hot ductility curves for Nb and Nb/V steels after solution treatment.

Comparison with the curves from the previous work [2] on the niobium free steels with similar V and N levels also indicates, that the presence of both Nb and V results in a significant deterioration in ductility for temperatures > 850°C, Figs. 1 and 3.

The occurrence of dynamic recrystallisation can be detected from the stress v total elongation curves by either an abrupt decrease or oscillations of the flow stress. It would appear from these curves (not shown) that the temperature for the onset of dynamic recrystallisation increases with V and N concentration as well as Nb level. It was found that the estimated temperatures for the onset of dynamic recrystallisation corresponded closely to when full recovery in ductility occurred indicating the importance of dynamic recrystallisation in influencing ductility at the high temperature end of the trough.

Raising the product of [V] x [N] in V-N steels from 2.55 to 11.1, (0.05%V,0.0051%N to 0.11%V,0.01%N) increases the temperature for the onset of dynamic recrystallisation from 900°C to 950°C. Nb containing steels as expected have in general higher temperatures for the onset of dynamic recrystallisation. In the Nb containing V free steels raising the Nb level from 0.015 to 0.028% Nb increases the temperature for dynamic recrystallisation from 950 to 1000°C. For the 0.030% Nb/V steels, dynamic recrystallisation occurred also at the high temperature of 1000°C.

Metallography and electron microscopy

The γ grain size was found to be 400um for all the solution treated steels and ~1mm when as-cast.

SEM examination of fracture surfaces gave high temperature rupture failure at temperatures > 1000°C. As the temperature was lowered to the minimum ductility temperature, 800°C, the fracture appearance changed to intergranular failure. At temperatures above the Ae_3, the fracture appearance was mainly flat smooth facets, indicative of grain boundary sliding in the γ and below 800°C the fracture faces were covered with microvoids, indicative of the presence of the thin film of ferrite accompanied by voiding at MnS inclusions.

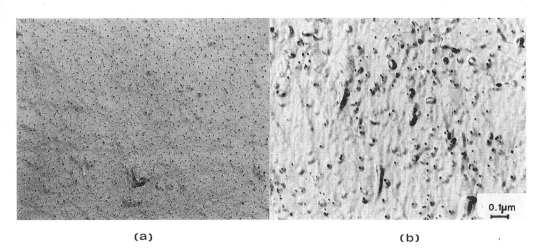

(a)   (b)

Fig. 4   (a) marked Nb(CN) precipitation present in steel 4.
            (b) Coarser VN and Nb(CN) precipitation found in steel 5.

TEM examination showed precipitation in the 0.028%Nb steel was the most extensive of all the steels examined. A fine matrix precipitation, average size 6nm, was observed, Fig. 4a and a coarser grain boundary precipitation (average size 11nm).

For the lower Nb containing steel the average size remained the same but precipitation was much reduced.

The precipitation in the V/N steels has been discussed previously in Ref. 2. It was found in this work that raising the product of [V] x [N] increased the amount of VN precipitated. Precipitate size seem to be little influenced by the V and N levels. However, the precipitate size was found to be coarser than in the Nb containing steel, the average size of the particles being 16nm in the matrix.

Only the higher Nb, 0.03% Nb, V containing steels were examined. Precipitation was found to be extensive, Fig 4b but not as extensive as in the V free Nb containing steel, and in accord with there being little difference in hot ductility behaviour little difference in precipitation behaviour was noted between the two V/Nb steels, steels 5 and 6.

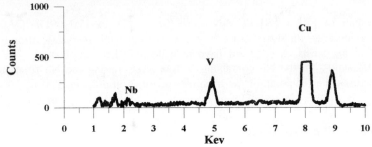

Fig. 5 EDX analysis of typical particles present in Nb/V steels showing the presence of both Nb and V.

Both VN and Nb(CN) were probably present as can be seen from the Energy Dispersive X-ray Analysis, Fig. 5. In consequence, the mean particle size came out between that for the V steels and that for the Nb steels at 10nm.

DISCUSSION

Influence of Nb

For Nb containing steels, intergranular failure in the higher temperature range (900°C) invariably occurs by grain boundary sliding in the austenite. Nb(CN) is particularly effective in reducing the hot ductility for the following reasons [3]:

1) The very fine matrix precipitation of Nb(CN) increases the stress required for deformation and therefore increases the stress in the grain boundary regions.

2) Often precipitate free zones are present at the grain boundaries concentrating the strain into the boundary regions. The situation is then very similar to that prevailing when thin films of ferrite are present.

3) There is always marked precipitation at the austenite grain boundaries and this will encourage voiding and extension of cracks formed by grain boundary sliding. If the precipitation is fine with close spacing of particles, as is found in these Nb containing steels, cracks can readily link up.

Previous work on hot-rolled samples heated up to 1330°C [5] has tended to indicate that all the Nb(CN) will have gone back into solution and it may all precipitate out in a very fine form during deformation at the test temperature. It has been shown that dynamic precipitation occurs very rapidly at temperatures close to the nose in the precipitation-temperature-transformation (PTT) diagrams, i.e. at 900°C for C-Mn-Al-Nb steels and it can often be assumed that the rate of reaction is sufficiently rapid to maintain equilibrium [6].

At lower temperatures (800°C), deformation induced ferrite is also formed surrounding the γ grains, and this with the precipitation leads to continued poor ductility. However, fracture now is by microvoid coalescence at the MnS inclusions present in these thin films, and this is exacerbated by the Nb(CN) precipitation. Ferrite although having excellent ductility is softer than γ and all the strain concentrates in these regions giving rise to poor ductility.

Full recovery at the high temperature end always corresponds to the onset of dynamic recrystallisation. Dynamic recrystallisation can not take place during the straightening operation in continuous casting but some improvement in ductility does occur at higher temperatures without recrystallisation either by reducing the degree of precipitation or coarsening existing precipitates.
In the case of the Nb containing steels, the fine precipitation of Nb(CN) produced on deformation even in the low Nb steel, delays the onset of dynamic recrystallisation to higher temperatures. However, the reduced precipitation in this lower Nb steel does result in a general raising of the trough, Fig. 3 and it is this part of the trough which is most relevant to continuous casting.
Ductility recovery at the low temperature end generally corresponds to when there is a significant volume fraction of ferrite present so that strain concentration does not occur.

Comparison between Nb and V containing steels.
Previous work [2] has clearly shown that V steels give better ductility than the conventional 0.03%Nb containing steel. This is due to the V containing steels producing coarser precipitates than Nb and a greater volume fraction of precipitate is required to cause the ductility to deteriorate [2]. Coarser precipitation has been shown to both reduce transverse cracking and give improvements in hot ductility [1]. This arises because a coarser matrix precipitation decreases the stress concentration at the boundaries, and because of the greater interparticle spacing makes it less easy for cracks formed by grain boundary sliding to join up [6].

Influence of Nb/V in combination on hot ductility.
In the case of V + Nb steels, the addition of Nb to a V containing steel would be expected to reduce the hot ductility. Not only would there be precipitation from VN but there would be additional precipitation from NbCN. As expected, Fig.3, ductility does generally deteriorate particularly so for the lower Nb containing steels. However, it should be noted that for the more conventional Nb containing steel, the presence of V leads to significantly better ductility at least in the temperature range 850 to 900°C, than the ductility obtained when Nb alone is the precipitating element. Coarser and less extensive precipitation was noted in these steels in accord with their better ductility.

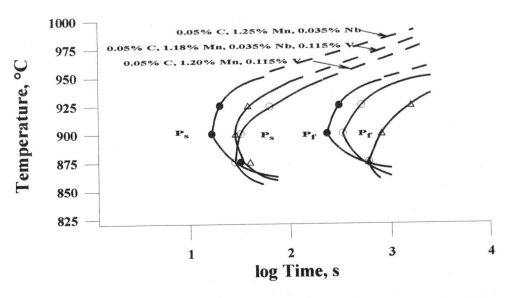

Fig. 6 Dynamic PTT curves for Nb, V, and V in combination. After Akben et al. ref. 6.

These observations are in accord with the work of Akben et al [6]. They examined a 0.05%C, 1.2%Mn steel at the 0.035%Nb and 0.115%V levels with 0.006%N. They showed that the kinetics of dynamic precipitation of Nb and V are similar and fast, Fig.6, with the V steel in general showing the slower kinetics. When present singly the nose temperature was approximately 900°C for the Nb containing steel and slightly lower at 880°C, for the V containing steel. Nb precipitated out more rapidly than V at temperatures above 880°C.

At the nose it needed about 5 min to complete precipitation for the Nb containing steel and 8 min for the V steel. However, when the 0.035%Nb and 0.115%V where together in combination the rate of precipitation at the nose (900°C) was considerably slowed down, the time taken for completion of precipitation being 15 minutes. The time to complete a test in the present exercise in the trough was about 3 min so it is expected that substantial precipitation will have occurred as has been noted from the replica examinations [2] but it is unlikely that precipitation would have gone to completion, and therefore it is likely that precipitation will be reduced when both Nb and V are present together. Akben et al [6] suggest that V slows down the precipitation of Nb because it increases the solubility of Nb(CN). Hence, not only will there be less Nb(CN) precipitation but their will be proportionally more of the coarser VN precipitation and less of the finer Nb(CN).

As-cast hot ductility

Hot ductility behaviour was similar to that given by the solution treated steels in that the conventional Nb containing steel gave the worst ductility, but changes in ductility were less marked. This arises because in the as-cast state due to segregation a significant amount of the microalloying elements are precipitated out prior to testing as coarse eutectics so there is less Nb and V available for precipitating out during deformation as the fine precipitation which is so detrimental to ductility [4].

CONCLUSIONS

1. In Nb containing, V free steels, lowering the Nb level from 0.03% to 0.015% improves hot ductility, both the width and depth of the trough being reduced.
2. Adding Nb to a V containing steel generally causes the ductility to deteriorate but the presence of V in a 0.03%Nb containing steel improves the hot ductility in the temperature range 850-900°C. This is believed to be due to the V slowing down the precipitation of the Nb(CN).
3. Hot ductility behaviour was similar for both as-cast and solution treated steels but changes in ductility with composition were more marked in the latter.
4. It would appear from all the results that V steels are likely to show reduced transverse cracking compared to Nb containing steels.

ACKNOWLEDGEMENTS

The authors would like to thank Vanitec for help in funding this work.

REFERENCES

[1] B.Mintz and J.M.Arrowsmith: Met. Technol.6 (1979),p. 24.
[2] B.Mintz and R.Abushosha: Ironmaking and Steelmaking, 20 (1993),p.445.
[3] B.Mintz,S.Yue and J.J.Jonas: Int. Materials Reviews, 36 (1991),No.5, p.187.
[4] B.Mintz and R.Abushosha: Mat. Sci. and Technol. 8,(1992), p.172.
[5]B.Mintz and Z.Mohamed: 7th Int.Conf. on Fracture, Houston, Texas, ed. by
    Salama,Ravi-Chandos,Taplin,Rama Rao,vol.4,(1989),p.2545.
[6] M.G.Akben,B.Bacroix and J.J.Jonas, Acta. Metall.,vol.31,(1983), p.161.

Materials Science Forum Vols. 284-286 (1998) pp. 469-476
© *1998 Trans Tech Publications, Switzerland*

The Influence of Composition on the Hot Cracking Susceptibility during Casting of Microalloyed Steels Processed to Simulate Thin Slab Casting Conditions

D.N. Crowther[1], M.J.W. Green[1] and P.S. Mitchell[2]

[1] British Steel Swinden Technology Centre, Moorgate, Rotherham, S60 3AR, UK

[2] Vanitec, High Street, Westerham, Kent, TN16 1AQ, UK

Keywords: Thin Slab Casting, Ductility, Hot Deformation, Microalloyed Steels, Transverse Cracking, Precipitation, Transformation, Vanadium, Niobium

Abstract

One of the problems associated with the thin slab casting process route is the formation of surface defects such as transverse cracks in the continuously cast slab. The objective of this work was to determine the likelihood of developing transverse surface cracks for various microalloyed steel types under thin slab casting conditions. A technique was developed to simulate the cooling conditions experienced during the casting of thin slabs 50mm in thickness in small laboratory ingots. To simulate the slab straightening procedure, the cast ingots were deformed in a three point bend mode using strain rate conditions expected during the straightening of thin slabs. To generate significant cracking, surface strains greater than those expected during continuous casting had to be applied. The tests were conducted over a range of temperatures, and the degree of surface cracking assessed. Five steel types were examined based on 0.06%C, 1.5%Mn and with additions of V, Nb and N. Small cracks, up to 2mm long, were observed at austenite grain boundaries for all the steels across a wide temperature range. However, the Nb, V-N, and V-Nb steels exhibited cracks which were longer than this in some parts of the temperature range. The Nb steel exhibited the longest cracks (17mm at 820°C) over the widest temperature range (725-875°C) while the V-N and V-Nb steels exhibited cracks which were up to 15mm and 12mm long, respectively, in the temperature range 775-825°C. The most severe cracking was associated with the austenite-ferrite transformation, together with detrimental precipitate distributions. It was concluded that transverse cracking would be most likely to occur in the colder regions of the slab such as the slab corner, or in areas with a high level of water impaction.

Introduction

The use of thin slab casting technology is becoming increasingly widespread throughout the world due to the reduced initial investment costs and reduced production costs associated with such plants [1]. However, there have been problems reported with the surface quality of some continuously cast thin slab, and surface quality considerations have led to restrictions on carbon content to avoid the peritectic range, and hence reduce slab longitudinal cracks [2]. There have also been reports of surface transverse cracks for some high strength microalloyed steels [2]. It is well known that some microalloyed steels exhibit a severe loss of ductility when tensile tested in the approximate temperature range 700 to 1000°C, and this loss of ductility has been associated with transverse cracking during conventional continuous casting [3]. The loss of ductility in this temperature range during hot tensile testing has been attributed to two mechanisms: 1/ strain concentration in the softer ferrite following the onset of the phase transformation, or 2/ intergranular fracture promoted by the presence of fine microalloy precipitates which can harden the matrix and retard recrystallisation [4].

There have been many reports of the use of the hot tensile test to assess the likelihood of transverse cracking for different steel types under conventional casting conditions [3-5]. However, this test technique has the disadvantage of not being able to accurately simulate the as-cast grain structures of continuously cast slab, and as transverse cracks usually follow austenite grain boundaries, this is a major shortcoming. Similarly, most hot tensile tests have not melted samples in-situ, and as such cannot accurately simulate certain aspects of high temperature precipitation. There have been some attempts to overcome the drawbacks of the hot tensile test by using alternative techniques. One such technique is to cast a small laboratory ingot and deform it at high temperature, thus preserving the as-cast grain structure and precipitate distribution [6]. The work carried out so far using both hot tensile testing and ingot bending has concentrated on simulating conventional slab casting, and there have been few reports of any laboratory test techniques being used to assess the likelihood of transverse cracking under thin slab casting production conditions. There are a number of important process variables which distinguish thin slab casting from conventional casting, and which could potentially alter the hot ductility behaviour. One of the most significant differences of the thin slab process route compared with conventional casting is the faster cooling rate of thin slab. This faster cooling rate leads to a reduced as-cast austenite grain size [7], and can lead to alterations in the size distributions of some precipitate types [8].

The objectives of this work were to compare the hot cracking behaviour of a number of different microalloyed steels using a laboratory ingot bending test designed to simulate thin slab casting conditions, and hence identify temperature ranges and steel compositions which would be prone to transverse cracking.

Experimental Techniques and Materials

To simulate thin slab cooling rates and solidification structures, a special ingot mould was designed to give cooling rates in the ingots similar to those experienced during thin slab casting of 50mm slabs. The approximate size of the ingots was 50x150x350mm, and the typical cooling rate over the temperature range 1470-1100°C at the slab quarter thickness position was $8°Cs^{-1}$.

A total of five steel types were examined, and the compositions are shown in Table 1. All the steels were based on 0.06% C, 0.50% Si, and 1.45% Mn. Three of the steels contained 0.1% V with variations in N (0.005-0.020%) and Nb (0-0.03%). A fourth steel contained 0.03% Nb with no V, and for comparison purposes, a C-Mn steel with no microalloying additions was also included. It should be noted that the nitrogen level in the V-N steel is at the top end of the likely useable range.

Steel	C	Si	Mn	P	S	Cr	Mo	Cu	Ni	Al	N	Nb	V
C-Mn	0.067	0.52	1.43	0.012	0.005	0.15	0.03	0.07	0.09	0.022	0.005	<0.005	<0.005
V	0.063	0.49	1.45	0.013	0.005	0.08	0.03	0.07	0.09	0.029	0.007	<0.005	0.10
V-N	0.069	0.57	1.47	0.013	0.005	0.09	0.03	0.07	0.09	0.027	0.020	<0.005	0.10
V-Nb	0.065	0.50	1.46	0.013	0.005	0.12	0.03	0.07	0.09	0.031	0.007	0.031	0.10
Nb	0.060	0.55	1.48	0.014	0.005	0.09	0.03	0.07	0.08	0.025	0.006	0.037	<0.005

Table 1 Average compositions of steels investigated, wt. %

Following casting, the ingots were allowed to solidify for approximately two minutes before being broken out of the moulds whilst still hot, typically with surface temperatures in excess of 1100°C. The ingots were insulated to reduce heat loss during transportation to the testing machine and to maintain as constant a temperature as possible during testing.

To simulate the unbending operation, three point bend tests were performed on the hot ingots. Calculations of the surface strain expected during the thin slab casting of 50mm slab with a machine radius of 3000mm using the approach of Lankford [6] indicated a surface strain of 0.8%. Strain rates were also estimated using the approach of Lankford, and for casting speeds in the range 3.5-6m/min, strain rates in the range 8×10^{-4} to 1.9×10^{-2} s^{-1} were estimated depending on the assumptions made for the length over which the maximum strain is developed. A surface strain rate of 9×10^{-4} s^{-1} was chosen as being a reasonable value. The temperature at which unbending takes place is also of critical importance, and it has been reported that the temperature prior to unbending is maintained at above 1000°C [2]. However, there are likely to be large variations in slab temperature around the cross section of the slab, with the slab corners likely to be significantly colder than the middle of the broad face. Three point bend testing was carried out over the temperature range 680 to 1030°C, to encompass the likely temperature range at all points over the slab cross section. Initial testing using a strain rate of 9×10^{-4} s^{-1} and surface strains of 1.8-3.6% revealed no significant surface cracking in a Nb steel, which would have been expected to be prone to cracking. To generate significant cracking, and hence allow the various steel types to be differentiated, it was found necessary to increase the surface strain to 19.2%.

After testing, the deformed ingots were water quenched, then sectioned, and the cut face prepared for metallographic examination. Cracking was examined in the most severely deformed part of the ingot, and the size and number of the surface cracks measured for each ingot.

To examine precipitate distributions, carbon extraction replicas were taken from below the surface of selected samples and examined using a Philips transmission electron microscope operating at 100kV.

To estimate transformation temperatures during testing, samples of the five steel types were heated to 1300°C in a Theta dilatometer to produce a coarse austenite grain size, and then cooled at a rate similar to that experienced during the ingot bending test. Each steel developed an austenite grain size of approximately 300µm following the 1300°C heating cycle, which was similar to that found near the surface of the ingots. The dilatometry exercise determined the transformation temperature during cooling, i.e. the Ar$_3$ temperature. The software package Chemsage was used to calculate the equilibrium transformation temperatures, Ae$_3$.

As-Cast Structure

The general as-cast structure of the slab had relatively fine equi-axed grains near the slab surface, followed by long columnar grains, and a central region of coarse equi-axed grains. The surface equi-axed zone was confined to a relatively narrow layer adjacent to the slab surface, typically one or two grains in depth, and this depth was similar for all the steel types. Mean austenite grain sizes near the slab surface were in the range 240-340µm for all the steel types. The columnar region consisted of elongated austenite grains which increased in length on moving away from the slab surface, and were typically 5-10 millimetres in length near the slab mid-thickness position. The length of these columnar grains was similar for the different steel types. The size and shape of the as-cast grains was very similar to those expected for commercially produced 50mm slab.

Transformation Temperatures

The transformation temperatures measured during dilatometry (Ar_3), and the equilibrium transformation temperature calculated using Chemsage (Ae_3) are shown in Table 2.

Steel	Measured Ar_3 Temperature, °C	Calculated Ae3, °C
C-Mn	755	886
V	770	895
V-N	745	899
V-Nb	730	896
Nb	750	896

Table 2 Transformation temperatures

As would be expected, the Ar_3 temperatures are significantly lower than the Ae_3 temperatures. For the cooling rate used, all the steels had similar Ar_3 temperatures in the range 730 to 770°C.

Observation Of Cracks

Examination of the deformed ingots revealed the presence of numerous surface cracks for all the steel types across a wide temperature range. For the higher test temperatures, when deformation was occurring in the single phase austenite region, the cracks almost always occurred at austenite grain boundaries. An example of this type of crack is shown in Fig. 1a. For the lower test temperatures, when transformation had commenced, the cracks formed in the ferrite located at the prior austenite grain boundaries. An example of this type of cracking is shown in Fig. 1b. It was evident that the grain boundary crack within the ferrite was formed by the nucleation, growth and coalescence of separate micro voids at inclusions within the grain boundary ferrite.

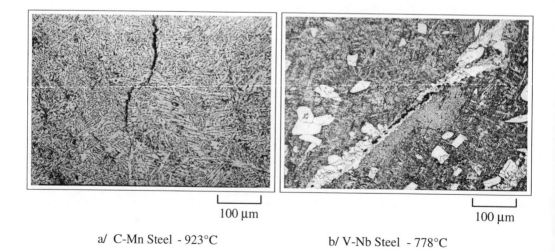

a/ C-Mn Steel - 923°C   b/ V-Nb Steel - 778°C

Figure 1 - Grain boundary cracks

Some of the steel types exhibited severe cracking over certain temperature ranges, whilst other steels exhibited little cracking across the entire range of temperatures. Examples of severe and minimal cracking are shown in Figure 2.

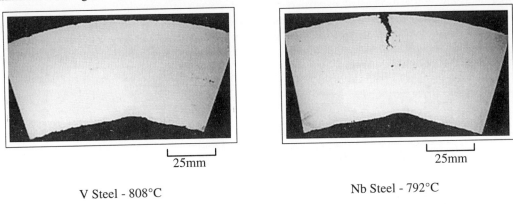

25mm

V Steel - 808°C

25mm

Nb Steel - 792°C

Figure 2 Comparison of cracking in V steel and Nb steel

Fig. 3 shows the variation in length of the longest crack with change in test temperature for all the steel types. The maximum length of crack in the C-Mn and C-Mn-0.1%V steels was similar, did not vary with temperature, and was generally less than 2mm. It is thought that such cracks are unlikely to result in defects in the final product. Cracks which were longer than this were observed in the other three steels with the Nb steel exhibiting both the longest cracks (up to 17mm at 820°C) and the widest range of temperature over which cracking occurred (725-875°C). The V-N and V-Nb steels also exhibited relatively long cracks, 15mm and 12mm respectively, but the range of temperature over which cracking occurred (775-825°C) was narrower than that exhibited by the Nb steel. It should be noted that all the steel types examined had similar numbers of cracks across the entire range of test temperatures, typically 1-2 per mm.

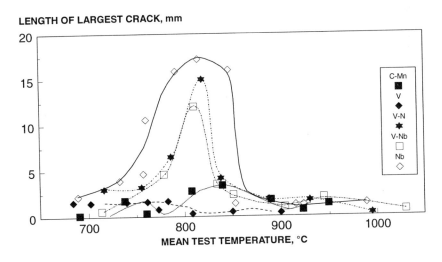

Figure 3 Variation of length of largest crack with test temperature and steel type

Electron Metallography

Examination of carbon extraction replicas from the V-N, V-Nb and Nb steels tested at approximately 800°C revealed extensive precipitation both at austenite grain boundaries and within the grains. A precipitate free zone was sometimes observed adjacent to the grain boundary precipitates. Precipitates in the Nb steel were generally finer than those observed in the V-N and the V-Nb steels , and this is illustrated in Fig. 4. Little precipitation was observed in the V steel, and no precipitates were observed in the C-Mn steel.

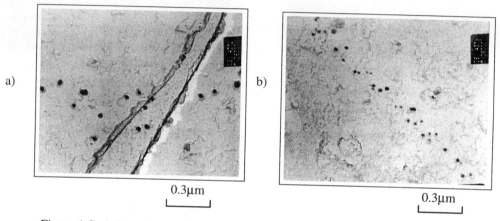

a)

b)

0.3μm

0.3μm

Figure 4 Grain boundary precipitates in a) V-Nb (811°C) and b) Nb steels (792°C)

Discussion of Results

All the steels examined contained cracks at austenite grain boundaries, or within the ferrite formed along austenite grains, across the entire range of test temperatures. This is consistent with results from hot tensile tests, in which grain boundary fracture is commonly observed within certain temperature ranges, and also with observations of transverse cracks in continuously cast slabs, which are also associated with austenite grain boundaries [3]. Comparing the numbers of cracks observed with the austenite grain size at the slab surface suggests that for the test conditions employed, approximately 60% of austenite grain boundaries exhibited cracking of some level of severity, for all the steel types at all test temperatures. However, although grain boundary cracks were common for all steel types across the entire temperature range, only for the Nb, V-N, and the V-Nb steel did these cracks propagate to any great extent, and even then only in a relatively narrow temperature range.

Two mechanisms have been proposed to account for the formation of these grain boundary cracks over the temperature range of interest. In the single phase austenite region, strain is concentrated along grain boundaries due to grain boundary sliding, and grain boundary failure can occur at triple points, or at second phase particles situated at the grain boundary. This mode of failure is favoured by the presence of fine microalloy precipitates, which retard recrystallisation and may further concentrate strain along grain boundaries [4]. The second mode of low ductility failure occurs at lower temperatures, when the austenite-ferrite transformation has commenced. During transformation, ferrite is formed along the austenite grain boundaries, and during testing, strain is concentrated in this softer ferrite phase, and fracture occurs by the nucleation, growth and coalescence of micro-voids at second phase particles within the ferrite. Examination of fracture surfaces from hot tensile tests has indicated that MnS type inclusions are commonly found on intergranular fracture surfaces [10], and it is these inclusions that are probably responsible for nucleation of the majority of the micro-voids within the ferrite films at austenite grain boundaries. Austenite grain size also has an important influence on the levels of ductility in the austenite-ferrite region [4].

The observation of cracking in grain boundary ferrite (Fig. 1b) in the Nb, V-N, and V-Nb steels suggests that it is this second mode of failure which is responsible for the development of the longest cracks. The measured Ar_3 temperatures are somewhat lower than the temperatures at which the most severe cracking was observed, but there have been reports that the deformation experienced during testing can substantially raise the transformation temperature until it approaches the Ae_3 temperature [9]. The calculated Ae_3 temperatures were all slightly higher than the temperature at which the most severe cracking was observed. The observation that the most severe cracking occurs in the Nb, V-N, and V-Nb steels, which would be expected to have the largest volume fractions of micro-alloy precipitates, suggests that it is these microalloy precipitates which are influencing ductility, as the austenite grain size, and S levels (and hence MnS volume fractions) was similar for all the steels. One possible mechanism by which microalloy precipitates could reduce ductility in the austenite-ferrite structure is by hardening the austenite matrix, and hence further concentrating strain in the grain boundary regions. In support of this idea, maximum loads during bend testing were slightly higher for the Nb, V-N and V-Nb steels than for the other steels. As well as hardening the austenite matrix, the microalloy carbides may retard softening processes in the ferrite itself, and hence reduce the ductility of the ferrite for a given austenite grain size and MnS volume fraction.

The differences between the Nb steel, which showed the worst cracking, and the V-N and V-Nb steels which showed less cracking, may be due to differences in the precipitate distributions. Indeed, electron metallographic examination showed that precipitates were generally finer in the Nb steel than the other two steels, and could have a greater effect in retarding softening processes in the ferrite. Conventional hot tensile testing has also shown that V-Nb steels and V-N steels have better levels of hot ductility than Nb steels [11].

An extensive body of data exists derived from hot tensile tests showing the existence of a ductility trough associated with the austenite-ferrite transformation even in C-Mn steels, and also in the single phase austenite phase field, which is particularly prominent for microalloyed steels. The present work indicated only relatively minor cracking which was similar for all steel types in the single phase austenite region. Recent work using hot tensile tests has suggested that the levels of ductility in the single phase region are slightly higher than in the austenite-ferrite region [11]. It is possible that had higher strain levels been applied in the current tests, severe cracking would also have been observed in the austenite region for the microalloyed steels. Similarly, higher levels of strain may have promoted severe cracking even in the C-Mn steel in the two phase region.

Commercial Implications

Due to the larger amounts of strain which had to be incorporated in the laboratory test, an exact correpondence between the laboratory test and defects in continuously cast slab cannot be made, but the trends evident in the laboratory tests are likely to be relevant to continuous caster operation. The C-Mn steels, which had mean crack lengths of approximately 2mm, are not usually associated with transverse cracking, and this would suggest that cracks of depth 2mm or less will have little likelihood of developing into a significant defect in the final product. Cracks with lengths significantly in excess of 2mm were only observed in the Nb, V-N and V-Nb steels, suggesting that these steels are the more likely to exhibit transverse cracking under thin slab casting conditions. Of these three steels, the Nb steel had the most severe cracking, and is most likely to suffer from transverse cracking. The temperature at which slabs are straightened prior to entry into the equalisation furnace is reported as typically being in excess of 1000°C [2]. Severe cracking was not observed for any of the steels at this temperature. However, it is likely that at certain parts of the slab, for example slab corners or areas of severe water impaction, the temperatures will be considerably lower, and the results would suggest that it is these colder regions which will be most prone to transverse cracking.

Conclusions

1. The C-Mn and C-Mn-0.1%V steels showed only small surface cracks, typically less than 2mm in depth, for all test temperatures examined.
2. The C-Mn-0.03%Nb steel exhibited both the largest cracks (17mm at 820°C) and the widest range of temperatures over which cracking occurred (725-875°C). The C-Mn-0/1%V-0/02%N and the C-Mn-0.1%V-0.03%Nb steels also exhibited relatively long cracks (15mm and 12mm, respectively), but these developed over a narrow range of temperature (775-825°C).
3. The cracking in the Nb, V-N, and V-Nb steels was probably related to the onset of the austenite to ferrite transformation, and the presence of microalloy precipitates.
4. Transverse cracking is most likely to occur in the colder parts of the slab such as the slab corners, or in areas where water impaction has been severe.

References

[1] G. Flemming et. al., Metall. Plant Technol. Int., 16, (1993), p. 84.
[2]. P. J. Lubensky, S. L. Wigman and D. J. Johnson, Microalloying 95, Pittsburgh, (1995), p. 225.
[3]. B. Mintz and J. M. Arrowsmith, Metals Technology, 6, (1979), p. 24.
[4]. B. Mintz, S. Yue and J. J. Jonas, Int. Mat. Rev., 36, (1991), p. 187.
[5] G. Suzuki et. al., Phys. Sim. of Welding, Hot Forming and Cont. Casting, Ottowa, (1988).
[6]. W. T. Lankford, Metall. Trans., 3, (1972), p. 1331.
[7] R. Kaspar and O. Pawelski, METEC Congress 94, Dusseldorf, (1994).
[8] A. M. Sage et. al., Processing, Microstructure and Properties of Microalloyed and Other Modern High Strength Low Alloy Steels, Pittsburgh, (1991), p. 443.
[9] B. Mintz, Mat. Sci. and Technol., 12, (1996), p. 132.
[10]. N. Hannerz, Trans. ISIJ, 99, (1985), p. 149.
[11]. B. Mintz and R. Abushosa, Ironmaking and Steelmaking, 20, (1993), p. 445.

Acknowledgements

The authors would like to thank Vanitec for funding the work, and Dr K. N. Melton, Research Director, Swinden Technology Centre, British Steel plc for permission to publish.

Correspondence

Please send any correspondence to D. N. Crowther, Fax. 0044 (0)1709 825419 or e-mail dave_crowther@technology.britishsteel.co.uk.

Materials Science Forum Vols. 284-286 (1998) pp. 477-484
© *1998 Trans Tech Publications, Switzerland*

Hot Ductility of Certain Microalloyed Steels under Simulated Continuous Casting Conditions

L.P. Karjalainen[1], H. Kinnunen[2] and D. Porter[2]

[1] University of Oulu, Department of Mechanical Engineering,
P.O. Box 444, FIN-90571 Oulu, Finland

[2] Rautaruukki Oy, Research Centre, P.O. Box 93, FIN-92101 Raahe, Finland

Keywords: Continuous Casting, Hot Ductility, Transverse Cracking, Tensile Testing, *in-situ* Melting, Microalloying, Sulphur Content, Calcium Treatment

Abstract

The hot ductility of steels containing 0.014-0.017%Ti, 0.34-0.72%Ni and various combinations of Nb, V and Cu has been investigated by tensile testing over the temperature range of 600-1050°C. The in-situ melting procedure was employed, followed by subsequent cooling at three different rates to the selected testing temperature. Two strain rates, 0.005 and 0.05 s^{-1} were used. The results show that a ductility trough exists in all the steels in the temperature range of 700-950°C, and the lowest ductility in terms of reduction of area (RA) is at 750°C. Markedly lower RA and elongation to fracture values were obtained at the lower strain rate. The ductility trough was quite similar in Nb bearing and V or Cu containing steels, especially at the high cooling rate of 4°C/s and with a Ti/N ratio of about 2.8. Reducing the cooling rate from 4 to 1 °C/s makes the ductility trough narrower and shallower in the Nb-bearing steels, but has little effect in the Nb-free steels. Increasing Ca/S ratios in the range 0 to 1.25 improve the ductility minimum especially at the lower cooling rate of 1°C/s. Low N and high Ti/N ratio may also improve the minimum ductility at lower cooling rates.

1. Introduction

The simple hot tensile test has been found to be very useful in tackling the problem of transverse and edge cracking of slabs encountered under certain circumstances in continuous casting [1-4]. The method has been employed in numerous investigations, and the state of the art was recently reviewed by Mintz et al.[3,4]. A comprehensive overview of the up-to-date information on the continuous casting of steel was given by McPherson and McLean [5]. Simple reheating of the specimen to temperatures around 1330°C is used in most tests and this procedure has been found to be successful in C-Mn(-Al) and Nb bearing steels to reveal the hot ductility behaviour [1,2,5], but in steels with Ti microalloying or to reveal the effects of S or C contents, the in-situ melting technique must be used to dissolve Ti and S and to provide a segregated, coarse-grained microstructure [2,4,6-8].

A ductility trough at temperatures 750-900°C is frequently observed. In C-Mn(-Al) steels, this is connected with the formation of thin films of deformation-induced ferrite on the austenite grain boundaries where strain is concentrated resulting in intergranular fracture at a low overall strain. Nb microalloying has been found to be detrimental to ductility, decreasing the minimum ductility and especially widening the ductility trough to higher temperatures [1,2,9]. Precipitation of fine (2-10 nm) NbCN particles in the matrix and subgrain boundaries and coarse eutectics as well as the precipitate free zones along grain boundaries result in an intensive grain boundary sliding and consequently in a low ductility until dynamic recrystallization is able to occur at higher temperatures [3,6]. The width of the ductility trough depends on the strain rate applied at the test temperature. Mintz and Arrowsmith [1] showed using the reheated specimens that the ductility trough is still wider at the strain rate of 0.0003 s^{-1}, a typical rate at the straightening of a strand, than at 0.003 s^{-1}. However, the minimum RA value in the trough was not influnced by the strain rate in this range.

The effect of Ti is quite complicated, depending on factors such as Ti/N ratio and cooling rate [4,9]. A beneficial influence of Ti has been observed at low Ti/N ratios ($\leq$ 2) and at low cooling rates ($\leq$ 1°C/s) in Nb-Ti steels, but a highly detrimental effect is found at the stoichiometric ratio [4]. The influence of V is weaker than that of Nb and, in practice, problems have not been encountered in steels with V<0.1% [4]. Cu may have a detrimental effect by forming fine CuS particles when testing under an oxidizing atmosphere, while Ni has been reported to be slightly beneficial [4].

Sulphur reduces ductility, but its effects can only be detected in as-cast specimens [6,10]. Both mill experience and simulation tests show that Ca treatment is beneficial, as a result of a reduction in the total amount of S in the steel thereby reducing the amount of S in solution or precipitated in a finer form during cooling after solidification [2,10,11].

Mill experience during the development of new steel grades with improved weldability has shown that low carbon steels containing Ni are more prone to surface defects caused by slab surface cracking than other steel grades. In order to try to elucidate the compositional and processing factors controlling the cracking a number of continuously cast Ni containing steels were chosen for a preliminary investigation of hot tensile ductility using the in-situ melting technique. The steels were Ti microalloyed low carbon, Ni alloyed grades containing Nb, V and Cu in various combinations.

2. Experimental

The chemical compositions of the steels used in the investigation are given in Table 1. Four of the steels were Ti microalloyed 460 MPa grade steels with a low carbon content of 0.07-0.10% and about 1.6%Mn. The fifth steel, a 13MnNi63 type, designated here as Nb-V, was a Ni alloyed HSLA pressure vessel steel, selected for comparison. All the steels contained 0.014-0.017%Ti, 0.34-0.72%Ni and 0.03-0.06%Al and various combinations of Nb, V and Cu. The Ti/N ratio is also given in Table 1. The S levels were below 100 ppm. Apart from the V grade, the steels also contained some Ca, even though the Ca/S ratio was relatively low, as indicated in Table 1.

Table 1. Chemical compositions (mass-%), Ti/N and Ca/S ratios of the steels tested.

Steel	C	Si	Mn	P	S	Cu	Ni	Ti	Nb	V	Al	Ca	N	Ti/N	Ca/S
Nb	0.079	0.23	1.50	0.015	0.009	0.009	0.36	0.017	0.034	0.007	0.031	0.003	0.006	2.8	0.33
Cu	0.096	0.27	1.60	0.017	0.006	0.295	0.34	0.014	0.001	0.021	0.036	0.004	0.005	2.8	0.67
V	0.100	0.23	1.58	0.015	0.006	0.008	0.35	0.016	0.001	0.091	0.026	0.000	0.006	2.7	0.00
Ni-Nb	0.067	0.19	1.51	0.009	0.004	0.232	0.72	0.017	0.022	0.007	0.061	0.002	0.003	5.7	0.50
Nb-V	0.132	0.35	1.53	0.012	0.004	0.016	0.49	0.014	0.039	0.064	0.040	0.005	0.005	2.8	1.25

For hot tensile testing round rods of 9.98 mm diameter and 115 mm in length were machined from mill rolled plates about 2 mm below the surface, parallel to the rolling direction. A Gleeble 1500 thermomechanical simulator was employed for determining the reduction of area and elongation to fracture. The free span of the specimens was 35 mm. The specimens were heated at 25°C/s in argon shielding gas to the melting temperature, while a length of 10-15 mm in the middle part of the specimen was supported by a slitted silica tube 30 mm in length. Melting was detected visually and the final temperature adjusted manually. The specimen was held for 5 min at this maximum temperature and subsequently cooled at cooling rates of 4, 1 or 0.4°C/s to the deformation temperature, held for 15 s and then tensile tested at a strain rate of 0.005 or 0.05 s^{-1}.

3. Results and discussion

3.1. Ductility trough

The reduction of area (RA) values in the temperature range of 600°C to 1050°C are shown in Fig. 1(a-e). All specimens were tested at the cooling rates of 4 and 1°C/s. In addition, the Nb steel was also tested at 750°C at the cooling rate of 0.4°C/s, and RA is given Fig. 1a. The strain rate in these tests was 0.005 s^{-1}. The RA values at the higher strain rate of 0.05 s^{-1} are plotted in Figs. 1c and 1d. The behaviour of all the steels is qualitatively similar, revealing a ductility trough in the temperature regime of around 700-900°C. The total elongation behaved in a similar manner. The minimum RA values are close to 20% in most cases. At the higher strain rate of 0.05 s^{-1}, RA values are distinctly higher and the trough is very shallow.

The A_{e3} and A_{r3} temperatures of the steels (the start of austenite decomposition under equilibrium and continuous cooling conditions, respectively) were calculated using the equations taken from the literature [12,13]. These temperatures and the equations are given in Table 2 and the temperatures are also marked in Fig. 1. The equation for A_{r3} is originally given for a heavily deformed austenite under continuous cooling, which may have a higher transformation temperature than that of the melted and cooled undeformed specimen in an isothermal test. However, straining in tensile testing also increases the A_{r3} temperature. The difference between A_{e3} and A_{r3} is almost constant, 108-131°C, for the present steels. For C-Mn-(Al) steels this range determines the width of the ductility trough [4,14], but in the present microalloyed steels the trough is somewhat wider extending to higher temperatures. In all cases the minimum RA occurs at the testing temperature of 750°C. This is well below the A_{e3} temperature for the steels concerned (about 850°C). Thus, it may be concluded that the ductility minimum is associated with the existence of ferrite films on austenite grain boundaries, although quite evidently precipitation also has its detrimental influence.

Table 2. A_{e3} and A_{r3} temperatures (°C).

Steel	Ae3	Ar3
Nb	853	737
Cu	848	732
V	845	737
Ni-Nb	855	724
Nb-V	842	719

$A_{e3} = (911-29Mn+70Si-10Cr)-(418-32Mn+86Si+Cr)C+232C^2$; Ref. [11]
$A_{r3} = 911-310C-80Mn-20Cu-15Cr-55Ni-80Mo$; Ref.[12]

In Nb bearing grades (Figs. 1a,1d and 1e) reducing cooling rate improves ductility. This is also shown in Fig. 2, where the effect of cooling rate on the minimum RA and the total elongation to fracture values are shown for the Nb, Ni-Nb and Nb-V steels. Data from Abushosha et al. [9] (for a Nb-Ti steel) are plotted for reference. In the Cu and V steels the cooling rate has no influence. Both RA and the total elongation to fracture reveal the same trend. Abushosha et al. [9] observed that the decreasing cooling rate had a beneficial effect in Nb-Ti steel at cooling rates below 1.7°C/s, but not in plain Nb steel. This was suggested as being a result of the precipitation of NbCN on the surface of TiN particles, thus reducing the amount of soluble Nb which could precipitate dynamically in a form of fine NbCN particles. This explanation is in accordance with the present results, as all steels are microalloyed with Ti.

3.2. Role of chemical composition

Several alloying elements varied in the present series of experiments and the number of tested steels was small so that the role of a single element is difficult to establish. One can, however, notice that the Ti microalloyed steels without Nb, i.e., the V and Cu steels, did not possess any higher minimum RA values (Figs. 1b and 1c) than the Nb bearing steel. In these steels, the Ti/N ratio is

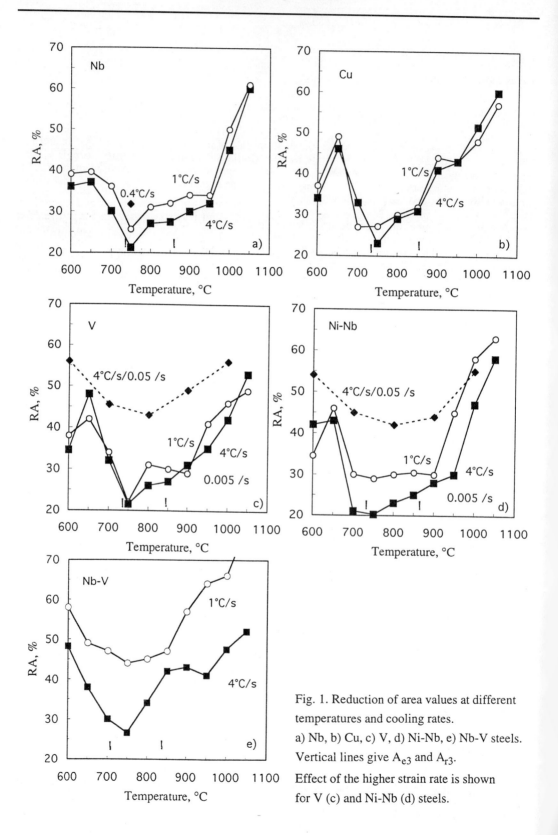

Fig. 1. Reduction of area values at different temperatures and cooling rates.
a) Nb, b) Cu, c) V, d) Ni-Nb, e) Nb-V steels. Vertical lines give A_{e3} and A_{r3}.
Effect of the higher strain rate is shown for V (c) and Ni-Nb (d) steels.

2.8, which is rather near the stoichiometric ratio, which may mean fine TiN precipitation, which is detrimental to ductility [4,9]. The cooling rate did not affect ductility in these grades, which suggests that it does not affect the size of fine Ti precipitates and that V precipitation has no significant effect on ductility.

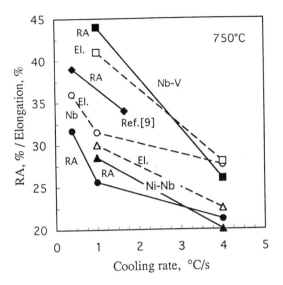

Fig. 2. Effect of cooling rate on minimum RA
and total elongation to fracture values.

Residual levels of Cu have been shown to increase the tendency to transverse cracking by forming fine Cu sulphides or oxysulphides under oxidising conditions, but not in a protective atmosphere [15]. Hence, such an effect cannot be expected in the present tests using a protective argon atmosphere (Fig. 1b). Moreover, the Cu steel contains Ni (Ni/Cu ≈ 1.15), and it has been shown that Ni restores ductility in Cu bearing steels by increasing the solubility of Cu in austenite [15]. Nevertheless, the minimum RA is rather low, ≈22%, and the reason for this may be connected to the near stoichiometric amount of Ti (Ti/N=2.8).

Ni in steels without Cu is found to improve ductility in simulation tests in an oxidising atmosphere, though only to a small extent (≈1-2%) [15]. Contradictory to this, according to mill experience Ni alloying seems to promote cracking [16]. In the Ni-Nb steel, the Ni content is quite high (0.72%), but on the other hand, the high Ti/N ratio (5.7) is obtained due to low nitrogen content (30 ppm compared with 50-60 ppm in the other steels). This is known to encourage precipitate coarsening and to improve ductility [4]. Low nitrogen, in turn, should reduce the precipitation temperature for TiN and lead to a smaller volume fraction of finer precipitates. These various factors may have either parallel or opposing effects. Anyhow, the minimum RA, 29%, at the low cooling rate of 1°C/s is relatively good. The role of Ni, Ti, Nb and N should be investigated more systematically.

At the low cooling rate of 1°C/s, the Nb-V steel had clearly the best ductility among the steels tested, the minimum RA value being as high as 44%. The C content is highest (0.13%) of the steels and the Ni content is also rather high. Generally, the width of ductility trough has been found to increase with increasing C content and the cracking tendency to be highest at 0.1-0.15%C. These are connected with the peritectic reaction leading to a coarse columnar austenite grain size and, perhaps, with an intense P segregation [5]. This structure cannot be simulated in tensile test specimens. However, the higher C content is hardly the reason for better ductility in the present tests.

The S content varied in the steels tested between 0.004-0.009% and the Ca/S ratio from 0 to 1.25 (see Table 1). The minimum RA values are plotted as a function of S content and Ca/S ratio in Fig. 3a and 3b, respectively. Similar dependences of RA values at 900°C are also found though not shown. The correlation between the minimum RA and S content is poor, (Fig. 3a), but Fig. 3b reveals a distinct relationship, particularly at the lower cooling rate, with ductility increasing with an increasing Ca/S ratio. The Ca/S ratio may be a factor explaining the better ductility of the Nb-V steel in the present case. Coleman and Wilcox [11] reported experimental observations that the addition of Ca or Ce produced crack-free slabs of C-Mn-Nb-Al steels. Mill data from Rautaruukki also indicate that the cracking tendency becomes lower with increasing Ca/S ratio between 0.5-1.6. It is recommended that the Ca/S ratio should be 2 to obtain fully modified calcium sulphide-calcium aluminates [10]. It was simply concluded by Coleman and Wilcox [11] that Ca is beneficial by reducing the total amount of S. Mintz et al. [10] explained the observed improvement of ductility of C-Mn-Al and C-Mn-Nb steels obtained by Ca treatment, when tested in the as-cast state, to be a result of S precipitation and removal by flotation in the liquid state. The present results, however, indicate that there is also an effect due to the presence of Ca per se. Fig. 3b shows that Ca/S ratio is able to explain the relative magnitudes of the minimum RA values very well, especially considering the range of compositional variables covered by these Ti microalloyed steels. Microstructural reasons for this correlation have not been studied in this work and at present it is only possible to speculate about possible explanations in the light of other observations reported in the literature.

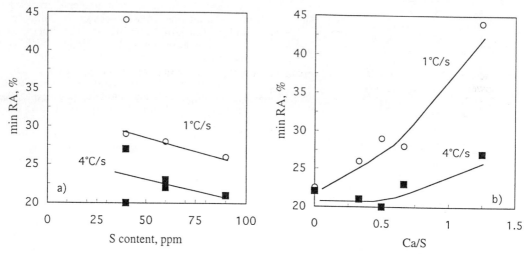

Fig. 3. Effect of S content (a) and Ca/S ratio (b) on minimum RA values at two cooling rates.

Firstly, regarding the poorer correlation between the ductility minimum and total S content shown in Fig. 3a, it should be remembered that the total S content mainly reflects the quantity of coarse sulphide inclusions in the steel. These inclusions are either MnS, (CaMn)S or CaS depending on the Ca content of the steel together with the Al and O contents. Of greater importance for the ductility trough are fine solid state sulphide precipitates or sulphur segregation to boundaries and interfaces [6,17]. It has been shown by Abushosha et al.[6] that Fe oxysulphides are able to precipitate and strengthen the matrix in a similar manner to Nb(CN). Maehara and Nagamichi [17] have proposed that unprecipitated S segregates to grain boundaries and Nb(CN)/matrix interfaces thereby facilitating decohesion and microvoid formation at grain boundary precipitates during straining. While this mechanism was proposed to explain loss of ductility in the austenite region, it may also operate after ferrite films have formed, i.e. at 750 °C, where the minium RA values were observed in the present case. The formation of fine precipitates or interface segregation is dependent on there being sufficient S in solution in the austenite immediately after solidification which, although dependent on the total S content, is quite different to it. It will also vary with position in the dendritic structure.

During the solidification of calcium treated, aluminium killed steel Ca is usually present as aluminates $(CaO)_x(Al_2O_3)_y$, and some Ca in solution both of which are then able to react with S to form CaS as solidification proceeds [18]. Finally any remaining, uncombined interdentritic S forms MnS. In the absence of detailed information on the aluminate and sulphide inclusion compositions for the present steels, the extent to which Ca is able to fully react with S, thereby preventing the formation of MnS, can only be estimated from their Ca/S ratios. The higher this ratio the lower the content of MnS inclusions and for Ca/S > 1.5 - 2.0 no MnS is generally encountered [19]. Similarly, due to the greater stability of CaS compared to MnS, increasing Ca/S ratio will also reduce the amount of S remaining in solution after solidification, thereby lowering the amount of S available to precipitate or segregate at lower temperatures.

The reason why the influence of Ca/S ratio is clearer at the lower cooling rate (Fig. 3b) might be explained by S segregation to TiN precipitate surfaces reducing their ability to act as nucleation sites for Nb(CN). To test the above hypotheses microstructural studies and thermodynamic calculations are of course required.

4. Conclusions

The hot tensile tests carried out for Ti microalloyed steels on a Gleeble thermomechanical simulator using the in-situ melting technique indicate that:

1. A ductility minimum in terms of the reduction of area (RA) or total elongation to fracture values is present at 750°C at the low deformation rate of 0.005 s^{-1} in all the steels investigated.
2. The minimum RA values are not higher in V or Cu bearing steels without Nb than in Nb bearing steels, with Ti/N ≈ 2.8.
3. Reducing the cooling rate from 4 to 1 °C/s makes the ductility trough narrower and shallower in the Nb-bearing steels, but has little effect in the Nb-free steels.
4. Increasing Ca/S ratios (0...1.25) improve the ductility minimum at 750 °C especially at the lower cooling rate of 1 °C/s.
5. Low N (30 ppm) and high Ti/N ratio (5.7) may improve the minimum RA at the low cooling rate of 1 °C/s.

5. Acknowledgements

The financial support from the Technology Development Centre of Finland (TEKES) is acknowledged with gratitude. The steels were provided by Rautaruukki Raahe Steel. Rautaruukki Oy is also thanked for the permission to publish these results.

6. References

[1] B. Mintz and J.M. Arrowsmith, Met. Technol. 6 (1979), p. 24.
[2] B. Mintz and R. Abushosha, Mater. Sci. Technol. 8 (1992), p. 171.
[3] B. Mintz, S. Yue and J.J. Jonas, Intern. Materials Reviews 36 (1991), p. 187.
[4] B. Mintz, R. Abushosha, O.G. Cominelli and S. Ayyad, The hot tensile test as a means of assessing the susceptibility of steel to cracking during continuous casting, Proc. 7th Intern. Symp. on Physical Simulation of Casting, Hot Rolling and Welding, 21-23 January, 1997, Tsukuba, Japan, p. 449.
[5] N.A. McPherson and A. McLean, Continuous Casting, vol. 8, Transverse cracking in continuously cast products, The Iron and Steel Soc., Warrendale, PA, USA, 1997, p. 1.
[6] R. Abushosha, R. Vipond and B. Mintz, Mater. Sci. Technol. 7 (991), p. 1101.
[7] B. Mintz and Z. Mohamed, Hot ductility of directly cast microalloyed steels, Intern. Symp. on Physical Simulation of Welding, Hot Forming and Continuous Casting, Ottawa, Ca, 1988, p. II39.
[8] H.G. Suzuki, S. Nishimura and S. Yamaguchi, Physical simulation of the continuous casting of steels, Intern. Symp. on Physical Simulation of Welding, Hot Forming and Continuous Casting, Ottawa, Ca, 1988, p. II1.
[9] R. Abushosha, R. Vipond and B. Mintz, Mater. Sci. Technol. 7 (1991), p. 613.
[10] B. Mintz, Z. Mohamed and R. Abushosha, Mater. Sci. Technol. 5 (1989), p. 682.
[11] T.H. Coleman and J.R. Wilcox, Mater. Sci. Technol. 1 (1985), p. 80.
[12] B. Donnay, F. Manassis, P. Fabregue and U. Lotter, Computer assisted modelling of metallurgical aspects of hot deformation and transformation of steels. ECSC project 7210-EC/250, p. 22.
[13] C. Ouchi, T. Sampei and I. Kozasu, Trans. ISIJ 22 (1982), p. 218
[14] B. Mintz, Mater. Sci. Technol. 12 (1996), p. 132.
[15] B. Mintz, R. Abushosha, and D.N. Crowther, Mater. Sci. Technol. 11 (1995), p. 474.
[16] N. Torvela, Rautaruukki Oy, personal communication.
[17] Y. Maehara and T. Nagamichi, Mater. Sci. Technol. 7 (1991), p. 915.
[18] S. Kitamura, K. Miyamura and I. Fukuoka, Trans. ISIJ 27 (1987), p. 344.
[19] T.T. Lee and M. Ichikawa, Optimization of calcium treatment at Dofasco, Proc. 2nd Canada-Japan Symp. on Modern Steelmaking and Casting Techniques, 20-25 August, 1994 (eds. J.J. Jonas, J.D. Boyd and N. Sano), 33rd Annual Conf of Metallurgist of CIM, p. 29.

Materials Science Forum Vols. 284-286 (1998) pp. 485-492
© *1998 Trans Tech Publications, Switzerland*

The Effects of Ti and S in Direct Rolled Thin Slab Cast Low Carbon Steel

L.D. Frawley[1] and R. Priestner[2]

[1] BHP Research Melbourne Laboratories, Materials Science Centre,
Clayton, Victoria, Australia/Manchester

[2] University of Manchester /UMIST Materials Science Centre, Manchester M1 7HS, UK

Keywords: Thin Slab Casting, Direct Rolling, Low Carbon Steel, MnS, Surface Defects, Edge Cracking

Abstract

The effects of sulphur on the metallurgy of mild steel produced via the Thin Slab Direct Rolling (TSDR) and Cold Charge Rolling (CCR) routes were compared. Particular attention was given to the precipitation behaviour of MnS and its influence on the microstructure, mechanical properties and susceptibility to cracking of the hot band. The sulphur content was varied from 0.003wt% to 0.04wt% and three levels of Mn (0.10, 0.24 and 0.64 wt%) were used. A number of casts with Ti additions of 0.024 and 0.036wt% were also made. CCR produced an homogenous, polygonal grain structure while TSDR product displayed banding of the microstructure into coarse and fine ferrite grains. The banding was less severe at low S contents. The average grain size of the TSDR product became finer than that of the CCR product as the S content was increased. Yield strength was greater in the TSDR material and increased to a maximum at a sulphur content which was inversely related to the level of Mn. Mechanisms responsible for the observed change in mechanical properties with S content include ferrite grain refinement, due to both the refinement of the austenite grain size through precipitation of MnS on the austenite grain boundaries, and strain induced precipitation of MnS during rolling, which retarded recrystallisation and increased retained strain prior to transformation. Surface defects in the form of edge cracking were observed in all TSDR strip with the general exception of low S and high Mn steels. The cracking was most severe in the low Mn TSDR steels where it occurred at very low S levels. It was found that an isothermal hold at 1150°C for 20 minutes prior to rolling eliminated cracking. The severity of the cracking appeared to be dependent on the amount of S in solution prior to rolling. Small additions of Ti significantly reduced the degree of edge cracking, due to the formation of a Ti-based sulphide and heterogeneous nucleation of MnS on Ti-based oxide particles.

Introduction

The current world trend in steel production is towards optimising processing and minimising costs. The capital cost per tonne of hot rolled strip can be reduced by 50%, compared with a fully integrated BF-BOS steelworks, by directly rolling the product of a thin-slab caster supplied by electric arc steelmaking[1]. Considerable work has been published on thin-slab direct rolling (TSDR) process and plant[2-4], but less is known about the new microstructural phenomena that result, during processing, from differences in temperature and deformation histories compared with conventional, thick-slab cold-charge rolling (CCR). In the conventional CCR route the, usually thick, cast slabs are allowed to cool to ambient temperature and are then reheated to above the A_{c3} for austenite conditioning prior to rolling. Although TSDR plants differ significantly, their common feature is that thinner slabs, nearer to net shape and requiring less hot rolling reduction, are rolled directly after casting without being allowed to cool below the A_{r3}. The most typical procedure[2] involves normalising the temperature of the slab to 1100°C before final entry to the hot mill for rolling to a hot band gauge and coiling at 650°C.

The constitution of just-cast austenite prior to direct rolling is different from that of reheated austenite prior to CCR, in that it is coarser grained, retains a dendritic substructure and segregation pattern inherited from the casting process, and may be more highly supersaturated with respect to precipitating species[5-6]. It does, however, have a finer cast structure than thick slab castings[5]. These differences can affect subsequent microstructure development during processing. Recent reports indicate that increasing sulphur content increases the yield strength[7] and hardness[8] of mild steel produced by TSDR, albeit with a reduction in ductility. Surface defects, such as edge cracks, are more severe in hot strip from the direct rolling process[9]. Such defects may cause steel to be scrapped, resulting in poor production performance. It has been well established that the hot ductility of mild steel is adversely affected by the presence of S and improved by the addition of Mn. The improvement has been closely attributed to the precipitation of MnS

The purpose of the present research is to investigate these effects of sulphur by TSDR simulated in the laboratory.

Experimental

<u>Melting and Casting</u>. Laboratory simulation of TSDR requires a freshly made ingot for each experiment. The facility developed at the Manchester Materials Science Centre has been described in more detail elsewhere[6] and is shown schematically in Fig.1. Approximately 1kg of low C, low

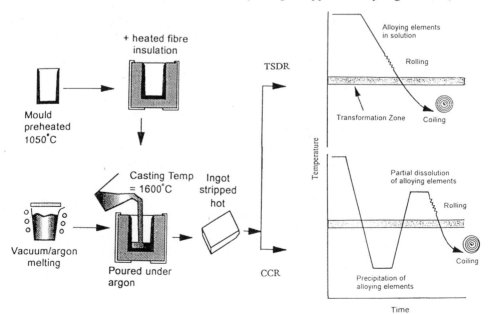

Figure 1. Schematic illustration of laboratory simulation of CCR and TSDR processing

Mn, low S melting stock that had been previously vacuum melted in 50kg lots was induction melted under 300mbar of argon, and additions made to adjust the composition to the following constant base composition (wt%) with varying sulphur and manganese contents:

C	Mn	Si	Al	P	Cu	N	O	S
0.04	0.10/0.24/0.64	0.20	0.005	0.005	0.01	0.0026	0.0028	0.005-0.04

The hot metal was cast under argon into a split, re-usable ceramic mould that had been previously heated to 1100°C and wrapped in ceramic fibre blanket preheated to 300°C. The procedure controlled the cooling rate to be 1.9°C/s between 1400°C and 1200°C, as measured by a thermocouple located on the centre line of the ingot and at about 1/3 height. This simulated the cooling near the centre of 50mm thick continuous cast slab, and resulted in an as-cast austenite grain size of approximately 1mm, with a secondary dendrite arm spacing of 150μm. The ingots were 75mm wide, 15mm thick and approx. 130mm long, and bevelled on their bottom edge to aid entry to the roll gap.

Thermomechanical Treatment. The ingots were either allowed to cool to room temperature and then reheated for 1 hour at 1200°C and rolled to simulate CCR, or were stripped from the mould at approximately 1150°C, allowed to cool in air to 1100°C, and then rolled, Fig.1. The three pass rolling schedule was as follows:

Roll Pass Number	Temp (°C)	Entry (mm) Thickness	Exit (mm) Thickness	Reduction (%)
1	1100	15	8.8	41.3
2	1000	8.8	5.7	35.2
3	900	5.7	3.7	35.1

The nominal rolling speed was 0.25m/s and the time between passes approximately 25s. After rolling the resulting plate was air cooled to 650°C, held in a furnace at this temperature for 1 hour and then furnace cooled to simulate coiling. A number of ingots were vigorously quenched into iced brine at -18°C at different stages of processing in order to investigate progressive changes in constitution.

Examination and Testing. In addition to conventional optical metallography, precipitates were extracted onto carbon replicas and examined using TEMs with EDS capabilities.

Tensile specimens had the gauge length of 25mm, width of 10mm and were 3mm thick. The strain rate was 0.04/minute.

To quantify the severity of edge cracking in the form of transverse cracks radiating inwards from the edge of the hot band, a 75mm long section of hot band was ground to its half thickness and the length of the 3 longest transverse edge cracks were measured and averaged. This value is referred to as the 'cracking index'. A cracking index was determined for each TSDR sample.

Results and Discussion

Ferrite Microstructure. All the CCR samples contained an homogenous, polygonal ferrite grain structure, Fig.2(a). The TSDR samples exhibited banding of coarse and fine ferrite, as observed by others after direct hot rolling[6,7]. The banding was minimal at 0.005wt%S, Fig.2(b), but increased with sulphur content, Fig.2(c). Cooling to room temperature and reheating through the α/γ transformation range for CCR refined the austenite grain size to approximately 150μm before rolling[6]. This increased the kinetics of recrystallisation so that the austenite was fully recrystallised after the first pass, to a finer uniform grain structure. This was repeated between successive passes, so that the final ferrite grain structure after CCR was generated from a uniform austenite grain structure. Examination of the 0.014wt%S TSDR steel quenched after the first roll pass showed only partial recrystallisation of the austenite, large unrecrystallised regions being observed adjacent to finer grained, recrystallised regions. The finer grains were located in bands along the former austenite grain boundaries. This inhomegeneity was not removed by successive passes and was inherited in the final ferrite grain structure.

In part, the slow recrystallisation of the austenite was due to its large grain size, but the degree of heterogeneity increased with sulphur content, and strain induced precipitation of MnS may also have been a factor inhibiting recrystallisation during TSDR. The solubility equation, Eq 1 [10],

$$\log[\text{Mn}][\text{S}] = 2.929 - 9020/\text{T} \qquad\qquad \text{Eq.1}$$

was used to calculate the supersaturation ratio (SR) for the precipitation of MnS. For the CCR steels SR was 2.8 at the start of rolling. For the 0.005wt% and 0.014wt%S TSDR steels SR was 5.2 and 14.7, respectively, assuming MnS was wholly dissolved in the austenite immediately after solidification. It has been suggested that dislocations introduced during rolling provide nucleation sites for precipitation of NbC in microalloyed steel ("strain induced precipitation") when SR reaches about 7 [11].

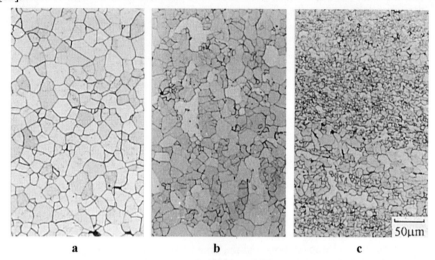

| a | b | c |

Figure 2. Ferrite microstructures: (a)CCR, 0.015wt%S; (b)TSDR, 0.005wt%S; (c)TSDR0.014wt%S

Effect of Sulphur on Strength and Ferrite Grain Size. In the CCR steels, as the S content was increased from 0.005 to.0.040wt%, the ferrite grain size became smaller, Fig.3a, and the yield strength increased, by 17.6 MPa, Fig.3b. Grain refinement increases yield stress according to the Hall-Petch relationship ($\sigma_y = \sigma_0 + k_y d^{-1/2}$). Assuming $k_y = 17$ MPa.mm$^{1/2}$, the observed maximum changes in grain size would account for an increase in yield stress of 16.9MPa in the CCR material. Clearly, the strengthening effect of S is wholly explained by the grain refinement.

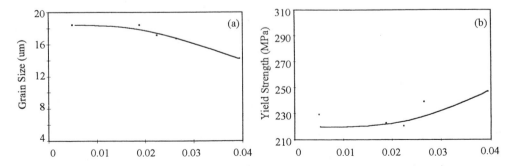

Figure 3. Effect of sulphur on (a) ferrite grain size and (b) yield strength in CCR steels.

In the TSDR steels the ferrite grain size was similar to the CCR steels at low S levels. However, as the S content was increased the ferrite grain size became significantly finer in the TSDR material. Again, the yield strength closely followed these trends as predicted by Hall-Petch. This is shown in Fig.4a and Fig.4b, for the 0.10wt%Mn steels, where the grain size continuously decreased from 15.2 to 7.6 µm and the yield strength increased by 47 MPa over the same S range. For the 0.24 and 0.64wt% Mn series steels the grain size decreased to a minimum, after which it became slightly coarser, Fig.4c and Fig.4e. The grain size minimum corresponded with the peak in the yield strength, Fig.4d and Fig.4f. In the 0.24wt% Mn steels the grain size decreased from 19.7 to 9.6µm as the yield strength increased by 60 MPa. Similarly, a ferrite grain refinement from 10.6 to 7.6µm and a corresponding increase in yield strength of 32.8 MPa was measured in the 0.64wt% Mn series. Again, using the Hall-Petch relationship the increase in strength due to grain refinement is 53 and 30 MPa respectively for these two series of steels. The strengthening effect of S for all the TSDR steels is wholly explained by the grain refinement. Comparing the three series of TSDR steels, the grain size minimum and yield strength maximum moved from low S levels (0.014wt%) in the 0.64wt%Mn steels to a higher S level (0.027wt%) in the 0.24wt%Mn steels. For the low Mn (0.10wt%) steels a peak in strength was not evident, however, it appears likely that it would occur at higher S levels than used here.

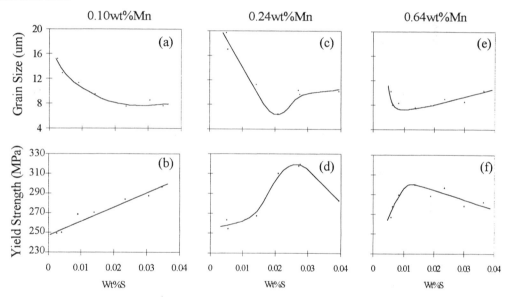

Figure 4. Effect of sulphur on ferrite grain size and yield strength in TSDR steels

Edge Cracking During Rolling, Transverse cracks radiating inwards from the edge of the hot band were observed in TSDR samples. The cracks, located at austenite grain boundaries, appeared during the first roll pass, and became more severe during subsequent passes. The severity of the edge cracking measured by a cracking index, as a function of S and Mn content, is shown in Fig.5. For the high Mn series (0.64wt%) of TSDR steels, where the minimum Mn/S ratio was 19, and all CCR steels, no edge cracking was observed. In the 0.24wt% Mn series of TSDR steels, no edge cracking was observed until the S content reached 0.015wt%, after which it increased with the level of S. In terms of the Mn/S ratio, cracking occurred when Mn/S was 14 or lower. However, in the low Mn TSDR (0.10wt%) steels, cracking occurred at very low S contents. The steels with the two lowest S levels in this series (0.0025 and 0.0040wt%S) had high Mn/S ratios of 32 and 33. An as-cast ingot containing 0.018wt%S was held isothermally for 20 minutes after reaching 1100°C before the commencement of rolling. This halt in processing eliminated edge cracking.

The effect of Ti on the susceptibility to edge cracking of TSDR steels was examined. Additions of 0.024 and 0.036wt%Ti were added to a 0.02wt%S/0.24wt%Mn TSDR steels and 0.024wt%Ti was added to a 0.02wt%S/0.10wt%Mn TSDR steel. The Ti was found to have a dramatic effect on the severity of cracking. Reductions in the cracking index from 2.5 to 1.4, 2.5 to 1.9 and 11 to 2.7 were measured. Carbon extraction replicas revealed the presence of complex (TiFeMn)S precipitates both randomly distributed and on austenite grain boundaries. MnS precipitates were also found precipitated on Ti-based oxide particles. Analysis of the very fine strain induced precipitates revealed only low levels of S. These particles were predominantly Ti(C.N) with some S present. The Ti apparently increased the kinetics of sulphide precipitation, thereby reducing the free sulphur at the time of rolling.

S and Mn levels are known to have a very marked influence on the hot ductility of low carbon steel. Increasing S or decreasing Mn decreases the hot ductility. Proposed mechanisms include segregation of S to the austenite grain boundary reducing the cohesive strength of the grain boundary, formation of a low melting point FeS compound on the grain boundary or precipitation of fine sulphides at the grain boundaries providing nucleation sites for micro voids. Clearly, from this study, cracking was associated with sulphur at the austenite grain boundary. Carbon extraction replicas have shown the presence of large (100 - 800nm diameter) sulphide precipitates on the boundaries but it appears that sulphur in solution has a greater effect on ductility because the hold at 1100°C significantly reduced the occurrence of cracking by reducing the free sulphur through precipitation at the boundary. These results indicate that edge cracking is of considerable importance to plants performing direct rolling, particularly when the time between casting and rolling is short.

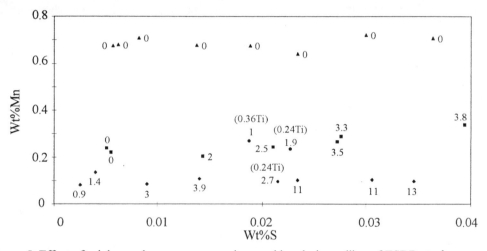

Figure 5. Effect of sulphur and manganese on edge cracking during rolling of TSDR steels.

Precipitation. Carbon extraction replicas from a 0.015wt%S/0.24wt%Mn CCR material contained MnS precipitates in two size ranges. Larger diameter particles, 100-500nm, formed during cooling and reheating that preceded rolling, and finer, more closely spaced precipitates, 5-30nm, during hot rolling. These fine MnS particles appeared to outline a dislocation cell structure within the γ grains, and it is assumed that they are strain induced precipitates and occurred either during a roll pass, or soon after a roll pass, and that they preceded recrystallisation.

To study MnS precipitation in the TSDR process a series of steels with 0.24wt%Mn and increasing S content were cast, allowed to cool at the normal thin slab cooling rate to 1100° C and then quenched. Another series of steels with 0.015wt%S and 0.24wt%Mn were quenched after the first and third roll pass and after the coiling treatment.

In the cast ingots, quenched immediately prior to rolling, carbon extraction replicas revealed the presence of precipitates, identified as (MnFe)S. In steels with low S content (0.005wt%), few particles were present, whereas in the case of high S (0.040wt%), many precipitates were found. Precipitation occurred preferentially on the austenite grain boundaries and were easily visible as lines of precipitates across the replicas, Fig.6a. Table 1 shows the density of MnS precipitates (n/0.25mm²), their average radius (r), their area fraction (f) and the ratio of r/f for the steels with various S contents. It is evident that the density, radius and area fraction of precipitates were larger at the higher S levels. This indicates that both the nucleation and growth rates of precipitates increased with S content. Measurement of the austenite grain size in these samples quenched at

Table 1 Effect of S content on MnS precipitation in austenite

S (wt%)	n (/0.25mm2)	r (nm)	f (%)	r/f (nm/%)
0.005	26	63	0.0001	489,708
0.015	612	102	0.0079	12,811
0.025	1,360	202	0.0694	2,904
0.040	1,282	239	0.0919	2,599

1100°C revealed that this increase in S content, and subsequent increase in MnS precipitation kinetics, significantly refined the austenite grain size. Fig.6b. It appears that MnS precipitates have a pinning effect on grain boundary migration of the austenite phase. The solubility of MnS in austenite at equilibrium is represented by Eq.(1). For the 0.005wt%S steel, MnS will begin to precipitate below 1250°C in the γ phase, and so there are no precipitates available to pin grain boundary migration from 1440°C, the end of the δ/γ transformation, to a temperature below 1250°C when precipitation occurs. Consequently, MnS will have little chance of retarding grain growth. However, at higher S levels, 0.025 and 0.040wt%, precipitation will occur immediately after the δ/γ transformation. Therefore, γ grain boundary migration can be suppressed by MnS precipitates. The relationship between the average grain size R and the pinning force exerted by particles on grain boundaries was defined by Zener, Eq.(2), where β is a constant.

$$R = \beta/(r/f) \qquad \text{Eq.2}$$

As shown in Table 1, the r/f ratio becomes significantly smaller as the S content is increased to 0.025wt%S, thereby increasing the pinning force on the grain boundary. Increasing the S content further to 0.04wt% does not significantly increase the pinning force as the average radius (r) of precipitates has increased but their density decreased. Precipitation may have occurred in the δ phase thereby reducing the total S available for precipitation on the γ grain boundary.

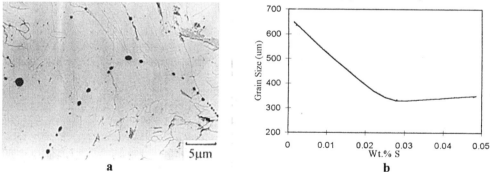

a   **b**

Figure 6. (a) MnS precipitates on γ grain boundaries (b) γ grain refinement due to sulphur

Extraction replicas taken from TSDR steels quenched after the first and third roll passes and after the coiling treatment were used to determine the sequence of precipitation and growth of fine, strain induced MnS precipitates. The volume fractions of precipitate particles varied greatly even on a single replica, and were not measured quantitatively. It was observed qualitatively that the particles were more numerous and were visible over a larger fraction of the replica as processing progressed. The diameter of precipitates extracted from material quenched after the first pass, third pass and coiling treatment were measured. After the first pass very fine precipitates with an average diameter of 8.4nm were found. After the third pass the average precipitate diameter had increased to 13.7nm, indicating precipitate growth. However, fine precipitates in the 2-4nm range were still present indicating that further nucleation of precipitates was occurring in the latter stages of the rolling sequence. After the one hour coiling treatment at 650°C the average precipitate diameter had increased, only slightly to 14.7nm. The virtual absence of very small precipitates, less than 8nm in diameter, indicated that some coarsening had occurred. However, as the average precipitate diameter only increased marginally it is apparent that these fine MnS precipitates were very stable at this coiling temperature. Ageing experiments conducted on the TSDR steels confirmed this as the hardness of the steels remained constant after several hours of isothermal heat treatment at 650°C.

Comparisons between TSDR and CCR material showed that the fine, strain-induced MnS precipitates were more numerous in the TSDR steels and increased with S content. It is considered that the fine MnS particles observed here could have contributed to retardation of recrystallisation if a sufficient volume fraction were present, but the possibility that solute drag was responsible should not be discounted.

Summary and Conclusions

Increasing the sulphur content of thin slab cast and direct rolled mild steel refined the ferrite grain size of the product and increased strength. The sulphur had two principal effects. The high supersaturation with respect to MnS present in the just-cast austenite caused precipitation of MnS preferentially on the austenite grain boundary. These precipitates have a pinning effect on the grain boundary resulting in a finer austenite grain size prior to rolling. The other effect is that the strain induced precipitate dispersion, or the sulphur in solid solution, or both, retarded austenite recrystallisation, augmenting a similar tendency arising from the coarseness of the just-cast austenite grain size. The retained strain in the banded, unrecrystallised fraction of the austenite caused that fraction to transform to fine grained ferrite. Yield strength was greater in the TSDR material and increased to a maximum at a sulphur content which was inversely related to the level of Mn. This concentration of sulphur where peak strength occurs corresponds to its maximum solubility in austenite at the peritectic temperature. Edge cracking originated in the first pass at 1100°C, and was related to sulphur in solid solution rather than to precipitated MnS. The addition of Ti significantly reduced the severity of edge cracking.

Acknowledgements.

One author (LDF) is grateful to BHP Research for supporting his PhD studies at the Manchester Materials Science Centre. The research on thermomechanical processing at The Manchester Materials Science Centre is supported by EPSRC grant GR/J80191.

References.

1. J.Edington, 'Hatfield Memorial Lecture - The New World of Steel', Ironmaking and Steelmaking, (1997), Vol.24, No.1, p19.
2. F.K.Iverson and K.Busse, 'A Review of First Year CSP Operations at Nucor Steel's New Thin Slab Casting Facility', Metallurgical Plant and Technology International 1, (1991), 40-51.
3. G.Gosio, M.Morando, L.Manini, A.Guindani, F.P.Pleschiutschnigg, B.Kruger, H.D.Hoppmann, I.v.Hagen, 'The Technology of Thin Slab Casting, Production and Product Quality at the Averdi I.S.P. Works, Cremona', METEC Congress '94, 2nd European Continuous Casting Conference, 6th International Rolling Conference, Dusseldorf, June 20-22, (1994), Proceedings, Volume 1, 344-351.
4. M.Espenhahn, R.Bertram, G.Flemming, D.Kruger, G.Pietzko, H.Streubel, 'CPR - A Combined Casting- Rolling Process for the Production of Steel Strips', ibid ref.3, Proceedings, V1, 359-366.
5. R.Kaspar and O.Pawelski, 'Thermomechanical Treatment of Direct Charged Thin Slabs', ibid ref.3, Proceedings, Volume 1, 390-395.
6. R.Priestner and C.Zhou, 'Simulation of Microstructural Evolution in Nb-Ti Microalloyed Steel in Hot Direct Rolling', Ironmaking and Steelmaking, (1995), Vol.22, No.4, 326-332
7. T.Senuma, S.Sanagi, K.Kawasaki, S.Akamatsu, T.Hayashida and O.Akisue, 'Precipitation Control of Cold Rolled Mild Steel Sheets in Thin Slab-Direct Hot Rolling Process', Tetsu-to-Hagane, (1993), 79(2), pp. 64-70.
8. S.Wilmotte, 'New Techniques for Casting Thin Slabs at High Speeds', Int.Conf. on New Smelting Reduction and Near Net Shape Casting Technologies for Steel, Vol 2, Pohang, Korea, (1990).
9. M.Suzuki, C.H.Yu, H.Shibata and T.Emi, 'Recovery of Hot Ductility by Improving Thermal Pattern of Continuously Cast Low Carbon and Ultra Low Carbon Steel Slabs for Hot Direct Rolling', ISIJ International, Vol.37, (1997), No.9, pp.862-871.
10. E.T.Turkdogan, S.Ignatowicz and J.Pearson, 'The Solubility of Sulphur in Iron and Iron-Manganese Alloys', Journal of the Iron and Steel Institute, August, (1955), 349-354.
11. B. Dutta and C.M. Sellars, Mat. Sci. and Technology, 3, (1987), pp.197-206.

For Correspondence: Fax L.D.Frawley (UK) 0161 2003590

Materials Science Forum Vols. 284-286 (1998) pp. 493-500
© *1998 Trans Tech Publications, Switzerland*

Precipitation of TiN and (Ti,Nb)(C,N) during Solidification, Cooling and Hot Direct Deformation

A. Köthe, J. Kunze, G. Backmann and C. Mickel

Institute of Solid State and Materials Research Dresden,
Postbox 270016, D-01171 Dresden, Germany

Keywords: Simulation, Thin Slab Casting, Direct Hot Rolling, Microalloying, TiN Precipitation, Precipitation of (Ti,Nb)(C,N) (-during Solidification, - in Undeformed Austenite), Strain Induced Precipitation, Microsegregation, Fractional Softening, HSLA Steels

Abstract

The precipitation behaviour of TiN and (Ti,Nb)(C,N) in low carbon HSLA steels has been studied by simulation of the casting and hot direct rolling of thin slabs. At ferritic solidification TiN particles 0.5 ... 1.0 µm in size were formed only in regions of maximum segregation where the supersaturation was higher than 10fold. During cooling, in the range of delta ferrite fast homogeneization of the segregated Ti takes place, and no TiN nuclei were formed. In undeformed austenite some large particles precipitated their arrangement reflecting the solidification structure. In the 1200 to 1100 °C range particle sizes between 20 and 70 nm were reached. A high density of deformation induced particles smaller than 10 nm could be obtained at or below 1050 °C. The density of the particles increased and their size decreased with decreasing deformation temperature. After fast cooling, below 1000 °C also in undeformed material finely dispersed precipitates were found. In TiNb microalloyed steels, complex precipitates are formed their Nb:Ti ratio being low at high temperature and increasing with decreasing precipitation temperature. Separate nucleation of Nb(C,N) particles is observed deformation-induced below 1000 °C and without deformation below 950 °C. Based on these results a number of thin slab casting and direct rolling experiments have been carried out on the thin slab pilot facility of Thyssen Krupp Stahl AG at Dortmund. With steels of 0.046...0.086%C–0.06%Si–0.92...1.05%Mn–0.03%Al-0.016...0.020%Ti and partly 0.035% Nb 50 mm slabs were rolled to 10 mm plates which after air cooling had the following mechanical properties: R_{eL} = 368 ... 394 MPa, R_m = 442... 464 MPa; A_5 = 34...37.4 %; $T_{tr}(50\%)$ = -38...-45 °C.

1. Introduction

When producing plate and hot strip of HSLA steels by the processing route „Casting and hot direct rolling of thin slabs" it is usually difficult to obtain high values of strength and toughness properties similar to those of thermomechanically processed conventional continuously cast thick slabs. This is mainly due to the facts that the rolling process starts with a very coarse initial austenite grain size, that the amount of total thickness reduction available for austenite grain refinement is small, and that an inhibition of austenite grain growth by precipitate particles is normally not available because all microalloying elements are in solid solution when the rolling process commences. Therefore, the present investigation is aimed at investigating the precipitation behaviour of microalloying elements during solidification, cooling and hot direct deformation under conditions pertinent to thin slab casting and rolling. The results shall allow to make an efficient use of the microalloying elements Ti and Nb for austenite conditioning under the circumstances of direct rolling of thin slabs which is necessary to obtain a fine-grained ferrite-pearlite microstructure of the hot rolled product.

The solubility of TiN in low-alloyed steel is small. The solubility product in pure liquid iron was reported [1] as

$$L_{TiN} = -17040/T + 6.40 \qquad \text{Eq. 1.}$$

Considering the stoichiometric composition at 1800 K, 0.0542% Ti and 0.0158% N can be dissolved in pure Fe-Ti-N alloys. Technical steels microalloyed with Ti for austenite grain growth inhibition normally contain 0.010 to 0.020% Ti and 0.005 to 0.010% N. Although this amount is soluble at the liquidus temperature, the solubility limit can be surpassed locally due to micro- and macrosegregations in the residual melt between dendrites or equiaxed grains [2].

Information about the kinetics of TiN precipitation in the as-cast solid state are relatively scarce. Subramanian et al. [3] calculated the thermodynamic equilibrium between (Ti,Nb)(C,N) particles and the solid solution. From the experimentally determined Ti:Nb ratio in the particles they estimated the temperature at which nucleation and growth of the particles had occured. In an alloy with 0.020% Ti and 0.009% N Abushosha et al. [4] observed particles of 100 nm diameter after solidification and cooling with a rate of 25 K/min pertinent to thick continuously cast slabs. More recently they detected particles in the 10 nm range at this cooling rate but smaller than this after cooling with a rate of 200 K/min characteristic of thin slab casting [5]. Zhou et al. [6] published results of (Ti,Nb)N precipitation at direct solidification and cooling experiments. They observed particles larger than 100 nm in all alloys where the Ti content was higher than 0.012% if the samples were quenched from any temperature up to 1408 °C. At less than 0.011% Ti, particles were detected only when the ingots were slowly cooled down to 1000 °C and then quenched.

The deformation-induced precipitation of Ti and Nb in reheated steels has been studied in a large number of publications. At 1200 to 1300 °C only TiC and Nb(C,N) can be totally dissolved. The solubility of TiN at 1300 °C amounts to $[\% \text{ Ti}][\% \text{ N}] = 3 \cdot 10^{-6}$ [7]. Thus, with respect to Ti only the precipitation of TiC has been investigated e.g. by Liu and Jonas [8...10] or by Medina et al. [11].

Therefore, here at first the conditions of TiN precipitation in the liquid segregation zones during solidification of pure iron and low-carbon steel are studied [2]. Then the precipitation of TiN and (Ti,Nb)(C,N) is investigated in as-cast samples where the cooling process was interrupted for some 10 seconds in the temperature range 1350 to 800 °C [12] and the size and distribution of the particles formed during holding determined. Deformation-induced precipitation is studied in samples in-situ melted, re-solidified and directly deformed in a torsional plastometer. The investigation is completed by the characterization of 10 mm air cooled plates of Ti and TiNb microalloyed low carbon steels produced by thin slab casting and hot direct rolling on a pilot plant [13].

2. Experimental details

Experiments without deformation were performed on a Fe-0.77%Mn-0.051%Ti alloy. 0.5g samples suspended on a thin pure iron wire were molten in an induction coil for 15 minutes (the liquid sample did not fall due to its surface tension) in a circulating atmosphere of N_2-H_2-Ar-CH_4 containing 33 mbar N_2 and a defined amount of methane. After dissolving 0.08 to 0.10%C and the equilibrium concentration of 0.0080% nitrogen the high frequency power was switched off. The samples solidified in $t_e = 6$ s and cooled down in the gas with a cooling rate of 21 K/s between 1300 and 800 °C. The cooling process was interrupted at pre-defined temperatures for a given time in order to allow nucleation and growth of precipitates to occur.

Deformation-induced precipitation was studied by means of a torsional plastometer described in [14]. A zone of about 15 mm in length of the middle part of a 6 mm diameter rod was inductively molten in an Ar-N_2 atmosphere and re-solidified with a local solidification time of 3...4 s. The molten zone solidified dendritically with a mean secondary dendrite arm spacing of 30 to 60 μm. The as-cast zone cooled off with a mean rate of 15 K/s between 1300 and 800 °C. As soon as the temperature reached a defined value a torsional deformation (surface strain $\varphi = 0.3$ or 1.0, surface strain rate $\dot{\varphi} = 1$ s^{-1}) was applied followed by an isothermal holding (in most cases of 30 s duration) and by a second deformation step of $\varphi = 0.3$. Then the sample was quenched into water or cooled off in the gas stream. The second deformation served to determine the fractional softening during isothermal holding after the first deformation step. The fractional softening FS is defined as

$$FS = 100 \cdot (\sigma_{max} - \sigma_2) / (\sigma_{max} - \sigma_1) \qquad \qquad \text{Eq. 2}$$

where σ_1 and σ_2 are the initial flow stresses of the first and second deformation step, respectively, and σ_{max} is the maximum flow stress attained at the end of the first deformation step.

A third group of experiments could be carried out at the experimental thin slab casting and rolling facility of Thyssen Krupp Stahl AG at Dortmund. Details of this facility are described in [13]. 50 kg heats molten in a vacuum induction furnace were cast in vacuum into a water-cooled copper mould to ingots of 50 mm in thickness. From the secondary dendrite arm spacings of 130 µm in the outer zone and 160 µm in the centre of the slabs a local solidification time of $t_e = 50$ s and 100 s, respectively, has been deduced. The slabs were cooled with a mean rate of 3 K/s between 1300 and 1100 °C (a value typical of thin slab casting) either to the equilibration temperature of 1250 °C (holding time 10 min) or to the rolling start temperature of either 1150 °C (TiNb steels) or 1050 °C (Ti steels). The 50 mm slabs were rolled to 10 mm plate in 4 passes with finish rolling temperatures of 800 to 820 °C and then air cooled to room temperature. Steel composition and processing parameters are given in **Table 1**.

Table 1: Chemical composition (wt.-%) and processing parameters of the materials of thin slab casting and rolling experiments carried out at Thyssen Krupp Stahl AG in Dortmund. Direct hot rolling of 50 mm slabs to 10 mm plate, air cooling from FRT. RST: rolling start temperature; FRT: finish rolling temperature

| Steel no. | Type | Chemical composition / wt-% | | | | | | | Rolling parameters | | |
		C	Si	Mn	Al	Ti	Nb	N	Hold at	RST °C	FRT °C
720	Ti	0.064	0.057	1.05	0.033	0.020	-	0.0060	-	1060	800
752	TiNb	0.086	0.081	0.98	0.056	0.020	0.039	0.0107	-	1160	825
762	TiNb	0.046	0.047	0.92	0.036	0.016	0.031	0.0071	-	1160	820
776	TiNb	0.054	0.058	0.98	0.046	0.017	0.038	0.0077	1250°C /10min	1150	820

Samples for metallographic investigations by optical microscopy were prepared by mechanical grinding followed by a 30 s electrolytic polishing in a 10:1 mixture of CH_3COOH and $HClO_4$ and a 10 s etching in 2% alcoholic HNO_3 (nital). The quantitative characterisation of grain size and grain size distribution was carried out by linear analysis of about 400 grains in both rolling and normal direction. Particles down to 2...3 nm were detected on carbon extraction replicas by means of a high resolution transmission electron microscope with field emission cathode. Using an EDX spectrometer particles ≥ 7 nm in size could be analysed.

3. Results

3.1. TiN precipitation during solidification

Pure and low carbon iron solidifies as delta ferrite. The thermodynamic distribution coefficients of Ti and N

$$k_{0,Ti}^{\delta/l} = [\% \text{ Ti}]_\delta / [\% \text{ Ti}]_l \qquad \text{and} \qquad k_{0,N}^{\delta/l} = [\% \text{ N}]_\delta / [\% \text{ N}]_l \qquad \text{Eq. 3, 4}$$

have been determined as 0.40 and 0.285, respectively [7]. With these values the segregation factors of Ti and N during dendritic solidification using the Brody and Flemings model of plate-like secondary dendrite arms [15] in the form improved by Ohnaka [16] with a local solidification time of 6 s and a secondary dendrite arm spacing of 70 µm have been calculated. In **Fig. 1a and b** the segregation factors of Ti and N and the TiN supersaturation η are plotted as a function of the solidified fraction f_s. The supersaturation η is defined as

$$\eta = [\% \text{ Ti}][\% \text{ N}] / L_{TiN} \qquad \text{Eq. 5.}$$

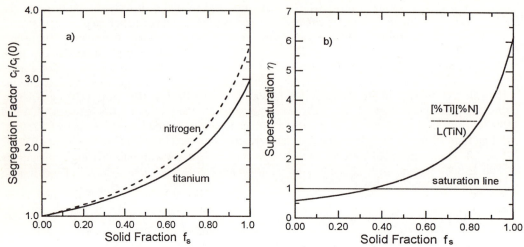

Fig. 1: a) Segregation factor, b) supersaturation in the interdendritic melt of a ferritically solidifying Fe-0.08%-0.77%Mn-0.051%Ti-0.008%N alloy as a function of the solidified fraction, calculated with a local solidification time of 6s, secondary dendrite arm spacing of 70 μm, and solidification temperature 1800 K.

The calculated enrichment of Ti in the interdendritic residual melt of up to a factor of 3 was experimentally confirmed by secondary ion mass spectrometry (SIMS; [2,17]). At spots where 3 or more microsegregation zones meet a Ti enrichment of up to 9 was observed.

In samples containing 0.08...0.10%C–0.77%Mn–0.051%Ti-0.016%N solidified within t_e = 30s TiN particles ≥ 0.5 μm in diameter have been observed. These particles are arranged in rows marking the boundaries between dendrite arms or equiaxed grains and the residual melt and correspond with the points of maximum Ti concentration. The particle spacing within the rows is about $2 - 3$ μm. This arrangement may be explained assuming that the nucleation starts at $f_s \cong 0.95$ in zones were the enrichment of Ti attains a value of $c_l/c_l(0)$ = 10. Each particle contains an amount of Ti which was formerly concentrated in a volume of about 3x3x2 μm^3. After a faster solidification with t_e = 6 s the particles were less numerous. In samples containing 0.008% N particles could be detected only at very few points in the centre of the sample.

From theoretical and experimental investigations described in [2] it may be concluded that a nucleation of TiN is possible if the concentration product [% Ti][% N] exceeds the solubility product $L_{TiN} \cong 8\cdot10^{-4}$ by more than a factor of 10. If by microsegregation in maximum a 20 to 30fold enrichment can be reached, in alloys containing a concentration product of more than [% Ti][% N] $\cong 3\cdot10^{-4}$ the formation of coarse TiN precipitates in liquid segregation zones is possible at a solidification time of t_e = 6s. If the microsegregation is superimposed by any macrosegregation, or if the solidification rate is slower, the critical concentration product may be even lower.

3.2. TiN precipitation during cooling in undeformed austenite

An interruption of the cooling process in the temperature range 1350 to 1250 °C for $\geq$ 30s leads to the formation of cuboidal particles having edge lengths between 50 and 200 nm. They are mostly arranged in rows having a particle distance of 0.5 to 2 μm. They belong to the same type of rows formed in the residual melt which mark the primary solidification structure. The distance between the rows is in the order of some ten μm.

Nucleation has been observed in undeformed upper austenite only in the zones were the last melt had solidified. Possibly very small titanium oxides or sulphides acted as nucleation sites since in the segregation zones increased concentrations of the very intensively segregating elements oxygen and sulphur are to be expected. Co-precipitation of MnS and TiN has been shown by Zhou et al. [6].

If the cooling from solidification temperature is interrupted between 1200 and 1100 °C, rows of TiN particles having diameters between 20 and 50 nm and mutual distances between 100 and 500 nm are formed. From the distance of these rows which is in the order of some μm it is assumed that the nucleation could occur at dislocations (a mean distance between dislocations of 3 μm corresponds to a dislocation density of 10^{11} m^{-2}, a value typical for undeformed as-grown metals).

Below 1000 °C, the supersaturation reaches values above η = 10000. At such high driving forces nucleation of new phases should take place also homogeneously in the matrix. Accordingly, at 1000 °C in addition to the rows mentioned above, statistically distributed particles in the 3 to 5 nm range have been observed. After a holding time of 30 to 60 s at 900 or 800 °C, 10 to 15 nm particles arranged in rows, and randomly distributed 3 to 7 nm particles were detected. In case of an overstoichiometric Ti:N ratio, at these low temperatures the additional formation of titanium carbonitrides or carbides has to be taken into account.

In **Figure 2** the supersaturation of TiN in a 0.1%C-0.77%Mn-0.051%Ti-0.008%N alloy as a function of temperature is plotted together with some observations made with respect to TiN formation. According to fig.2, nucleation at dislocations requires a supersaturation of about 500fold.

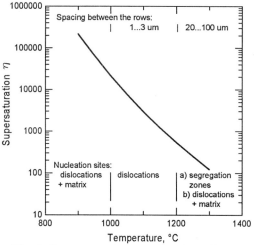

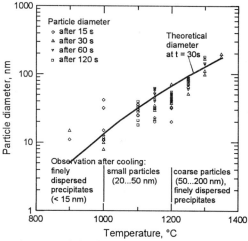

Fig. 2: Supersaturation as a function of temperature calculated for a homogeneous solution of 0.051%Ti + 0.008%N in a Fe-0.1%C-0.77%Mn alloy

Fig. 3: Particle diameter measured after interruption of the cooling at distinct temperatures (points) and calculated for holding time of 30s (solid line)

The particle size depends on the temperature at which the cooling process is interrupted. For diffusion-controlled growth of a single particle without interaction of the diffusion fields of neighbouring particles, the radius r can be described [18] by

$$r = \sqrt{2Dt\frac{c_w - c_a}{c_p - c_a}}$$

Eq. 6,

where D - diffusion coefficient, t - time, c_w, c_p, c_a - concentrations far away from the particle, inside the particle and in the solution adjacent to the particle, respectively. Since the particle growth is determined by the diffusion of Ti, the Ti concentration c_a near the particle surface is [%Ti]= L_{TiN}/[%N]. The particle radius can be calculated by this and the diffusion coefficient

$$D_{Ti}^{\gamma-Fe} = 1.5 \cdot 10^{-5} \exp(-25100/RT), \, m^2 s^{-1} \qquad\qquad \text{Eq. 7}$$

into Eq. 6. The results obtained for an isothermal holding of 30 s at various temperatures in the austenite region are shown in **Fig. 3**. The points inserted in this figure represent the particle diameters determined experimentally after various holding times. It can be concluded that the particle size is determined by the Ti diffusion at all temperatures. The Ti-depleted spheres surrounding the particles are about 1 μm in diameter at 1200 °C and about 3 μm at 1350 °C. That means that in 30 s the equilibrium concentration can be reached when dislocations act as nucleation sites, but only a small part of the Ti can precipitate if the nucleation takes place along secondary dendrite arms. Accordingly, finely dispersed particles may be formed at low temperatures in samples the cooling of which was interrupted at 1300 °C and above or which were cooled without interruption, but not when nucleation took place at dislocations at 1200 to 1100 °C.

3.3. Deformation-induced precipitation of TiN

1200 °C: After torsion at 1200 °C only a small number of cuboidal particles having edge lengths between 20 and 45 nm were detected. Eq. 6 yields a size of 50 nm if a diffusion time of 30 s (equal to the holding time after deformation) is assumed. The small number of particles points to a nucleation of TiN after the effect of deformation had been eliminated by recrystallisation, i. e. the deformation had no effect on TiN nucleation.

1100 °C: After a small deformation (less than the critical deformation for recrystallisation) a deformation-induced nucleation of TiN has been observed. According to Eq. 6 a particle size of 15 nm can be reached after a diffusion time of 15 to 20 s. In zones where the nucleation density was high, the equilibrium precipitation has been reached after this time. In zones with few nuclei the particles reach a size of about 25 nm. Applying a higher degree of deformation again the recrystallisation proceeds faster than the nucleation of TiN thus no deformation-induced precipitation took place.

1050 °C: In most parts of the torsion samples a high density of cuboidal particles of 6 to 15 nm edge length has been detected, whereas at a few places a small number of particles > 20 nm has been found. In these areas no finely dispersed particles had been formed.

1000 °C: A deformation at 1000 °C led to very fine particles in all parts of the sample. Frequently two kinds of particles have been detected: bands of 10 to 12 nm particles (probably nucleated at slip bands) and randomly distributed particles in the 3 to 7 nm range.

The size of the particles decreases and their number increases with decreasing deformation start temperature. **Figure 4** shows the particle size measured on experimentally rolled thin slabs [12] as a function of the reciprocal of the rolling start temperature varied between 1160 and 980 °C. The linear dependence points to a nucleation and a nearly total precipitation of the Ti during or immediately after the first rolling pass. The number of particles increases with decreasing rolling start temperature.

3.4. TiNb microalloyed steels

A combined microalloying with 0.01...0.02 % Ti and about 0.03 % Nb is often applied in HSLA steels in order to make use of TiN precipitates for inhibition of austenite grain growth and of deformation-induced precipitation of Nb(C,N) for retardation of austenite recrystallisation (and eventually for some precipitation hardening of ferrite). However, it is now well established that in TiNb microalloyed steels complex (Ti,Nb)(C,N) precipitates are formed (see e.g. [3,6]). At high temperature in the austenite region their composition is close to TiN with small Nb content. The Nb content increases with decreasing precipitation temperature. A separate nucleation of small Nb(C,N) particles has been observed only deformation-induced in the lower austenite region. In order to study the influence of a microalloying with Ti and/or Nb on the austenite structure during simulated hot direct deformation, some fractional softening experiments have been carried out with the torsional plastometer on steels of base composition 0.08%C-0.35%Si-1.44%Mn-0.005%S-

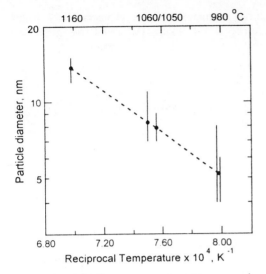
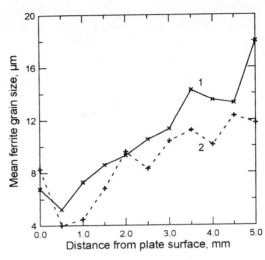

Fig. 4: Particle diameters measured in experimentally rolled thin slabs (composition similar to table 1) in dependence of the rolling start temperature

Fig. 5: Variation of the ferrite chord length in rolling direction as a function of the distance below the plate surface of two 10mm hot rolled plates (each point is the mean value of 300...400 grains)

0.027%Al-0.005%N and up to 0.019%Ti and/or 0.039%Nb. A two-step deformation ($\varphi_1 = 1$, $\varphi_2 = 0.3$, interpass time 30s) has been applied at 1150, 1100, 1050, 1000, 950 and 900 °C, respectively. The dependence of the amount of fractional softening determined according to eq. 2 from temperature, Ti and Nb content has been determined by means of multivariate linear regression analysis. The influencing factors turned out to be in decreasing order of significance: T_{def}, [Ti], [Nb] and [Ti]·[Nb]. The [Ti]·[Nb] interaction term vanished at $T_{def} < 1000$ °C which points to a separate nucleation of Nb(C,N) precipitates below 1000 °C.

3.5. Thin slab casting and rolling experiments

Chemical composition and processing parameters of the steels used in these experiments are given in Table 1. The rolling was subdivided into one pass above the recrystallisation stop temperature and 3 passes at relatively low temperature in order to obtain a fine-grained ferrite-pearlite microstructure of good strength and toughness. Therefore, in the Ti steel the rolling passes were applied at 1050 – 900 – 850 – 800 °C, and in the TiNb steels at 1150 - 920 – 870 – 820 °C. The mechanical properties and some microstructural characteristics determined on the air-cooled plates are presented in **Table 2**.

The mechanical properties are quite satisfying. The plates have fine-grained ferrite-pearlite structure which shows, however, a characteristic inhomogeneity in the through-thickness direction as shown in **Fig. 5** which is not uncommon in hot rolled flat products. A region of coarser grains at the surface is followed by a very fine-grained area down to about 1 mm below the surface, and a gradual increase in grain size towards the centre of the plates. The ferrite grains in the fine-grained region are nearly globular and relatively homogeneous. In the centre the structure is more inhomogeneous and elongated in the rolling direction. The factors leading to these microstructural features are not yet investigated in detail. It is assumed that they are caused by a number of factors: a certain surface decarburization (confirmed by microhardness measurements), a gradient of the rolling deformation over the cross section including differences in shear and compression components, and the local temperature-time regime in through-thickness direction. Some local grain

Table 2: Mechanical properties and microstructural parameters of the experimentally rolled 10 mm plates cooled on air. d: mean ferrite chord length; D_{95}/d: relative grain size inhomogeneity. D_{95}: 95% quantile of the ferrite chord length distribution; d_l/d_t: grain elongation in rolling direction

Steel no.	Type	Mechanical properties				Microstructural characteristics					
						Fine-grained zone			Centre of plate		
		R_{eL} MPa	R_m MPa	A_5 %	$T_{tr}(50\%)$ °C	d µm	D_{95}/d	d_l/d_t	d µm	D_{95}/d	d_l/d_t
720	Ti	368	446	37.4	-45	4.0	1.70	1.08	10.4	2.06	1.28
752	TiNb	386	464	34.0	-38	4.7	1.76	1.04	10.8	2.28	1.63
762	TiNb	380	442	37.4	-42	4.2	1.92	1.00	10.2	2.22	1.23
776	TiNb	394	454	37.1	-40	3.2	1.95	1.06	9.3	2.36	1.21

size inhomogeneities observed in the plates are assumed to be due to a possible extension of the finish rolling into the $\gamma+\alpha$ two phase region without a sufficient prior austenite grain refinement. All in all, the results obtained should be considered as a promising platform for further optimisation of steel composition, processing parameters and microstructure leading to consistently good mechanical properties.

Acknowledgements

The financial support granted by the Commission of the European Communities (ECSC Steel RTD Programme) under contract no. 7210 – CA/176 is greatly acknowledged. The authors would like to express their sincere thanks to Dr. B. Engl, Dr. Chr. Klinkenberg and R. Mempel of Thyssen Krupp Stahl AG Duisburg for the thin slab casting and rolling trials described in chapter 3.5.

References

[1] E.T. Turkdogan, ISS Steelmaking Proc. 70, Pittsburgh, Pa., March 29 – April 08 1987, p. 399

[2] J. Kunze, Chr. Mickel, M. Leonhardt and S. Oswald, steel res. 68(1997)9, p. 403

[3] S.V. Subramanian, S. Shima, G. Ocampo, T. Castillo, J.D. Embury and G.R. Purdy, in J.M Gray, T. Ko, Zhang Shouhua, Wu Baorong and Xie Xishan (Eds.), Proc Int. Conf. HSLA Steels '85, Nov. 4 – 7 1985, Beijing, China, ASM Intern.1986, p. 151

[4] R. Abushosha, R. Vipond and B. Mintz, Mater. Sci. Technol. (1991),7, p. 613

[5] B. Mintz and R. Abushosha, private communication

[6] C. Zhou and R. Priestner, ISIJ Intern. 36(1996) 11, p. 397

[7] J. Kunze, B. Beyer, S. Oswald and W. Gruner, steel res. 66(1995)4, p. 161

[8] W.J. Liu and J.J. Jonas, Metall. Trans. 19A(1988) 6, p. 1403

[9] W.J. Liu and J.J. Jonas, Metall. Trans. 19A(1988) 6, p. 1415

[10] W.J. Liu and J.J. Jonas, Metall. Trans. 20A(1989) 4, p. 689

[11] S.F. Medina and J.E. Mancilla, Acta Met. Mater. 42(1994) 12, p. 3945

[12] J. Kunze, C. Mickel, G. Backmann, B. Beyer, M. Reibold and C. Klinkenberg, steel res. 68(1997), 10, p. 441

[13] B. Engl and C. Klinkenberg, this conference

[14] R. Kaspar, P.M. Müller, F.-D. Philipp and G. Backmann, steel res. 64(1993) 12, p. 61

[15] H.D. Brody and M.C. Flemings, Trans. Met. Soc. AIME 239(1967)3, p. 426

[16] I. Ohnaka, Tetsu to Hagané 70(1984), p. 913

[17] J. Kunze, Chr. Mickel, S. Oswald and K.-H. Spitzer, Z. Metallkd. 88(1997)3, p. 182

[18] G. Sauthoff, Ausscheidung und Alterung, in W. Pitsch (ed.), Grundlagen der Wärmebehandlung von Stahl, Verlag Stahleisen GmbH, Düsseldorf, 1976

Materials Science Forum Vols. 284-286 (1998) pp. 501-508
© *1998 Trans Tech Publications, Switzerland*

Influence of Thin Slab Casting on the Properties of Microalloyed Steel for High Cold Forming Requirements

B. Engl and C. Klinkenberg

Thyssen Krupp Stahl AG, D-47161 Duisburg, Germany

Keywords: Thin Slab Casting, Direct Rolling, TiN Precipitation, Grain Size

ABSTRACT

The grain refining influence of a TiN-precipitation during conventional continuous casting is already well established /1/. During direct hot processing of thin slabs, TiN-precipitation should occur in the liquid phase during solidification /2, 3/ and in the austenite /4/. Various experiments have been carried out in order to examine the effect of a TiN-precipitation before and during hot rolling of thin slab casted Ti- and TiNb- microalloyed high strength structural steels.

The precipitation behaviour has been determined by TEM investigations. The samples show a large variety of TiN and Ti(N,C) precipitates. Number and size of the precipitates are depending on the holding and rolling-start temperatures. The impact of the individual hot rolling conditions on precipitation state, grain size as well as strength- and toughness-properties have been determined.

1. INTRODUCTION

This contribution concerns with the influence of different rolling and alloying parameters on the microstructure evolution and their impact on the mechanical and technological properties of direct hot rolled microalloyed structural high strength steels.

For a given hot band thickness, thin slabs receive lower total rolling reductions than conventional slabs. In thin slab casting the $\gamma-\alpha-\gamma$ phase-transformation, which takes place during cooling down of the conventional as cast slab to ambient temperature and reheating before hot rolling, does not occur. Both effects, higher total rolling reduction and $\gamma-\alpha-\gamma$ phase-transformation have a grain refining influence on the final hot band microstructure. Therefore direct rolling of thin slab cast materials requires well suited cooling and hot rolling schedules, in order to get best processing and application properties of steels, being produced in this way.

As direct rolled thin slabs do not cool down below the austenite range before hot rolling, TiN-precipitation should occur in the liquid, δ-ferritic or austenitic state. According to /3, 4/, TiN-precipitation does not take place in the δ-ferrite, but in the liquid phase and in the austenite phase. The experiments have been carried out in order to examine the role of a TiN-precipitation in the austenite phase before and during hot rolling of Ti- and TiNb- microalloyed steels. For comparison, a Nb-alloyed steel variant has been examined too.

2. MATERIALS AND EXPERIMENTAL PROCEDURES

Variant	C	Si	Mn	P	S	Al$_{sls}$	N	O	Nb	Ti	Ti/N	Ti$_{Rest}$
Ti	0.053	0.048	0.926	0.009	0.005	0.029	0.007	0.005	0.002	0.044	6.4	0.018
Nb	0.064	0.044	0.937	0.018	0.006	0.029	0.005	0.007	0.044	0.003	0.6	0.000
Ti +Nb	0.078	0.047	0.900	0.009	0.006	0.031	0.004	0.005	0.037	0.027	6.4	0.013

Table 1 Chemical compositions of the examined microalloyed structural steels (mass-%)

Table 1 gives the chemical compositions of the examined steels, which correspond to Ti-, Nb- and TiNb-variants of a hot rolled microalloyed high strength structural steel.

50 mm thin slabs have been produced by means of a special water cooled copper mould in a laboratory vacuum induction furnace and directly hot rolled on a 6.6 MN pilot mill. Experimental parameters as rolling force, torque and temperature are automatically recorded in order to provide exact experimental data. The variable parameter settings enable a detailed examination of the individual effects. **Table 2** shows the single experimental parameters.

Cast No.	Variant	Holding Temp.[°C]	WAT [°C]	WET [°C]	HKT [°C]	HT [°C]	KG1 [°C/s]	KG2 [°C/min]
752	Ti	Cooling down to ambient temperature in the mould.						
804	Ti	1200	1150	880	Air-Cooling			
805	Ti	1100	1050	870	Air-Cooling			
806	Ti	1150	1080	860	Air-Cooling			
860	Ti	1100	980	850	Air-Cooling			
861	Ti	-	1050	855	Air-Cooling			
862	Ti	-	980	855	Air-Cooling			
897	Ti	1100	1060	885	690	620	14.4	0.5
899	Nb	1100	1015	920	630	620	17.0	0.5
901	Ti +Nb	1100	1050	865	630	620	17.4	0.5

WAT = rolling start temperature   WET = rolling finish temperature

HKT = temp. at end of run out table   HT = coiling temperature

KG1 = cooling rate on run out table   KG2 = cooling rate during simulated coil cooling

slab thickness: 50 mm, hot band thickness: 10 mm

Table 2 Casting and hot rolling parameters of the examined microalloyed structural steels

The local cooling rate in the solidification temperature range has been calculated from Secondary Dendrite Arm Spacing (SDA) for the centre of the slab. As can be seen in **Figure 1**, the simulation mould's conditions are in good agreement with those given in literature /5/. In order to have a homogeneous and most complete TiN-precipitation according to /2, 6/, as cast slabs were held during 5 minutes at a temperature between 1100 and 1200°C before direct hot rolling. All slabs were hot rolled at rolling start temperatures between 980 and 1150°C and rolling finish temperatures of about 870°C. The slabs were rolled to a final thickness of 10 mm, corresponding to a total rolling reduction of 80%.

3. RESULTS

3.1. Transmission electron microscopical (TEM) investigations and grain size

TEM investigations have been carried out at the Institut für Festkörper- und Werkstofforschung (IFW) Dresden in order to determine the effects of the different experimental parameters on the precipitation behaviour in a Ti-variant steel. All examined samples showed a large variety of TiN and Ti(N,C) precipitates. Number and size of the precipitates are depending on the holding- and rolling start temperatures. The main part of the observed particles has a size in between 5 to 70 nm. This particle fraction subdivides into a d >10 nm division (**Figure 2**) which is missing when no holding is done before hot rolling and a 5-10 nm fraction **(Figure 3)**, which is always present when the rolling start temperature is below 1100°C.

Particles of about 2- 4 µm and in the range from 300 to 500 nm have been observed too. The 300 - 500 nm particles are arranged in a network structure when the as cast slab is cooled down to ambient temperature in the mould without deformation (Cast No. 752). This same slab exhibits a large number of particles < 5 nm, whereas the 5 to 70 nm particle fraction is less important /2, 3/.

The backdriving force P_R on a grain boundary may be expressed by the equation /7/:

$$P_R = -3/2 \cdot \gamma \cdot f/r$$

γ = grain boundary energy $\qquad$ f = volume fraction of dispersed phase

r = mean particle radius $\qquad$ f/r = dispersion degree

It is obvious, that a given volume fraction f of particles is more effective in retarding grain growth, when it is dispersed into a high number of small particles. Thus, the mean grain size will shrink with rising dispersion degree f/r. According to this, grain size should be depending on the precipitation state of TiN and Ti(N,C), which can be expressed in terms of the *dispersion degree f/r*. The volume fraction f of the dispersed phase was determined by evaluation of the TEM results giving number and size of the precipitates.

The results for different treatments of the Ti variant steel exhibit a close relationship between ferrite grain size and dispersion degree (**Figure 4**). As there is a well-established relationship between the austenite and ferrite grain sizes it may be supposed, that the variation of ferrite grain size is coupled with a change of austenite grain size. A detailed survey of the literature data is given in /8/. The mean grain size decreases with rising dispersion degree f/r. The highest dispersion degree and smallest grain size are obtained with a rolling start temperature of 1050 °C. Rolling start at 980°C without holding creates an inhomogeneous grain structure, consisting of mixed regions of large and small grains. As the dispersion degree of this sample is quite inhomogeneous, it is not given in Figure 4.

3.2. Mechanical tests

Figure 5 shows the results of the mechanical tests and the grain sizes as a function of the dispersion degree for the Ti-variant steel from Figure 4. All samples were air-cooled after hot rolling. The results indicate a strong influence of the dispersion degree on the grain size and on the mechanical properties of the hot rolled strip. A dispersion degree > 5 $*10^{-5}$/nm leads to a smaller grain size and to the highest yield and tensile strength. Ductile and toughness properties, which are given in terms of elongation at fracture and $A_{v50\%}$ transition temperature, are somewhat lower but still satisfying.

In order to rise the yield and tensile strength, experiments were carried out with accelerated water-cooling on the run-out table after hot rolling. The applied cooling rates were in the range of 12 - 23°C/sec. These quickly cooled down samples were subsequently transferred into a furnace and cooled down from 620 °C to ambient temperature with a cooling rate of 0,5°C/min, simulating coil-cooling conditions. The results are given in **Figure 6**. For comparison, the scattering band of conventional production is given too. Beside the Ti variant, a Nb- and a TiNb-alloyed variant have been examined. The Nb-bearing variants have been elaborated in the same way as the Ti-variants with a 5 min holding at 1100°C before hot rolling, accelerated water cooling and simulated coil cooling. As given in Figure 6, the TiNb- and Nb- variants exhibit somewhat higher yield and tensile strength than the Ti variant. This was primarily referred to lower temperatures at the end of the water cooling device, leading to more bainitic components in the microstructure.

4. DISCUSSION AND CONCLUSIONS

The observed precipitation state is according to the literature /2, 3, 4, 6, 9/. Coarse Ti-particles in the 2- 4 µm range precipitate during ferritic solidification in interdendritic residual melts, if a supersaturation $\eta > 10$ is exceeded /3/.

$$\eta = [\%Ti][\%N] / L_{TiN} \text{ , with } L_{(T)} = \text{solubility product}$$

Particles having diameters of 25-70 nm precipitate around dislocations during holding before hot rolling and 5-10 nm precipitates are deformation induced at rolling temperatures equal or below to 1100°C /4/. The very finely dispersed < 5 nm precipitates in the non deformed slab seem to be formed at lower temperatures in the ferritic state as Ti-carbides.

The impact of the variation of the dispersion degree on the materials mechanical and technological properties may be discussed in terms of its grain refining effect and of dispersion hardening. Grain refining leads to elevated strength and ductile properties whereas dispersion hardening only gives rise to strength but reduces ductility and toughness properties. Dispersed phases generally act to reduce ductility, because of the increased tendency towards crack nucleation and growth coupled with rapid work hardening /7/. In the case of the examined Ti-microalloyed structural steel, a dispersion degree of $15*10^{-5}$/nm should not be exceeded, in order to have an optimal combination of strength and toughness properties. As the results show, direct hot rolling of thin slabs with an optimised parameter set, including accelerated water-cooling on the run-out table gives the same hot band properties as conventionally produced material (Figure 6).

ACKNOWLEDGEMENTS

The authors like to thank Prof. Dr. A. Köthe, Dr. J. Kunze, C. Mickel and G. Backmann of the Institut für Festkörper- und Werkstofforschung (IFW) Dresden for performing the TEM experiments and various helpful discussions. They also gratefully acknowledge the financial support by the European Community (ECSC Steel Research Programmes).

LITERATURE

1   C. Lang, G. Stich, L. Meyer, "Verbesserung der Eigenschaften von Warmbreitband aus weichem unlegiertem Stahl", Stahl u. Eisen 106 (1986), Nr. 3, pp. 122 - 128

2   J. Kunze, Schlußbericht zum AiF-Vorhaben 358D/III, "Erarbeitung von Grundlagen-kenntnissen für die Erzeugung mikrolegierten Stähle durch Dünnbrammen-oder Bandgießen bzw. -gießwalzen.", 1993

3   J. Kunze, C. Mickel, M. Leonhardt, S. Oswald, "Precipitation of Titanium Nitride in low-alloyed Steel during Solidification", Steel Research 68 (1997), No. 9, pp. 403 - 408

4   J. Kunze, C. Mickel, G. Backmann, B. Meyer, M. Reibold, C. Klinkenberg, "Precipitation of Titanium Nitride in low-alloyed Steel during Cooling and Deformation", Steel Research 68 (1997), No. 10, pp. 441 - 449

5   W. Bleck, I. von Hagen, H. Pfeifer, U. Schriever, "Werkstofftechnische Entwicklungsmög-lichkeiten mit neuen Gießverfahren für Stahlflachproduktion", Stahl u. Eisen 118 (1998), No. 5

6   J. Kunze, B. Beyer, S. Oswald, W. Gruner, "Thermodynamic Data on the Formation of Titanium Nitride in Iron", Steel Research 66, 1995, No. 4, pp. 161-66

7   "Physical Metallurgy", ed. by R.W. Cahn, North-Holland Publishing Company Amsterdam-London, American Elsevier Publishing Company INC. New York, Second revised Edition, 1970, pp. 1178

8   B. Engl, J. Kruse,"Laborwalzungen als Ergänzung und Hilfsmittel für die Entwicklung von Simulationsmodellen zur Gefügeberechnung", Stahl u. Eisen 118 (1998), to be published

9   A. Köthe, J. Kunze, G. Backmann, C. Mickel, contribution to this conference

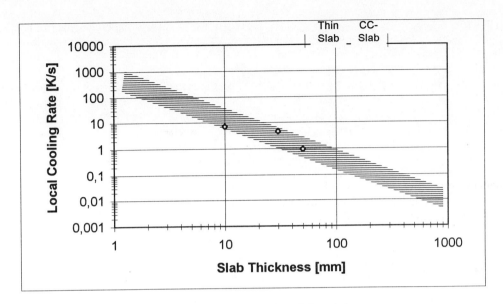

Figure 1   Local cooling rate in the center of the slab.
Calculated from secondary dendrite arm spacing (SDA).

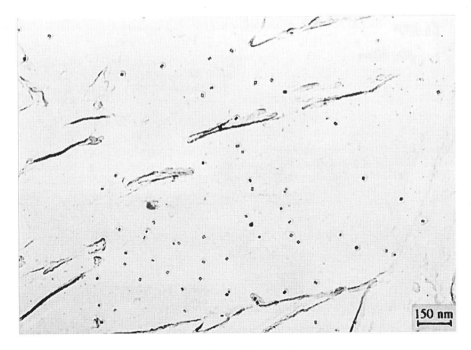

Figure 2   TiN-particles in the range d > 10 nm
TEM of a carbon extraction replica

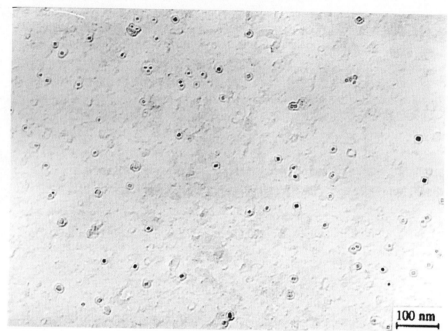

Figure 3 TiN-particles of the 5 - 10 nm fraction
TEM of a carbon extraction replica

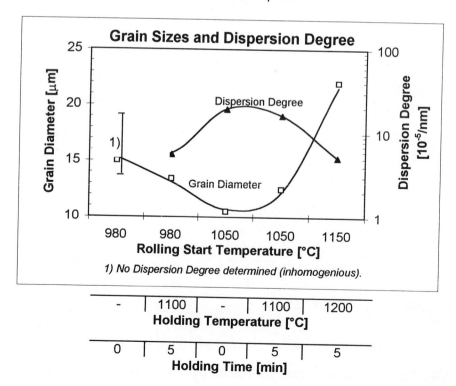

1) No Dispersion Degree determined (inhomogenious).

-	1100	-	1100	1200
		Holding Temperature [°C]		
0	5	0	5	5
		Holding Time [min]		

Figure 4 Grain sizes and precipitation state (Ti-variant, air-cooling)

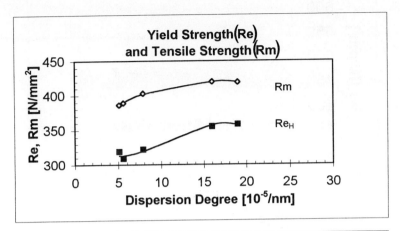

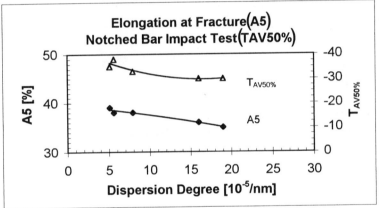

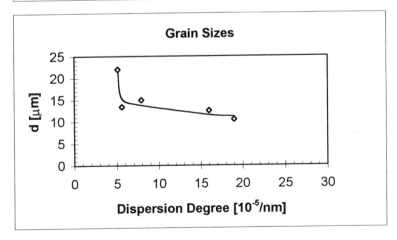

| Fig. 5 | Ti-microalloyed structural steel Influence of dispersion degree on hot strip properties after air-cooling | |

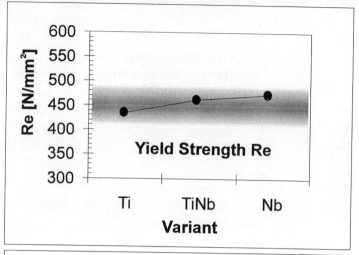

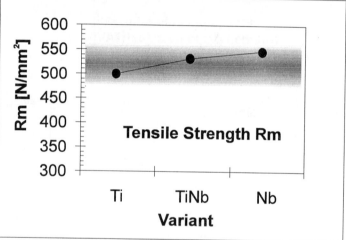

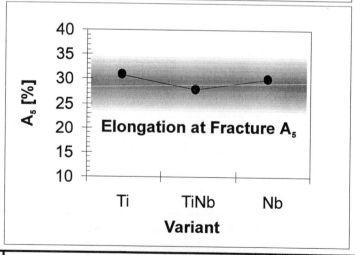

| Fig. 6 | **Mechanical properties of the microalloyed structural steel** Ti-, TiNb- and Nb-variant, water cooling *grey area: conventional production* | |

Materials Science Forum Vols. 284-286 (1998) pp. 509-516
© *1998 Trans Tech Publications, Switzerland*

Effect of Additional Elements on the Morphology of TiS
in Steels during Solidification

Y. Ouchi[1], K. Oikawa[2], I. Ohnuma[1] and K. Ishida[1]

[1] Department of Materials Science, Graduate School of Engineering, Tohoku University,
Aoba, Aramaki, Sendai, 980-8579, Japan

[2] Materials Engineering Division, Tohoku National Industrial Research Institute,
Nigatake, Sendai, 983-0036, Japan

Keywords: Titanium Sulfide, Inclusion, Morphology, Phase Diagram, Eutectic Reaction, Monotectic Reaction, Solidification

Abstract

The morphology of TiS in steels formed during solidification and the modification of this morphology by the addition of alloying elements of Mn, Zr, Cr and C were investigated by means of optical and scanning electron microscopy. It is shown that the morphologies can be classified into three types: (i) globular (ii) rod-like and (iii) plate-like sulfides. The morphology of TiS in the Fe-0.5%Ti-0.3%S base alloy is the rod-like sulfide while additions of Mn and Zr change the morphology from rod-like to globular. The tendency of enrichment of alloying elements in the sulfide is in the order Zr>(Ti)>Mn>Cr, which is similar to that in MnS. The formation of different TiS morphologies is discussed based on phase diagram information.

1. Introduction

Since non-metallic inclusions in steel are generally detrimental to its material properties, considerable efforts has been directed towards the development of alloys and processes to reduce their presence to a minimum. However, of late, the utilization of inclusions in steel as a means of controlling microstructures and thereby improving their mechanical properties, is becoming a subject of considerable interest [1]. In this sense, an understanding of the formation of different morphologies and distributions is essential for arriving at the conditions necessary for achieving the optimum morphology and distribution, commensurate with the required properties in a given steel.

It is well known that titanium, one of the most important micro-alloying elements, forms not only NaCl-type carbide and nitride but also hexagonal-type sulfide, TiS [2]. It has also been pointed out that TiS acts as precipitation sites of TiC in interstitial-free (IF) steels [3,4]. These facts suggest that the precise control of the morphology and distribution of TiS would be important in further development of micro-alloyed steel. However, very few studies [2,5,6] on TiS have been reported. In the present study, the morphology of TiS in steel during solidification was investigated by means of optical and scanning electron microscopy. The effect of Mn, Zr, Cr and C on the morphology of TiS was also investigated and the formation of the sulfide is discussed based on phase diagram information.

(a) Fe-0.25%Ti-0.3%S   (b) Fe-0.5%Ti-0.3%S

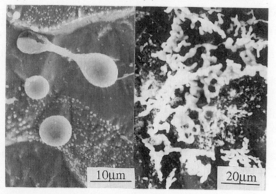

(c) Fe-0.75%Ti-0.3%S   (d) Fe-1.0%Ti-0.3%S

Fig.1 Scanning electron micrographs of sulfide
in the Fe-Ti-S as-cast alloys.

The composition of sulfide determined by EDS analysis is shown in Fig.4. According to the phase diagram of the FeS-TiS pseudo-binary system [10], a miscibility gap appears between FeS and TiS where maximum solubility of Ti into FeS is about 25 at%Ti. The present results show that the solubility of Ti in FeS of an as-cast specimen is comparable with the maximum solubility.

mass%Ti	0.25	0.5	0.75	1.0
morphology	⣿	⣿ψ	⣿ψ	▦
sulfide type	FeS	TiS	TiS	TiS

⣿	ψ	▦
globular sulfide	rod-like sulfide	plate-like sulfide

Fig.2 Morphology of sulfide in the
Fe-Ti-S as-cast alloys.

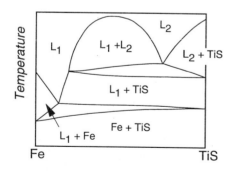

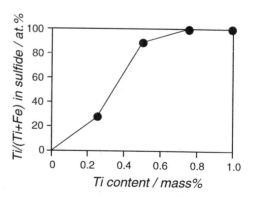

Fig.3 Schematic phase diagram of
Fe-TiS pseudo-binary system
proposed by Kaneko et al. [6].

Fig.4   Composition of sulfide in the
Fe-Ti-S as-cast alloys.

2. Experimental procedure

Each alloy (about 300 g) was melted by induction heating in a magnesia crucible under an Ar atmosphere. The procedure adopted in the case of each alloy was to melt the electrolytic iron first and then add the other elements namely, Mn, Zr, Cr and C, to the molten iron. After melting, the bath was allowed to solidify in the crucible. The cooling rate in the solidification range was about 30°C/min. The base composition of specimens was selected as Fe-0.5Ti-0.3S(mass%); typical examples of oxygen and nitrogen contents in selected alloy specimens are shown in Table 1.

Microstructural examination of specimens was conducted with an optical microscope and a scanning electron microscope (SEM). Specimens for microstructural examination were taken from the region 10 mm from the bottom of the ingots. The sections to be examined by SEM were mechanically polished and then etched electrolytically in a 1% tetra-methyl-ammonium chloride-10% acetyl-acetone-methanol solution. The composition and crystal structures of sulfide were determined by energy dispersion spectroscopy (EDS) and X-ray diffraction (XRD), respectively.

Table 1 Content of oxygen and nitrogen in typical specimens (mass%).

Ti	S	M	O	N
(0.5)	(0.3)	—	0.0053	0.0005
(0.5)	(0.3)	Cr(5.0)	0.0035	0.0027
(0.5)	(0.3)	Mn(3.0)	0.0073	0.0008
(0.5)	(0.3)	Zr(0.5)	0.0053	0.0003

( ) charged value

3. Results and discussion

3.1. Morphology of sulfide in the Fe-Ti-S system

Typical SEM micrographs of sulfide in Fe-(0.25~1.0)%Ti-0.3%S ternary alloys are shown in Fig.1 and the types of sulfide are summarized in Fig.2. Three types, namely, globular, rod and plate-like sulfide, were observed, which is quite similar to the well-known classification of morphology of MnS [7]. It can be seen from Fig.1 that the morphology changes from spherical to plate-like sulfide with an increase in Ti content. The sulfide of Fe-0.25%Ti-0.3%S specimen is FeS with bimodal globular shape of about 10μm in diameter and fine particles as shown in Fig.1 (a). Two types of morphology of the FeS have been reported as fine particle and cell wall structure formed by remelting reaction ($\alpha \rightarrow \gamma + L$) and eutectic reaction ($L \rightarrow \gamma + FeS(s)$), respectively [8]. The fine particle sulfide observed in this study is formed by remelting reaction. On the other hand, large globular sulfide which is similar to that of MnS in steel [9] is considered to be formed through metastable monotectic reaction ($L_1 \rightarrow Fe(s) + FeS(L_2)$) because the two-liquid region in the Fe-Ti-S ternary system extends to Fe-rich portion [10].

The rod or plate-like TiS has been observed in the case of high Ti content (Ti/S>0.5) [5], which is in good agreement with the present results. Figure 3 shows a schematic phase diagram of Fe-TiS pseudo-binary system proposed by Kaneko et al. [6], which suggests that the rod or plate-like sulfide would be formed through the eutectic reaction of $L \rightarrow Fe(s) + TiS$.

3.2. Effects of Additional Elements on the Morphology of Sulfide

Effects of alloy additions of Mn, Zr, Cr and C on the morphology of sulfide of Fe-0.5%Ti-0.3%S base alloys were investigated. Typical micrographs are shown in Fig.5(a)-(g). and the effects of alloying on the morphology of sulfide are summarized in Fig.6. The compositions of alloying elements in the sulfide of as-cast specimens were also determined by EDS analysis and are shown in Fig.7. The effect of individual elements on morphology is described below.

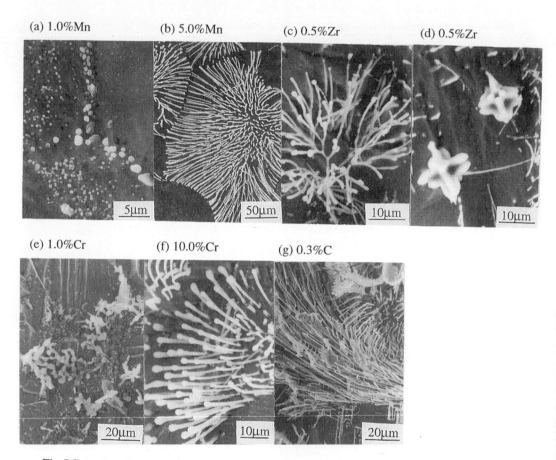

(a) 1.0%Mn   (b) 5.0%Mn   (c) 0.5%Zr   (d) 0.5%Zr

(e) 1.0%Cr   (f) 10.0%Cr   (g) 0.3%C

Fig.5 Scanning electron micrographs of sulfide in the Fe-0.5%Ti-0.3%S base as-cast alloys.

Addition of Mn and Zr

The addition of small amounts of these elements results in the formation of globular-like sulfide, while rod-like sulfide is formed with increasing alloying content as shown in Fig.5(a)-(c). In the 0.5%Zr alloy, tetrakaidecahedral shape of angular sulfide is formed and rod-like sulfide develops from the angular sulfide as shown in Fig.5(d). According to XRD analysis, the crystal structure of sulfide changes from hexagonal to NaCl-type of ZrS or MnS with increases in the content of Zr or Mn. The addition of 0.5%Zr results in the formation of NaCl-type ZrS, while both sulfides of MnS and TiS appear with the addition of 1%Mn and only MnS is formed by the addition of 5%Mn. These elements have a strong tendency to distribute in sulfide as shown in Fig.7 and form NaCl-type sulfide. These tendencies are very similar to the alloying behavior seen in Fe-0.5%S alloys [12]. The present results suggest that the solubility of Ti in MnS and ZrS is high and that of Mn and Zr in TiS is low, although the previous data [2] on the solubility of Ti in MnS is rather low in comparison with the present study. Further studies are needed to determine the solubility limit of these sulfides.

Since the phase diagrams of Fe-ZrS and Fe-MnS pseudo-binaries exhibit the eutectic type [6,9], the rod-type sulfide observed in the specimen of alloying Zr or Mn is formed through eutectic reaction. In 0.5%Zr alloys, the angular sulfide is formed by primary crystallization and then eutectic sulfide nucleates on the angular sulfide. This suggest that the solubility limit of ZrS for liquid iron is very small in comparison with that of TiS and MnS. On the other hand, the droplet-like sulfide formed in the specimens to which small amounts of Zr or Mn have been added may be due to the metastable monotectic reaction in analogy with the case of Fe-MnS base alloys [9], where the difference between the stable eutectic temperature and metastable monotectic temperature is so small that a metastable reaction leading to the formation of globular sulfide becomes a distinct possibility.

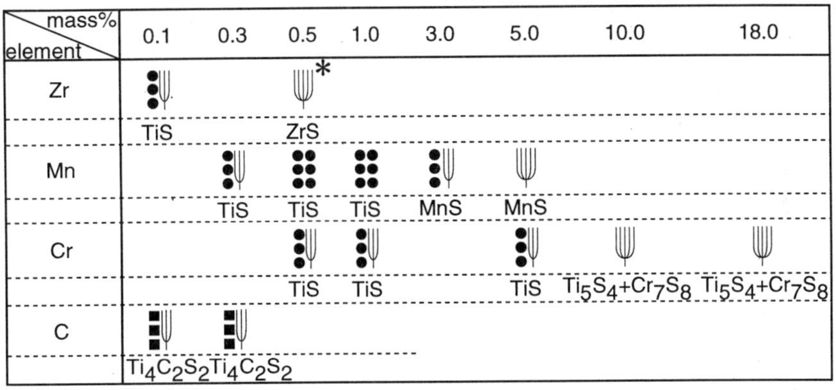

∗ : Primary + eutectic ZrS

Fig.6 Alloying effect on the morphology of sulfide in the Fe-0.5%Ti-0.3%S base as-cast alloys.

Addition of Cr

The morphology of sulfide in the specimens containing 0.5~5.0%Cr exhibits a mixture of globular and rod types, which is essentially the same morphology as that of Fe-0.5%Ti-0.3%S base alloy. This can be understood from the fact that TiS and CrS have similar hexagonal structures and that Cr has less tendency to concentrate into sulfide as shown in Fig.7. However, the morphology of sulfide of 10%Cr and 18%Cr alloys changes to the rod-like sulfide as shown in Fig.5(f), which has also been observed in the ferritic stainless steel [11]. Two types of sulfide, Cr_7S_8 and Ti_5S_4, are found in ferritic stainless steels with low Mn content [11].

The present results on the tendency of the enrichment of alloying elements in sulfide are in the order of Zr>(Ti)>Mn>Cr, which corresponds well with the data on Fe-S base alloys [12].

Addition of C

The morphology of sulfide by the addition of 0.3%C addition shows a mixture of rod-like and irregular plate-like sulfides. This type of morphology has often been observed in MnS [7,9], which is formed through eutectic reaction. The sulfide was identified by XRD measurement as $Ti_4C_2S_2$. This carbosulfide appears in ultra-low carbon steel with the addition of Ti (Ti-IF steel), where TiS and $Ti_4C_2S_2$ are formed at higher and lower temperatures, respectively [3,13]. In the present alloys, however, the carbon content is rather high compared with IF steel which stabilizes $Ti_4C_2S_2$ rather than TiS. The present results on the rod and irregular plate-like morphology resulting from the addition of C suggests that the $Fe-Ti_4C_2S_2$ quasi-binary phase diagram may be a eutectic type similar to that in the Fe-TiS system as shown in Fig.3.

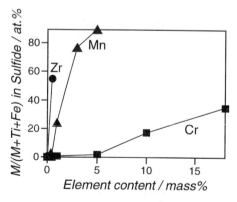

Fig.7 Composition of alloying
element M in sulfide.

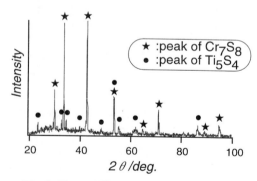

Fig.8 X-ray diffraction pattern of sulfide
in Fe-0.5%Ti-0.3%S-10.0%Cr
as-cast alloy.

4. Conclusions

The morphology of TiS and the effect of the addition of elements of Mn, Zr, Cr and C on the modification of morphology of sulfide in the Fe-0.5Ti-0.3S (mass%) base alloys have been studied. The results obtained are as follows:

(1) There are three types of sulfides in the Fe-Ti-S as-cast alloys: globular, rod-like and plate-like.

(2) The addition of a small amount of Mn or Zr results in the formation of globular sulfide, while rod-like sulfide of MnS or ZrS with a NaCl-type structure is formed with increasing Mn or Zr content.

(3) The addition of Cr results in little modification of sulfide morphology.

(4) The morphology of sulfide is discussed in terms of eutectic and monotectic-type phase diagrams in the Fe-sulfide quasi-binary system.

(5) The tendency of alloying element in sulfide is in the order of Zr>(Ti)>Mn>Cr.

Acknowledgments

This work was supported by Grants-in-aids for Scientific Research from the Ministry of Education, Science, Sports and Culture, Japan, and by Proposal-Based Advanced Industrial Technology R&D Program.

References

[1] J.Takamura and S.Mizoguchi, CAMP-ISIJ, 3 (1990), p.276.

[2] R.Kiessling, Non-metallic inclusions in steel, The Metals Society, London, (1978), p.128.

[3] N.Yoshinaga, K.Ushioda, S.Akamatsu and O.Akisue, ISIJ Int., 34 (1994), p.24.

[4] T.Obara, S.Satoh, S.Okada, S.Masui and K.Tsunoyama, Proc.Int.Conf.on physical Metallurgy of Thermomechanical Processing of steels and Other Metals, ISIJ, Tokyo, (1988), p.676.

[5] H.Kaneko,T.Nishizawa and K.Tamaki, J. Jpn. Inst. Met., 25 (1961), p.328.

[6] H.Kaneko,T.Nishizawa and K.Tamaki, J. Jpn. Inst. Met., 27 (1963), p.312.

[7] C.E.Sims and F.B.Dahle, Trans.Am. Foundrymen's Ass., 46 (1938), p.65.

[8] A.S.Keh and L.H.Van Vlack, Trans. AIME, 206 (1956), p.950.

[9] K.Oikawa,H.Ohtani,K.Ishida and T.Nishizawa, ISIJ Int., 35 (1995), p.402.

[10] R.Vogel and G.W.Kasten, Arch.Eisenhüttenw., 19 (1948), p.65.

[11] R.Kiessling, Sulfide inclusions in steel, Ed. by J.J.deBarbadillo and E.Snape, American Society for Metals, (1975), p.104.

[12] H.Kaneko,T.Nishizawa and K.Tamaki, J. Jpn. Inst. Met., 27 (1963), p.299.

[13] X.Yang, D.Van Der Schueren, J.Dilewijns, C.Standaert and Y.Houbaert, ISIJ Int., 36 (1996), p.1286.

Materials Science Forum Vols. 284-286 (1998) pp. 517-524
© 1998 Trans Tech Publications, Switzerland

Eutectic Precipitation of (TiNbV)(CN) in Cast, Microalloyed Low-C Austenite and Effects of Reheating

P.H. Li, A.K. Ibraheem and R. Priestner

University of Manchester/UMIST Materials Science Centre,
Grosvenor Street, Manchester, M1 7HS, UK

Keywords: Microalloyed Steel, Carbonitrides, Precipitation, Eutectic Carbonitrides, Modelling, Direct Rolling

Abstract

The constitution of austenite in thin-slab cast, low-C, Mn steel microalloyed with 0.007wt%Ti, 0.04wt%Nb and 0.011wt%V was simulated in small, laboratory castings. Two carbon levels, 0.007wt% and 0.22wt%C, each at two nitrogen levels, 0.003wt% and ~0.013wt%N, were investigated. After solidification the castings were quenched rapidly from ~1400°C in order to investigate the eutectic carbonitrides. Samples of the castings were reheated to a range of temperatures in the austenite phase field to investigate the fine precipitates that then appeared, and the re-solution of the eutectic carbonitrides. The compositions of precipitates were determined using high resolution electron microscopy with EDX and PEELS, and compared with the results of a computer model of the solution thermodynamics of $(Ti_xNb_vV_{1-x-v})(C_yN_{1-y})$ in austenite.

The eutectic carbonitride formed only in the higher-C steels independently of N content. However, increasing the nitrogen content of the steel increased the N:C ratio in the eutectic carbonitride. At both nitrogen levels the carbonitride was almost pure Nb(CN). After allowing for strong segregation of Nb and weak segregation of Ti to the liquid during freezing, these results agreed with the thermodynamic model, assuming the eutectic carbonitride to be in equilibrium with austenite. The presence of the eutectic carbonitride had little effect on subsequent precipitation in reheated austenite. The new precipitates were Nb-rich at low and intermediate temperatures, with some V at the lowest reheating temperatures. With increasing reheating temperature, first the V content and then the Nb content of these precipitates decreased, and their Ti content increased, as they also became N-rich, in approximate agreement with the computer model. At temperatures above approximately 1200°C the Ti tended to form Ti-rich Ti-Mn-oxides. The eutectic precipitates partially dissolved, but remained Nb-rich on reheating.

Introduction

In the conventional controlled rolling (CCR) of strip and plate products, continuously-cast slabs are allowed to cool to ambient temperature. They are then reheated, typically over a period of 6-8 hours to a temperature of 1250°C. Cooling and heating through the austenite-to-ferrite and ferrite-to-austenite transformation temperature ranges breaks down the original cast microstructure, and the time at temperature results in a significant degree of homogenisation. In the case of microalloyed steels the time at temperature is assumed to be sufficient to dissolve all or most of the microalloy carbonitrides, and to bring them close to equilibrium with their solid solution in austenite. The constitution of the austenite is then reasonably well defined, providing a sound basis for the prediction of recrystallisation and precipitation phenomena during subsequent hot rolling.

In Hot Direct Rolling (HDR) the temperature of the cast slab is maintained in the austenite phase field during transfer to the rolling mill. Thus, the constitution of the austenite at the start of rolling is inherited directly from the casting process. It is expected that the constitution of the just-cast austenite will be significantly different from that of reheated austenite, with respect to segregation, grain size and grain structure. In the case of microalloyed, low carbon steel, two additional, potential effects need to be considered.

One of these is that the austenite microstructure that is rolled in HDR practice is generated at the peritectic transformation temperature, in excess of 1450°C, at which the equilibrium solubility of microalloy carbonitrides is very much greater than at the soaking temperature used in conventional controlled rolling (CCR). This could result in the austenite being more supersaturated with respect to microalloying constituents during the rolling of just-cast austenite than during the rolling of reheated austenite, with consequential effects on the development of microstructure during thermomechanical processing. For steels containing about 0.01wt%Ti, it has been reported [1,2] that HDR resulted in higher yield strength of the product than CCR. This was attributed to increased supersaturation of the austenite with respect to TiN at rolling temperatures, giving rise to an increase in the recrystallisation-stop temperature and a consequential refinement of ferrite grain size, and to an increase in precipitation strengthening.

The second difference between just-cast austenite and reheated austenite of microalloyed steels might result from segregation of microalloying elements during freezing. Zhou *et al* [3] and Khan *et al* [4] investigated the HDR of a 0.04wt%Nb, 0.1wt%C steel and found that segregation of Nb to the liquid phase, during freezing at cooling rates that simulated 50 mm thick slab casting, led to the eutectic precipitation of NbC. Even though these precipitates should dissolve at equilibrium in austenite at temperatures above about 1120°C, they persisted throughout HDR processing. Thus, less Nb was available in solution to affect microstructure development than would be the case after conventional reheating, and the effect of the Nb that was dissolved was modified by its inhomogeneous distribution. It was found [5-7] that in steels containing less Nb (0.027wt%Nb, 0.08wt%C) the eutectic carbonitride was avoided; however, adding more than 0.012wt%Ti to those steels led to the eutectic precipitation of a Nb-rich carbonitride.

It is emphasized that these examples are of Nb-rich carbonitrides that were present in austenite after solidification and at temperatures at which they could not be in equilibrium with the austenitic solid solution: the precipitates should dissolve at equilibrium. It has also been shown by Kunze *et al* [8,9], although at higher Ti contents (upto 0.051wt%) that segregation of Ti in Nb-free compositions during freezing led to precipitation of TiN in the enriched liquid before complete solidification. However, TiN would be expected to be precipitated in their austenites at equilibrium. Similarly, in the cases of some other observations of Nb-rich carbonitrides precipitated interdendritically during freezing [10-13] it is not clear that the precipitates observed should dissolve at equilibrium at normal reheating temperatures.

In view of the differences that have been conjectured between the constitutions of just-cast austenite prior to HDR and reheated austenite prior to CCR, in the case of microalloyed, low carbon steels, a study of the microalloy precipitates present in austenite just after casting, and of their response to reheating, was initiated. The work reported here was concerned with four steels having approximately constant additions of nominally 0.006wt%Ti, 0.04wt%Nb and 0.011wt%V, but in which the carbon and nitrogen concentrations were varied independently. During the research the results of quantitative analyses of the composition of precipitates were compared with thermodynamic calculations of equilibrium compositions.

Calculation of Equilibrium Compositions of Microalloy Precipitates.

The solubility of an individual microalloy carbide or nitride of formula MX in austenite may be expressed in terms of the solubility product:

$$[M].[X] = 10^{A - \frac{B}{T}}$$

Eq. 1

where $[M]$ and $[X]$ are, respectively, the concentrations of microalloy element and of carbon or nitrogen in solution at the temperature T, and A and B are experimentally determined constants. A mass balance calculation is required to obtain the weight fraction of precipitate and the individual concentrations, $[M]$ and $[X]$, in equilibrium with the precipitate at temperatures below the solvus. The carbides and nitrides of titanium, niobium and vanadium that precipitate in microalloyed steels have the rock salt crystal structure, are isomorphous, and are mutually intersoluble. The metal ions (Ti, Nb, V) reside on one of the interpenetrating f.c.c. lattices and C and N reside on the other. In general, therefore, when Ti, Nb, V, C and N are all present in a steel austenite the equilibrium precipitate will be a compound expressed by the formula $(Ti_xNb_vV_{1-x-v})(C_yN_{1-y})$. in which, assuming stoichiometry, $0 \le x \le 1$, $0 \le v \le 1$, $(x+v) \le 1$ and $0 \le y \le 1$. Hillert and Staffansson [14] proposed a thermodynamic analysis for calculating the equilibrium between a mixed microalloy carbonitride and its solid solution from a knowledge of the equilibrium solubility products of the individual carbides and nitrides, which has been used as the basis of such calculations by a number of authors [15-21]. Accordingly, a computer programme was written to calculate the values of x, v and y, and of the concentrations of Ti, Nb, V, C and N in solid solution in the austenite, [Ti], [Nb], [V], [C] and [N], at equilibrium, as a function of temperature and as a function of steel composition. The values of *A* and *B* for the solubility products of individual carbides and nitrides used in the calculations are given in Table 1.

Compound	*A*	*B*	reference
TiC	5.33	10475	Narita [22]
NbC	3.42	7900	Narita [23
VC	6.72	9500	Narita [23]
TiN	0.322	8000	Matsuda [24]
NbN	2.8	8500	Narita [23]
VN	3.02	7840	Roberts [25]

Table 1. Data for solubility products of carbides and nitrides in austenite.

The computer model requires the concentrations of Ti, Nb, V, C and N in the steel to be input in wt%. It calculates the atomic fractions, x, v and 1-x-v, of Ti, Nb and V on the metal sublattice and the atomic fractions y and 1-y of C and N on the interstitial sublattice, the weight or volume fraction of precipitate, and the concentrations [Ti], [Nb], [V], [C] and [N] in equilibrium solution in the austenite as functions of temperature.

Experimental Details.

Four steels were made by vacuum/argon induction melting and cast into a heated ceramic mould. The ingots weighed approximately 1 kg and were 15 mm thick and 75 mm wide. The cooling rate between 1400°C and 1200°C was approximately 2 K/s, in order to simulate that in thin-slab casting. The mould was designed to allow the ingots to be stripped at approximately 1400°C and quenched in iced brine. The compositions of the ingots and the quench temperatures are listed in Table 2.

	C	N	Ti	Nb	V	Mn	Si	Quench Temperature
Cast 1	0.22	0.0024	0.006	0.04	0.011	1.27	0.25	1401
Cast 2	0.075	0.0032	0.007	0.04	0.011	1.23	0.23	1421
Cast 3	0.218	0.0117	0.005	0.04	0.011	1.27	0.25	1347
Cast 4	0.074	0.0145	0.006	0.04	0.011	1.27	0.25	1396

P and S were 0.006-0.008 wt%, Cu was 0.007-0.008 wt%, Al was < 0.002 wt%.

Table 2. Compositions of the experimental steels (wt%) and quench temperatures (°C).

This series of steels was designed to have constant Ti, Nb and V contents, but with high and low nitrogen contents at both high and low carbon contents.

Samples of the quenched ingots were polished and etched in nital, and examined using optical and scanning electron microscopy, and high resolution transmission and scanning transmission electron microscopy of extraction replicas. Quantitative analyses of extracted particles were obtained using EDX and EELS. Samples of quenched ingots were also reheated in evacuated quartz capsules at temperatures between 800°C and 1400°C for various times and then quenched. Particles extracted on replicas from these samples were also analyzed using high resolution electron microscopy.

Results and discussion.

Eutectic precipitates in the high-C steels. Microalloy carbonitrides were observed only in the quenched, as-cast microstructures of the high carbon steels, casts 1 and 3,. Fig. 1 illustrates the

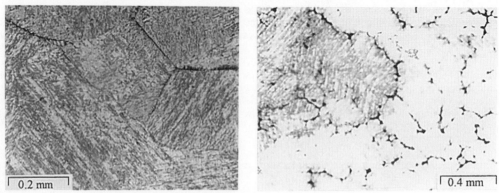

Fig.1. Distribution of eutectic precipitation in Cast 3.

distribution of microalloy carbonitrides (dark phase) in these steels at optical magnifications. In Cast 1 (high-C, low-N) the particles were virtually confined to austenite grain boundaries and grain corners. In Cast 3 (high-C, high-N) the particles were more profuse, and appeared in some interdendritic spaces within austenite grains as well as at grain boundaries. Fig.2 shows typical

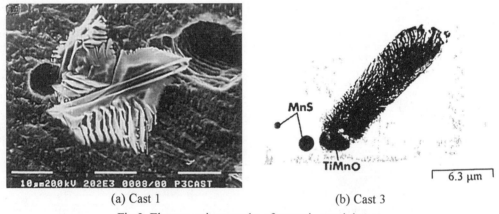

(a) Cast 1   (b) Cast 3

Fig.2. Electron micrographs of eutectic precipitates

examples of the particles at higher magnification, in the SEM after deep etching, Fig.2(a), and after extraction on a replica, Fig.2(b). They exhibited a mixed lamellar and rod-like morphology characteristic of eutectic particles and were usually associated with holes, which are interpreted as being microporosity and, therefore, associated with the last volume of liquid to freeze. They were nearly always associated with MnS particles and often with oxide particles.

The compositions of these precipitates are shown in Table 3

steel	x	v	1-x-v	y	1-y
	Ti/(Ti+Nb+V)	Nb/(Ti+Nb+V)	V/(Ti+Nb+V)	C/(C+N)	C/(C+N)
Cast 1	0.01	0.98	0.01	0.76	0.24
Cast 3	0.02	0.97	0.01	0.56	0.44

Table 3. Compositions of Eutectic Carbonitrides

The compositions of the metallic content of the carbonitrides are the averages of EDX analyses from at least 30 particles. The interstitial concentrations were obtained using EELS and are the averages of 7 analyses from the edges of particles. The eutectic particles were almost purely NbCN, the nitrogen content of which increased, but did not become the major interstial, in the higher nitrogen cast. The thermodynamic model (see Fig.5) predicts that the particles in equilibrium with the austenite at the quench temperature should be essentially Ti-rich nitrides. It is clear, therefore, that the composition of the eutectic carbonitride was determined by the greater segragation of Nb than of Ti during freezing.

If it is assumed that there is no diffusion in the solid and complete mixing in the liquid during freezing, the composition of the liquid when its volume fraction is f_L can be expressed as:

$$C_L = C_0 \cdot f_L^{k-1}$$

Eq.2

where C_0 and C_L are the original composition and the composition of the liquid during freezing and k is the equilibrium distribution coefficient. The volume fractions of eutectic were estimated by point counting, and found to be 0.1 for Cast 1 and 0.05 for Cast 3. Using values of $k = 0.3$ for Nb [10] and $k = 0.61$ for Ti [26], the compositions of the eutectic liquid at final solidification were calculated to be 0.22 wt% Nb, 0.0147 wt%Ti, 0.11 wt%V, 0.22 wt%C, 0.0024 wt%N and 0.325 wt% Nb, 0.016 wt%Ti, 0.11 wt%V, 0.218 wt%C, 0.0117 wt%N in Casts 1 and 3 respectively. (It was further assumed that V, C and N do not segregate.) The themodynamic model was then used to predict the composition of the carbonitride within these volumes, assuming it is in equilibrium with austenite at the quench temperature. The prediction gives the formula of the carbonitride in Cast 1 to be $(Ti_{0.08}Nb_{0.92}V_0)(C_0N_{0.2})$ and that in Cast 3 to be $(Ti_{0.09}Nb_{0.91}V_0)(C_{0.63}N_{0.37})$. These predictions are in fair agreement with the measured compositions presented in Table 3: despite the frailty of the assumptions, it is clear that the eutectic carbonitride should indeed be expected to be Nb-rich instead of nearly pure TiN because of segregation during freezing.

The eutectic precipitates were persistent and only partially redissolved, and became globuralized, on reheating into the austenite phase field. Fig. 3. The thermodynamic model (see Fig.5) indicates that

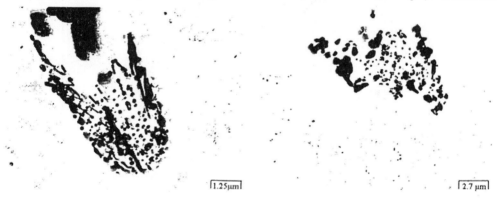

Fig.3. Eutectic precipitates after reheating at 900°C (left) and 1300°C.

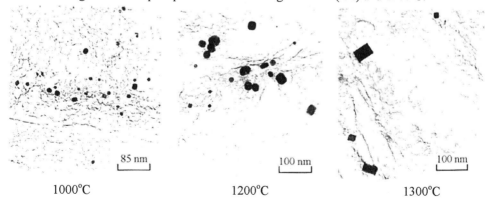

1000°C   1200°C   1300°C

Fig.4. New precipitation in austenite on reheating.

precipitate should be present even at 1400°C, but that it should be nearly pure TiN at high temperatures. However, the compositions of the eutectic precipitates were remarkably stable. Their Ti contents increased only very slightly on holding for upto 24 hours, except in the case of holding the high-C, low N steel, Cast 1, at 1300°C; after ½ hour its Ti content increased to about 5%, and after 24 houirs it increased to about 42%, still considerably less than expected at equilibrium.

Precipitation in the austenite on reheating. New carbonitride precipitates appeared on reheating the quenched steels into the austenite phase field. Fig.4 illustrates such precipitates in Cast. 1 (high-C, low-N). Broadly, these precipitates became larger and more cuboidal as the temperature was raised. Fig.5 shows the thermodynamic calculation of equilibrium compositions and volume fractions of precipitate as functions of temperature for each of the four steels. Fig.6 shows the compositions of the particles that were determined.

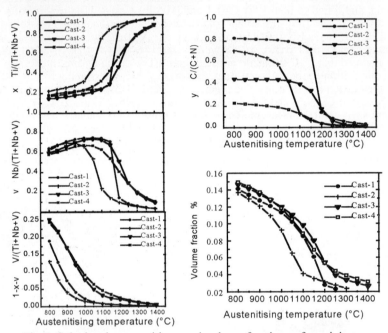

Fig.5 Calculated compositions and volume fractions of precipitates.

The thermodynamic model indicated that the equilibrium carbonitrides should be almost pure TiN at high temperatures, changing to be Nb-rich at low and intermediate temperatures. The carbon content of the Nb-rich precipitates at low and intermediate temperatures should increase with increasing ratio of C:N in the steel. The V content of the carbonitrides should be significant only at low temperatures. The analytical results were in approximate agreement with these trends. After reheating in the austenite phase field for only half an hour the precipitates were Nb-rich at temperatures up to 1300°C in Cast 1 (high-C, low-N) and up to approximately 1100°C in the other steels. In the high-N steels, Casts 3 and 4, The Ti content of the precipitates increased and their Nb content decreased at higher temperatures. However, the Nb content of the Nb-rich precipitates was significantly higher than predicted, and their Ti content correspondingly lower. The V content of the precipitates formed at low temperatures was lower than predicted. Holding at the austenitising temperature for 24 hours had little further effect on the microalloy content of the precipitates, except to increase their Ti content at high temperature in Cast 1, and to increase their V content a little at low temperatures. Samples of Casts 3 and 4 were held for 48 hours at 800°C, and this raised the V content of the precipitates further, close to the predicted equilibrium values.

Only a limited number of light element analyses using EELS were possible. The C:N ratio in precipitates formed in ½ hour at 900°C in Cast 2 (low-C, low-N) was 3:1, and in Cast 4 it was 1.2:1. After 24 hours at 900°C the C:N ratio in precipitates in Cast 4 was 1.5:1. The Ti rich carbonitrides in Cast 4 after 24 hours at 1200°C were found to be virtually pure nitrides.

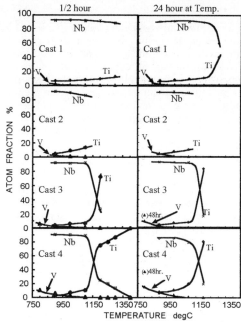

Fig.6 Measured compositions of precipitates.

It had been expected that the presence of the eutectic precipitate in Casts 1 and 3, and the fact that their compositions were so far from equilibrium, would affect the compositions of precipitates formed on reheating. This was clearly not observed, the low and intermediate temperature carbonitride precipitates being richer in Nb than predicted, rather than poorer, despite the loss of Nb from solution due to the presence of the Nb-rich eutectic carbonitrides.

At temperatures above about 1100°C in the case of the low-N steels, Casts 1 and 2, and above 1200°C in the high-N steels, TiMnO became the main Ti-containing particles precipitated in the austenite, instead of TiN. In the low-N steels the nitride precipitates were nearly always co-precipitated with oxide particles, whereas in the high-N steels less than half were associated with oxides.

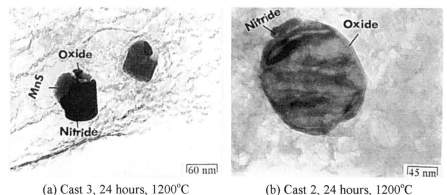

(a) Cast 3, 24 hours, 1200°C   (b) Cast 2, 24 hours, 1200°C

Fig.7. Co-precipitation of TiN with TiMnO

Fig.7(a) illustrates the co-precipitation of nitride and oxide particles (in this example with MnS also) in the high-N steel, Cast 3 after 24 hours at 1200°C. Here, the oxide is the smaller particle. Fig.7(b) shows nitride and oxide particles co-precipitated in the low-N steel, Cast 2, also after 24 hours at 1200°C. Here the nitride is the smaller particle. This figure suggests the possibility that the oxide may form *in situ* from the nitride.

Conclusions.

A thermodynamic model of the equilibrium between (TiNbV)(CN) and its solid solution in austenite suggests that the carbonitride should be nearly TiN at high temperatures, strongly rich in Nb at low and intermediate temperatures with a C:N ratio determined by the ratio of C:N in the steel, and V should dissolve in the NbCN to any significant extent only at low temperatures. These predictions were only approximately confirmed for carbonitride particles that precipitated in the matrix on reheating quenched castings into the austenite phase field. These precipitates were nearly pure TiN at high temperatures, except in the case of the low-C, low-N steel after a short reheating time. The carbonitrides present at low and intermediate temperatures were richer in Nb and poorer in Ti than predicted, and the V content of the precipitates only approached the equilibrium value after reheating for 48 hours at a low temperature.

In the two high-C steels a eutectic carbonitride was present after quenching from about 1400°C, which was NbCN virtually free of Ti and V. It was mainly carbide in which the C:N ratio depended on the nitrogen content of the steels. Although these carbonitride partially dissolved after reheating to high temperatures, their compositions remained Nb-rich. Their presence did not appear to affect the precipitation in the austenite on reheating. It was also observed that TiMnO tended to supplant TiN at temperatures above about 1200°C, and it is possible that initially precipitated nitrides transformed *in situ* to the oxide.

Acknowledgements. The authors are grateful to EPSRC for supporting this research through Grant GR/J37010.

References.

1. K. Kunishiga and N. Nagao, Trans. Iron Steel Inst. Japan., vol.25, (1985), p.315.
2. R.K. Gibbs, R. Peterson and B. Parker, Proc. Int. Conf. "Processing , Microstrucyure and Properties of Microalloyed Steels", Publ. ISS-AIME, Warrendale, Pa., (1992), p.201.
3. C. Zhou, A.A. Khan and R. Priestner, Proc. 1st. Int. Conf. on Modelling of Metal Rolling Processes, 21-23 Sept. 1993, London, publ. The Inst. of Materials, London, pp. 212-223.
4. A A Khan, R. Priestner, C. Zhou and A. K. Ibraheem, THERMEC'97, Wollongong, Australia, July, 1997, pp.
5. R Priestner and C Zhou, Ironmaking and Steelmaking, vol. 22, (1995), pp.326-332.
6. C, Zhou and R, Priestner, ISIJ International, vol.36, (1996), pp.1397-1405.
7. R. Priestner, C. Zhou and A.K. Ibraheem, "Titanium Technology in Microalloyed Steel", Ed. T.N. Baker, publ. The Inst. of Materials, London, (1997), pp.150-168
8. J. Kunze, C. Mickel, M. Leonhardt and S. Oswald, Steel Res. 68, (1997), No.9, pp.403-408.
9. J. Kunze, C. Mickel, G. Backman, B. Beyer, M. Reibold and C. Klinkenberg, Steel Res., (1997), No.10, pp.441-449.
10. V.K. Heikkinen and R.H. Packwood, Scand. J. Metallurgy, vol.6, (1977), p.170.
11. Z. Chen, M.H. Loretto and R.C. Cochrane, Mat. Sci. and Tech., vol.3, (1987), p.836.
12. D.C. Houghton, G.C. Weatherly and J.D. Embury, "thermomechanical Processing of Microalloyed Austenite", eds. A.J. DeArdo *et al*, TMS-AIME, Warrendale, Pa., (1982), p.267.
13. Zhang Hongtao, Guo Huawei and Wu Baorong, "THERMEC-88", ed. I. Tamura, (1988), p.170.
14. M. Hillert and I.L. Staffansson, Acta. Chem. Scand., vol.24, (1970), p.3618.
15. P.R. Rios, Mat Sci. Tech., vol.4, (1988), p.324.
16. W.J. Liu and J.J. Jonas, Metall. Trans., vol.20A, (1989), p.1361.
17. S. Okaguchi and T. Hashimoto, ISIJ International, vol.32, (1992), p.283.
18. H. Zou and J.S. Kirkaldy, Metall. Trans., vol.22A, (1991), p.1511.
19. H. Zou and J.S. Kirkaldy, Metall. Trans., vol.23A, (1992), p.651.
20. H. Adrian and F.B. Pickering, Mat. Sci. and Tech., vol.7, (1991), p.176.
21. H. Adrian, Mat. Sci. and Tech., vol.8, (1992), p.406.
22. K. Narite, J. Chem. Soc. Japan, vol.80, (1959), p.266.
23. K. Narita and S. Koyama, Japan Inst. Metals Journal, vol.52, (1966, p.292.
24. S. Matsuda and N. Okamura, Trans. ISIJ, vol.18, (1975, p.198.
25. W. Roberts and A. Sandberg, Tech. Rep. IM-1489, Swedish Inst. for Metals Research, Stockholm, (1985)
26. R.P. Elliot, "Constitution of Binary Alloys", McGraw-Hill Inc., (1965)

Correspondence: R.Priestner, e-mail: ronald.priestner@umist.ac.uk   fax: 0161 200 3586.

COLD DEFORMED AND ANNEALED STEELS

Materials Science Forum Vols. 284-286 (1998) pp. 527-534
© 1998 Trans Tech Publications, Switzerland

Recrystallization in Ultra Low Carbon (ULC) Steel: Influence of the As-Deformed Microstructure and Texture

B. Verlinden[1], I. Samajdar[1], P. Van Houtte[1] and L. Kestens[2]

[1] Department MTM, KU Leuven, de Croylaan 2, B-3001 Heverlee, Belgium

[2] Department Flat Rolling, CRM, Technologieperk 9, B-9052 Gent, Belgium

Keywords: Ultra Low Carbon Steel (ULC), γ (ND//<111>) Fibre, α (RD//<110>) Fibre, Recrystallization, Texture

ABSTRACT

Development of recrystallization texture in an ultra low carbon (ULC) steel was studied at four (50-90%) different cold rolling reductions. During deformation, a steady increase in α fibre (RD//<110>) was observed, while γ fibre (ND//<111>) increased from 0-50% reduction but then did not change significantly. In the recrystallization texture, however, a steady and significant increase of γ fibre, but no changes in α, was noted with increased reductions. Based on the as deformed state, two physical parameters were identified, which may explain the changes in recrystallization texture with strain. Those being: (1) spacings (as measured along ND) of γ/α oriented deformed bands and (2) relative ability of such bands to form recrystallized grains. An estimation of (2) for different orientations may be obtained from their respective nucleation factors (N_i, defined [6, 12] as the number of grains of a particular orientation per deformed band of the same orientation, measured/estimated along normal direction). Evidently, a decrease in γ band spacing and/or an increase/decrease in γ/α nucleation factors will strengthen the γ recrystallization texture. With increased reductions, spacings of the γ bands decreased - mainly from the geometrical considerations. On the other hand, above 70% reduction, nucleation factors for α bands, especially for I {112}<110> component, dropped significantly.

INTRODUCTION

The excellent deep drawing quality of low carbon steels is primarily due to the presence of strong γ-fibre (ND//<111>) texture, which, on the other hand, is typically obtained by a suitable combination of cold-rolling (deformation) and annealing (recrystallization) [1-10]. In spite of the existence of a relatively impressive amount of literature on the subject, the actual developments in the recrystallization texture remains a subject of intense scientific debate and of considerable industrial interest. It is generally agreed that the γ-fibre recrystallization texture evolves from the preferred nucleation and/or selective growth of the γ-oriented grains, presumably coming out from the deformed regions of near γ orientations [1-10]. Rather than solving the controversies of nucleation and growth, it is possibly far more important, at least technologically, to identify the exact parameters responsible for the development of recrystallization texture in low carbon steels. Evidently such parameters will be related to the as-deformed microstructure and microtexture [1-7].

Recent studies on IF (interstitial free) steel [6, 7] had shown that two features of the deformed structure may play a critical role in determining the final recrystallization texture and microstructure: (I) spacings of the α/γ oriented deformed regions [6] and (II) relative ability of such regions to provide recrystallized grains [6, 7]. In the present study, the development of recrystallization texture was studied at four different reductions and an effort was made to relate such developments with actual differences in deformed, but recovered, microstructure/microtexture.

EXPERIMENTAL METHODS

The ULC (ultra low carbon steel, for chemical composition see table I) was hot rolled (reheating, finishing and coiling temperatures being respectively 1250°, 920° and 750°C). Hot rolled material (with a weak transformation texture) was cold rolled to 50, 70, 82 and 90% reductions. Although the 50% deformed sample was recrystallized at 650°C, an annealing temperature of 550°C was selected for 70-90% reductions. Bulk texture measurements were obtained by inversion of four incomplete X-ray pole figures. X-ray ODFs (orientation distribution function) were calculated by using the software MTM-FHM, which uses the standard series expansion method [11]. For volume fraction measurements, X-ray ODFs were convoluted by appropriate model functions (with integrated ODF value of 1 and a Gaussian spread of 16.5°). Local orientation measurements were obtained by OIM (orientation imaging microscopy). Details of the OIM sample preparation in low carbon steels are given elsewhere [6, 7].

C	N	S	Mn	Al	Si	P
.0024	.0017	.003	.079	.021	.0045	.010

Table I. Chemical Composition (in mass %) of the ULC steel.

RESULTS

As in some of the previous studies [6-8], texture in the present ULC steel was modelled as α (RD//<110>) and γ fibres and random components (i.e. any orientation outside the α/γ fibres). F {111}<112> and E {111}<110> components were considered as part of γ-fibre, while I {112}<110> and H {001}<110> were considered as part of α. For X-ray volume fraction measurements a 16.5° gaussian spread was used, but local orientation measurements were characterized by 20° misorientation from the ideal components. Any orientation misoriented by more than 20° from the ideal F/E/I/H was considered random in the present study. Note that for the same samples, approximate values of the volume fractions (as measured from X-ray texture for 16.5° spread) and area fractions (as measured from OIM for 20° misorientation criteria) were nearly the same.

Figures 1 and 2 show the $\phi_2 = 45°$ ODF sections (as obtained by X-ray diffraction) describing changes in deformation (figure 1) and recrystallization (figure 2) texture at four different reductions (50-90%). For full recrystallization (as in figure 2) samples were annealed at 650° (50% reduction) and 550°C (70-90% reduction) and for times sufficient to cause complete recrystallization, but no noticeable grain growth. Note that the starting hot rolled material showed a weak texture. After 50% reduction (see figure 1a), noticeable presence of γ and α fibres was observed. Subsequent reductions

(70-90%) strengthened α further, although γ did not change significantly (see figures 1b to d). On the other hand, recrystallization texture of 50% deformed ULC steel is somewhat randomized (see figure 2a), while a gradual increase in γ fibre recrystallization texture was observed at higher reductions (see figures 2b-d). Volume fractions of the two fibres and their individual components may provide a narrower, but more quantitative estimation of the textural changes during deformation (see figure 3) and also during recrystallization (see figure 4). As described in figure 3, between 50-90% reduction the volume fraction of α fibre increased steadily with deformation, although no significant change in γ was observed. During recrystallization, however, the phenomenon was just reversed, see figure 4. On recrystallization, γ volume fractions increased with strain, while α remained almost the same. The changes in fibre volume fractions are reflected in their respective components.

Given the fact [6, 7] that γ/α oriented recrystallized grains are expected/observed to form/nucleate from the deformed regions of corresponding orientations, the recrystallization texture (and its evolution with strain) is also expected to be related to the as deformed microstructure or

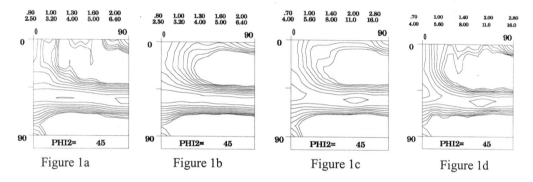

Figure 1a   Figure 1b   Figure 1c   Figure 1d

Figure 1. Developments in the deformation texture with strain. $\phi_2 = 45°$ sections of the X-ray ODFs after 50% (figure 1a), 70% (figure 1b), 82% (figure 1c) and 90% (figure 1d) reductions.

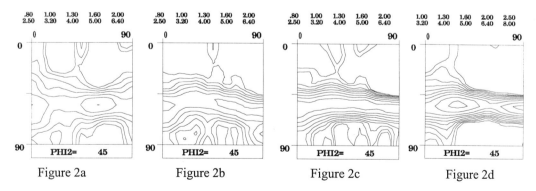

Figure 2a   Figure 2b   Figure 2c   Figure 2d

Figure 2. Developments in the recrystallization texture with strain. $\phi_2 = 45°$ sections of the X-ray ODFs for 50% (figure 1a), 70% (figure 1b), 82% (figure 1c) and 90% (figure 1d) deformed, but fully recrystallized, ULC steels.

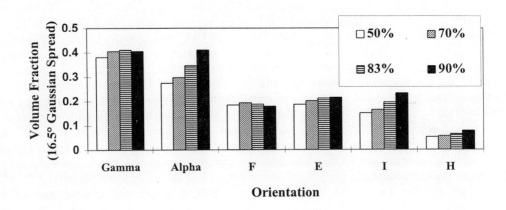

Figure 3. Volume fractions (after different reductions) of γ and α fibres and of F/E/I/H components in as deformed ULC steel.

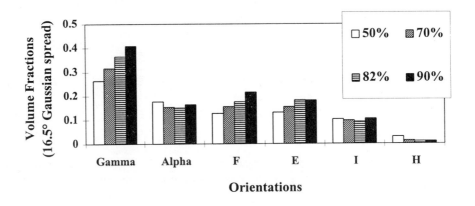

Figure 4. Volume fractions (after different reductions) of γ and α fibres and of F/E/I/H components in fully recrystallized ULC steel.

microtexture. Keeping this in mind, extensive OIM measurements were carried out from the as deformed, but recovered, ULC samples (typically the samples were given prolonged annealing at 400-450°C). Figures 5a-d show the OIM images of after 50 to 90% reduction in thicknesses of such ULC samples. The main observations from the OIM scans are as followed:

1) Band Thicknesses: Even at 50% reduction, grains were pancaked - see figure 5a. As in figure 5b-d, at larger reductions even more pancaking was observed. Average thicknesses (along ND) of the deformed grains/bands (irrespective of the orientations) were measured from the OIM scans and are listed in table II. Note that no significant differences between average thicknesses were observed between bands of different orientations.

2) Band Spacings: For each individual orientation (F/E/I/H), the relative spacings between bands were also measured from the OIM scans, see table III. As shown in table III, such spacings decreased with increasing reductions.

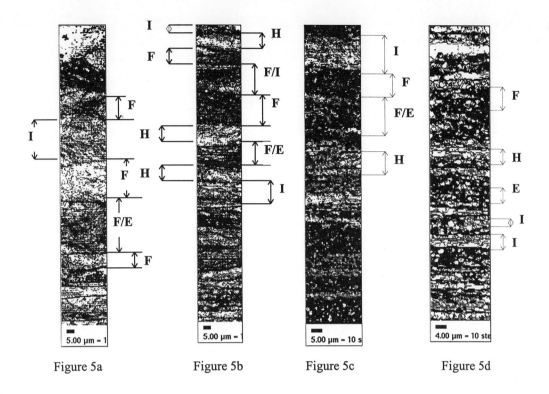

Figure 5a   Figure 5b   Figure 5c   Figure 5d

Figure 5. OIM images of a) 50%, b) 70%, c) 82% and d) 90% deformed (but recovered) samples. Grain boundaries were drawn from 1-20° (light boundaries) and >20° (dark boundaries) misorientation. Some (but not all) of the deformed α/γ grains are maked.

3) <u>Stored Energies:</u> As in earlier studies [6, 7], an effort was made to estimate the stored energies of different orientations from the OIM measurements of cell size (d) and cell misorientation (θ). Note that such estimations (although statistically very much representative) have two major drawbacks. Firstly, in such an estimation of stored energy only geometrically necessary dislocations (and not the statistically stored ones) are taken in account. Secondly, absolute accuracy of the OIM measurements are subject to the resolution (typically of the order of 0.5 μm, so some of the smaller cells might be missed) and also to the relative extent of recovery (note that for different samples different extent of recovery treatments were needed and used). Inspite of such limitations, the OIM scans do provide a distinct trend of change in the stored energies for different orientation. After 50% reduction, measured (θ/d)s were similar between deformed grains/bands of all orientations. At 70% reduction, however, stored energies of F/E was distinctly larger than I/H. The stored energy advantage for γ fibre orientations continued (and even slightly expanded) at higher reductions.

50%	70%	82%	90%
25	7	4.4	2.4

Table II. Average thicknesses (in μm and measured along ND) of the deformed bands at four different reductions.

Reduction	F	E	I	H	Random
50%	125	124	153	442	54
70%	31	30	37	107	16
82%	23	21	23	68	13
90%	14	11	10	30	8

Table III Average distances (or spacings, as measured along ND) in μm between deformed bands of different (e.g. F/E/I/H/Random) orientations.

Interestingly, the spreads of stored energies were noticeably <u>higher</u> in orientations with higher stored energies (especially for >70% reductions).

One probable cause for the increased stored energies in γ bands at and above 70% reduction is a somewhat more preferred appearance of 'strain localizations' in γ oriented deformed regions. The strain localizations are plastic instabilities or deformation inhomogenities and are visible in the OIM scans in the form of 'more frequent grain boundaries of relatively higher misorientations'. Such strain localizations (as in figure 5a) were often observed at an angle of ≈ 37° with the rolling direction (RD). At 50% reduction, their appearance (as observed in the OIM scans) was relatively independent of the orientation of the deformed grain/band. In the ULC deformed to 70% reduction (and more so at even higher reductions), however, the strain localizations (which were definitely more intense) appeared on a preferential basis in the deformed bands of γ fibre, which possibly brought the changes in stored energies, as shown in table IV.

Reduction	F	E	I	H
50%	1.4	1.3	1.43	1.36
70%	2.78	2.6	2.12	1.38
82%	3.1	3.2	2.1	1.36
90%	3.23	3.16	2.2	1.39

Table IV. Average (°/μm) values of θ/d (cell misorientation/cell size) in F/E/I/H and after 50-90% deformation (+ the recovery treatment).

DISCUSSION

The investigation of the deformed structure (figures 5a-d) clearly showed that the main components of γ/α fibres exist as deformed grains/bands. The initial hot rolled grain size was approximately 50 μm. Without the formation of any new grain boundaries, each original grain will form an average deformed band thickness of $50(1-F_d)$ μm along ND (where F_d is the fraction deformation/reduction). For 50% reduction, the measured and estimated values were similar, while for 70-90% reductions the measured values were approximately 1/2 of the estimated values, which indicate (statistically) that each grain, on average, has split into two [6, 12]. Note that the appearance of the relatively more intense (and appearing somewhat more preferentially in deformed γ regions) strain localizations accompanied the observed grain splittings.

The pancaked deformed grains/bands of α/γ orientations were observed to serve as the <u>only</u> sources of respective recrystallized grains. Evidently, the final recrystallization texture should depend on two factors - (I) availability and (II) potency of such sources. Relating these two somewhat abstract factors to more physically based parameters possibly holds the key to any successful semi-quantitative prediction of the final recrystallization texture. In rolling (or in an idealized plane strain compression), spacings of the deformed grains/bands may change significantly only along ND, as the deformed grains may be considered as infinitely long ribbons along RD with constant spacings along TD. Hence the factor (I) or availability of recrystallized α/γ grains should relate to their respective spacings along ND. Such spacings were measured from the OIM scans (as in table III). They may also be estimated from a combination of X-ray and microstructural data (as the spacing of an orientation i is $\lambda_i = t_i/V_i$, where t_i and V_i are the deformed band thickness and volume fraction for the orientation i). Note that the present OIM results closely approximate such estimated values. The second factor (the potency of α/γ bands) is more difficult to assess, as it may involve nucleation as well as growth. However, a simple quantification of factor (II) is possible, if one considers the concept of so called nucleation factor (N_i) [6, 12]. This nucleation factor has been conveniently defined [6, 12] as the number of recrystallized grains of i orientation coming out of an average i band, as measured or estimated along ND. The volume/area fraction of fully recrystallized i, as obtained from simple geometry, will be $N_i d_i/\lambda_i$, where d_i is the recrystallized grain size. Using this simple formula (and using area/volume fractions measured by OIM/X-ray ODFs), N_is were estimated for different orientations and at different reductions, see table V. The numerical values of N_is may depend on the nucleation/growth potentials and also on the relative thicknesses of the deformed bands , see[12]. Although thicknesses of the deformed bands decreased with increasing reductions (see table II), for any particular reduction they were similar for all orientation types. Hence the numerical values of N_is at any particular reduction may give an index of the relative nucleation/growth advantage/disadvantage.

As given in table V, N_is for all orientations dropped, as the percentage reduction increased from 50 to 70%. This was primarily because of the fact that recrystallization at 50% reduction was mainly from the relatively few grain-interior strain localizations and hence the recrystallization mechanisms (or kinetics) was quite different than at the other reductions. Between 70 and 90% reductions, N_is for F remained almost the same, but N_i of E dropped (at 90% reduction). At 90% reduction, an average band contained 3 dislocations cells - which may give a maximum N_i of 1; while for other reductions bands were thicker and correspondingly a maximum $N_i > 1$ may be possible [12]. Hence a drop in N_i of E at 90% reduction does not neccesarily indicate a reduced nucleation/growth advantage for E, but an increased advantage for F. Another interesting development was the estimated dramatic drop in the N_i of I component above 70% reduction, see table V. There are two apparent causes for this drop. Firstly, the absolute values (as well as the spread) of the stored energy of F/E was significantly more than I above 70% reduction. This was apparent caused by somewhat more preferred formation of strain localizations in F/E bands. Secondly, some (and not all) of the I bands (above 70% deformation) were observed to be separated by only 1 or 2 bands of other orientations. Newly recrystallized I grains were often stopped by such neighborhood I bands, in somewhat classic observations of orientation pinning [6,12]. This also contributed to the reduced N_i for I.

Reduction	F	E	I	H	Random
50%	0.69	0.71	0.68	0.56	1.43
70%	0.28	0.27	0.21	0.09	0.55
82%	0.31	0.29	0.07	0.024	0.54
90%	0.28	0.181	0.095	0.03	0.355

Table V. Nucleation factors (Ni) estimated for the deformed bands of different orientations.

SUMMARY

1) Between 50 and 70% reduction, the strength of the γ fibre deformation texture did not change significantly, although a steady increase in α was observed. On recrystallization, however, γ increased steadily with increased reductions, while α did not change.

2) Two physical parameters of the as deformed state were identified, which may relate the as deformed microtexture to the final recrystallization texture. These are: (I) spacings (as measure along ND) of different deformed bands and (II) relative ability of such bands to form recrystallized grains. Nucleation factor (N_i, number of recrystallized grains of a particular orientation per deformed band of the same orientation, as measure/estimated along ND) may serve as an index of (II).

3) At higher (>50%) reductions, spacing of the γ bands reduced, while N_is for α and random bands dropped. A combination of these two defined the final γ recrystallization texture.

4) The drop in γ band spacing was purely geometric, as γ fibre volume fractions remained unchanged with reductions above 50%. Two causes for the drops in N_is of α and random bands were identified: (a) stored energy advantage for γ bands or disadvantage for α (possibly caused by the preferred formation of strain localizations in γ oriented deformed regions) and (b) orientation pinning in some of the closely spaced α.

Acknowledgement

Financial support from IUAP contract no P4/33 and supply of the rolled materials from CRM are well appreciated.

REFERENCES

[1] W.B. Hutchinson and L. Ryde, 16th Risø symp. on Mater Sci., eds. N. Hansen, D. Juul Jensen, Y.L. Liu and B. Ralph, Risø National Lab, Roskilde, Denmark, 1995, p. 105.

[2] R.K. Ray, J.J. Jonas and R.E. Hook, Int. Mater. Rev., **39** (1994) 4, p. 129.

[3] W.B. Hutchinson, Int. Mater. Rev., **29** (1984) 1, p. 25.

[4] L. Kestens and J.J. Jonas, Metall. and Mater. Trans., **27A** (1996), p. 155.

[5] W.B. Hutchinson and E. Lindh, Int. Forum on Phys. Met. in IF steel, Tokyo, ISIJ, 1994, p. 127.

[6] I. Samajdar, B. Verlinden and P. Van Houtte, Acta Mater., In Press.

[7] I. Samajdar, B. Verlinden, P. Van Houtte and D. Vanderschueren, Mater. Sci. Eng, In Press.

[8] I. Samajdar, B. Verlinden, P. Van Houtte and D. Vanderschueren, Scripta Mater., **37** (1997) 6, p. 869.

[9] F. Emren, U. von Schlippenbach and K. Lücke, Acta Metall., **34** (1986), p. 2105.

[10] N. Rajmohan, Y. Hayakawa, J.A. Szpunar and J.H. Root, Acta Mater., **45** (1997) 6, p. 2485.

[11] P. Van Houtte, Manual of MTM-FHM, ed. MTM-KULeuven, Belgium, 1995.

[12] R.D. Doherty, I. Samajdar, C.T. Necker, H.E. Vatne and E. Nes, ibid [1], p. 1.

Bert.Verlinden@mtm.kuleuven.ac.be

Materials Science Forum Vols. 284-286 (1998) pp. 535-542
© *1998 Trans Tech Publications, Switzerland*

Effect of Al Content on r-Value and Recrystallization Texture of B-Bearing Al-Killed Steel Sheets

Y. Funakawa[1], T. Inazumi[1], Y. Hosoya[1] and T. Murayama[2]

[1] Materials and Processing Research Center, NKK Corporation,
1 Kokan-Cho, Fukuyama-Shi, Hiroshima 721, Japan

[2] Applied Technology Research Center, NKK Corporation,
1 Minamiwatarida-Cho, Kawasaki-Shi, Kanagawa 721, Japan

Keywords: Boron, Al-Killed Steel Sheet, Continuously Annealing, r-Value, Texture, Carbide Morphology

Abstract

It has been demonstrated that boron (B) had a beneficial effect on the mechanical properties of continuously annealed Al-killed steel sheets. However, the effect of Al content on r-value in the B-bearing steels has not been well discussed yet. In this study, the effect of Al content on r-value and recrystallization texture formation was investigated for laboratory-melt B-bearing low C Al-killed steel sheets.

The r-value of continuously annealed B-bearing steel increased with decreasing Al content under the same grain size. In the steels with lower Al (Al=0.005%) content, the carbide coarsening in hot-rolled sheet was more significant as compared to those with higher Al (Al=0.055%) content. The recrystallization temperature was lowered with decreasing Al content, and the development of <111>//ND components was accelerated while that of <110>//ND components was retarded. Since the increase in Al content accelerates AlN precipitation in hot-rolled sheets, B might remain as in solution and segregate to grain boundaries. On the other hand, the reduction in Al content causes preferential precipitation of BN in hot-rolled sheets. The improvement of r-value by reducing Al content in B-bearing low C steels was caused by the development of <111>//ND components which reduced by solute B.

1.Introduction

It has been reported that the B addition decreased the yield strength and increased the total-elongation and r-value of Al-killed steels by improving the grain growthability by substituting larger BN for AlN [1] . The effects of B/N ratio and Al content on the mechanical properties of B-bearing Al-killed steels have been widely studied. It has been reported that deep drawability of B-bearing steels could be improved either by optimizing the B/N ratio between 1.0 and 1.5 in wt% or by reducing Al content [2,3]. The decrease in r-value under the B/N ratio of more than 1.5 was considered to be attributed to the insufficient grain growth caused by fine $Fe_{23}(CB)_6$ precipitation. As for the effect of Al content, however, the mechanism has not been well discussed yet. In this study, the effect of Al content on r-value and recrystallization behavior of B-bearing Al-killed steel sheets was investigated.

2.Experimental procedures

B-bearing low carbon Al-killed steels with different levels of Al contents were prepared by vacuum induction melting and cast into 50kg ingots. The chemical compositions of the steels are shown in Table 1.

Table 1 Chemical compositions of steels used in this study (wt%).

No.	C	Si	Mn	P	S	Al	N	B	B/N
A	0.019	0.01	0.14	0.013	0.010	0.005	0.0028	0.0016	0.74
B	0.019	0.01	0.15	0.013	0.011	0.055	0.0023	0.0011	0.62
C	0.013	0.01	0.13	0.014	0.009	0.010	0.0028	0.0020	0.93
D	0.012	0.01	0.14	0.014	0.012	0.014	0.0026	0.0024	1.20

Steels A and B had B/N ratio of about 0.7 and their Al contents were 0.005% and 0.055% , respectively. Steels C and D were prepared to investigate the effect of B/N ratio on recrystallization behavior of the steels with lower Al content. After hot-rolling to 30 mm thick bars, they were cut to 120 mm long and 100 mm wide specimens. They were reheated at 1350℃ for 30min in vacuum followed by furnace cooling to simulate the thermal history of commercial slabs. The specimens were reheated at 1080,1120,1150,1250℃ for 1 hour to change grain size of annealed sheets under the same annealing temperature and hot-rolled to 4.5 mm thick sheets. The finishing temperature was controlled within 880℃-900℃. They were coiled at 700℃ and cooled in a furnace after holding at 700℃ for 1 hour to simulate the hot-coiling. After reducing the thickness by grinding, they were cold-rolled to 0.7 mm thick sheets, and annealed at 750℃ for 90 seconds followed by overaging at 350℃ for 120 seconds by using salt bath. The r-value was evaluated under the tensile strain of 12% by using JIS 5-modified type specimens(GL:25 mm,GW:12.5 mm) after 1.7% temper rolling. For the investigation of recrystallization behavior, cold-rolled sheets were heated to the temperature ranging from 570℃ to 750℃ under the heating rate of 20℃/sec followed by N_2 gas quenching without soaking. The developments of cold-rolling texture and recrystallization texture were evaluated by X-ray integrated intensities of <111>//ND, <100>//ND, <110>//ND fiber textures.

The microstructures were observed by optical and scanning electron microscopes for the cross section parallel to longitudinal direction. The measurement of the ferrite grain size was carried out by lineal analyses.

BN^- and C^- secondary ion images were detected by the SIMS using Cs^+ primary ion bombing. The accelerating voltage and the diameter of analyzed area were 10kV and 150 μ m, respectively.

3.Results and discussion

3.1 Effect of Al content on the correlation between r-value and grain size

Correlation between the r-values and the ASTM grain size numbers is shown in Fig. 1. R-value of steel A (Al=0.005%, B/N=0.74) was higher than that of steel B (Al=0.055%, B/N=0.62) under the same grain size. This result suggests that the Al content dominates the preferred orientation in the recrystallization texture formation. In addition, the plot of steel C (Al=0.010%, B/N=0.96) is on the line of steel A, while that of steel D (Al=0.014%, B/N=1.20) is on the line of steel B. This suggests that the increase in Al content has similar effect as increasing solute B content on the recrystallization texture formation.

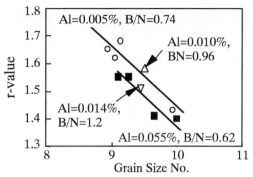

Fig.1 Effect of Al content and B/N ratio on r-value

3.2 Effect of Al content on recrystallization behavior

Figure 2 shows the change in <111>//ND, <100>//ND and <110>//ND fiber textures during recrystallization in all the steels for which soaking temperature was 1250℃. While there was no difference in the cold-rolling texture, <111>//ND components of steels A and C started increasing at lower temperature as compared to steels B and D. On the other hand, the development of <110>//ND components of steels A, B and C were retarded as compared to that of steel D. These results suggest that the increase in Al content suppresses the development of <111>//ND components and promotes the development of <110>//ND components. The effect of Al content on the development of <111>//ND components is similar to that of B in solution. Intensities of <111>//ND components in steels C and D were decreased at 675℃ before starting development. It has been reported that recrystallized grain of <111>//ND components nucleates both near grain boundaries as {111}<011> and in matrix as {111}<112>. Recrystallization of {111}<112> grain was suppressed by recrystallization of {110}<001> from the deformation bands due to grain boundary constraint by solute C, N and B and, then, such decrease in intensities of <111>//ND components was observed in IF steel [4]. The similar phenomena is thought to occur in this case. However, the effect of B segregation on grain boundary constrain in low C Al-killed steel has not been examined yet and the further investigations are necessary.

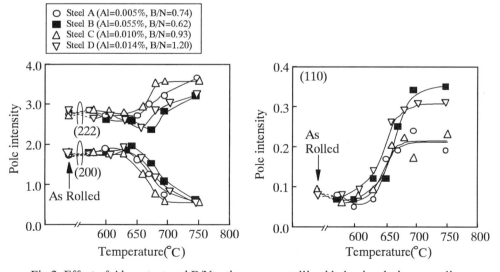

Fig.2. Effect of Al content and B/N ratio on recrystallized behavior during annealing.

It has been reported that the nucleation behavior of recrystallized grains is influenced by not only the ferrite grain size [5] but also the amount of solute C. In a low C Al-killed steel, the coarsening of the ferrite grain of hot-rolled sheet decreases the r-value since the recrystallized grains of <111>//ND components preferentially nucleate near grain boundaries [6]. Increase in solute C also decreases the r-value caused by the development of deformation bands during cold-rolling process which enhances the nucleation of <110>//ND components during annealing. Figure 3 shows the optical micrographs of the hot-rolled sheets and the annealed sheets of steels A and B. The grain sizes of both annealed steels were almost same. Since the grain size of the hot-rolled sheet of steel B was smaller than that of steel A, the grain size of the hot-rolled sheet can hardly dominate the r-value to be lower in steel B. Table 2 shows the amount of solute C and N measured by internal friction. Because the amount of solute C in steel B was less than that in steel A, change in solute C in hot-rolled sheet is not the cause for the retardation in nucleation of recrystallized grains in steel B.

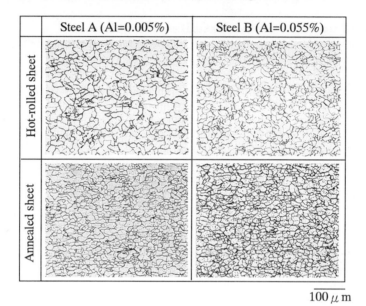

$\overline{100\,\mu\,\mathrm{m}}$

Fig.3. Optical microstructures of steels A and B (reheated to 1250℃).

Table 2 Amount of solute C and N in hot-rolled sheets.

	C(ppm)	N(ppm)
Steel A	11	5
Steel B	7	4

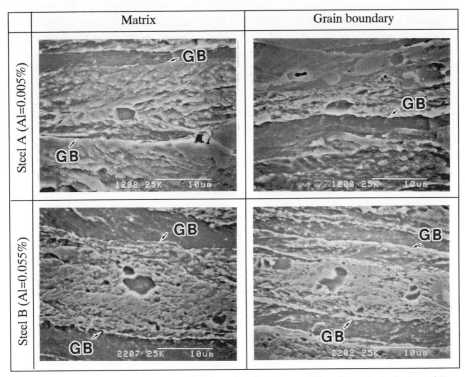

Fig.4. Nucleation sites of recrystallized grains during rapid heating of steels A and B.

Figure 4 shows the SEM micrographs of the specimens annealed at 600℃. Nucleation of the recrystallized grains were observed near grain boundaries only for steel A. The difference between steels A and B suggests that the nucleation of recrystallized grains near grain boundaries is suppressed by increasing the Al content. As shown in Fig. 5, the coarsening of the grain boundary carbides was insufficient in steel B as compared to steel A. In low C Al-killed steels, coarsening of hot-rolled sheet's carbides is effective to improve the r-value of continuously annealed sheets by reducing the dissolution of carbides during recrystallization [7]. The insufficient coarsening of the hot-rolled sheet's carbides is considered to be one of the causes for the lower r-value in steel B.

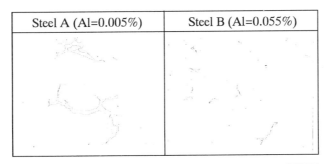

$\overline{25 \mu m}$

Fig.5. Optical carbide morphology of steels A and B (reheated to 1250℃).

Figure 6 shows the fraction of solute B in hot-rolled sheets evaluated by chemical analyses. Free B was hardly retained in steel A, while about 20% of total B was retained as in solution in steel B. The increase in solute B with increasing Al content may be attributed to the accelerated AlN precipitation which decreases BN precipitation. It has been reported that the solute B segregated to ferrite grain boundaries during $\gamma \rightarrow \alpha$ transformation in the B-bearing IF steel [8]. The segregation of B at grain boundaries in steel B without forming BN competes with the C segregation and carbide nucleation at grain boundaries. The effect of solute B at grain boundaries, such as the suppression of nucleation at grain boundaries by solid solution strengthening as well as solute C, may also have to be taken into account. Despite the effect of $Fe_{23}(CB)_6$ on the r-value has been considered in the B-bearing steel [9], $Fe_{23}(CB)_6$ was not observed in this study.

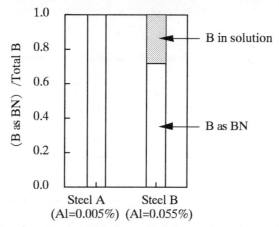

Fig. 6. Fraction of solute B in hot-rolled sheets of steels A and B.

3.3 Mechanism for change in carbide morphology

Figure 7 shows the morphological change of carbides for the hot-rolled sheets of steel B. The fine carbides precipitated not only at grain boundaries but also in matrix. Figure 8 shows BN⁻ and C⁻ secondary ion images by SIMS. Almost the same distribution was observed for BN⁻ and C⁻. Since the BN precipitation is thought to start at higher temperature, this result suggests that BN can become the nucleation sites for carbide precipitation in matrix.

Fig.7. Carbide distribution in hot-rolled sheet of steel B.

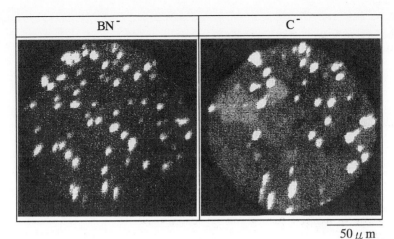

50μ m

Fig.8. BN⁻ and C⁻ secondary ion images showing distribution of BN and carbide using by SIMS.

4. Conclusions

The effect of Al content on r-value and recrystallization texture of B-bearing Al-killed steel sheets was investigated and the following conclusions were obtained.

(1) R-value of B-bearing Al-killed steel was reduced by increasing Al content even at the same grain size.

(2) The development of <111>//ND components was retarded by increasing Al content, which was similar to the case of increasing solute B content.

(3) Nucleation of recrystallized grains at grain boundaries was suppressed by increasing Al content, and it is considered to be the cause for the retardation in development of <111>//ND components.

(4) In the B-bearing Al-killed steel sheets, carbide precipitation occurred not only at grain boundaries but also in matrix of hot-rolled steels because BN could become the nucleation sites for carbide precipitation. Carbide coarsening was suppressed by increasing Al content because solute B was retained at grain boundaries by the precipitation of AlN and retard C segregation at grain boundaries.

(5) The decrease in r-value with increasing Al content for cold-rolled B-bearing Al-killed steels could be attributed to the increase in solute B at grain boundaries to suppress carbide coarsening in hot-rolled steel sheets and to retard the development of <111>//ND components. B segregation at grain boundaries resultantly causes deterioration in r-value of continuously annealed sheets.

References

[1] N.Takahashi, M.Shibata, Y.Furuno, H.Hayakawa, K.Kakuta and K.Yamamoto, Metallurgy of Continuous-Annealed Sheet Steel (1982), p.133.

[2] P.Messien and V.Leroy, Steel Research 60 (1989)7, p.320.

[3] M.Sudo and I.Tsukatani, Technology of Continuously Annealed Cold-Rolled Steel Sheet (1984), p.203.

[4] Y.Nagataki and Y.Hosoya, ISIJ International, 36(1996), p.451.

[5] T.Senuma and M.Kameda, J. Japan Inst. Metals, 60(1996), p.717.

[6] H.Inagaki, Tetsu-to-Hagane 62(1976), p.50.

[7] H.Kubotera, K.Nakaoka, K.Araki, K.Watanabe, and K.Iwase, Tetsu-to-Hagane 62(1976), p.60.

[8] Y.Hosoya, S.Hashimoto and A.Nishimoto, TMS-AIME, Colorado, (1987), p.273.

[9] N.Takahashi, M.Shibata, Y.Furuno, H.Hayakawa, K.Kakuta and K.Yamamoto, Metallurgy of Continuous-Annealed Sheet Steel (1982), p.133.

E-mail address:yfunakaw@lab.fukuyama.nkk.co.jp
Fax:81-849-45-4059

Materials Science Forum Vols. 284-286 (1998) pp. 543-550
© *1998 Trans Tech Publications, Switzerland*

Effect of Boron Addition on the Dynamic Strain Aging Behavior of a Medium Carbon Steel Wire Rod

B.H. Cao[1], J.J. Jonas[1], P.R. Hastings[2] and N. Nickoletopoulos[2]

[1] Department of Metallurgical Engineering, McGill University,
3610 University Street, Montreal, H3A 2B2, Canada

[2] Ivaco Rolling Mills, Box 322, L'Orignal, Ontario, K0B 1K0, Canada

Keywords: Dynamic Strain Aging, Cold Heading, Boron, Nitrogen Scavenging, Dislocations, Work Hardening, Medium Carbon Steel

ABSTRACT

The effect of boron addition on the dynamic strain aging behavior and elevated temperature mechanical properties has been investigated in a medium carbon steel. For this purpose, four steels with additions in the range 0 to 63 ppm. by wt. of boron were tensile tested at temperatures from ambient to 350 °C and at a strain rate of 1×10^{-3} s^{-1}. The presence of serrated yielding, work hardening peaks, and minima in ductility was observed over certain ranges of temperature. These characteristics are attributed to dynamic strain aging. Boron addition is shown to be effective in controlling such aging, where higher boron contents lead to lower degrees of dynamic strain aging. These results are discussed in terms of recent models of dynamic strain aging.

1. INTRODUCTION

Strain aging involves changes in the properties of alloys that occur as a result of the interactions between dislocations and solute atoms [1-3]. Two types of strain aging can be characterized: static strain aging (SSA), where the aging process takes place after prestraining, and dynamic strain aging (DSA), where the aging occurs during plastic deformation [2-5].

Extensive research has been carried out to investigate DSA in steels because of its significant effects on the mechanical properties, leading as it does to the presence of serrated stress-strain curves, high work hardening rates, negative strain rate sensitivities of the flow stress, and reductions in ductility [2-11]. These arise because the temperature is high enough to allow interstitial atoms, principally C and N, to diffuse to dislocations during deformation and thus to hinder their movement through the formation of atmospheres [1,12-14]. Moreover, due to the higher solubility of nitrogen than of carbon, it is generally accepted that DSA in steels is caused mainly by nitrogen rather than carbon [4,9,15]. The most important variables determining this behavior are the interstitial concentration and cooling rate after hot rolling, followed by the temperature and strain rate of testing [3,7,8,11].

The production of fasteners by the cold heading of steel wire rod usually involves severe deformation and multiple forming operations applied over short time intervals. The heating of the workpiece resulting from these forming conditions increases the drift rates of the interstitial C and N to sinks and thus accelerates the aging kinetics. Depending on the level of mobile C and N in the steel, DSA can occur so that the volume being deformed becomes less tolerant of the high strains imposed upon it. Such aging can lead in turn to head bursts, shear cracking, and increased machine/tool wear [10,16].

Microalloying additions have been successfully used to prevent both static and dynamic strain aging and one approach has been to add a carbonitride former, such as Al, B, Nb, Ti and V, for this purpose [10,15, 17-20]. Cold heading quality wire rods, in either alloy or carbon grades, are generally process annealed so as to render them sufficiently soft and ductile for upsetting. In low carbon steels, the addition of B converts the free nitrogen into BN [10,15,17,18,21-24], which leads to improved formability and ductility, decreased work hardening and the elimination of nitrogen-related strain aging [15]. As a result, boron-treated steels can often be cold headed in the hot rolled condition, thus eliminating the annealing operation usually employed prior to cold heading. Furthermore, boron-treatment has also proven to be useful in decreasing the deleterious effects of DSA in low carbon steels that are subjected to cold deformation during production [10,15].

To date, B addition has largely been restricted to low carbon steels for the prevention of strain aging. For example, Ivaco Rolling Mills have developed B-containing low carbon (0.04 to 0.15 %C) steels for wire drawing down to 95% reduction without annealing [26,27]. By contrast, little attention has been given to the use of B in *medium* carbon steels. In our previous paper [28], the effects of B addition on the static strain aging behavior, mechanical properties and cold formability were described in a series of medium carbon steels. The present investigation was undertaken to assess further the effect of B additions on the DSA behavior and elevated temperature mechanical properties of a medium carbon steel. The results are discussed in terms of recent models of DSA.

2. EXPERIMENTAL APPROACH

Four steels with additions in the range 0 to 63 ppm by wt. of boron were selected for investigation. The chemical analyses of these materials are given in Table 1. The steels were continuously cast into 120 mm square billets and then hot rolled into 8.5 to 16.5 mm diameter wire rod at Ivaco Rolling Mills, L'Orignal, Ontario. The steels that contain 0, 26, 41 and 63 ppm of B will be referred to as the Base, Low-B, Med-B, and High-B steels, respectively. Note that these B levels are equivalent to 0, 0.37, 0.54, and 0.96 times the N content on an atomic or stoichiometric basis, respectively.

Cylindrical ASTM subsized specimens with threaded ends were machined from the wire rods for tensile testing purposes. The dimensions were 40 mm gauge length and 5 mm diameter. In order to provide comparable initial structures, all four steels were normalized before testing, by heating to 900 °C, holding for 20 minutes at temperature and air cooling. The optical micrographs indicated that the microstructures consisted of ferrite and pearlite and that the B additions resulted in finer grain sizes after normalizing, compared with B-free wire rods.

Table I   **Chemical compositions of the experimental steels (in wt.%)**

Grade	Diameter (mm)	C	Mn	Si	P	S	N	B	B/N-ratio (atomic)
Base	16.5	0.31	0.73	0.22	0.014	0.012	0.0087	0.0000	0
Low-B	8.5	0.30	0.74	0.23	0.014	0.014	0.0090	0.0026	0.37
Med-B	10.5	0.30	0.72	0.26	0.014	0.007	0.0098	0.0041	0.54
High-B	13.5	0.32	0.73	0.23	0.013	0.010	0.0085	0.0063	0.96

All the tensile tests were carried out on a servohydraulic computerized MTS machine, with a microprocessor-controlled tungsten lamp radiant furnace mounted on the machine frame. A chromel-alumel thermocouple placed against the center of the specimen gauge length was used for temperature measurement and control. Temperature variations during testing were found not to exceed ± 2 °C. The tests were performed in the range of temperature from ambient to 350 °C and at a strain rate of 10^{-3} s^{-1}. For elevated temperature testing, the specimens were heated at 1 °C/s and held at the testing temperature for 5 min prior to testing. In order to minimize oxidation, argon gas was passed through the quartz tube enclosing the specimen and grips.

The load vs displacement information was converted into engineering stress vs engineering strain curves. The following conventional mechanical properties were derived from the foregoing information: yield strength or lower yield strength, flow stress at a given strain, UTS, fracture stress and total elongation. In the case of the continuous curves, the yield strengths were determined from the stress-strain curves by the 0.2% strain offset method.

3. RESULTS

Flow curves of the four steels tested between room temperature and 350 °C at a strain rate of 10^{-3} s^{-1} are shown in Figs. 1a to 1d. Serrated flow, one of the characteristics of DSA was clearly

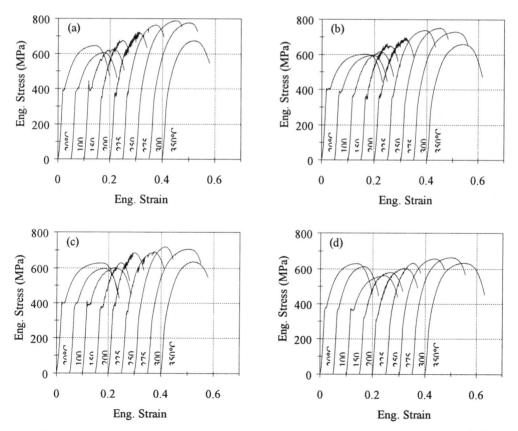

Fig. 1   Stress-strain curves at various temperatures determined at a strain rate of 10^{-3} s^{-1}.
(a) Base steel, (b) Low-B steel, (c) Med-B steel, and (d) High-B steel.

observed at certain temperatures in all four steels. Serrated flow first appeared at about 150 °C in the Base, Low-B and Med-B steels, whereas it was delayed to about 200 °C in the High-B steel. Nevertheless, the serrations are clearly observed to disappear beyond 250 °C in all four steels. At temperatures of 275 °C or higher, yielding is continuous. For all four steels, the work hardening rate (slope of the stress-strain curve) increases to a maximum then decreases sharply with increasing temperature. By increasing the amount of boron addition up to a B/N ratio of about 1.0, the magnitude and frequency of serrations decrease significantly. Also, the work hardening rate decreases markedly with B addition.

The effect of B addition on the temperature dependence of the yield strength is presented in Fig. 2. This property generally decreases with increasing temperature, although there is a plateau extending over the temperature range where serrated flow behavior is observed. It can be seen that all levels of B addition resulted in a decrease in yield strength, especially at the higher temperatures. Furthermore, the higher the B/N-ratio, the lower the yield strength.

The effect of B addition on the temperature dependence of the work hardening behavior is shown more clearly in Fig. 3, where the flow stress at 5 pct strain is plotted as a function of the temperature. In the temperature range where serrated flow occurs, the flow stress increases gradually with increasing temperature and reaches a maximum between 250 and 275 °C. The flow stress then decreases with further increase in temperature. The magnitude of the peak decreases as the B/N-ratio is increased; however, the temperature of the peak does not change.

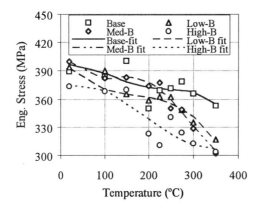

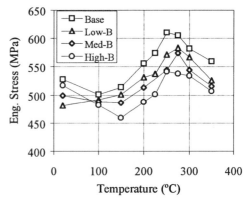

Fig. 2   Effect of B addition on the temperature dependence of the yield strength.

Fig. 3   Effect of B addition on the temperature dependence of the flow stress at 5 pct strain.

The evolution of the ultimate tensile strength (UTS) with temperature for the four steels is illustrated in Fig. 4. The UTS first decreases slightly with increasing temperature, reaching a minimum between 100 and 150 °C; it then increases with increasing temperature and reaches a maximum at around 275 °C, before decreasing with still further increase in temperature. The increase in the UTS in the temperature range between 150 and 275 °C coincides with the occurrence of serrations in the flow curves (see Fig. 1). At temperatures above 100 °C, the magnitude of the UTS decreases significantly with B addition, especially in the temperature range where the maximum UTS is observed. In addition to the serrated nature of the stress-strain curves, the appearance of a UTS peak at a particular temperature is also a manifestation of DSA [7].

The evolution of the fracture stress with temperature for the four steels is displayed in Fig. 5. The behavior observed is similar to that of the UTS (as seen in Fig. 4). Again, there is a peak in fracture stress, with a maximum situated around 275 °C. The addition of B results in a decrease in the magnitude of the fracture stress over nearly the whole investigated temperature range

The dependence of the total elongation on test temperature is shown in Fig. 6. The lines drawn through the experimental points are based on best fit quadratic equations. It is evident that there is a considerable reduction in ductility at intermediate temperatures and that the minimum in the total elongation is located at around 200 °C for all four steels. At sufficiently high temperatures, the total elongation returns to its room temperature level. It can be seen that all levels of B addition lead to increases in total elongation and that the greatest increase is observed when the B/N-ratio is the highest.

4. DISCUSSION

4.1 General characteristics of the elevated temperature mechanical behavior of the present medium carbon steel

It is now well accepted that the flow stress discontinuities and serrated yielding displayed by a material being deformed originate on a dislocation scale from interactions between diffusing solutes and mobile dislocations, i.e. from DSA [30,32]. The tensile behavior of the present material over the temperature range ambient to 350 °C is consistent with the general features of DSA observed in low carbon steels [2-11]. Serrated stress-strain curves, typical of DSA, were present at temperatures between 150 and 250 °C during deformation at a strain rate of 10^{-3} s^{-1}. Such serrated behavior was accompanied by increases in the rate of work hardening, peaks in flow stress, UTS and fracture stress, and a reduction in ductility.

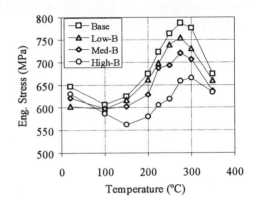

Fig. 4 Effect of B addition on the temperature dependence of the UTS.

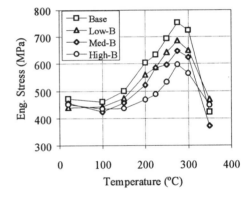

Fig. 5 Effect of B addition on the temperature dependence of the fracture stress.

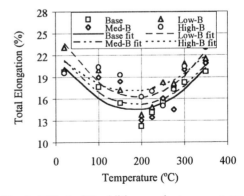

Fig. 6 Effect of B addition on the temperature dependence of the total elongation.

These characteristics are typical of the "blue-brittle" behavior of mild steel [2,4,14,31], which is attributed to the occurrence of DSA.

In the model initially proposed by Cottrell [1,12], it was assumed that serrated flow is initiated when solute atmospheres are formed about, and then dragged along by dislocations during a quasiviscous type of motion. More recent extensions of this theory [13,32-35] suggest that, during deformation, moving dislocations are temporarily held up at local obstacles, then break away and advance to further obstacles. Serrated yielding occurs when the waiting time is sufficiently long for solute atmospheres to form around the arrested dislocations; the latter results in an increase in the friction component of the flow stress. In the light of these recent models, it appears that the DSA that develops in medium carbon steels with ferrite-pearlite structures results from the interaction of carbon and nitrogen with moving dislocations in the ferrite phase.

According to the present results, the yield strength generally decreased with increasing temperature, except for the presence of a plateau in the DSA temperature range (Fig. 2); such observations have also been reported by other researchers [7-9,37]. It is of interest that the concentration of nitrogen in dislocation atmospheres first decreases with increasing temperature before approaching a minimum at around 150 °C, and then increases with temperature until it attains its maximum [38]. Thus, while the decrease in yield strength can be attributed to the softening that accompanies temperature increase, the appearance of the plateau results from superimposing the effect of DSA on the softening influence of heating.

The appearance of peaks in work hardening rate in the serrated flow region can be correlated with the increase in the rate of dislocation multiplication due to solute pinning during testing (Figs. 3 to 5). This enhanced multiplication persists until just beyond the disappearance of the serrations. The appearance of a minimum at around 200 °C in the total elongation vs. testing temperature curve is associated with the presence of the largest serrations in the stress-strain curves (see Figs. 1 and 6). This can be attributed to the optimal locking of dislocations by solute atoms in the DSA region, which leads in turn to non-uniform deformation. In addition, it can be related to changes in the strain for the onset of plastic instability and to the rate of localization of plastic flow during necking and fracture [5,7].

4.2   Effect of B addition on the DSA behavior and the elevated temperature mechanical properties

Since DSA originates from interactions between mobile solutes and dislocations, the solute concentration of the alloy can be expected to influence the serrated flow characteristics [30]. The magnitude of the serrations in fact reflects the degree of locking and of dislocation multiplication produced by DSA [9,36]. In the present case, the magnitude and the frequency of the serrations both decrease significantly with boron addition, as seen in Fig. 1. When B is added to a steel for nitrogen scavenging, and is consumed for this purpose, there should be no increase in hardenability. The elimination of interstitial N is accomplished by reacting the N with B to produce BN, a compound that has no significant effect on steel properties [15].

The marked decrease in work hardening rate and considerable improvement in ductility that accompanied the addition of B in the present experiments provide direct evidence of the affinity of B for the interstitial N present in wire rods; this was shown in Figs. 3 to 6. These results therefore suggest that, in the present material, DSA is largely due to interstitial nitrogen, as in mild steel. Thus, the decrease in the work hardening rate can be directly associated with the decrease in the interstitial N concentration. The fact that the serrations are absent at about 150 °C in the high-B steel indicates that the N is almost entirely scavenged and that insufficient solute atoms are available to fully lock the dislocations (see Fig. 1d).

4.3   Implications for cold heading quality wire rod

One of the most promising applications for the addition of B to medium carbon steels is to prevent DSA from occurring and in this way to reduce the effects of DSA during cold heading operations. For this objective to succeed, the amounts of solute C and primarily N must be reduced below levels that affect the process or the properties of the final products. The elimination of interstitial N is accomplished by reacting the N with B to produce BN. It is clear from the present results that B addition improves the ductility considerably and also reduces the work hardening rate significantly in the elevated temperature range. Thus, B addition can be an efficient way to prevent or reduce DSA in medium carbon steel wire rods, as in low carbon steels [10,15].

In terms of the B content, it has been shown that the amount of B required for scavenging depends on the N content and that the B/N atomic ratio must be greater than 1.0 to remove the N totally from solid solution [15]. With regard to the present materials, the further addition of B (beyond the 63 ppm B level employed here) is expected to eliminate N-related DSA totally. Conversely, from our previous investigation [28], it appears that non-aging steels can also be produced with lower B/N-ratios; in this case, lower cooling rates after rod rolling must be employed to lower the solute interstitial levels in conjunction with a BN precipitation treatment to promote N scavenging.

5. CONCLUSIONS

The effects of B addition on the DSA behavior and on the elevated temperature mechanical properties have been assessed on a medium carbon steel. The main conclusions from this investigation are as follows.

(1)   Dynamic strain aging occurs in the tested steels during uniaxial tensile testing in the temperature range 150 to 250 °C at a strain rate of 10^{-3} s^{-1}; this phenomenon has a considerable effect on the elevated temperature mechanical properties of this material.

(2)   The main features that can be attributed to dynamic strain aging are: (i) serrated flow between 150 and 250 °C, (ii) a yield strength plateau in the serrated flow range, (iii) peaks in the flow stress, UTS and fracture stress vs testing temperature curves at 275 °C, and (iv) a minimum in the total elongation vs testing temperature curve at 200 °C.

(3)   The addition of up to 63 ppm B by weight (B/N atomic ratio=0.96) decreases significantly the magnitude and frequency of serrations, the flow stress, UTS and fracture stress peak heights, and improves the total elongation. This can be related to the decrease in the amount of free nitrogen that results from BN precipitation.

(4)   More complete DSA control can be achieved either by raising the B level so as to convert all the interstitial N into BN, or by employing slower cooling rates after hot rolling to promote more BN precipitation.

ACKNOWLEDGMENTS

The authors acknowledge with gratitude the financial support received from the Natural Sciences and Engineering Research Council of Canada (NSERC) and the Canadian Steel Industry Research Association (CSIRA). They also thank Mr. Edwin Fernandez and Ms. Lorraine Mello for technical and administrative assistance.

REFERENCES

[1] A.H. Cottrell, Dislocations and Plastic Flow, Oxford University Press, Oxford, 1953.
[2] J. D. Baird, Metall. Rev., 16, (1971), p. 1.
[3] W. C. Leslie, The Physical Metallurgy of Steel, McGraw-Hill, New York, NY, 1983, p. 79.
[4] J.D. Baird, Iron & Steel, 36 (1963), p. 186, p. 326, p. 368, p. 400, p. 450.
[5] J.D. Baird, The Inhomogeneity of Plastic Deformation, ASM, Metals Park, Ohio, (1973), p. 191.
[6] B.J. Brindley and J.T. Barnby, Acta Metall., 14 (1966) p. 1765.
[7] A.S. Keh, Y. Nakada and W.C. Leslie, Dislocation Dynamics, ed. by W.A. Owen and M.J. Roberts, McGraw Hill, New York, (1968), p. 381.
[8] C.-C. Li and W.C. Leslie, Metall. Trans., 9A (1978) p. 1765.
[9] S. Lou and D.O. Northwood, Can. Metall. Quart., 33 (1994), p. 243.
[10] C. Weidig, M. Espindola, B. Gonzalez, P. Rodrigues and M. Andrade, Wire Journal International, 28 (1995), p. 82.
[11] A. Karimi Taheri, T.M. Maccagno, and J.J. Jonas, ISIJ International, 35 (1995), p. 1532.
[12] A.H. Cottrell, Phil. Mag., 74 (1953), p. 829.
[13] A. van den Beukel, Acta Metall., 28 (1980), p. 965.
[14] R.W. Honeycombe and H.K Bhadeshia, Steel: Microstructure and Properties, Arnold, London, 1995.
[15] G. M. Faulring, Electric Furnace Conference Proceedings, 1989, p.155.
[16] I.G. Ritchie and Z. Pan, 33rd MWSP Conf. Proc., ISS-AIME, XXIX (1992), p. 15.
[17] F. Boratto, C. Weidig, P. Rodrigues and B.M. Gonzalez, Wire Journal International, 26 (1993), p. 86.
[18] I.D. McIvor, Ironmaking and Steelmaking, 16 (1989), p. 55.
[19] W.B. Morrison, Ironmaking and Steelmaking, 16 (1989), p. 123.
[20] D.T. Llewellyn, Ironmaking and Steelmaking, 20, (1993), 35.
[21] E.R. Morgan and J.C. Shyne, J. Iron Steel Inst., 185 (1957), p. 156.
[22] E.R. Morgan and J.C. Shyne, Trans. of AIME, 209 (1957), p. 781.
[23] J.F. Butler, Trans. of AIME, 224 (1962), p. 84.
[24] P. Messien and V. Leroy, Steel Research, 60 (1989), p. 320.
[25] R.M. Guthrie, 51st Annual Convention, the Wire Association International, 1981, p. 195.
[26] T.A. Perlus and J.T. Griffiths, U. S. Patent, No. 4,168,181, September, 1979.
[27] T.A. Perlus, J.T. Griffiths and M. Coburn, Canadian Patent, No. 1,102,588, June, 1981.
[28] B.-H. Cao, J.J. Jonas, P.R. Hastings and N. Nickoletopoulos, 39th MWSP Conf. Proc., ISS-AIME, XXXV (1997), p. 783.
[29] S. Kinoshita, P.J. Wray and G.T. Horne, Trans. Met. Soc. AIME, 233 (1965), p. 1902.
[30] J.M. Robinson and M.P. Shaw, Inter. Mater. Rev., 39 (1994), p. 113.
[31] J.S. Blakemore and E.O. Hall, J. Iron Steel Inst., 204 (1966), p. 817.
[32] L.P. Kubin and Y. Estrin, Acta metall.mater., 38 (1990), p. 697.
[33] P.G. McCormick, Acta metall., 20 (1972), p. 351.
[34] R.A. Mulford and U.F. Kocks, Acta metall., 27 (1979), p. 1125.
[35] J. Schlipf, Scripta metall., 29 (1993), p. 297.
[36] D. Park and J.M. Morris, Scripta metall., 29 (1993), p. 363.
[37] J.W. Kim and I.S. Kim, Nucl. Eng. Des., 172 (1997), p. 49.
[38] S. Lou and D.O. Northwood, J. Mater Sci., 30 (1995), p. 1434.

For further information on this paper, please contact J.J. Jonas. E-mail address: johnj@minmet.lan.mcgill.ca, Fax: 1-514-398-4492.

Materials Science Forum Vols. 284-286 (1998) pp. 551-558
© *1998 Trans Tech Publications, Switzerland*

Towards a Controlled Bake Hardening by Numerical Calculation of the Precipitate Behaviour during Continuous Annealing of TiNb and Nb Alloyed IF BH Steels

R.A. Hubert[1], S. Vandeputte[1], A. Van Snick[2] and C. Xhoffer[1]

[1] OCAS N.V., Research Centre of the SIDMAR Group, J. Kennedylaan 3, B-9060 Zelzate, Belgium

[2] Laboratory for Iron and Steelmaking, University of Ghent,
Technologiepark 9, B-9052 Zwijnaarde, Belgium

Keywords: IF, NbC, Bake Hardening, Precipitation Kinetics, Numerical Modelling

Abstract

Numerical modelling of the NbC precipitates' behaviour during coiling and continuous annealing was used to calculate the amount of free solute carbon after processing of TiNb and Nb IF Bake Hardening (BH) steels. The simulations were confronted with experimental BH results as a function of coiling temperature, soaking time, soaking temperature and cooling speed. The approach of combining annealing experiments, modelling and (S)TEM/EDX for determination of the precipitates allowed to obtain deeper insights in the NbC behaviour and the influence of the different processing parameters. Using the simulation tool, the final goal is to estimate the BH level from the calculated free carbon content.

Introduction

For automotive applications, where a reduction of steel sheet thickness is searched for through the use of high strength steels, Bake Hardenable (BH) steels exhibit interesting properties as to both strength and deformability. Whereas the material gets its final yield strength increase during paint baking, its lower initial yield strength before press forming is advantageous for the car manufacturer. For the steel producer, it is important to be able to guarantee the minimum desired BH-level, mostly 40 MPa. The BH-process itself is dependent on the amount of free carbon in solution, the dislocation density and the carbon mobility. Yet, the combination of these factors does not facilitate the realisation of a stable, reproducible BH-value. From the steelmaker's point of view, the carbon content is probably the most regulable parameter.

Since a low free carbon content results in low BH and a too high content has to be avoided because of ageing, the goal of the present research is the control of the BH-level through a more careful adjustment of the solute carbon level after final processing.

In general, two metallurgical routes can be followed to produce IF-BH steel grades: one, in which carbon remains free in the whole production process from hot rolling to final annealing (typical 20 ppm C) [1] and, a second one, in which carbide dissolution during continuous annealing is aimed at [2]. In this work, BH is achieved by following the second route, i.e. dissolution of NbC during annealing of TiNb and Nb alloyed vacuum degassed steels.

In order to limit the number of industrial and laboratory trials, computer simulations were performed to monitor the NbC precipitate evolution during the processing. Attention was focused on the last processing stages (coiling following hot rolling and annealing following cold rolling) since these steps are determining for the precipitate behaviour. By comparison with experimental BH results and using information from TEM/STEM and EDX measurements, a better understanding of the precipitation phenomena and the influence of the processing factors is searched for. Afterwards, optimised settings have to lead to a more controlled BH level.

Materials and experimental procedure

The steel chemistry of the investigated alloys is given in Table 1.

Table 1:   Steel compositions (all concentrations in mass ppm)

Steel	C	Mn	P	Nb	Ti	N	Nb/C$_{at}$
Nb1	26	9737	687	219	40	22	1.09
Nb2	28	745	150	175	45	16	0.81
TiNb1	18	1080	85	35	100	19	0.25
TiNb2	23	960	95	210	110	22	1.20

No Ti is available for carbide formation if one assumes that TiN and TiS are formed at high temperatures (Ti-3.4N-1.5S <0). In the Nb steels, N is bound by both Ti and Al and in case of steel Nb1 also by B (11ppm).

After reheating at 1250°C for 1 h, hot rolling was performed in the austenitic region, followed by water jet cooling (20°C/s) to the coiling temperature. The latter was varied between 540 and 720°C. The cooling pattern to room temperature was computer controlled and corresponded to that of the warmest point of a 20t coil, modelled with the following equation :

$$T(t) = (CT)^{\frac{100}{100+t}} \tag{1}$$

where t is the elapsed time in hours and CT is the coiling temperature.

After wards, a cold reduction of 75% was given in case of TiNb1/2 and Nb2, and 69% in case of Nb1. The continuous annealing was simulated using a Gleeble 1500/20. The heating rate was always 10°C/s whereas the soaking temperature was varied between 750-900°C , the soaking time between 5 and 60s, and the cooling speed after soaking between 5 and 40°C/s. To eliminate stretcher strains, the skin pass elongation was either 1% (Nb1), 1.3% (Nb2) or 1.5% (TiNb1/2).

The bake hardening (BH$_2$) was measured following the SEW 094 specification. BH$_2$ is equal to the difference between the stress at 2% prestrain and the lower yield strength after subsequent annealing at 170°C for 20 min.

TEM/STEM measurements were performed on a JEOL 2010 transmission electron microscope operated at 200kV and equipped with an EDX analysis and an image processing system. The characterisation of the precipitates as to nature and size was done on carbon extraction replicas. That way, it was possible to resolve ultrafine precipitates of ≥3-5 nm equivalent spherical diameter.

Numerical model

The OCAS developed MODIPS software [3] has been used to calculate the equilibrium NbC pre-cipitated fractions as a function of the temperature as well as the kinetic evolution of the precipitates as a function of the time and the temperature. The TEM investigations revealed the presence of Ti$_4$C$_2$S$_2$ beside NbC in the TiNb steels [4] but this was not taken into account in the simulations as the NbC fraction was much larger than the Ti$_4$C$_2$S$_2$ fraction. In the Nb steels titanium was fully precipitated at high temperature as TiN only and was thus not available anymore to bind carbon. Hence, only the NbC, precipitating at temperatures typically below 900°C, needed to be modelled.

The equilibrium calculations based on the solubility product of Petrova [6] show that the steels studied do not contain NbC in their austenite phase. The Petrova equation was chosen because it is close to the average of all equations available in the literature [5-10]. However, the uncertainty on this type of equation is quite large, especially in ferrite.

As the steel processing is often too fast for the equilibrium to be reached, the kinetic aspects of precipitation and/or dissolution must be taken into account. Our model is capable of predicting the kinetics of precipitation and/or dissolution of binary precipitates not taking into account the presence of singularities such as grain boundaries. The precipitates are assumed to be identical, spherical and homogeneously distributed in the matrix. Each precipitate is associated with a closed cell in such a way that the modelling of just one precipitate in its cell is sufficient to represent the whole sample. The precipitation/dissolution kinetics can be followed by looking at the evolution of the matrix concentration profiles of niobium and carbon, which can be calculated by solving Fick's second equation (in case of spherical symmetry):

$$D\left[\frac{\partial^2 c}{\partial r^2} + \frac{2}{r}\frac{\partial c}{\partial r}\right] = \frac{\partial c}{\partial t} \tag{2}$$

where c is the concentration (in this case c^{Nb} or c^C), r, the position in the cell, t, the time and D, the diffusion coefficient. Diffusion coefficients in ferrite were taken from [11] and [12] for Nb and C respectively. Local equilibrium is assumed at the precipitate-matrix interface and precipitates are supposed to remain stoichiometric. More details about the model and some examples of applications can be found elsewhere [3, 13, 14].

As no reliable nucleation model is available, the software used is unable to create new precipitates. This means that the calculation always starts with a given distribution of precipitates and that the precipitate density is not modified during the calculation.

If the initial situation corresponds to complete dissolution, the temperature must first be lowered until the precipitation can start. Undercooling can also be taken into account by lowering again the temperature. To effectively start the calculation, a given density of precipitates is injected in the system at the chosen temperature. In this case the precipitate size is chosen very small (typically 0.1nm). The density is determined from the distance D between 2 precipitates which is a parameter of the model. As the density does not change during the calculation, not all the parameters can be chosen independently: the precipitate size ϕ, the distance D and the carbon concentration in solution [C] are linked by the following relation:

$$[C] = C_{tot} - 6 \cdot 10^4 \frac{\phi^3}{D^3} \quad (in\ ppm) \tag{3}$$

where C_{tot} is the total carbon concentration. This equation which is deduced from simple geometric considerations - assuming both NbC and ferrite densities to be equal - is valid for spherical NbC precipitates only.

Calculations and comparison with experimental results

First some equilibrium calculations were performed to evaluate the amount of free carbon at equilibrium as a function of the temperature for the different steels. In most cases, precipitation starts at or just below the γ-α transformation temperature as indicated in Fig. 1. The results confirm that Nb1 and TiNb2 are superstoichiometric while Nb2 and TiNb1 are substoichiometric.

Influence of the coiling temperature (CT). For this calculation, we assume that the precipitation starts at the beginning of the coiling. The thermal cycle is described by equation (1). Both the coiling temperature and the distance between precipitates (i.e. the precipitate density) were varied. Those two parameters are of course related with each other: for low coiling temperatures, the undercooling is larger and more nuclei are "activated".

Fig. 2a shows the evolution of the amount of carbon in solution after cooling as a function of CT and of D, the distance between precipitates, for steel Nb1. As expected, the precipitation is favoured by a high CT and a high precipitate density (short D's).

Complete precipitation, corresponding to the equilibrium, is reached only for high CT's and short D's. This calculation confirms the TEM observations which reveal that no visible precipitation occurs in the ferrite grains after coiling at 600°C. Actually, NbC sizes varied between not visible (<2-

3nm?) at 540°C to 3-30nm at 680-720°C and the precipitate density after coiling is also depending on CT. Furthermore, yield point elongations of hot rolled non-aged Nb1 material are 0%, 2.1% and 3.1% at CT's of 680, 600 and 540°C respectively. This is another confirmation of the fact that free solute carbon can still be present after coiling. As a consequence, the simulation of the continuous annealing section must be done starting from a non equilibrium situation in the case of low CT's.

Fig. 2b shows the evolution of the precipitate size after cooling as a function of CT and D for steel Nb1. Interesting to notice is the fact that for a given CT the precipitate size is almost constant over a wide range of densities: as can be seen from the figure, the different lines merge with each other when CT decreases. This phenomenon can be explained by the slow diffusion of niobium: when D

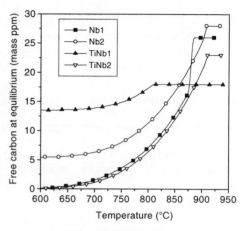

Figure 1   Free Carbon at equilibrium as a function of the temperature for all steels used in this study.

increases, the amount of atoms available in one cell also increases which should lead to a larger precipitate. However only the niobium atoms which are in the vicinity of the precipitate can reach it and the size "saturates" at a level which depends on the temperature but not on the distance.

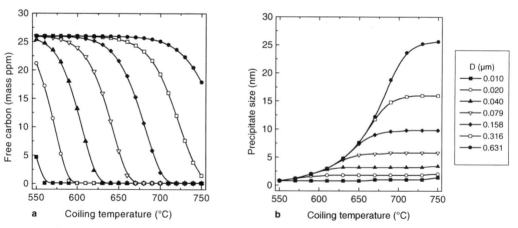

Figure 2   Free Carbon in solution (a) and precipitate size (b) after cooling as a function of the coiling temperature and of the distance between the NbC precipitates for steel Nb1.

Simulation of the continuous annealing section. As indicated above, the coiling temperature defines the starting situation. For high CT's, all the "precipitable" carbon is precipitated and the equilibrium is established at the beginning of the annealing cycle; the precipitate density is low and the precipitate size is rather large. Conversely, low CT's lead to non equilibrium situations with a lot of carbon remaining in solution and numerous small NbC precipitates. In this case, the precipitate size after coiling does almost not depend on the density and the calculations are performed with a constant starting size ϕ_i. This is of course a simplification as the precipitate size after coiling is in the reality not uniform.

Fig. 3 shows some calculation examples for steel Nb2 with a soaking temperature of 850°C and a soaking time of 60s. The equilibrium carbon concentration at 850°C is 17ppm. Hence, during the soaking, the system tends towards this value. This implies that the concentration either decreases or increases depending on the start concentration. For systems initially at or near equilibrium, the concentration increases. Conversely, when the start concentration is higher than 17ppm, it decreases but it effectively reaches the equilibrium concentration only if the soaking time is long enough and/or if the distance is short enough. For distances of 0.05 and 0.3µm and supposing the same precipitate starting size, the equilibrium is reached after 5 and 50 s respectively (Fig. 3). The initial situation (cold state after coiling) has also a large influence for large D's and short soaking: the free carbon difference between the two curves with D=0.3µm (starting and not starting at equilibrium) decreases from 13 to 1 ppm when the soaking time increases

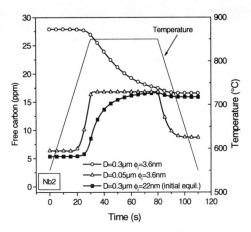

Figure 3 *Evolution of the amount of free carbon during annealing for steel Nb2 in three typical situations (see text for details)*

from 5 to 50 s. During the cooling, the concentration always decreases due to reprecipitation and the amplitude of the drop depends only on the density: the shorter the distance, the larger the drop.

Influence of the soaking time. Fig. 4 shows the evolution of the bake-hardening level with the soaking time for steel Nb2. The BH level decreases as the soaking time increases, which means that the free carbon concentration also decreases supposing some constant dislocation density in the recrystallised steel. This behaviour corresponds to a non equilibrium situation at the beginning of the annealing cycle as shown in Fig. 3. The coiling temperature (670°C) was not high enough to completely precipitate carbon as NbC. The corresponding simulation is shown in Fig. 5 where the free carbon evolution is given as a function of the soaking time and of the distance between precipitates. The experimental BH results are compatible with D's greater than 0.2µm.

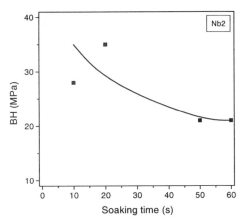

Figure 4 *Bake-hardening as a function of the soaking time for steel Nb2 (CT=670°C, soaking temperature=850°C; cooling rate=10°C/s)*

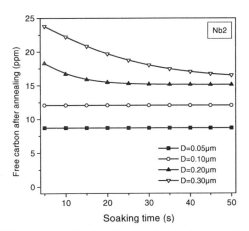

Figure 5 *Calculated carbon in solution after annealing as a function of the soaking time and of the distance between the NbC precipitates for steel Nb2 (soaking temperature=850°C; cooling rate=10°C/s; ϕ_i=3.6nm)*

<u>Influence of the soaking temperature.</u> Experimental BH results concerning the influence of the soaking temperature have been obtained for steels TiNb1 and TiNb2 (Fig. 6). The experimental conditions were: CT=720°C, soaking time=60s, cooling rate=20°C/s. The TiNb1 BH level is almost constant and equal to 30MPa for temperatures lower than 825°C and increases to 37MPa for a soaking temperature of 900°C. The explanation of this behaviour is obvious as this steel has always at least 13 ppm free carbon because of its low niobium content (see Fig. 1). A numerical simulation for a soaking temperature of 900°C shows that the 18 ppm carbon contained in TiNb1 is in solution after annealing (Fig. 9). The influence of the soaking temperature is much larger for steel TiNb2: the BH level increases from 0 to about 40MPa when the soaking passes from 750 to 900°C. Fig. 7 shows the corresponding numerical simulation. As the CT was 720°C, we may assume that the equilibrium was reached at the beginning of the annealing. The calculation compares favourably with the experiment when the distance between precipitates is larger than 0.2μm with corresponds to a NbC size of 15nm after coiling in agreement with the TEM observations. For higher precipitate densities (e.g. D=0.05μm), the final amount of free carbon saturates due to reprecipitation during cooling in spite of the higher dissolution at higher soaking temperatures.

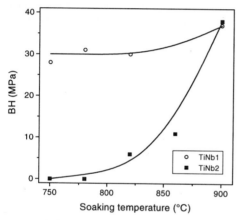

Figure 6 *Bake-hardening as a function of the soaking temperature for steels TiNb1 and TiNb2 (CT=720°C; soaking time=60s; cooling rate= 20°C/s)*

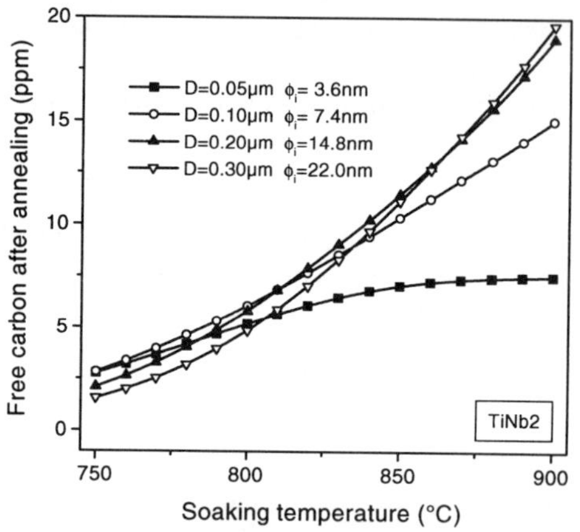

Figure 7 *Calculated carbon in solution after annealing as a function of the soaking temperature and of the distance between the NbC precipitates for steel TiNb2 (soaking time=60s; cooling rate=20°C/s; equilibrium assumed before annealing.*

<u>Influence of the cooling rate.</u> This simulation was performed in order to compare the results with those obtained by Lips et al. [4] who published BH values obtained after laboratory processing of steels TiNb1 and TiNb2. They used the following cycle: heating rate: 10°C/s, soaking time: 60s, soaking temperature: 900°C, cooling rate: varied between 5 and 40°C/s. Their results are depicted in Fig. 8. The dissolution temperatures of NbC in TiNb1 and TiNb2 are ≈ 800 and 910°C respectively. This means that the dissolution is almost complete at the end of the soaking for both steels. Precipitation starts again during cooling but one has to take the nucleation time into account. This can be done by introducing an undercooling which was assumed to be 50°C in this case. Hence, the simulation has been limited to the cooling section and the start temperature has been chosen 50°C under the dissolution temperature.

The simulation results are shown in Fig. 9 and indicate that almost no precipitation occurs for steel TiNb1 as all the carbon remains into solution as indicated in the previous section. For steel TiNb2 they find a BH level of 15 MPa for a cooling rate of 10°C/s. This level increases up to 30 MPa at 20°C/s where it saturates. The calculation is compatible with the observation if the distance

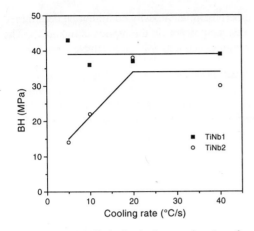

Figure 8 Measured bake-hardening as a function of the cooling rate for steels TiNb1 and TiNb2 (CT=720°C; soaking time=60 s; soaking temperature=900°C)

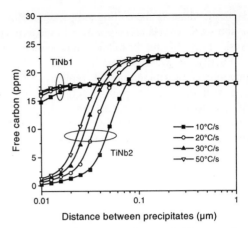

Figure 9 Calculated carbon in solution after annealing as a function of the cooling rate and of the distance between the NbC precipitates for steels TiNb1 and TiNb2 (soaking time=60s; soaking temp.=900°C; undercooling=50°C)

between precipitates is close around 0.03-0.05μm. In this range, the carbon concentration varies between 5 and 20 ppm when the cooling rate passes from 10 to 50°C/s: the higher the cooling rate, the less reprecipitation occurs.

Influence of the precipitate size distribution. Some preliminary calculations have also been performed with a distribution of some 500 precipitates. In this case, precipitates are allowed to interact with each other due to Ostwald ripening. This results in a growth of the biggest precipitates at the expense of the smallest, even in isothermal conditions.

Fig. 10 shows the size distributions before and after annealing for a typical continuous annealing cycle. The number of precipitates decreases from 500 to 150 during the annealing, while the mean diameter increases. Most of the precipitates disappear at the end of the heating stage and during the soaking. A comparison between the ripening model and the simple model indicates that more free carbon is found in the multi-particle case because most of the small particles vanished leaving their carbon into solution.

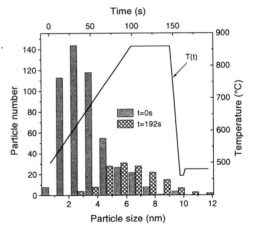

Figure 10 Evolution of a distribution of 500 precipitates during an annealing cycle (black line), modelled with the ripening model (soaking temperature=860°C)

Discussion

According to the results in Table 2, for steels Nb1 and Nb2, the lowest BH corresponding to the lowest free carbon at the end of the processing is obtained for the lowest CT. In this case, a high precipitate density is expected, which means that the equilibrium will be rapidly reached during the soaking but also that much reprecipitation will occur during cooling (see Fig. 3). This means that beside the size information, some idea of the precipitate density is necessary as input data for the numerical modelling. After annealing, in all steels fine NbC were found with typical sizes between

3-30nm, whatever the coiling temperature. However the precipitate density could not be determined as the steel volume corresponding to the extraction residue present on the replica is unknown. The work of quantifying the extraction process is still in progress and data will be available soon.

Finally, it was noticed that the precipitate-containing carbon replica film still showed the grain structures of the steel matrix after its removal through etching. Therefore, it was checked if important differences between grain boundaries and grain interior should be considered. In many cases, there were indeed more NbC precipitates found in the grain boundaries, especially in the hot rolled material. However, the size frequency distributions were nearly equal. Because the difference in number of precipitates per image field with the grain interior was not that high (factor two times max. difference) and considering the surface versus volume ratio for grains of typical 20 µm, it was concluded that for the calculations, segregation effects of Nb and C could indeed be neglected.

Table 2: BH values as a function of coiling temperatures for steel Nb1 (T_{soak} = 860°C during 47s, cooling rate 33°C/s) and for steel Nb2 (T_{soak} = 850°C during 60s, cooling rate 10°C/s). Dashes indicate non-measured values.

CT (°C)	540	600	620	670	680	720
BH, Nb1	16	22	-	-	31	-
BH, Nb2	-	-	23	21	-	32

Conclusions

Modelling provided a deeper insight in the precipitation behaviour of NbC in IF-BH steels, as a function of different processing parameters: coiling temperature, soaking time and temperature, cooling rate, steel chemistry.

Using this tool the free carbon content at the end of the processing can be calculated, from which the BH can be estimated (supposing some dislocation density). If quantitative information as to initial precipitate sizes and densities becomes available from analytical chemistry and microscopy, the methodology can lead to a real prediction of the behaviour.

References

[1]   W. Bleck, R. Bode, O. Maid, L. Meyer, Proceedings Symposium on High-Strength Sheet Steels for the Automotive Industry, (ed. R. Pradhan, ISS) (1994), p. 141.

[2]   S. Satoh, S. Okada, T. Kato, O. Hashimoto, T. Hanazawa, H. Tsunekawa, *Kawasaki Steel Techn. Rep.* n° 27 (1992), p. 31.

[3]   R. Hubert, *ATB Metallurgie*, 34-35 (1995) pp. 5-10.

[4]   K. Lips, X.H. Yang and K. Mols, *Steel Research*, 67(1996), pp. 357-363.

[5]   S. Akamatsu, M. Hasebe, T. Senuma, Y. Matsumura and O. Akisue, *ISIJ International*, 34 (1994) pp. 9-16.

[6]   E. F. Petrova et al., *Soviet Phys. Doklady*, 3(1958) p. 876.

[7]   A. Pichler, M. Mayr, G. Hribernig, H. Presslinger and P. Stiaszny in *Proc. Int. Forum for Physical Metallurgy of IF Steels (IF-IFS-94) Tokyo, May '94* , publ. The Iron and Steel Institute of Japan & Nissho Iwai Corporation, (1994) pp. 249-268.

[8]   R.C. Sharma, V.K. Kakshmanan and J.S. Kirkaldy, *Metall. Trans.* A15(1984), pp. 545-553.

[9]   K. A. Taylor, *Scripta Metal. Mater.*, 32(1995), p. 7.

[10] R.C. Hudd, A. Jones and M.N. Kale, *J. Iron and Steel Inst.*, 209(1971), pp.121-125.

[11] J. Geise and C. Herzig, *Z. Metallk.*, 76 (1985) pp. 622-626.

[12] S.N. Tewari and J.R. Cost, *J. Mater. Sci.*, 17(1982), pp. 1639-1648.

[13] G.P. Krielaart, R.A. Hubert and Y. Houbaert, in *Proc. Moving Boundaries IV, Ghent, Belgium,* ed. R. Van Keer and C.A. Brebbia, Publ. Computational Mechanics Publications, (1997), pp. 327-339.

[14] B.C. De Cooman and R.A. Hubert, *Steel Research,* 68(1997), pp. 487-494.

E-mails: roger.hubert@ocas.be, sven.vandeputte@ocas.be, chris.xhoffer@ocas.be
Web site: www.ocas.be

Materials Science Forum Vols. 284-286 (1998) pp. 559-566

Simulation of Hot-Dip Galvanizing for Automotive Applications: Requirements for Modern Steel Product Design

A. Pérez[1], L.J. Andrés[1], B. Fernández[2], I. González[2] and J.M. Puente[2]

[1] Instituto Tecnológico de Materiales, Parque Tecnológico de Asturias, 33428 Coruño (Asturias), Spain

[2] ACERALIA Corporación Siderúrgica, Centro de Desarrollo Tecnológico, P.O. Box 90, 33480 Avilés (Asturias), Spain

Keywords: Zinc Coated Steel Sheet, Hot-Dip Galvanizing, Galvannealing, Continuous Processing, Physical Simulation

Abstract

With the trend toward increased continuous processing, batch annealing has been partially replaced by continuous annealing during the last years. The prime advantages of continuous annealing are the considerable increase in surface cleanliness and product uniformity along the length of the coil. Also, the versatility of continuous annealing processing lines (CAPL) and continuous galvanizing lines (CGL) has made possible the development of sophisticated processing routes leading to the introduction of a wide variety of new automotive cold rolled steel grades. As an example, the continuous production of hot-dip galvanized steels for automotive applications comprises both thermal (annealing and cooling stages), mechanical (final temper rolling) and chemical treatments (alloying of the Zn coating with the steel substrate). These stages take place continuously within a few seconds, thus demanding excellent on-line control.

This situation has led to an increasing need for simulation techniques to reproduce accurately the CA processing conditions. The hot-dip process simulator (HDPS) available at ITMA is being used in conjunction with ACERALIA for the design of new steel products for automotive applications. This equipment has been designed to meet strict control requirements concerning the abrupt changes of cooling conditions within a non oxidizing atmosphere, temperature gradients and thickness of the Zn coating. In this testing machine a cold rolled specimen (120x200 mm^2) is annealed in a radiation furnace under a H_2-N_2 atmosphere, then controlled cooled by He-blowing and conducted pneumatically into the Zn pot. Additional wiping for thickness control of the deposited Zn layer together with alloying treatment of the coating (galvannealing) by means of infrared or induction furnaces complete the simulation.

This article is focused on the description of the processing conditions achievable with the HDPS compared to those of a modern CGL, together with a presentation of the results obtained to date in the field of microalloyed ULC bake-hardening steels.

Continuous processing versus batch annealing

After the first suggestions towards continuous processing of cold rolled sheet in the 30's, almost forty years went by until the first two production lines were installed in Japan in 1972 [1]. Thereafter this technology has experienced a significant growth. During continuous annealing, uncoiled steel sheet is passed through a multi-stage heating and cooling equipment, with a residence time on the order of a few minutes. This has a first influence on the productivities reached with this technology compared to those achievable with the classical batch annealing, where a set of coils are placed under a cover with reducing atmosphere (typically 95%N_2 5%H_2) and heated for a time period that may involve several days (**table 1**). In general, thicker gauges of the strip and higher throughputs increase the net savings in operating and capital costs of a CA-line compared with a BA-line.

The control of temperature of the single strand of sheet in continuous lines is better than for batch annealing. Thus more uniform properties are obtained along the coil length, obviating the need to crop head and tail sections and so improving yield. This is specially important in steels where the strengthening mechanisms have a great dependence on annealing temperature and cooling rates. With respect to coated products, the adhesion and corrosion resistance of the deposited layer is highly dependent on the sheet surface cleanliness. The introduction of in-line cleaning units (electrolytic or mechanical) in continuous annealing lines followed by annealing at high temperatures provides a cleaner surface than is attained for batch annealed sheet [2]. CAP-lines offer these advantages but do so at the expense of flexibility (fixed production capacity and sheet width). Furthermore, recent results by using 100% H_2 reducing atmospheres in BA-lines have shown better consistency in mechanical properties, cleaner strip surface and higher productivity, and its adoption has been recommended for manufactures where flexibility is an important issue [3,4].

As already mentioned, the specific requirements of each product play an important roll in the kind of technology to be adopted. For instance, during continuous annealing recrystallization takes place at a high rate, resulting in harder qualities which do not allow the cold reduction necessary for certain applications. As another example, the isotropic grade ZstE 250i tipically undergoes a recrystallization annealing in batch-type furnaces [5]. On the other side, modern CAP-lines offer a wide range of cooling rates and heating patterns that are not possible on batch annealing furnaces because the large mass of steel involved in a usual industrial annealing cycle. This has permitted the development of a range of high-strength-steels which cannot be produced through the BA route, and also of additional means of controlling strip shape through optimized cooling rates.

During 1997, 43% of the tinplate and 80% of the cold rolled

BATCH ANNEALING	CONTINUOUS ANNEALING
✓ flexibility (small orders, special steels) ✓ excellent ductility ✓ 100%H_2 technology ✓ more simple control	✓ increased productivity ✓ better yield ✓ appropiate for high-strength grades ✓ product uniformity ✓ better surface cleanliness and shape

Table 1 *Comparison of main advantages of batch and continuous sheet processing*

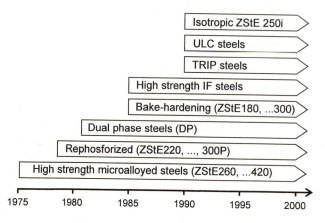

Fig. 1 *Historical evolution of HS cold rolled steel grades for automotive sheet after [7]*

coil at ACERALIA-Avilés was pro-duced through batch annealing. Predictions for the next future estimate that around 75% of the tinplate world production will be caried out on continuous lines, even down to thicknesses of 0,08 mm [6].

The new generation of HS-steels and coatings for automotive applications

Today more than twenty different steel grades are used for the manufacturing of indoor and outdoor panels in automotive applications [7]. The traditional deep drawing qualities with low levels of C and N (FePO4, 5 and 6) have been extended with SULC qualities where carbon is further reduced down to about 20 ppm and Ti and N are designed stoichiometrically to provide the necessary amount of TiN precipitates (**figure 1**). This reduction of C and N has made it possible to minimize the microalloying additions, thus improving coating quality. For the development of high-strength grades several strategies have been proposed, for example the addition of Mn, P or even the microalloying with Nb. The bake-hardening grades reach a high level of strength "in service" due to an aging effect during the baking treatment of the panels after press-forming and painting. This requires the presence of a certain amount of interstitial carbon dissolved in the ferritic microstructure after annealing and hot-dipping, which can only be achieved through an adequate design of chemical composition and heat treatment patterns in a CAP-line. Other possible choices for high-strength applications are the dual-phase steels and the promising TRIP steels, whose development is also connected to stringent requirements in process control only achievable at continuous lines.

The automotive industry has clearly adopted the zinc coated steel sheet as an appropriate corrosion resistant material for the outer panels of the car body [8]. The use of deep drawing steel grades to match the high local strains exerted in modern press-forming lines must be accompanied by a significant forming capacity and adherence of the coating to the steel substrate. In this sense it is of fundamental importance to keep an adequate Al-content in the molten bath (0.1-0.2%), leading to the formation of an intermetallic Fe-Zn-Al layer between coating and substrate during hot-dipping. Apart from the classical hot-dipping with almost 100% Zn, other coatings as *Galfan*© (95% Zn, 5% Al), *Lavegal*© (70% Zn, 30% Al) and *Galvalume*© (43,4% Zn, 55%Al, 1,6% Si) have been developed for specific applications and require different bath compositions (**figure 2**).

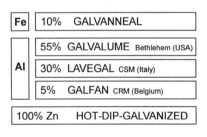

Fig. 2 *Different commercial Zn-based sheet coatings produced by hot-dipping*

The *galvanneal* coating represents another alternative to improve the coating quality by means of an additional annealing treatment after hot-dipping. A schematic view of the resulting thermal treatment is shown in **figure 3**. As a result, a multilayer of intermetallic phases with 7-11% Fe is obtained by diffusion of Fe into the Zn deposit: Γ-phase at the interface, δ at the bulk and ζ at the surface. This product shows improved paintability

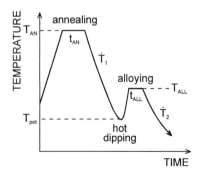

Fig. 3 *Schematic representation of the thermal treatment during continuous processing of galvannealed sheet*

and weldability compared with the conventional galvanized sheet.

Finally, other techniques leading to better performance of zinc coated parts for certain applications make use of organic films (*Durasteel*©, with improved anti-chipping behavior) and chemical (chromating, phosphating) and electrochemical treatments (Fe-flash).

Industrial production of galvanized and galvannealed sheet

The development of Zn coated products during the last years as described above has demanded the progressive introduction of new technologies, specially for the continuous production

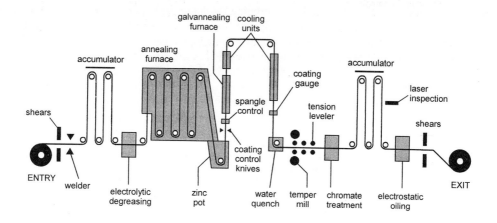

Fig. 4 *General layout (schematic) of the CGL at GALMED (Sagunto - Spain)*

Fig. 5 *View of the hot-dip process simulator (HDPS)*

CONTINUOUS ANNEALING	
Heating rate	5 - 50 °C/s
Cooling rate	5 - 60 °C/s
Maximum temperature during annealing	950 °C
Atmospheres:	
Inert	Pure nitrogen
Reducing	0-25% H_2
Dewpoint	-60 a +5 °C
HOT-DIPPING	
Dimensions of the sample	120 x 200 x 0,1 - 3,2 mm
Thickness of the coating	40 - 300 g/m^2 each face
Speed of the sample	10 - 1000 mm/s
Control of coating's thickness	N_2 - wiping
ALLOYING - INDUCTION FURNACE*	
Maximum temperature during annealing	760 °C (Curie's temperature)
Heating rate	0 - 150 °C/s
Speed of the sample	200 mm/s
MOLTEN BATH	
Maximum temperature	900 °C ± 2 °C
Capacity	25-30 kg Zn

Table 2 *Ranges of variation of different working parameters of the HDPS (*induction furnace had not been yet comissioned by the time of the preparation of the manuscript)*

The hot-dip-process simulator (HDPS)

Figure 5 shows a view of the hot-dip process simulator (HDPS) available at the Instituto Tecnológico de Materiales and ACERALIA. It has been designed to serve as a physical simulation tool for the development of new steel grades as well as for the optimization of the thermal cycles during continuous processing. A vertical mechanical actuator holds the 120 x 200 mm sample and moves it along the different modules of the simulator under a reducing atmosphere (**figure 6**). The temperature of the sample is monitored and recorded by means of a welded thermocouple. The facility allows an accurate simulation of all the stages involved in the industrial continuous processing of cold rolled sheet: annealing inside a radiation furnace suitable for thicknesses in the range 0,1 – 3,2 mm; hot dipping in a graphite crucible resistance heated; gauge control of the deposited Zn through N_2 wiping; and controlled cooling by means of inert gas jets. Optionally, the coated sample can undergo a low temperature annealing in order to achieve a *galvanneal* alloy. A pilot cold rolling mill can also be used to skin-pass the processed samples. Other coatings as *Galfan* and *Galvalume* as well as the influence of tramp elements on coating quality may also be studied if the chemical composition of the molten bath is properly changed.

A PC controls the operation of the equipment. For each sample a complete set of parameters (linear speed for displacement of the sample, cooling and heating rates, soaking times, chemical composition of the reducing atmosphere) together with the actual thermal profile during the test are recorded for later analysis and validation. **Figure 7** shows the thermal profiles during annealing in the radiation furnace, measured at three different points on the sample's surface. The temperatures lay in a range of ±5°C around the set point and seem to be homogeneous enough to guarantee the existence of a central part of the sample processed under nominal conditions. **Table 2** lists the available ranges of variation of different working parameters of the simulator.

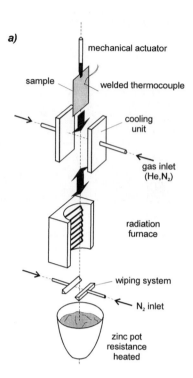

a)

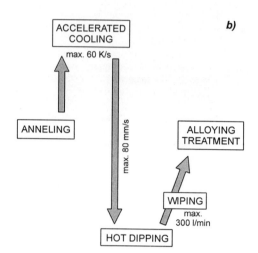

Fig. 6 *(a) Schematic view of the hot-dip process simulator and (b) operating stages during each test*

b)

of sheet. Regarding the design of the lines (**figure 4**), it has been since the adoption of vertical radiant furnaces that the CGLs have been able to reach the soaking times and cooling rates required by the new HS steel grades. Electrolytical and mechanical cleaning and degreasing units as a first step in the production lines guarantee the elimination of coating defects as a consequence of surface residues. Regarding the surface quality of the sheet, the actual trend towards minimum spangle through Zn-powder projection on the band offers a better paintability as well as environmental benefits derived from the elimination of Pb in the molten bath. Finally, the use of on line skin-pass rolling mills improves the surface roughness of the sheet and eliminates Lüders strain without decrease of productivity.

The increased standard quality has been made possible through a growing degree of automation in the modern lines (laser detection of surface defects). Together with this, the implementation of mathematical models allows an on line control of important parameters as the deposited mass, its homogeneity along the transverse direction or the alloying degree in the induction furnace.

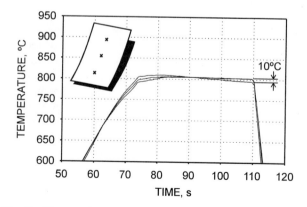

Fig. 7 *Temperature profiles at three different points on the sample's axle during annealing*

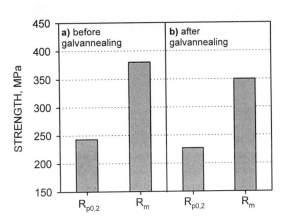

Fig. 8 *Coating appearance and change in tensile properties after a typical galvannealing treatment (T_{AN}=850°C, dT/dt=60K/s, T_{GAN}=510°C)*

Research has been carried out so far in the field of galvannealed IF high-strength steels and Ti-microalloyed ULC bake-hardening steels. Besides the physical simulation of the hot-dip process, complete characterization of the samples has been done by means of metallography, mechanical testing (tensile test, hardness, powdering, etc.) and advanced techniques for coating quality assessment. The studies have been focused in the determination of optimal annealing and galvannealing cycles. Special attention has been paid to the relationship between composition of the steel substrate and the coating quality, and to the influence of the galvannealing treatment on the mechanical properties of the sheet (**figure 8**).

Acknowledgements

Mr. J.R. Ovies (Depto. Programación), and Mr. P. Blanco and Mr. L.O. Valera (Depto. Diseño Técnico) from ACERALIA are gratefully acknowledged by the authors for their help with the preparation of the manuscript.

Bibliography

[1] "Metalurgia física. Aplicaciones al recocido continuo", CSI Planos/CEIT, Vol. 2 (1997)

[2] *Pradham R.:* "Continuous annealing of steel", Metals Handbook ASM, Vol. 4 (Heat treating of steel), p. 56

[3] *McManus G.J.:* "Batch annealing stages come back, but continuous annealing gains ground too", Iron and Steel Engineer, Sept. 1997, p. 58-59

[4] *Wittler P., Scheuermann W., Bovalina T.:* "Recent developments in high performance hydrogen annealing", 38th MWSP Conf. Proc., ISS, Vol. XXXIV, p. 383-389

[5] *Kruska J.U., Freier K.:* "A new developed HSS-grade", HSS Sheet for the Automotive Industry, ISS, Baltimore 1994, p. 159-169

[6] "100%H2 improves batch annealing but the CAL still offers more", Steel times international, Nov. 1996, p. 22-23

[7] *Bleck W.:* "Anforderungen an Karroseriewerkstoffe", Stahl u. Eisen 115 (1995), 4, p. 55-63

[8] *Teulon H.;* "Coated steel in the automotive industry", 3rd International zinc coated sheet conference INTERGALVA, Barcelona (1991), S5F/1-S5F/4

Materials Science Forum Vols. 284-286 (1998) pp. 567-574
© *1998 Trans Tech Publications, Switzerland*

Influence of Bake Hardenability on the Properties of the IF, BH and Isotropic Sheet Steels

J. Janovec[1], F. Mohelsky[2] and J. Ziegelheim[1]

[1] Czech Technical University Prague, Karlovo náměsti 13, 121 35, Czech Republic

[2] Autoškoda Volkswagen Group, 293 60, Mladáb Boleslav, Czech Republic

Keywords: Interstitive Free - IF Steels, Bake Hardenability - BH Steels, Isotropic Steels, BH Effect, Precipitation and Substitution Hardening, Mechanical Properties, Δr&r Values, Microstructure Analysis

1. Introduction

Automotive steel sheets, inclusive the coating sheets, are final coachworks pieces. After pressing, point welding and consecutive lackering a number of many antagonistic demands must be realized. One of the most important demands is to increase the strenght characteristics, because reducing the sheet ply of the body brings the reduction of the weight of the automobile, which results in the total weight being reduced by one quarter. With the consumption, the ratio of harmful gases decreases as well [1]. Increasing yield strenght can solve the body weight reduction, preserving prosperous plastic properties. Thus up to 40 % weight savings comparable to leading world producers Mazda, Audi, BMW etc can be achieved.

2.Steels used for automotive sheets

An assortment of the steels used for automotive sheets is unusually rich. Its strenght level moves between 280 up to 1500 MPa. A very important part of these steels occupy new developed low carbon steels for deep drowing used for inside and outside parts. It is important to preserve a stable pressing of the complicated shapes without demaging them, high dimensional precision and low grade of elasticity [3].

2.1 Steels used in Autoškoda Mladá Boleslav

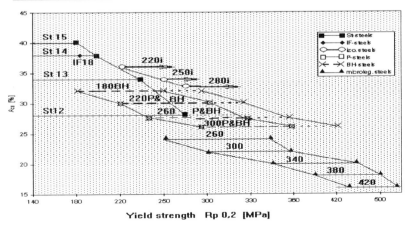

Fig 1. Types of steels used in Autoškoda Mladá Boleslav [3]

Fig. 1 shows the characteristics of new developed automotive steels used in Autoškoda MB in co-operation with the whole Volkswagen Group production.

In case of all the sheets, zinc galvanization was made on both sides, coating thickness was not more than 7,5 um.

Proportion of steels with high strength characteristics applied to today´s Škoda MB production has not rised according to the original hypotheses. The reasons are mainly technological ones: higher consistence and higher elasticity. Target task is to achieve 40 % of this rate, as in the case of top world producers. The complex physical - metallurgical analysis of used steels investigates the effect of particular technology factors to final properties.

Introduction of isotropical steels into the production relates to the lappetness reduction. Tab.5 shows values of the grade of the plane anisotropy, with respect to ISO 11 113. This parameter gives changes of deformation rate in various directions in the plane of sheet. For isotropical sheets $\Delta r = 0$, on contrary, the rise of Δr value signalizes an inclination to the creation of the lappets.

2.2 Precipitation and substitution hardening of automotive steels.

Original theory about precipitation in IF steels originated in the idea of a simple precipitation reaction [4]. Later reports [5] showed the importance of influence of Nb (C, N) and mainly $Ti_4C_2S_2$ for the stabilization process. Ti content required for full stabilization in these steels has been given in per cents of weight as follows:

$$Ti_{stab.} = 4\ C + 3,42\ N + 1,5\ S$$

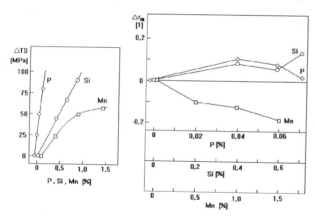

On the contrary, stabilization content of Nb agrees with the total Nb content because the occurrences of NbS and NbN are minimal. Typical IF steel consits of 0,003 % C, 0,003 % N and 0,006 % S, the atomic weight ratio is C : N : S = 12 : 1,2 : 1. Carbon represents only 35 % of atoms forming precipitates, analogically N makes 35 % ratio, and S 30 % ratio precipitating atoms [5].

Substitution hardening provide P, Si or Mn, with respect to Ti, Nb or Ti + Nb content. The effect of these elements is shown in Fig 2.

Fig 2. Effect of substituting hardening elements

The effect of Cr, Ni, Cu, Mo and Sn elements is often not taken into account, although their content is very low. Nevertheless it can also cause substitution hardening - see Tab. 2.

3. BH effect

The complex content of elements increasing the strength described above cause not only problems with the coating process, but also formability becomes hard (r_m value decrease).

The growing strength properties in case of steels with the minimal content of elements and solved carbon ($\approx$ 0,001 %) are made possible by so called BH (bake hardenability) effect. The steel supplied has a low level of yield strenght. After pressing, the plastic deformation brings higher dislocation density. This deformation hardening, WH - work hardenability, also brings the rise of yield strength above 20 MPa [2].

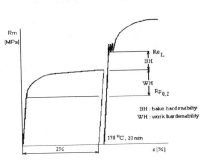

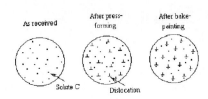

Fig 3 Schematic illustration showing mechanism of
strain age- hardening and measuring method
of bake - hardenability (BH) in tensile test

BH effect is reached during the lacker baking in 20 minutes and 170 °C. Yield strength can generally rise above 60 up to 80 MPa [2]. Resistance to a deformation of the final products increases too.

Fundamental mechanism of the BH effect is the deformation ageing caused by the interaction of the dislocation substruction and intersticial carbon [2].

The low content of solved carbon in BH steels causes its easy difusion even at room temperature. It means these steels incline to ageing even before forming [6]. N can be the cause of the deformation ageing as well. The N difusion is faster than the C difusion. Concurrently a temperature row and delay in case of BH steels is being investigated, because high energy savings can be achieved.

180 BH sheets gave positive results. Not only because the level of hardening ΔBH is the best, but also the influence of time period is defined. On the contrary, the level of temperature between 125 and 170°C does not in fact influence the bake-hardenability process. The strongest control is indicated by the rise of deformation grade. This result has been observed with all materials.

Experimental measurement observed not only total yield strength value in dependence on temperature and time period but also other deep drawing characteristics - see Tab. 1.

Tab. 1

Material	$Rp_{0,2}$ [MPa]	Rm [MPa]	$Rp_{0,2}$/Rm [1]	A_{80} [%]	r [1]	n [1]	th. [mm]
RR St 15 - 03	172,0	288,4	60,03	42,21	2,11	0,208	0,754
IF	151,0	312,0	48,40	39,98	2,23	0,280	0,697
ZStE 180 - BH ZE 75/75	231,0	331,0	69,80	35,16	2,17	0,187	0,754
250 i	255,3	349,6	73,03	37,50	1,12	0,189	0,720

For BH effect application it is best to use extra low carbon steels sheets with the Ti or Nb addiction. After hot rolling a continual annealing can follow, while annealing of the whole rolls is not used any more. Moreover continual annealing lines can be combined with the lines for galvanic coating - zinc.

Annealing temperatures are higher (900 C) so that NbC could be solved and carbon could start precipitating. Atomic ratios of Nb/C and Ti/C play a very important role in affecting BH and r_m values. Ti is a Ti content which can be obtained if the Ti content bound to S and N is obtained from the total Ti content. Low ratios of Nb/C = 0,3 (0,05 % C and 0,012 % Nb) lead to high carbon solving and it means increasing the BH effect, but with higher ratios of Nb/C = 0,7 (0,005 % C and 0,029 % Nb) the r_m value rises. If the atomic weight ratio of Nb/C is reduced, hardening can be reached in an easier way even at low annealing temperatures. Optimal BH effect is obtained with 6 ÷ 8 ppm solved carbon, but this value is pretty hard to be controlled in an operation.

To achieve the equal effect in case of the Ti steels, the ratio Ti*/C must minimally be of the unit value. From this results that is the temperature has increased by other 50°C. Higher affinity of Ti to S and N means problems for exact checking of Ti*/C ratio. S content is necessary to be reduced as much as possible, minimally under 0,003 %.

4. Chemical content

The results of a chemical analysis of the basic elements of the investigated sheets and the ratio values of Mn, Ti and Al are shown in Tab. 2.

These content´s values are usual in case of the basic elements Si, P, S, Al or N. In case of 180 BH steels a higher C content is surprising, because minimal Ti content means to reach very low Ti/C and Ti/N values, suitable for an effective BH effect.

Tab.2

Steel	Chemical content								Rate values			
	C	Mn	Si	P	S	Ti	Al	N	Mn/S	Ti/N	Ti/C	Al/N
	weight % . 10^3							ppm	-			
P05/St15	39	173	12	9	2	13	56	42	86,5	3,1	0,33	13,3
IF	17	236	7	22	25	135	31	26	9,4	51,9	7,94	11,9
180BH	28	21	5	19	5	7	60	74	4,2	0,9	0,25	8,1
250i	16	185	17	17	41	43	63	45	4,5	9,6	2,69	14,0

The alternatives of Mn, Ti and Al rate values are high. From this aspect the best balance is given by 180 BH and 250 i steels.

For the content of minor elements and contaminations including summarization of secondary substituting hardening elements see Tab. 3.

Tab.3

Steel	Chemical content									$\sum Cr + Ni + Cu + Mo + Sn$
	Cr	Ni	Cu	Mo	V	Sn	As	Pb	Ca	
	weight % . 10^3									
P05/St15	41	21	20	3	0	3	4	2	0,2	88
IF	32	64	61	41	18	0	95	86	11	198
180BH	13	21	27	2	5	5	3	2	0	68
250i	37	72	77	32	23	0	107	90	14	218

Higher value of the sum of the minor elements is common at isotropical steels. The incidence of nearly the same value at the IF steels is surprising.

5. Microstructural analysis

Grain size was evaluated according to czech standard. From the materials supplied cuttings were recovered in directions α = 0; 45 and 90° in respect to rolling. The results are given in Tab. 4. In this table **m** means the number of grains in 1 mm^2 and d_m expresses the value of the grain diameter.

From this table the minimal influence orientation to axe of rolling direction on grain size is evident.

For details see microstructural view (Picture 1). In thickness the size of feritic grain is uniform, mostly it is moving in one size class. The smallest grains were observed in 250i steels, the greatest ones in 180 BH and St 15 steels. Plane grind shows a higher heterogenity of grain size. Grains range between two and three size classes. Grains of IF, 180 BH and St 15 materials even after annealing have their typical elongate shapes suitable for deep drawing demands. In case of 250i practically one axe symmetrical grains were observed.

Picture 1 shows for each material the pictures of microstructures taken from sheet thickness, in rolling direction, as follows from Tab. 5 - the minimal size difference in various directions of recovering the cuttings.

Tab.4

Material	α [°]	m [mm^{-2}]	d_m [µm]
P 05/St 15	plane of rolling	2 872	18,7
	0	2 872	18,7
	45	2 746	19,1
	90	3 338	17,3
BH 180	plane of rolling	2 158	21,5
	0	3 932	15,9
	45	4 568	14,8
	90	3 852	16,1
250i	plane of rolling	6 945	11,49
	0	11 558	9,30
	45	11 513	9,32
	90	12 413	8,98
IF	plane of rolling	4 449	14,99
	0	7 475	11,6
	45	8 084	11,1
	90	8 408	10,9

The determination of the deformation size was effected by the metallography method according to czech standard. This method was used because local values of deformation can be very different from the total nominal deformation. The principle is in the deformed grain evaluation. Dimensionless rate of the longest and the shortest grain axes is determined. So called parallel-axeless coefficients $\alpha = \dfrac{a}{b}$ or $\delta = \ln \alpha$ are provided.

Dispersion of magnitude is calculated $(\sigma^2(\delta))$. δ value characterizes the true deformation and $(\sigma^2(\delta))$ characterizes a grade of deformationof the non-homogeneity in individual grains. Measuring was taken on ten typical places for the material and from the obtained values arithmetical diameter was calculated. For results see Tab. 5 incl. the grade of plane orientation.

Tab.5

Material	speciment orientation	α [1]	δ [1]	$\sigma^2(\delta)$ [1]	grade of plane orientation
BH 180	plane of rolling	1,140	0,131	0,081	6 %
	0°	3,319	1,199	1,336	54 %
	45°	2,411	0,880	0,412	42 %
	90°	2,915	1,069	0,336	38 %
250i	plane of rolling	1,069	0,068	0,058	3 %
	0°	1,580	0,457	0,149	23 %
IF	plane of rolling	1,460	0,378	0,141	19 %
	0°	2,208	0,792	0,547	38 %
P05/St15	plane of rolling	1,320	0,278	0,116	14 %
	0°	2,429	0,887	0,265	42 %

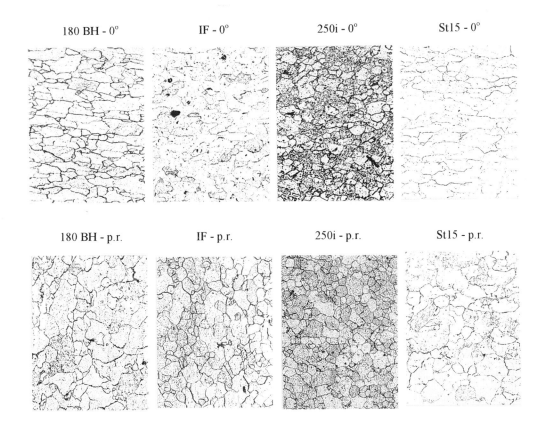

180 BH - 0°　　　IF - 0°　　　250i - 0°　　　St15 - 0°

180 BH - p.r.　　　IF - p.r.　　　250i - p.r.　　　St15 - p.r.

Picture 1 Microstructure of investigated steel sheets, p.r.-plane of rolling. (250x, Nital 2%)

6. Changes of mechanical properties caused by BH effect

Mechanical properties of investigated sheets in supplied sort are presented in Tab.1. Values were calculated as an arithmetical mean from 9 speciments. Constantly three of them had an orientation of 0, 45, 90°. The dependence of mechanical properties of the orientation angle was minimal, and in case of 250i steels they approached a zero.

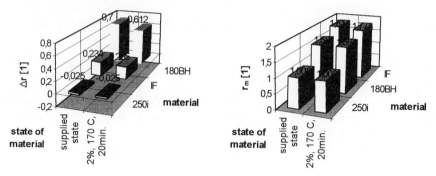

Fig 4. Changes of r_m and Δr values depending on bake process for each material.

At first, dependence of the BH effect was investigated with respect to the speciment orientation under model conditions. It means, the bars were elongated by 2% of permanent deformation. Then they were annealed at 170°C for 20 minutes and then they were tested for tension.

Besides classical deep drawing characteristics r, n and A_{80}, the rate Re/Rm was espressed. Dependence A_{80} and Re/Rm values on orientation is zero. First, attention was dedicated to total rise of yield strength - it is BH value - see Fig 5.

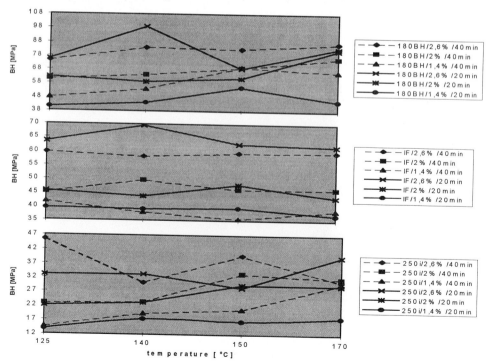

Fig 5. Changes of ΔBH value depending on bake process.

7.Conclusions

Some results of material properties of sheet steels used in Autoškoda Mladá Boleslav after BH effect were presented. The various baking times and baking temperatures for simulation of bake - painting were tested.

1) Increasing strength characteristics can lead even up to sheet destruction during forming. It is mainly the problem of 250i and 180BH steels. Moreover elasticity increases too.

2) In case of BH steels, ΔBH value reaches in average 70 MPa. But final ΔBH values were also found at IF steels $\approx$ 45 MPa and against expectance even at iso. steels ($\approx$ 25 MPa). The reason is, that C or N migration is realized not only at higher temperatures, but also at room and lacquer baking temperatures as well. High rise of the ΔBH value led to a higher crack frequency during press forming and to higher elasticity.

3) Lacquer baking temperatures of 125, 140, 150,170°C, and times 20 and 40 minutes at these temperatures do not influence final mechanical properties at all. Nevertheless, the deformation has a considerable influence on all the investigated properties.

4) It is optimal to investigate the microstructure in speciments recovered from the rolling plane, because minority phases can thus be observed pretty fine.

References

[1] Janovec J., Mohelský F.,: Růst užitných vlastností tenkých automobilových plechů, 5. Mezinárodní metalografické symposium Metal' 96, Ostrava, Květen 14-16,1996, 3.p. 205 - 214

[2] Ushioda K., Yoshinaga N.,: Recent Progress in physical metallurgy of cold - rolled sheets steels, Proc. of Physical metallurgy' 96, Stockholm, September10 - 12, 1996, 120 - 144

[3] Beneš F.,: Mezní stavy zpracovatelnosti ocelí se zvýšenými pevnostními charakteristikami. Rozměrová a tvarová nestabilita výlisků, Hutní projekt Praha, 1996

[4] Tither G., Garcia C. I., Hua M., De Andro A.J.,:Precipitation behaviour and solute effects in IF steels, Proceeding of Int. Forum For Physical Metallurgy of IF steels, Tokyo, 1994, p.293 - 322

[5] Yoshinaga N., Ushioda K., Akamatsu S., Akisue S.,: ISIJ International, 34, 1, 1994, p.24

[6] Satoh S., Okada S., Kato T.,et al.: Davalopment of Bake - Hardening High - Strength Cold - Rolled Sheet Steels for Automobile Explosed Panels, Kawasaki Steel Technical Report, No.27, November 1992

Materials Science Forum Vols. 284-286 (1998) pp. 575-582
© *1998 Trans Tech Publications, Switzerland*

Microalloying with Titanium to Improve Drawability in Low Carbon Wire-Rod Steels

M.P. Staiger[1], C.H.J. Davies[1], P.D. Hodgson[2], B. Jessop[3] and A. Brownrigg[4]

[1] Department of Materials Engineering, Monash University, Wellington Road, Clayton, Australia

[2] School of Engineering and Technology, Deakin University, Waurn Ponds, Australia

[3] BHP Newcastle, Rod and Bar Division, Newcastle, Australia

[4] BHP Research, Melbourne Laboratories, 245 Wellington Road, Mulgrave, Australia

Keywords: MA Microconstituent, Nitrogen, Microalloying with Titanium, Titanium Nitride, Drawability, Wire Drawing, Dilatometry, Torsion, Hardenability, Strain Aging

Abstract The effect of nitrogen and titanium on the phase transformations of low carbon manganese silicon steels was assessed using deformation dilatometry. The role of nitrogen in the formation of the martensite/retained austenite microconstituent is discussed. Three alloys of constant base composition were prepared; one without added titanium, and the others with 0.018 wt% and 0.043 wt% Ti. Samples were heated to 900°C, deformed, and then cooled at rates of 0.2, 1.2, 3.5 and 5.2°C/s. The effect of titanium and nitrogen on microstructural evolution was assessed using optical microscopy. Microstructural changes in the titanium-added steels were investigated, particularly any changes in the development of the martensite/retained austenite microconstituent and/or pearlite phase, and the effects of these on ductility. Torsion testing was used in an attempt to deconvolute the factors which limit the cold drawability of the steels investigated.

Introduction The hardenability of some low carbon manganese silicon steels which have relatively high amounts of austenite stabilizing elements, may be near the point at which austenite or lower transformation phases such as martensite may appear as microconstituents if processing conditions are not carefully controlled. In these steels, a microstructure containing ferrite is preferred for optimizing drawability. The presence of martensite/retained austenite microconstituents (MA phase) in wire rod having a ferrite microstructure is detrimental to drawability and can account for wire breaks, especially in finer wires.[1]

Previous investigations into the effects of free nitrogen on the microstructural evolution in low carbon manganese silicon steels (without nitride-formers) has shown that when nitrogen is present it will encourage the formation of MA phase. It was found that even relatively low amounts of nitrogen could influence phase transformation behaviour by depressing the pearlite transformation start temperature. In effect, the nitrogen enhances the hardenability of the steel and the likelihood of MA phase formation. The presence of excess free nitrogen can not only reduce drawability through strain aging but also indirectly by increasing the MA phase content of the steel. Figure 1 clearly shows the difference in MA phase content for two steels having the same base compositions and varying free nitrogen levels while the difference in ferrite grain size is negligible.[2]

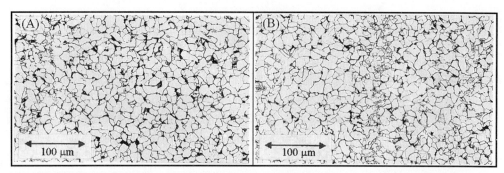

Figure 1: The absence of MA phase in a "low" nitrogen steel (A) having 30 ppm N is compared with a "high" 100 ppm nitrogen steel (B) containing around 10 vol% MA phase. Also note the difference in pearlite content. Cooling rate: 5.2°C/s.[2]

The work presented here shows that in these particular steels it is possible to completely eliminate MA phase caused by excessive free nitrogen by alloying with microadditions of titanium. Microstructural parameters such as volume fraction, hardness, and grain size were measured to reveal the significant effects that relatively small amounts of nitrogen and titanium can have on the final microstructure. Torsion testing is used to compare the drawability of Ti-containing and Ti-free steel. The benefits derived from eliminating MA phase are highlighted by the torsion results.

Experimental Procedure

Materials preparation Steel grades were prepared having the compositions shown in Table I. The first steel is the reference steel and contained no added Ti, while the other two steels had different levels of titanium added. The Ti-High alloy contains the stoichiometrically required amount to completely tie up the free nitrogen as TiN, while the Ti-Low alloy was designed to do this only partially. The steels were produced under laboratory conditions using a vacuum induction furnace. The ingots were homogenised for two hours at 1100°C, and then rolled down to 15mm thick plates in two passes.

Table I: Base compositions of alloys							
Alloy	**C**	**Mn**	**Si**	**Ti**	**N**	**C$_{eq}$***	**Mn$_{eq}$***
Ti-free	0.07	1.41	0.86	-	0.010	0.31	1.45
Ti-low	0.08	1.42	0.86	0.018	0.011	0.32	1.46
Ti-high	0.08	1.45	0.87	0.043	0.011	0.33	1.47

$$\text{*derived from[3] } C_{eq} = C + \frac{Mn}{6} + \frac{Cu + Ni}{15} + \frac{Cr + Mo + V}{5},$$

$$Mn_{eq} = Mn + Cr + Cu + Mo + \frac{Ni}{2} + 10(Nb - 0.02)$$

Deformation dilatometry, Torsion testing An MMC dilatometer was used to austenitise and deform solid cylindrical samples which were then cooled at rates of 0.2, 1.2, 3.5 and 5.2°C/s. The schedule used is depicted in Figure 2. A longer hold time after deformation (30 s) was used for the Ti-containing steels in an attempt to increase the austenite grain size, and compensate for the grain refining effects due to TiN precipitation.

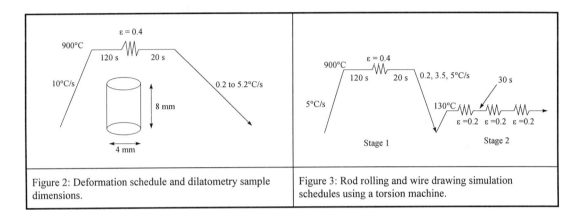

Figure 2: Deformation schedule and dilatometry sample dimensions.

Figure 3: Rod rolling and wire drawing simulation schedules using a torsion machine.

Hot torsion was used to simulate rod rolling so as to develop the desired rod microstructure, while warm torsion was used to simulate the subsequent wire drawing process. The two stages of torsion testing are shown schematically in Figure 3. The sample gauge length and diameter were 20 and 6.7 mm, respectively. The parameters for hot torsion testing were chosen so as to obtain similar microstructures as to those developed by the dilatometry work while the temperature and interpass times used for Stage 2 were taken from studies of actual wire drawing conditions.[4] A strain rate of 3 s^{-1} was used.

Optical microscopy, Hardness, Volume fraction, Grain size measurements Samples for optical microscopy were etched using 2% Nital. Klemms reagent I was used to tint samples so as to highlight the MA microconstituent. Tinting was achieved after 1-2 minutes soaking at room temperature. The hardness was measured using a Vickers indenter and load of 10 kg. The volume fraction of MA and pearlite phases was determined using point counting and area fraction measurements, respectively. The ferrite grain size was measured using the mean linear intercept method.

Experimental Results
Deformation dilatometry The CCT diagrams for the alloys tested are shown in Figure 4. The Ar$_3$ temperature remained approximately the same for all alloys at the highest cooling rate but tends to increase in the Ti-free steel relative to the Ti-containing steels as the cooling rate decreases. The pearlite transformation start temperature was found to be 25°C higher in the Ti-containing alloys compared with the Ti-free alloy at the highest cooling rate of 5.2°C/s. As the cooling rate decreases so too does the difference in pearlite start temperatures between Ti-free and Ti-containing steels. The pearlite transformation start temperature was essentially the same for both of the Ti-containing steels for all cooling rates.

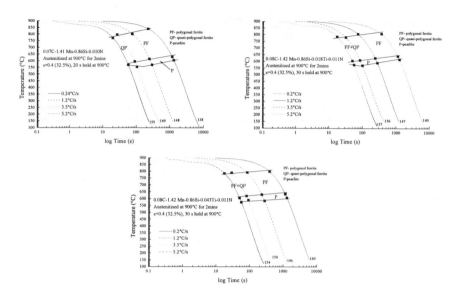

Figure 4: CCT diagrams used to compare phase transformation behaviour (Vickers hardness values are attached).

Optical microscopy; MA and pearlite phase volume fraction measurements The optical micrographs in Figure 5 show the typical microstructures found in an alloy with and without added Ti. The MA phase is easily identified by its morphology: it is made up of jagged islands, etching light gray using 2% Nital, which are often arranged in stringers and associated with pearlite.

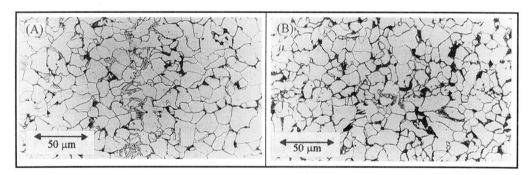

Figure 5: The presence of 10 vol% MA phase is clearly seen in the Ti-Zero steel (A), while Ti-Low (B) shows the absence of MA microconstituents. Cooling rate: 5.2°C/s.

The volume fraction of MA phase was found to be substantially higher for Ti-free steel compared to the Ti-containing alloys (Figure 6). The amount of MA phase measured in the steel containing 0.018 wt% Ti was less than 0.3 vol%, while no MA microconstituents could be detected in the steel containing 0.043 wt% Ti. At a cooling rate of 5.2°C/s, the Ti-containing alloys remained essentially MA-free while the Ti-free alloy contained 10 vol% MA phase (Figure 5). The presence of TiN cuboids was also observed in the Ti-containing alloys, especially in Ti-High. The TiN cuboids were normally less than 2 μm. The amount of pearlite in the Ti-containing steels was found to be significantly higher than in the Ti-free steel especially at higher cooling rates (Figure 7). The difference in pearlite content between the Ti-containing and Ti-free steels increased as the cooling rate increased.

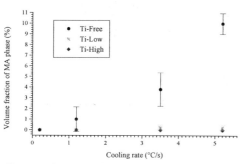

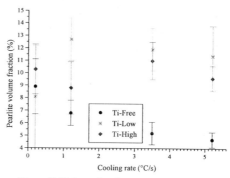

Figure 6: The volume fraction of MA phase was very low for Ti-containing steels while significantly higher levels were measured in the Ti-free steel

Figure 7: Volume fraction measurements reveal the significant difference in pearlite content between the Ti-free and Ti-containing steels

Grain size and Hardness measurements The ferrite grain size of the Ti-containing steels was smaller than the Ti-free steel at the lower cooling rates. The grain size of the Ti-containing steels remained relatively constant over the range of cooling rates tested, while the grain size of the Ti-free steel was strongly dependent on the cooling rate as shown in Figure 8. As the cooling rate was increased the difference in grain size between the Ti-free and Ti-containing steels also decreased. At a cooling rate of 5.2°C/s the smallest difference in grain size between all of the steels was at most 3-4 μm.

In general, it was observed that the hardness of the Ti-containing steels was similar to the hardness of the reference steel. For some cooling rates the hardness of the alloy was decreased by the addition of titanium (Figure 9).

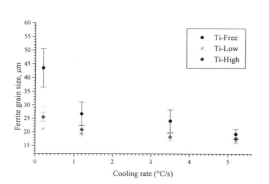

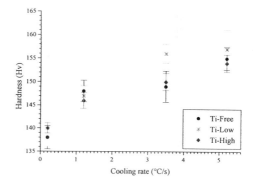

Figure 8: Ferrite grain size measurements

Figure 9: Vickers hardness measurements

Torsion testing The simulations carried out using torsion were performed on Ti-free and Ti-High. The wire drawing simulations indicate that the number of twists to failure for Ti-High was approximately doubled under all of the cooling conditions compared to that of Ti-free (Table II, Figure 10).

Table II: Stage two wire drawing simulation results		
Cooling rate (°C/s)	Twists to failure	
	Ti-Free	Ti-High
0.2	13	21
3.5	9	16
5.0	8	20

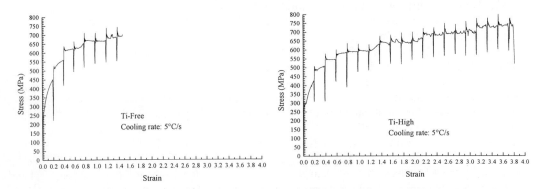

Figure 10: Wire drawing simulation results showing differences in ductility for Ti-free and Ti-High steels at 5.0°C/s

Discussion These current results reinforce the findings from previous work [2] that indicate free nitrogen is responsible for the formation of MA phase. They show that in steels having similar grain size and morphology, the occurrence of the MA microconstituent in their microstructures depends on whether or not nitrogen is dissolved in solid solution or tied up by a nitride former such as titanium.

To test the hypothesis that free nitrogen is responsible for increased MA formation in the steels investigated, the austenite grain size of the steels should be kept constant as it has a strong effect on hardenability. For the Ti-containing steels, a low reheat temperature of 900°C was chosen so as to retain the TiN cuboids. If reheat temperatures are high enough, the TiN could redissolve, and on cooling would precipitate as fine TiN precipitates, which could lead to a much smaller austenite grain size.

The strongest indication that free nitrogen is affecting the phase evolution during transformation are the results obtained from cooling these steels at 5.2°C/s. At this cooling rate, the grain size converges for the Ti-free and Ti-containing steels to within 3-4μm. But the very large differences in both MA and pearlite volume fraction for the samples cooled at 5.2°C/s is good evidence for nitrogen's ability to locally increase hardenability so that in a Ti-free steel the pearlite volume fraction is half that of the Ti-containing steels and MA volume fraction is 10%. Furthermore, the 0.043 wt% Ti steel, which has sufficient Ti to completely tie up the 100 ppm of free nitrogen, contains no MA phase while the 0.018 wt% Ti steel has very small amounts of MA phase as the Ti level only partially ties up the free nitrogen. It could be argued that for the Ti-free steel a larger austenite grain size, which is indicated by the measurement of a slightly larger ferrite grain size in this steel, will contribute to an increase in hardenability but unpublished work has shown that an increase in grain size of 3-4 μm caused an MA increase of around only 2 vol%. At the cooling rate of 5.2°C/s, there is thus 8 vol% of MA phase unaccounted for in the Ti-free steel. Previous work [2] shows that two steels containing 30 ppm and 100 ppm N have a difference in MA content of around 7.5 vol% at a cooling rate of 5.0°C/s, while the difference in grain size was negligible. From these observations it is concluded that the MA content may be increased by around 8 vol% at 5.0°C/s by a critical free nitrogen level which must lie somewhere between 60 and 100 ppm nitrogen.

Being able to compare microstructures for steels with and without Ti is made more complex by the route taken to arrive at their final ferrite grain sizes. The Ti-free steel has its grain size determined by the cooling rate and transformation temperature, while it is assumed the Ti-containing steels started with a smaller austenite grain size. This might be presumed to lead to a change in ferrite morphology such that for example the ferrite in the Ti-free steel is more acicular in nature. However, differences in ferrite morphology are not observed in the optical micrographs. This observation is in line with the Ar$_3$ temperatures which are also quite similar for a cooling rate of 5.2°C/s. At lower cooling rates the Ar$_3$ temperature for Ti-free steel increases above that for the Ti-containing steels, and this difference in Ar$_3$ temperature is reflected in the much larger grain size of Ti-free steel compared to the Ti-containing steels.

The addition of Ti does not alter the hardness of the base composition. In steels containing higher levels of Ti in solution, the occurrence of TiC can lead to unacceptable increases in strength through TiC precipitation strengthening which diminishes drawability. At these low levels of Ti, Ti-high should have only 0.005 wt% Ti in solution. At these Ti levels the strength has not been increased according to the hardness values measured and therefore the drawability should not be affected by small additions of Ti used in this study. That the hardness is not decreased more by removal of the MA phase would seem to be an anomaly. It may be that as the pearlite level increases simultaneously with decreasing MA volume fraction, the hardness is partially compensated for by pearlite. The presence of TiN cuboids may also increase the hardness lost by a diminishing MA content. In cases where the hardness is *decreased* by the addition of Ti this may be explained by the removal of free nitrogen from solid solution which is itself a strong solid solution strengthener.

The relationship among cooling rate, grain size, pearlite formation, and composition is complex. The results obtained from cooling the samples at 5.2°C/s suggest that while ferrite grain sizes differ by only a few microns, a change in free nitrogen content makes possible a variation in pearlite and MA levels. Obviously the lower pearlite content found in the Ti-free steel must be directly linked to its higher MA content. With less pearlite forming, this provides more carbon available to enrich the austenite pockets between the ferrite grains during the transformation. The implications of this are that nitrogen is influencing the pearlite reaction, and this is supported by previous findings that increasing free nitrogen suppresses the pearlite reaction in the same types of steel investigated in this work.[2]

The findings from the torsion tests show that the ductility of Ti-High steel is approximately double the Ti-free steel, based on the number of twists to failure. For cooling rates of 3.5 and 5.0°C/s, the large difference in ductility can be explained by the presence of MA microconstituents in Ti-free steel. The lower ductility in Ti-free steel at a cooling rate of 0.2°C/s compared with Ti-High steel cannot be explained in terms of MA content since the amount of MA should be negligible at this cooling rate for both steels. However, the lower ductility of Ti-free steel could perhaps be explained by a higher sensitivity to strain aging of Ti-free compared with Ti-High as the former should contain higher free nitrogen levels. It is most likely that the ductility will be decreased at low cooling rates by the effect of strain aging alone while at higher cooling rates the decrease in ductility can be attributed to the combination of strain aging and presence of MA microconstituents.

Conclusions The following points are the main findings drawn from this study of low carbon manganese silicon steels:

1) Free nitrogen has the potential to suppress the pearlite reaction, while simultaneously increasing the volume fraction of MA microconstituents.
2) Small additions of Ti are able to suppress the formation of MA microconstituents presumably by removing dissolved nitrogen, and forming TiN precipitates.
3) The addition of small amounts of Ti to tie up dissolved nitrogen improved ductility during wire drawing simulations using torsion testing.

Acknowledgments One of the authors (M.S.) would like to thank BHP Research and the Australian Government (ARC) for financial support, and B. Perrett at BHP-MRL for his help with torsion testing.

References

1. C.M. Bae and W.J. Nam., "Crack initiation during wire drawing of low carbon, manganese steel wire rods containing ferrite, pearlite, and martensite", Mater. Sci. Technol., **10**, 1994, p.971.

2. M.P. Staiger, C.H.J. Davies, B. Jessop, P.D. Hodgson, and A. Brownrigg., "The effect of nitrogen on the microstructural evolution of low carbon manganese steels", Thermec-97: International Conference on Thermomechanical Processing of Steels and Other Materials, July, 1997

3. "Atlas for Bainitic Microstructures", **1**, ISIJ, 1992

4. R. Louis., "Rod drawability- Part 2, Typical wire drawing parameters", BHPR Melb Lab., Report no. BHPR/OE/N/033

Correspondence All correspondence should be directed to Mark Staiger via email: Mark.Staiger@eng.monash.edu.au.

AUTHOR INDEX

KEYWORD INDEX